D0368845

BASIC PRINCIPLES IN NUCLEIC ACID CHEMISTRY

VOLUME II

CONTRIBUTORS

WILLIAM BAUER
D. M. BROWN
C. ALLEN BUSH
HENRYK EISENBERG
PAUL O. P. TS'O
JEROME VINOGRAD

QD433
.T 77
V. 2

BASIC PRINCIPLES IN NUCLEIC ACID CHEMISTRY

Edited by
PAUL O. P. TS'O

Division of Biophysics
Department of Biochemical
and Biophysical Sciences
The Johns Hopkins University
Baltimore, Maryland

VOLUME II

ACADEMIC PRESS   *New York and London   1974*

A Subsidiary of Harcourt Brace Jovanovich, Publishers

INDIANA
UNIVERSITY
LIBRARY

NORTHWEST

COPYRIGHT © 1974, BY ACADEMIC PRESS, INC.
ALL RIGHTS RESERVED.
NO PART OF THIS PUBLICATION MAY BE REPRODUCED OR
TRANSMITTED IN ANY FORM OR BY ANY MEANS, ELECTRONIC
OR MECHANICAL, INCLUDING PHOTOCOPY, RECORDING, OR ANY
INFORMATION STORAGE AND RETRIEVAL SYSTEM, WITHOUT
PERMISSION IN WRITING FROM THE PUBLISHER.

ACADEMIC PRESS, INC.
111 Fifth Avenue, New York, New York 10003

United Kingdom Edition published by
ACADEMIC PRESS, INC. (LONDON) LTD.
24/28 Oval Road, London NW1

Library of Congress Cataloging in Publication Data

Ts'o, Paul On Pong, Date
        Basic principles in nucleic acid chemistry.

        Includes bibliographical references.
        1.   Nucleic acids.   I.   Title.
QD433.T77   547′.596   72-13612
ISBN 0−12−701902−2 (v. 2)

PRINTED IN THE UNITED STATES OF AMERICA

CONTENTS

1. Chemical Reactions of Polynucleotides and Nucleic Acids

D. M. Brown

2. Ultraviolet Spectroscopy, Circular Dichroism, and Optical Rotatory Dispersion

C. Allen Bush

3. Hydrodynamic and Thermodynamic Studies

Henryk Eisenberg

4. Circular DNA

William Bauer and Jerome Vinograd

5. Dinucleoside Monophosphates, Dinucleotides, and Oligonucleotides

Paul O. P. Ts'o

LIST OF CONTRIBUTORS

Numbers in parentheses indicate the pages on which the authors' contributions begin.

WILLIAM BAUER* (265), Department of Chemistry, University of Colorado Boulder, Colorado

D. M. BROWN (1), University Chemical Laboratory, Cambridge, England

C. ALLEN BUSH (91), Department of Chemistry, Illinois Institute of Technology, Chicago, Illinois

HENRYK EISENBERG (171), Polymer Department, The Weizmann Institute of Science, Rehovot, Israel

PAUL O. P. Ts'o (305), Division of Biophysics, Department of Biochemical and Biophysical Sciences, The Johns Hopkins University, Baltimore, Maryland

JEROME VINOGRAD (265), Division of Biology and Division of Chemistry and Chemical Engineering, California Institute of Technology, Pasadena, California

* *Present Address*: Department of Microbiology, State University of New York, Stony Brook, New York.

PREFACE

About one hundred years ago, a young Swiss physician, Friedrich Miescher, published the first paper on "nuclein" (or nucleohistone in current terminology) and thus launched chemical research on nucleic acids. Nearly twenty-five years ago, nucleic acid was identified as the physical basis of genes, and since that time the quest for knowledge on genes rightfully has become a major thrust in modern biological research. In fact, the tremendous progress in nucleic acid research has raised the possibility that advancements in this field may exert a profound influence on the future of man.

We, as researchers in nucleic acid chemistry, have prepared this multi-volume treatise in honor of this historic event: the centennial anniversary of the discovery of nucleic acid. Our view is that progress in nucleic acid chemistry has been substantial and sufficient to justify an attempt to formulate certain basic principles in this field. We hope that these basic principles will not only endure the test of time but will serve as a foundation for further advancement in nucleic acid research as well. Not only have we critically examined the achievements of the past, we have also contemplated the future: the momentum of nucleic acid research and its contribution and influence on the destiny of man. Knowledge of nucleic acid chemistry will be utilized more extensively than ever in biomedical research areas such as cell biology, differentiation, microbiology, virology, oncology, genetic therapy, and genetic engineering. Hopefully, this treatise will serve as reference and resource material for many workers in biomedical research and as teaching material for instructors in institutions of higher learning.

In following the approach of Volume I, the first four chapters in Volume II are written by scholars who have expert knowledge in a particular area of research in nucleic acid chemistry. These are Chapter 1, Chemical Reactions of Polynucleotides and Nucleic Acids; Chapter 2, Ultraviolet Spectroscopy, Circular Dichroism, and Optical Rotatory Dispersion; Chapter 3, Hydrodynamic and Thermodynamic Studies; and Chapter 4, Circular DNA. Chapter 5, Dinucleoside Monophosphates, Dinucleotides, and Oligonucleotides, describing the current knowledge and concepts of nucleic acid chemistry at this level of complexity is written by the editor.

We wish to thank the following reviewers for their helpful suggestions in the preparation of this volume: Chapter 1, Freidrich Cramer, Max–Planck–Institut für Experimentelle Medizin, Göttingen, Germany; Chapter 2, Ignacio Tinoco, Jr., University of California at Berkeley; Chapter 3, John E. Hearst, University of California at Berkeley; Chapter 4, James C. Wang, University of California at Berkeley; Chapter 5, Charles Cantor, Columbia University. The valuable assistance of Carl Barrett and Jean Conley in the preparation of the Subject Index and the help from many colleagues and from the staff of Academic Press are gratefully acknowledged.

Onward to Volume III!

PAUL O. P. Ts'o

CONTENTS OF OTHER VOLUMES

1

CHEMICAL REACTIONS OF POLYNUCLEOTIDES AND NUCLEIC ACIDS

D. M. BROWN

1

I. Reactivity of Polynucleotides

A. Introduction

Ostensibly the subject of this chapter is the susceptibility of the nucleic acids toward attack by chemical reagents. This is indeed an important matter. It would, for example, be of great interest to know the principles which govern the reactivity of bases in DNA toward certain carcinogens or mutagens. What effect does nucleotide sequence or secondary structure have? What effect does tertiary structure in ribosomal, messenger, and transfer RNA have, in corresponding reactions? The fact is that very little is known. No *ab initio* estimates of steric hindrance in polynucleotide reactions have been made, nor is it likely that these could be usefully made at present, though the nature of ordered polynucleotide structures and the forces involved in stabilizing them have received much attention [1]. As a result much of the chapter will be devoted to a discussion of a variety of reagents important in nucleotide chemistry, and on the mechanistic principles which inform their reactions. Emphasis will be placed on those classes of compounds whose reactions with polynucleotides have been studied.

We may begin by introducing a generalization to the effect that the reactivity of bases or of the internucleotide linkages in polynucleotides will only differ from those of the corresponding monomeric systems *in rate*. This may appear a sufficiently obvious and naive suggestion as to warrant no mention, but, for example, it is not long since the view was seriously held that the difference between DNA and RNA in their susceptibilities to base hydrolysis was due to the secondary structure in the former. The generalization is introduced because it provides the justification for the view that a study of the chemistry of the monomeric species (be it base, nucleoside, or nucleotide) is a necessary and valid approach to the chemistry of the same residue in a polymeric system. The rate difference may of course be, and often is, very large. Thus, for example, reactions which depend on a specific approximation of the reacting species—photodimerization of pyrimidines—may be much faster (or the amount of product at equilibrium greater) in a polynucleotide duplex. On the other hand, reactions of the same polynucleotide with external reagents may be exceedingly slow.

The assumption behind many recent studies has been that tentative conclusions can be drawn in respect of secondary and tertiary structure on the basis of comparative rate data. This will be discussed in more detail later in the chapter.

B. Types of Nucleotides

The discussion will be confined, in the main, to the chemistry of the base residues adenine, guanine, cytosine, and uracil as it is seen in nucleosides and nucleotides or suitable analogs. The natural modified bases, e.g., dihydrouracil, the thiopyrimidines, the 5-substituted pyrimidines, pseudouridine, and certain alkylated purines will be treated, though in less detail. The discussion will be limited to certain classes of reagents which are considered to be important from the standpoint of biological activity or use in polynucleotide structure elucidation. For this reason reactions that involve vigorous conditions are not included. A number of topics included here are dealt with more exhaustively elsewhere [2].

II. Base Modification by Nucleophilic Species

A. HYDROLYSIS

The positions available for nucleophilic displacements and addition reactions in the common purine and pyrimidine bases are those indicated by arrows:

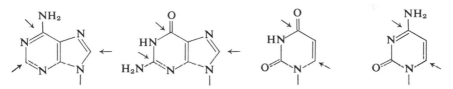

It is worth noting here that nucleophilic reactions may often be subject to electrophilic catalysis. For instance, hydrolysis or aminolysis displacement reactions may proceed much more rapidly on the protonated heterocycle than on the free base [3,4]. Alkylation, too, has marked effects on the hydrolytic stability of nucleosides. The ease with which purines are hydrolyzed, with opening of the 5-membered imidazole ring, varies enormously. Attack at C-8 is much facilitated by alkylation on N-7 in nucleosides (Section III,B). Reactions involving exchange of C-8 also occur and, for example, α-diketones may afford pteridines [4a].

Addition of nucleophilic reagents to the C-6 position in the pyrimidines leads to a variety of results which have important biological consequences. This will be discussed in Section II,C, but in broadest outline the initial consequence of nucleophilic attack at C-6 is activation of C-4 to nucleophilic and of C-5 to electrophilic substitution.

The hydrolysis of adenine and of guanine to hypoxanthine and xanthine, respectively, either acid- or base-catalyzed, appear to be reactions so slow as not to be significant [5]. Cytidine is very slowly deaminated to uridine in strong base [6]. It is also hydrolyzed in a reaction having a rate maximum at pH 4–5 [7,8]. This reaction is dependent on buffer ion catalysis, for example, by phosphate and citrate. The hydrolysis may result in part from direct attack by water at C-4 of the pyrimidine ring with general acid–base catalysis, but it is clear that a more complex mechanism is also involved, which is discussed below. In any event, the neutral hydrolysis of cytidine to uridine, though slow, is sufficiently fast to be biologically significant.

The displacement of sulfur from 2- and 4-thionucleosides can be effected, the order of reactivity being 2-thio < 4-thio [3]. The reaction rate is much increased if the thiono group is first modified. Thus osmium tetroxide [9], or

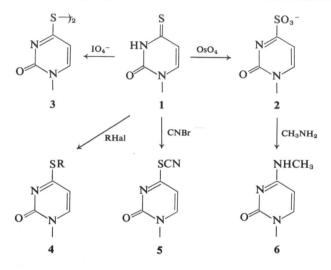

periodate oxidation [10], alkylation [3], or treatment with cyanogen bromide [11], lead to reactive intermediates (**1–5**, respectively) which can undergo hydrolysis or aminolysis (to **6**). Thus the thiouridine residue in yeast tRNA is converted to a uridine residue via **5** with retention of acceptor activity and its conversion to a cytidine or N_4-methylcytidine residue via **2** and **5** has also been demonstrated [10,12].

Dihydrouridines (**7**) suffer ring cleavage in strong base to give the corresponding β-ureidopropionate (**8**). The reaction would be much faster but for the formation at high pH values of the nonproductive anion (**9**; λ_{max}, 230 nm) [13,14]. When this cannot form, as in the case of 1,3-dimethyl-5,6-dihydrouracil (**10**; R = CH_3), hydrolysis is rapid [15].

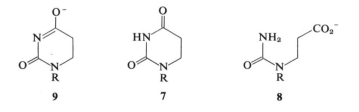

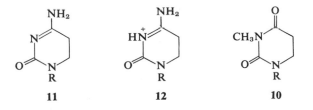

11   12   10

Dihydrocytosines (**11**) undergo extremely rapid hydrolysis under mild conditions. Thus the cation (**12**) has a $t_{1/2} = 15$ min at 37°C [16]. Dihydrocytosine nucleosides and nucleotides can be made by catalytic hydrogenation over rhodium-on-alumina, but purity is difficult to achieve due to deamination [16,17]. Part of the latter results from overhydrogenation to the tetrahydro derivative [18]. No dihydrocytosine derivatives have been observed to occur naturally. Their sensitivity to hydrolysis may have prevented detection, but it is probable that the dihydrouridine residues in tRNA's are modified uridines and not artifacts of dihydrocytidine hydrolysis [19].

B. REACTIONS WITH HYDRAZINES

It has long been known that hydrazine hydrate cleaves uracil and cytosine but not the purines. The former gives rise to pyrazolone (**15**) and the latter to aminopyrazole (**16**) as the major product [20,21]. Mechanistically the reaction is believed to proceed as shown for the uracil case, the essential point being that nucleophilic addition to the activated 5,6-double bond, to give **13**, initiates the cleavage reaction; a ureidopyrazolidone intermediate (**14**) has been isolated in one instance [22]. Discussion of the further consequences of

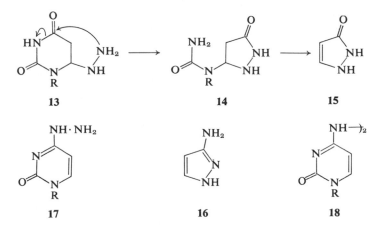

13   14   15

17   16   18

this cleavage reaction for polynucleotide degradation is deferred until later (Section IV,D,3). More dilute hydrazine solutions and those buffered in the

region of pH 8 are erratic reducing agents converting uridine to dihydrouridine [23]. The major reaction at pH 6 with cytosines is the displacement at
N-4 generating an N_4-aminocytosine (**17**) and even the *N,N'*-bispyrimidinyl-
hydrazine (**18**) [24], uridine and thymidine apparently being stable [25].
Other *N*-alkyl hydrazines also react in this way, but of particular interest are
a series of *N*-acyl derivatives, *inter alia* **19** (R = CH_3, NH_2, $CH_2CO_2^-$, and
4-pyridyl). These do not react with uridine or thymidine, nor with the
purine nucleosides, but with cytosine derivatives give the corresponding
4-hydrazides (**20**) [26–30]. The pH dependence (rate maxima in the pH range
4–5) suggests that the reaction may have value in polynucleotide modification,
though the reactions are disadvantageously slow, $t_{1/2} = 50$ hr with **19**
(R = $CH_2NC_5H_5^+$). On the other hand, it is noted that ribonuclease

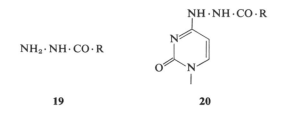

NH₂·NH·CO·R

19   **20**

hydrolysis of cytidine-2′,3′-phosphate after modification with **19** (R =
$CH_2NC_5H_5^+$) is one-thirtieth of that of the parent cyclic phosphate. The
application to polynucleotide sequence studies is obvious [31].

It is not clear whether the N_4-substitution products are the primary reaction
products or whether an intermediate 5,6-double bond adduct is formed
initially in these reactions. The significance of this point will become clearer
when the related reaction with hydroxylamine is discussed.

C. REACTIONS WITH HYDROXYLAMINES

Hydroxylamine resembles hydrazine in that, although it is a relatively
weak base (pK_a, 6.5), it is an exceptionally powerful nucleophile. In this
connection the so-called "α-effect" has been discussed extensively [32].
Alkoxyamines (pK_a of CH_3ONH_2 is 5.8), too, are very reactive, though less
so than the parent base. Since it is important that reactions with nucleic acids
should be conducted near neutrality, the low pK_a values of oxyamines allow
a high concentration of the free base to be present in these circumstances.
Hydrazine and hydroxylamine and their *N*-substituted derivatives are unstable. It is likely that many of their biological effects are to be ascribed to
reactions resulting from radical production and not only from the chemical
processes discussed here [33,34]. Alkoxyamines in general are much more
stable.

Hydroxylamine is highly specific for pyrimidine bases [35]. The rate of reaction with adenosine is about 1/200 that of cytidine. The first product is N_6-hydroxyadenosine and, in an even slower reaction, this is converted to adenosine-N_1-oxide; other minor products of unknown structure are also formed. Guanosine reacts extremely slowly [36,37].

When cytosine reacts with hydroxylamine solutions at neutral pH a fall in optical density is noticed and two products are obtained. These are N_4-hydroxycytosine (**21**) and N_4-hydroxy-6-hydroxylamino-5,6-dihydrocytosine (**22**) [38,39]. Kinetic analysis of the reaction using UV or NMR (in D_2O) spectroscopy shows that the two products are formed simultaneously [40,41]. These in turn equilibrate with each other at a much slower rate [42]. The preferred tautomeric states of **21** and **22** are those shown [43]. Analogous products are formed with methoxyamine [44]. N_4-Hydroxycytosine is clearly formed by a direct substitution reaction. The bishydroxylamino compound (**22**) is formed as a result of the addition of hydroxylamine to the 5,6-double bond, the resulting dihydrocytosine (**23**) then undergoing

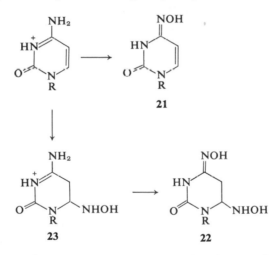

fast substitution of the amino group at C-4. There is a good analogy for the second step. 1-Methyldihydrocytosine (pK_a 6.5) as the cation (**24**) reacts extremely rapidly with hydroxylamine to give **25**; the free base reacts much more slowly [16]. The pK_a of **23** is probably about 6.0, i.e., between that of 1-methyldihydrocytosine [16] and cytidine hydrate (**26**) [45], so that **23** will be protonated under the conditions of the reaction. Cytosine hydrate residues (as formed photochemically in TMV nucleic acid, for example) also react rapidly with hydroxylamine to give the relatively stable **27** [46]. It is also probable that it is the protonated form of cytosine which reacts with hydroxylamine. The rate is maximal at pH 5–7 falling off on either side [47]. Had

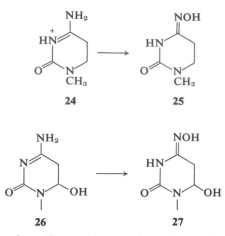

the free cytosine base been the reacting species an increasing rate with increase in pH would have been predicted. Protonation on N-3 is ideally suited to increase the electrophilic character of C-6.

In order to understand the mutagenic action of hydroxylamines and their use as probes for polynucleotide structure it is necessary to consider the reaction more closely. Consider the first addition reaction to give **23**. Is this a one-step process and is it reversible? What is the stability of **23**? Because the irreversible conversion of **23** to **22** supervenes, the answers can only be approached indirectly. The following points are relevant.

When 1-methyl-5,6-dihydrocytosine is dissolved in D_2O the C-5 protons are found to undergo exchange. As with hydrolysis and the reaction with hydroxylamine, the cation (**28**) undergoes deuterium exchange very rapidly,

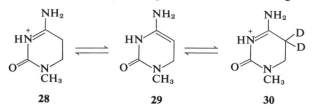

the free base very slowly, to give **30**. The reaction is conveniently followed by NMR spectroscopy. Presumably the exchange occurs via the enediamine (**29**), the C-5 protons being rendered acidic by the amidinium system [16]. Analogously photochemical hydration of [5-³H]cytidine 5′-phosphate (**31**) leads to the release of tritium and the reaction can be used to estimate the production of cytosine hydrate in nucleic acids [48]. The chemistry of the compound (**32**) is instructive [49]. In solution in D_2O at pD 6.8, the C-5 proton undergoes exchange ($t_{1/2}$ 10 min at 39°C). The compound also undergoes hydrolysis to the uracil (**36**) with intermediate accumulation of **35**, and with hydroxylamine

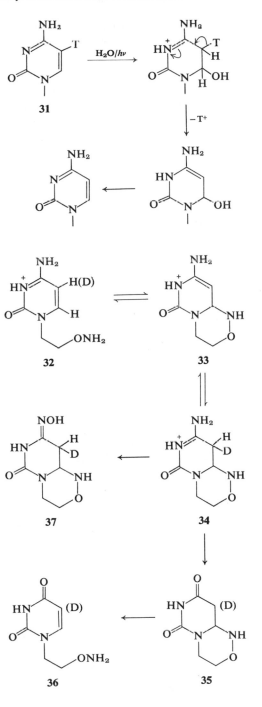

the product **37** is rapidly formed. Evidently the addition product **34** is formed reversibly giving rise to the various phenomena observed. From this it seems likely that the first step in the hydroxylamine reaction on cytosine should be reversible. Recently deuterium exchange into cytidine catalyzed by hydroxylamine at pH 7.5, but not 5.5, has been observed [50].

Two other observations are important in the elucidation of the mechanism. The ratio of the products depends on the reciprocal of the hydroxylamine (or methoxyamine) concentration and on the pH [51]. Finally when the hydroxylamine reaction on 1-methylcytosine is carried out in D_2O the product (**38**) carries only one deuterium at C-5 which is *trans* to the hydroxyamino group. An analysis of the NMR spectrum shows that the added groups are

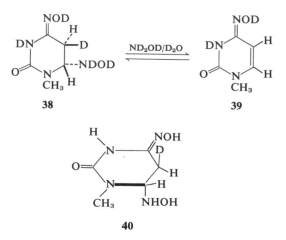

diaxial (**40**) [52]. Under hydroxylamine-catalyzed elimination to form an equilibrium mixture (**38** ⇌ **39**) the nondeuterated product alone is formed, so that the elimination is also stereospecific. It showed a primary isotope effect, K_H/K_D, of 5.5 [52]. These data may be rationalized in a reaction scheme in which the additional enediamine intermediate (**42**) is introduced [53], with the added assumption that hydroxylammonium ion and its conjugate base are involved in the proton transfers in **41** ⇌ **42** ⇌ **43**. If, too, $k_2 \gg k_{-2}$, as is universally true for the association and dissociation of carbon acids, the kinetic observations can be shown as being satisfied. The important conclusion is that the intermediate (**43**) should be stable under conditions where base-catalyzed elimination is prevented. Evidence has been given that this situation may prevail in polynucleotides, though this is not direct [54]. The question is important in deciding whether the species (**43**) can be involved in hydroxylamine mutagenesis. An earlier kinetic analysis [51], less satisfactory in that the reaction scheme involved a fast uncatalyzed elimination of hydroxylamine (**43** → **41**), led to the conclusion that **43** can have but fleeting existence

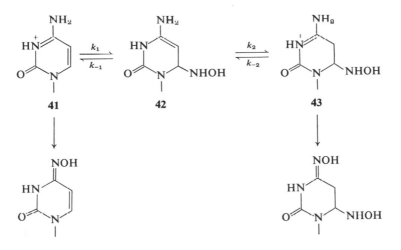

41   **42**   **43**

[50,51]. The closely related cytidine hydrate is stable enough for manipulation; its reversion to cytidine is general acid–base-catalyzed [55,56].

One general point emerges from the above discussion. It is that the addition reaction of the 5,6-double bond in cytosine is an equilibrium process, the proton transfers being subject to general acid–base catalysis. The principle

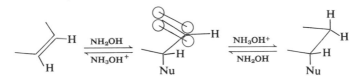

of microscopic reversibility is then upheld. In a later section it will be shown that the addition of another nucleophile, bisulfite, follows the same principles. In a number of cases nucleophilic addition proceeds axially and the axial conformer does appear to be thermodynamically the more stable [52,57]. The steric consequences of this must be important in reaction with polynucleotides.

The addition reaction is strongly inhibited in the case of 5-methylcytosine and 5-hydroxymethyl cytosine. The exchange reaction at C-4 to give the corresponding 5-substituted N_4-hydroxycytosine proceeds at a somewhat reduced rate [42,58].

Why then does uridine not appear to react with hydroxylamine in the region of neutrality? The answer is that the equilibrium is established, but is not in favor of the adduct at pH 5–7; the drop in optical density suggests this. There is no subsequent process (as in the case of cytosine) to remove the adduct irreversibly. 5-Nitrouracil shows no reaction again presumably because the equilibrium is unfavorable, but 5-bromouridine does react giving rise to uridine via the 5,6-adduct. The strong lethal effect of hydroxylamine on bacteria containing 5-bromouracil incorporated in their DNA is probably related to this phenomenon [35]. When uridine is treated with hydroxylamine at higher pH values (9–10) there is a subsequent irreversible step available— that of cyclization (**44**) [2,38,47]. Isoxazolone (**46**) is formed together with ribosylurea (**47**). The latter then forms *N*-hydroxyribosylamine (**48**) [59]. The rate of the reaction decreases at higher pH's due to the ionization of the

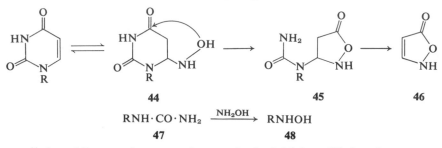

44   **45**   **46**

$$RNH \cdot CO \cdot NH_2 \xrightarrow{NH_2OH} RNHOH$$

47   **48**

uracil ring. Alkoxyamines can take part in the initial equilibrium but cannot undergo the ring closure. Therefore no reaction is observed. It is for this reason that their specificity in distinguishing cytosine from uracil (or thymine) is considered to be superior to that of hydroxylamine [2]. The application of these reactions to polynucleotides is discussed in Section IV,D.

A number of *N*-arylhydroxylamines, or compounds which are converted to these *in vivo*, are biologically active. It is certain that these are acting as electrophilic agents and are therefore dealt with in the corresponding Section III,H.

The concept of nucleophilic addition at C-6 in uracils and cytosine cations leads to an understanding of other aspects of pyrimidine chemistry. The hydrolysis of cytosine derivatives, though slow, shows a rate maximum at pH 4.5 together with buffer ion catalysis [7,8]. Moreover, during the hydrolysis and at a greater rate deuterium incorporation occurs at C-5 [60]. This is ascribed to the reversible addition of the buffer (e.g., citrate) at C-6 to give **49**, with the expected consequences.

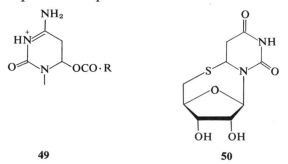

49   **50**

5'-Thio-5'-deoxyuridine is in equilibrium with the ring-closed form (50) which predominates at neutrality [61,62]. The addition of the 5'-hydroxyl group in isopropylidineuridine is not favored at equilibrium, but in basic solution it catalyzes deuterium exchange at C-5 [63]. Deuterium exchange is simply an electrophilic substitution. Formation of 5-hydroxymethyluridine

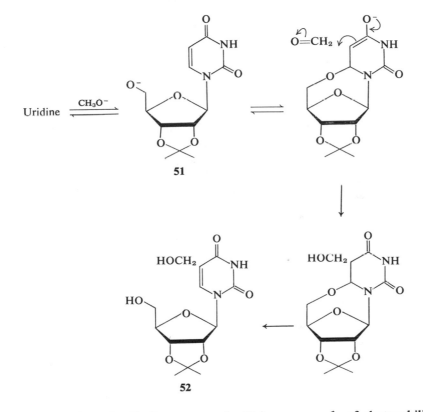

(52) catalyzed by the 5'-alkoxy group in 51 is an example of electrophilic substitution leading to C—C bond formation [63]. 1-Methyldihydrocytosine is extremely reactive in this respect [64]. In general one should expect a class of reactions of the type 53 → 54 with a corresponding set in the cytosine series.

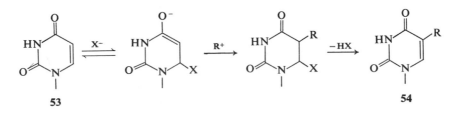

D. REACTIONS WITH BISULFITE

A particularly important nucleophilic addition to the pyrimidine 5,6-double bond is shown by bisulfite ion. The reaction is in essence closely similar to that with hydroxylamine, but it has features which make it intrinsically interesting. In the first place the equilibrium is established rapidly and under the right conditions favors the adduct. The equilibrium position is pH-dependent, formation of the uridine adduct (**55**) being favored at pH 6, rapid elimination occurring in more alkaline medium. The addition reaction has *trans* stereospecificity and the added groups in the case of the product formed in D_2O solution can be shown by NMR spectroscopy to take up the diaxial conformation (**56**). The adduct derived from uridine can be isolated in a pure state [65,66].

The adduct (**57**) derived from cytidine can also be isolated but is, as expected, less stable. It is prone to hydrolysis, like dihydrocytosines, although the pH-rate relationship appears to be complex [67,68]. Thus bisulfite addi-

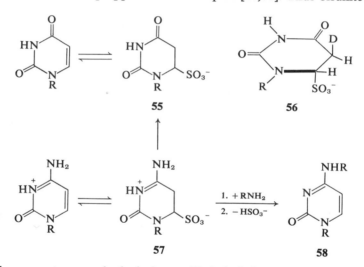

tion, then spontaneous hydrolysis at pH 5–6, followed by elimination at pH 8–9, is probably the mildest way yet devised for converting cytosine to uracil [69]. If, however, the pH is kept at 7.0 no hydrolysis occurs and cytosine is recovered intact from the adduct. The mutagenic activity of bisulfite may reasonably be ascribed to this process [70,71]. Present evidence suggests that reaction with bisulfite is confined to the pyrimidine nucleosides. Adenine and guanine derivatives are unreactive. Applied to polynucleotides, the cytosine-to-uracil conversion is effective but is inhibited by secondary structure as in DNA and in RNA [69]. The cytosine adduct (**57**) undergoes substitution reactions at C-4 other than hydrolysis. Reaction with a variety of amines, aliphatic and aromatic, leads to the corresponding N-substituted cytosine (58) [72], and, as expected, bisulfite catalyzes the reaction of cytosine at N-4 with hydroxylamine [73].

E. REDUCTION

It is unlikely that catalytic reduction will be extensively applied to poly-nucleotides, though it is effective with small oligomers. Room temperature and pressure hydrogenation of the pyrimidine bases can be effected over platinum or, more commonly, rhodium-on-alumina [74]. Uracil (and thymine) derivatives go easily to the 5,6-dihydro compound. Cytosine and its congeners afford the 5,6-dihydro product, but further reduction of the 3,4-double bond with subsequent elimination of ammonia can lead to derivatives of 2-pyrimidone [18]. Pseudouridine (59) undergoes allylic reduction affording 5-ribityluracil (60) [75].

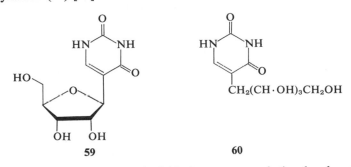

59   60

Of recent years sodium borohydride in aqueous solution has been investi-gated extensively [76]. On the grounds that uridine will react in a photo-chemically excited state with a variety of nucleophiles, borohydride, as a source of hydride ion, was studied [77]. 5,6-Dihydrouridine was formed in competition with the hydrate. Thymidine and pseudouridine were also reduced, though at one-tenth the rate, and cytidine even more slowly. The course of such reactions can be conveniently studied by the use of boro-deuteride; the structure of the product can then be defined by NMR spectroscopy. Thus the first reduction product of thymidine is 61 [78]. In nonphotochemical reactions 4-thiouridine gives 62, while N_4-acetylcytidine (63) affords the tetrahydro product 65, probably via 64 [79,80]. In the nonphotochemical reactions the deuteride ion is behaving in a way comparable to other nucleophiles.

Dihydropyrimidine nucleosides are also subject to borohydride reduction in solutions of pH 8–10, in a dark reaction. Thus dihydrouridine reduction followed by acid hydrolysis of the ribosyl residue gives ureidopropanol; using borodeuteride the product is (66) [79]. When uridine photohydrate is reduced

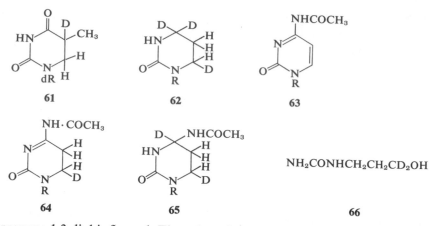

61   62   63

64   65   66

propane-1,3-diol is formed. The value of these reactions can at once be seen since if borotritiide ($[^3H]BH_4{}^-$) is used, a radioactive label can be introduced. Thus uracil photohydrate residues formed on irradiation of coliphage R17 RNA have been assayed in this way [81,82]. Dihydrouridine residues in transfer RNA are reductively labeled, as are, of course, uridine and *N*-acetylcytosine residues, though reaction is inhibited in structured regions of the tRNA molecule [83–85]. Among other minor bases in tRNA which are altered by borohydride reduction are 7-methylguanine, base Y in tRNA$_{yeast}^{Phe}$, 1-methyladenine, and 3-methylcytosine [86]. The reduction of 1-methyladenosine (**67**) gives the 1,6-dihydro compound (**68**) which, interestingly, is rapidly autoxidized to N_6-methyladenosine (**69**), essentially a mild Dimroth rearrangement; oxidation with nitrous acid gives back 1-methyladenosine [87].

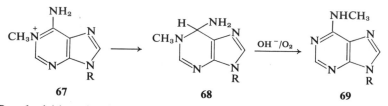

67   68   69

Borohydride reduction as a tool in the study of polynucleotide structure is considered in Section V,B,2.

III. Base Modification by Electrophilic Species

A. INTRODUCTION

In this section we discuss a heterogeneous group of reagents which can all be classified as electrophilic in character. A very important member of this class, the hydroxonium ion (and other acids) will not be discussed as it should be subsumed under the broader aspect of tautomeric equilibria. Nevertheless, the state of ionization of a heterocyclic base may control in large measure its rate of reaction with electrophilic agents. Unfortunately few definitive mechanistic studies have been made, so that one must fall back on analogies from related systems.

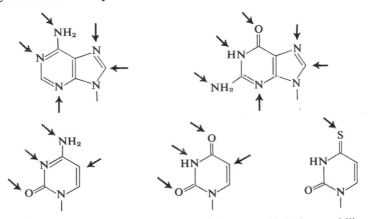

In the formulas, arrows indicate the positions at which electrophilic attack may occur. The heteroatoms represent positions of high electron density. The multiplicity of reactive centers precludes at present any overriding theory, certainly so far as relative rates are concerned. Thiol and thiono groups in organosulfur compounds are generally very reactive toward electrophilic reagents, so that the sulfur atom in the thiouracils and thiocytosine in tRNA will, in principle, be subject to attack by many of the reagents mentioned in this section.

B. ALKYLATION

A number of reviews in depth are available which deal with the very large amount of research on the biological effects of nucleic acid alkylation and the attempts to discern the underlying chemistry [88–90]. These should be consulted; here we only discuss the more general aspects of the chemistry [3,4,91,92]. Acylation is not dealt with in this chapter, since much of the large body of work is related to functional group protection in synthetic studies [92a].

The rate of alkylation at a given position in a base is affected by structure in the polynucleotide in which the base is present; it is also dependent on the nature of the alkylating agent. We may roughly classify alkylating species into four groups, of which some generalized examples are shown. These will either represent the reagent itself or intermediates generated from it, *in situ*.

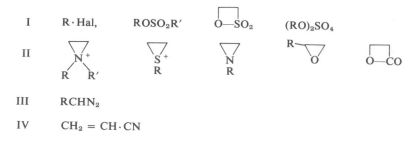

I   R·Hal,   ROSO$_2$R'   O—SO$_2$ (RO)$_2$SO$_4$

II

III   RCHN$_2$

IV   CH$_2$ = CH·CN

It is not possible here to do justice to the enormous range of reagents that have been studied, many of which can be brought under the classes shown in the scheme shown here. It is probable that many compounds in classes I and II alkylate by way of a bimolecular (S$_N$2) mechanism. For instance, there is kinetic evidence that methyl and ethyl methane sulfonates alkylate bases in DNA by an S$_N$2, but the isopropyl ester by an S$_N$1 process [93]. The molecularity of substitution reactions depends on the structure of the alkylating agent, as in the above examples, on the reactivity of the nucleophile and on the solvent. Detailed mechanistic interpretations are, in consequence, difficult to reach, the more so in cases where the reaction is carried out *in vivo* and where correlation with a biological phenomenon (e.g., mutagenesis or inactivation) is desired. Attempts to do this with a wide range of methane sulfonates have been made [94]. It may be worth pointing out, however, that many compounds in classes I and II will react with the poorly nucleophilic water (i.e., destruction of the reagent) by an S$_N$1 process, but with a base by a bimolecular substitution.

The 2-chloroethylamino and 2-chloroethylthio derivatives are exceedingly reactive. This is associated with two successive nucleophilic displacements, the first intramolecular by the fast formation of an ethyleneiminium or ethylenesulfonium ion, the second with the opening of the strained ring system [95].

R$_2$NCH$_2$CH$_2$Cl ⟶ R$_2$$^+$N◁ ⟶ R$_2$N·CH$_2CH_2$Nu

RSCH$_2$CH$_2$Cl ⟶ RS$^+$◁ ⟶ RS·CH$_2$CH$_2$Nu

ArNRCH$_2$CH$_2$Cl ⟶ ArNRCH$_2$CH$_2$$^+$ ⟶ ArNRCH$_2$CH$_2$Nu

70

It has been held that with the *N*-arylchloroethylamines an ethyleneimine intermediate is not formed; instead a rate-determining ionization to give a reactive carbonium ion (**70**) occurs [88]. This has always seemed most unlikely. The kinetic results are equally well accommodated on the view that the formation of the *N*-arylethyleneiminium ion is rate-determining, whereas the second step is rate-determining in the case of the *N*-alkyl compounds. Fortunately, the matter has recently been clarified [96]. In the acetolysis of the dideuteriated compound (71) deuterium is completely scrambled between the 1- and 2-position of the hydrolysis product. The intervention of the iminium

$$\text{Ar}\cdot\text{NRCH}_2\text{CD}_2\text{Cl} \longrightarrow \overset{R}{\underset{CH_2}{\overset{+}{Ar}N}}\overset{CD_2}{\underset{}{\big|}} \longrightarrow \text{ArNR}\cdot\text{CH}_2\text{CD}_2\text{OH} + \text{ArNRCD}_2\text{CH}_2\text{OH}$$

71   **72**

ion (72) and not 70 is therefore proven. With strong nucleophiles a direct displacement pathway competes.

The bis(2-chloroethyl) compounds, the nitrogen (73) and sulfur (74) mustards, the bisepoxides (75), and bismesylates (76) have been particularly widely studied as they are biologically very active, a result that may, in part, be ascribed to crosslinking (Section V). However, the monofunctional

$CH_3N(CH_2CH_2Cl)_2$ $S(CH_2CH_2Cl)_2$

73   **74**

$\overset{O}{\overset{/\backslash}{CH_2}}\cdot\text{CH}\cdot\text{CH}\cdot\overset{O}{\overset{/\backslash}{CH_2}}$ $CH_3SO_2\cdot O(CH_2)_4OSO_2\cdot CH_3 \cdot$

75   **76**

compounds are generally better mutagens. The relation between functionality and biological activity has been discussed in detail elsewhere [89].

Diazoalkanes present a somewhat different picture [97]. They do not react, per se, with nucleophiles (they are themselves nucleophilic in character, $CH_2^-\cdot N_2^+$). Alkylation of anions is therefore very slow. It is supposed that an alkyldiazonium ion is the reactive species, generated by abstraction of an acidic proton, e.g.,

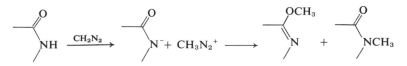

A number of compounds, such as *N*-nitroso-*N*-ethylurea and *N*-methyl-*N'*-nitro-*N*-nitrosoguanidine (NG), are of the type which generate diazoalkanes on reaction with strong base. They are also alkylating agents in neutral aqueous solution and potentiation of their alkylating ability by thiols has been noted and this may be important *in vivo* [97–99]. Though at one time NG was held to act via diazomethane [100], more recent evidence shows that the alkyl group is transferred intact [101].

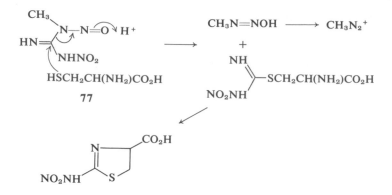

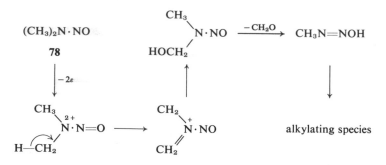

Another group of agents, particularly studied for their powerful carcino-genicity, are the *N,N*-dialkylnitrosamines (e.g., **78**). These are alkylating agents, *in vivo*, for RNA and DNA [102], from which the alkyl group is transferred intact, i.e., a diazoalkane is not involved [103,104]. Their activa-tion is believed to depend on oxidative removal of an alkyl group, e.g.,

In considering the position at which alkylation occurs on the heterocyclic bases, class I and II reagents broadly fall into the same category. As is general among *N*-heterocycles having amino groups in the α- or γ-relation to the heteroatom, the basic centers are the ring nitrogens. Thus N-1, N-3, and N-7 in adenosine, N-3 and N-7 in guanosine, and N-3 in cytidine derivatives are subject to alkylation. However, the order of basicity of these positions, as measured by pK_a is not directly related to their relative rates of alkylation. As has been pointed out [91], protonation is an equilibrium process, whereas alkylation is irreversible. The latter is subject to kinetic control and the product distribution is dependent on this; steric factors may therefore play a large role.

There have been many studies made on nucleosides, nucleotides, and other simple systems with a variety of alkylating agents. Toward esters of strong acids the order of reactivity is guanosine > adenosine > cytidine with

uridine essentially unaffected. The corresponding major products using, for example, dimethyl sulfate in neutral aqueous systems are **79** to **83** [90,105–112]. The N-7 position of guanine is the most nucleophilic of all sites in mono- and polynucleotides alike toward alkylation. The next is N-1 in adenine. 1-Methyladenosine (**80**) is essentially the sole product with dimethyl sulfate [113,114]; however, N-7 and N-3 alkylation (**82**) are observed as very minor processes, the latter position taking precedence over N-1 in alkylation of DNA [90].

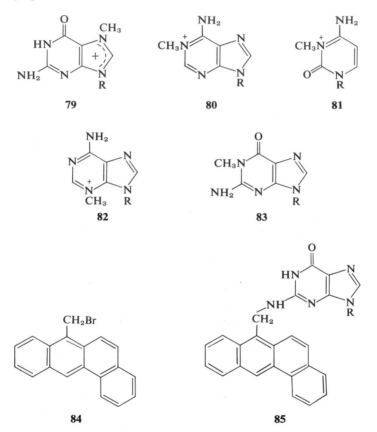

It is very interesting, by contrast with the above, that the highly reactive "benzylic" halide, 7-bromomethylbenz[*a*]anthracene (**84**), shows anomalous behavior. This carcinogen reacts preferentially at the amino group of guanosine (to give **85**), and of adenosine, when the reaction is performed in aqueous solution [115]. Its reaction with DNA is discussed in Section V,B,3.

When alkylation is carried out under more basic conditions the ionizable lactam system becomes methylated. Thus guanosine gives 1-methylguanosine

(83) [116]. In the same way inosine gives the N-7 and N-1 methyl derivatives under neutral and basic conditions, respectively. O-Alkylation has not been observed in neutral aqueous conditions. Under mildly basic conditions 1-alkyladenosine undergoes a Fischer–Dimroth rearrangement, via **86**, to give the N_6-alkyl derivative (**87**) [117]. Alkylation under these conditions can therefore lead to a 1-N_6-dialkyladenosine (**88**) [118].

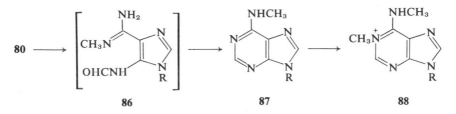

With diazomethane—often used by shaking an ethereal solution with a buffered aqueous solution of the substrate—the order of reactivity is uridine (to give **89**) > guanosine > cytidine > adenosine [105,106,119,120]. Thymidine is considerably less reactive than uridine. It is interesting that the major positions (but not rates) of alkylation are the same as those using esters of strong acids. In addition to the large reactivity of N-3 in uridine (also shown toward trimethyl- and triethyloxonium ions [121]) there is one important difference between diazomethane and dimethyl sulfate. The former effects O-alkylation by a minor reaction pathway. Inosine affords the N-1

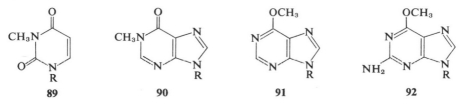

and O-6 products (**90** and **91**) in comparable amounts [122], although the tautomeric state of inosine strongly favors the lactam form [123, cf. 124]. Guanosine also shows O-6 alkylation to give the minor product **92** with diazomethane [125], and, importantly, with alkyl nitrosamides [126]. O_6-Alkyl guanine is formed in salmon sperm DNA by N-methyl- and N-ethyl-N-nitrosourea (NMU and NEU) and by N-methyl-N'-nitro-N-nitrosoguanidine (NG), but not by methyl sulfate [126,127]. The difficulties of defining the precise molecular target to explain a given biological activity is illustrated by these facts. Mutagenesis in T2 phage can be effected by NMU, NEU, and diethyl sulfate. Dimethyl sulfate is inactive [126]. NG and other nitrosamines are effective liver carcinogens [103]. These and dimethyl sulfate lead to alkylation of liver nucleic acid, although dimethyl sulfate is not carcinogenic. It has

generally been held that N-7 of guanine and N-3 of adenine are the major alkylation sites, although attack at N-3 of thymine is quantitatively more important in T2 with NMU and NEU [128]. It would appear, at least, that the earlier views concerning N-7 alkylation of guanine as the target and the corresponding explanations for GC → AT mutation transition must be modified. Have we then to consider explanations in terms of thymine alkylation, or of guanine O-6 alkylation—for which experimental tests and mutagenic theories could be devised—or is further search necessary for an even less evident reaction than the latter? We will discuss in Section IV the cleavage of internucleotide linkages resulting from base and phosphate alkylation. It should be pointed out here that as a result of model experiments with dinucleoside phosphates and with homopolynucleotides the consensus of opinion is that very little *P*-alkylation occurs [108], though this has been contested [129]. It seems reasonable to suppose that the biological results of alkylation, mutagenesis, inactivation, and carcinogenesis are to be ascribed to a variety of causes and not to one particular chemical process.

Mention was made earlier that sulfur is a potentially very nucleophilic site, so that the thionucleosides, e.g., those present in transfer RNA, are prone to *S*-alkylation. Thus ethyleneimine reacts readily with 4-thiouridine **(93)** to give **(95)** via the intermediate **(94)** [130].

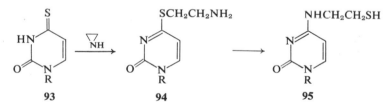

93   94   95

C. MICHAEL ADDITION REACTIONS

Another type of alkylation which has assumed importance in recent years is mechanistically different from those discussed above and warrants independent consideration. Compounds which have α,β-unsaturation with respect to a carbonyl or equivalent function (e.g., cyano, nitro, sulfonyl) undergo Michael addition reactions with nucleophiles. Such reactions may often be base-catalyzed, the function of the base being to generate the more powerfully

$$\text{Nu} \curvearrowright \overset{|}{\text{C}}\text{H}{=}\text{CH}{-}\overset{|}{\text{C}}{=}\text{O} \longrightarrow \text{NuCH}{-}\text{CH}{=}\overset{|}{\text{C}}{-}\text{O}^- \longrightarrow \text{NuCH}{-}\text{CH}_2{-}\overset{|}{\text{C}}{=}\text{O}$$

nucleophilic anion. Elimination to give the starting materials, initiated by base-catalyzed removal of the acidic α-proton, is often encountered. Potentially there are a very wide range of reagents in this class [131], though the number used successfully in aqueous solution is quite small.

Acrylonitrile has been extensively studied as a reagent for base modification. The reaction is base-catalyzed but at pH 8–9 a valuable degree of specificity is obtained [132–135]. Thus pseudouridine, inosine, and 4-thiouridine react rapidly to give **96–99**. 1-Cyanoethyl pseudouridine (**97**) is slowly converted reversibly to the 1,3-bis product (**97**). Uridine and ribothymidine react

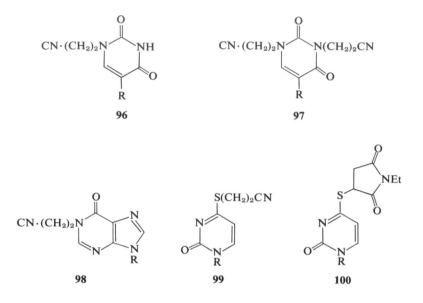

slowly. Thus $k_\psi/k_u \simeq 20$. Cytidine, adenosine, and guanosine are essentially unreactive. 4-Thiouridine, as the free nucleoside and in tRNA, is *S*-alkylated. The product (**99**) is formed reversibly, although 2-thiouridine is unaffected [134].

Another α,β-unsaturated carbonyl compound which is exceptionally reactive toward thiols is *N*-ethylmaleimide [136]. The reagent reacts specifically with 4-thiouridine to give **100**. As applied to transfer RNA it can, in effect, protect the 4-thiouridine residues while leaving 2-thionucleosides unaffected and available for oxidation or other manipulation [137].

D. Reactions with Aldehydes

1. *Formaldehyde* [137a]

The reactions of aldehydes with nucleophilic centers are, in general, subject to acid and base (very often general species) catalysis [32]. The initial addition reaction may be followed by dehydration and subsequent transformations. The reactions are potentially and very often realizably reversible, e.g.,

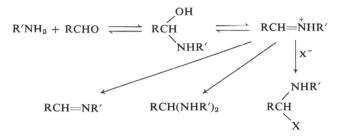

Formaldehyde in aqueous solution is present as the hydrate, $CH_2(OH)_2$, to the extent of 98%, but is nevertheless extremely reactive [138]. It reacts with the base residues of all nucleosides. An equilibrium position is established depending on the concentrations of the reactants, while slower, possibly irreversible processes may also occur [138–143].

The reaction with the nonbasic nucleosides uridine, pseudouridine, and inosine, is very fast in the forward and reverse directions [143]. The reaction therefore may become significant when nucleic acids are studied in presence of formaldehyde. An appreciable fraction of the uracil or thymine residues may, in these circumstances, be derivatized with consequent effects on higher structure, but the derivatives will not appear in any degradation products. The kinetics, pH optima, and structural specificity have been studied extensively both with mono- and polynucleotides, while rate and equilibrium constants for the reaction with nucleoside 5'-phosphates have been given [141,144,145].

The structures of none of the formaldehyde reaction products have been established rigidly. The product from adenosine, the most studied, has a UV spectrum consistent with the N_6-hydroxymethyl compound (**101**) and it is difficult to believe that this product would not be favored at equilibrium over the stronger bases (e.g., **102**) resulting from hydroxymethylation on N-1, N-3, or N-7. The highly reversible reactions with uridine and inosine presumably result from hydroxymethylation of the lactam-*N*, as in **103**.

It was noted at an early stage that the formaldehyde reaction appeared to be biphasic [140]. In an investigation of the second slower step (e.g., pH 4.8/15 days) with purine nucleotides and with RNA methylene bisnucleotides were isolated [146]. The latter, e.g., **104** are stable to base, to the extent that

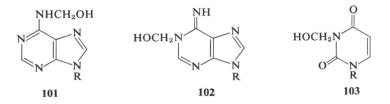

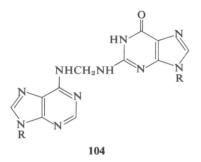

104

they can be isolated after alkaline hydrolysis of treated RNA, but they are less stable to acid. There is evidence for crosslinking in DNA [147] in addition to tRNA [148]. Development of this aspect of the formaldehyde reaction, with characterization of crosslinked products could contribute to tertiary structural studies, though the denaturing effect of the reversible reaction may vitiate this. The reaction with nucleic acids is discussed in Section V.

A slower reaction with formaldehyde under somewhat more vigorous conditions, of the pyrimidine nucleosides and nucleotides, gives 5-hydroxy-methyl derivatives [149]. It seems clear, now, that at least one reaction pathway involves activation by nucleophilic addition at C-6 (see Section II,C). However, the mechanistic question has certainly not been entirely clarified and, at least in acid solution, direct electrophilic attack (by $CH_2=O^+H$) seems likely. The base-catalyzed hydroxymethylation of uracil itself [151] probably depends on the process **105**, a view which correlates well with the instability of 5-trifluoromethyluracil by process **106** to give the carboxylic acid (**107**), compared with the stability of 5-trifluoromethyluridine [152].

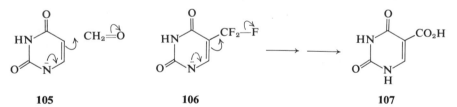

105   **106**   **107**

The 5-hydromethyluracil system is of interest in that it shows the reactivity characteristic of an allylic alcohol, for example, in its catalytic reduction to thymine and in other reactions [150, 151].

2. *Dicarbonyl Compounds*

The antiviral activity of β-ethoxy-α-ketobutyraldehyde (Kethoxal) (**108**) inspired the investigation [153] of its action and that of glyoxal and of related compounds on the nucleic acid bases. Of the compounds studied those that inactivated tobacco mosaic virus RNA also reacted rapidly with guanine, but not with the other base residues. These α-dicarbonyl compounds, therefore, show a satisfying degree of specificity. With Kethoxal a product could also be obtained from N_2-methylguanine and from isocytosine but not from 1-methylguanine, indicating that the 1- and N_2-positions were involved in the reaction. The adducts are in general quite labile, readily reverting to their components at pH values above neutrality; pH 7 appears to be optimum for the forward reactions.

The key to the structures of the compounds is to be found in the reaction with periodate [154]. In each instance an N_2-acylguanine is formed in a slow

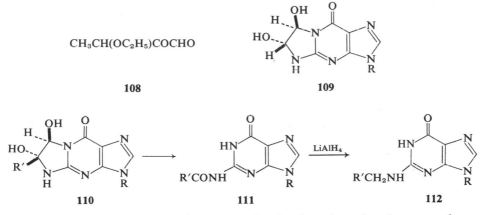

$$CH_3CH(OC_2H_5)COCHO$$

108   **109**

110   **111**   **112**

oxidation. A cyclic diol (**109**) structure for the glyoxal product is proposed, the two aliphatic protons showing no spin–spin coupling in the NMR spectrum, a result consistent with a *trans*-diol system [155]. Compounds derived from pyruvaldehyde and Kethoxal are formulated as **110** [R' = CH$_3$, CH$_3$·CH(OEt)] since the corresponding N_2-acyl derivatives [**111**; R' = CH$_3$, CH$_3$·CH(OEt)] are formed on periodate oxidation. Lithium aluminum hydride reduction of the *N*-acyl derivatives gives N_2-alkylguanines [**112**; R' = CH$_3$, C$_2$H$_5$, and CH$_3$CH(OEt)CH$_2$]. The orientation of the products derived from the ketoaldehydes warrants comment. In discussing the formaldehyde reactions it was noted that fast reversible addition to the imino nitrogen appeared to occur. Since in α-ketoaldehydes it is the aldehyde group which is the most reactive of the carbonyl functions, it is to be expected that this will interact first with the imino group, the subsequent ring closure then establishing the product in the observed orientation. The "denaturing" effect of formaldehyde on DNA [144,156] seems to find a parallel in Kethoxal reactions. The extent of reaction, at equilibrium, with tRNA is a function of the reagent concentration [157,158]. Likewise a reversible reaction occurs with the 6-amino group of adenine, its presence being shown by periodate oxidation of a mixture with glyoxal which affords N_6-formyladenine [155]. Nevertheless Kethoxal can be used for distinguishing guanine residues in tRNA which are not involved in higher structure [159]. At higher temperatures, all of the guanine residues in tRNA form a glyoxal adduct and this it is claimed protects against T1 nuclease cleavage at these sites [160].

Ninhydrin also reacts with guanine. The stability of the adduct is low, but its periodate oxidation product is more stable and protects the guanine residue against nitrous acid deamination [155]. Cytidine also forms a product (113) with ninhydrin, in this case stable to alkali, so that conceivably ninhydrin or another triketone could be used in nucleic acid modification [161]

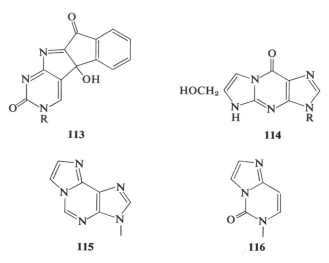

Other bifunctional aldehydes have been looked at in connection with specific base modification. Glycidaldehyde gives a product, probably 114 or an isomer, with guanosine [162]. Chloroacetaldehyde reacts specifically with adenosine and cytidine under mild aqueous conditions (pH optima 4.5 and 3.5, respectively) [163]. The fluorescence spectra of the products (115 and 116) are characteristic and the intense emission has been used to follow the reaction with tRNA [164].

E. REACTIONS WITH CARBODIIMIDES

Carbodiimides have played a large part in synthetic nucleotide chemistry, mainly in condensation reactions involving phosphate groups (see Chapter by Letsinger in Volume III). In these reactions, a transient intermediate (such as 117) is formed with the carbodiimide (often N,N-dicyclohexylcarbodiimide) which then may react, possibly by way of a polyphosphate (such as 118), with a variety of other nucleophiles, acids, amines, and alcohols. Dialkylphos-

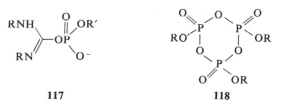

117   **118**

phoric acids themselves react to give tetraalkyl pyrophosphates, but neutral salts of these are unreactive. In effect this means that in reacting a polynucleotide with a carbodiimide at neutrality and above, the internucleotidic linkage will be unaffected, although a terminal monoesterified phosphate may be activated by the reagent. Indeed derivatization of the 5′-terminal phosphate of tRNA's can be effected in this way. However since carbodiimides undergo rapid acid-catalyzed hydration to the corresponding urea, both they and the derivatizing agent have to be supplied in large molar excess. Nevertheless, by this means the terminus of a tRNA(ROPO$_3$H$_2$) can be converted to a labeled methyl ester or phenylphosphoramidate [165].

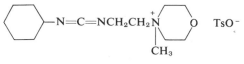

Bearing these considerations in mind, we can say that in aqueous solution at neutrality the sugar phosphate backbone or terminal residue of nucleic acids are unlikely to be effected by carbodiimide-type reagents.

It was observed that the water-soluble carbodiimide (**119**) reacted with uridylic acid under mild conditions (pH 8; 30°C; 4 hr) and that the adduct reverted to uridylic acid at pH 10.5. The product was given the general structure **120** on spectroscopic grounds. Guanine derivatives also reacted, though somewhat slower, and the products, thought to be **121**, also underwent reversion at the higher pH [166,167]. Forward and reverse reaction rates have

119

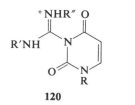   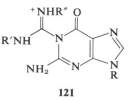

120   **121**

been measured [168]. No detectable reaction with adenine or cytosine derivatives was found [167]. Subsequently dihydrouridine, inosine [169], and pseudouridine were shown to react [170,171]. The latter gave a mono- and disubstitution product (**122, 123**; R′ = CMC residue).

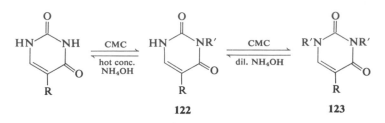

122   **123**

The value of these observations can be appreciated when it is added that modification of uridine or pseudouridine residues in a polynucleotide prevents pancreatic ribonuclease acting at these points so that the specificity of the enzyme can be increased [172]. The CMC residue, too, is cationic and the consequent alteration of electrophoretic mobility can be utilized to advantage [173]. Since the reaction with CMC appears to be confined to unstructured parts of a nucleic acid [174–176], the reagent has found much subsequent application. This is discussed in Section V.

F. REACTIONS WITH NITROUS ACID

The reaction of nitrous acid with nucleic acids has been widely studied since the observation that it inactivated viral RNA and led to mutations [177–179]. In the simplest terms it is clear that the deamination of cytosine to uracil must be mutagenic. It is not so certain what effect the changes adenine to hypoxanthine or guanine to xanthine will have. Moreover, at the pH's used for deamination (~4.2) slow loss of purine bases can also occur from DNA [180], and in addition nitrous acid leads to changes other than deamination.

Adenosine and cytidine as nucleosides or in nucleic acids are deaminated, without detectable side reactions, to inosine and uridine, respectively. Guanosine gives xanthosine (**124**) as major product, 2-nitroinosine (**125**) as a minor product and, it may be, some further unidentified product, when nucleic acids are deaminated [181–184]. A considerable number of estimates of deamination rates have been given, since it was felt that the variation of these with, for example, pH, might cast light on which bases are involved in mutations or inactivation. Some relative values are given in Table I [178, 185–187]. It is noticeable that the three bases (as nucleosides) only differ overall by a factor of 6, with G > A > C. Increasing the pH from 4.2 to 5.0 reduced the rate in TMV nucleic acid for A and C to 1/90 and that for

TABLE I

RELATIVE RATES OF DEAMINATION OF BASES WITH NITROUS ACID[a]

Substrate	G	A	C	Conditions	Ref.
Nucleoside	6.5	2.2	1.0	pH 4.5, 21.5°C	184
tRNA (*E. coli*)	2.1	1.4	1.0	pH 4.3	185
TMV–RNA	2.1	1.9	1.0	pH 4.3, 21.5°C	181
TMV	0.01	0.53	1.0	pH 4.2, 21.0°C	183
DNA (calf thymus)	2.2	0.44	1.0	pH 4.2, 20°C	182
DNA (heat-denatured)	1.3	0.63	1.0	pH 4.2, 20°C	182

[a] From Shapiro and Pohl [184].

G to 1/35. The mutation and inactivation rates were decreased by factors of 90 and 30, respectively. It was inferred that therefore hits on A and C were mutagenic and those on G were lethal [178]. This view can no longer be maintained as a general explanation in the light of newer understanding of mutagenic processes [188]. Among other types of mutagenic changes those resulting from large deletions are observed [189]. Of considerable interest, too, is the observation that short treatment of double-stranded DNA with nitrous acid renders it reversibly denaturable [190,191]. This is considered to depend on interstrand crosslinking. In fact this reaction is quantitatively important; in *Bacillus subtilis* transforming DNA [192] at pH 4.15 one crosslink is estimated to be formed per four deaminations [191], and it is suggested that it is the crosslinks which account for most of the deactivating events. About four deaminations contribute one inactivation, too, in the phages T2 and P22 [188].

The nitrous acid reactions presumably proceed through diazonium ion intermediates. That from guanosine (**124**) reacting with water gives xanthosine, and with nitrite ion, gives **125**. Halopurine nucleosides have also been formed

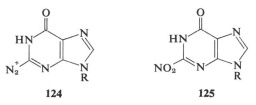

124   **125**

in this way (Goodman, Chapter 2, Volume I). The crosslinking reaction clearly depends on a proximity effect operating in double-stranded systems. If it is supposed, for example, that guanine is involved, many explanations in which crosslinking through an azo (—N=N—) group, a diazoamino (—N=N—NH—)

group or directly to C-2 by displacement of the diazonium group, can be conceived. All are common arlydiazonium ion reactions. Linkage to O-2 of the cytosine originally forming the base pair to the diazotized guanine residue with loss of N_2 is an obvious candidate. There is some evidence that this type of damage can be repaired *in vivo*, and, if this is so, mutational error induction of the point and deletion types could result in the process [188].

We should bear in mind that nitrous acid will react with secondary amines to generate nitrosamines. Aminopyrine with nitrite gives the carcinogenic dimethylnitrosamine, *in vivo*, and as we have seen this generates a methylating species when metabolized [193]. It is also known that dialkylnitrosamines undergo a transnitrosation reaction with other amines [194], so that alkylation may not be the only biologically important reaction of these compounds.

Finally, mutagenesis studies with the single-stranded S13 suggest strongly that nitrous acid reacts with thymine residues [195]; the nature of the reaction is quite unknown.

G. Reactions with Peracids

Over the years many more or less vigorous oxidizing agents have been applied to nucleic acid components [5]. A number of these are discussed later with reference to the cleavage of internucleotide links (Section D,3). Here we confine ourselves to *N*-oxidation, a process which is mild enough to preclude adventitious polynucleotide chain scission. The oxidation of heterocyclic bases to their *N*-oxides has been studied intensively [196]. This is so too for the nucleic acid bases, particularly as some of the *N*-oxides have been found to be carcinogenic [197]. A variety of methods have been used (Chapter 2, Volume I), but of these oxidation by peracids in neutral aqueous solution has been most successful. Although some destruction of all bases may occur, under optimal conditions uracil and guanine derivatives are essentially inert to monoperphthalic acid. Adenylic acid is oxidized smoothly to the N_1-oxide **(126)** and cytidylic acid to the N_3-oxide **(127)**. The oxidation of the former is somewhat more rapid though an estimate of relative rates is not available [198–200]. The use of *m*-chloroperbenzoic acid has been advocated for the synthesis of cytidine-3-oxide derivatives [201]. It oxidizes all of the common nucleotides [202], but its advantages have not been assessed in the wider

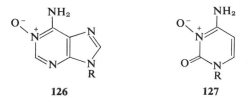

126   127

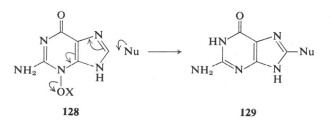

128   **129**

context. Monoperphthalic acid appears to have a degree of specificity for adenine derivatives sufficiently satisfactory for nucleic acid modification work and it has been so applied [200] (see Section V). The rate and extent of reaction with poly A and poly C are dependent on the reagent concentration.

The N-oxide of adenosine, but not of cytidine, is stable to base [203]. These compounds have characteristic UV spectra so that the quantitative aspects of N-oxide chemistry are readily dealt with. A novel substitution reaction is shown by the 3-oxides of guanine and xanthine, both of which are carcinogenic. On activation by O-acylation, e.g., **128** (X = $COCH_3$, SO_2CH_3) guanine 3-oxide undergoes reaction with many nucleophiles (Nu) in aqueous solution affording the 8-substituted purine (**129**) [204,205].

H. REACTIONS WITH ARYLHYDROXYLAMINES

A number of carcinogens are not in themselves effective but are converted *in vivo* into the active agent. Here we refer to 2-acetamidofluorene (**130**), 4-methylaminoazobenzene (**131**), and to 4-nitroquinoline-1-oxide (**132**). A large volume of work, recently reviewed [206], testifies to the fact that an important step in their activation is the conversion to the corresponding hydroxylamine (**133–135**). It is recognized that even these are probably not the "proximate" carcinogens. The latter may be the corresponding O-sulfonates, O-phosphonates (e.g., X = SO_3H, PO_3H_2) or other O-acyl or glycosyl derivative (**136**), since, at least in the case of the fluorene derivative, extensive studies have shown that the enzymic activities required for their generation are present [207–209].

We should recognize, first, that N-arylhydroxylamines will behave very differently from hydroxylamine itself or from O- or N-alkyl hydroxylamines. Nucleophilic reactivity at the nitrogen atom will be much reduced by conjugation to the aryl group. Moreover, heterolytic cleavage of the weak N—O bond will be accentuated by O-acylation, that is by making OX in **136**, a good leaving group. Phenylhydroxylamine undergoes acid-catalyzed conversion to p-aminophenol. The process generates an electrophilic species (**137**) which can undergo reaction at the o- or p-position or at the nitrogen center [210].

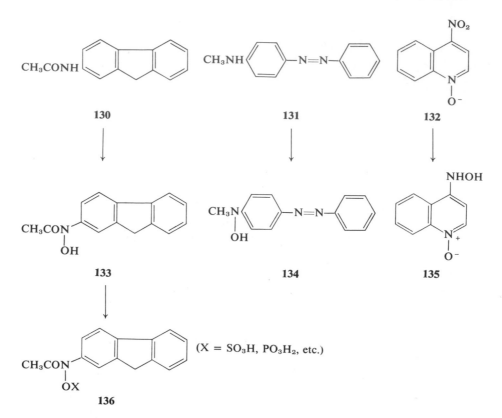

130	131	132
133	134	135

$(X = SO_3H, PO_3H_2, etc.)$

136

This essentially forms the basis of the important reactions of **130–132** in biological systems [211].

N-Acetoxy-2-fluorenylacetamide (**138**) reacts with proteins in part at methionine residues; reaction occurs at the *ortho*-position (via **139**) [206]. With nucleic acids, guanine is the major target *in vivo* [212] and this is confirmed by studies with the individual nucleotides [213]. With guanosine

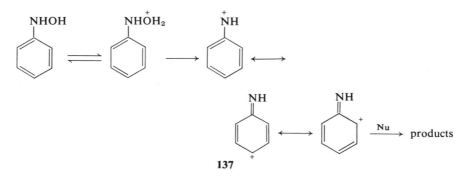

137

for example, electrophilic substitution at C-8 gives the product (**140**). The reaction proceeds in aqueous solution at pH 7. The structure of the product is proved by its hydrolysis to **141** and the independent synthesis of the latter, from 8-bromoguanine and 2-aminofluorene [213]. It is interesting that in the acid hydrolysis of **140** (R = deoxyribosyl 5′-phosphate) to **141** the lability of

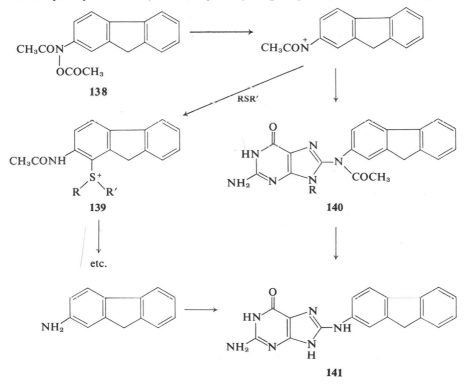

the glycosidic linkage is closely similar to that of deoxyguanylic acid, so glycosidic bond cleavage cannot be important for the observed biological effects. It may be that, as with 8-halogenation, the more stable conformation is *syn* (**142**; X = halogen or fluorenylamino) and not *anti* (**143**) as it is in the nucleoside itself [214,215].

Adenosine reacts in the same way as guanosine but at a much slower rate, though this minor reaction may account for the mutagenic activity of **138** for T2 bacteriophages [216] (but see Maher *et al.* [217]).

The specificity shown by *N*-acetoxyfluorenyl-2-acetamide and related compounds for guanine makes them potentially valuable reagents for nucleic acid modification studies. The *N*-acetoxy compound reacts with transforming DNA and with ribosomal and transfer RNA [217,218]. More work is necessary to show whether the reactivity of a guanine or adenine residue is

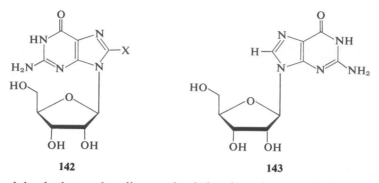

142   143

altered by hydrogen-bonding or by being in a base-stacked environment [215].

The reaction of guanine residues in mono- or polynucleotides with the diazonium salt (**144**) gives **145**, a further example of electrophilic substitution at C-8 [219,220].

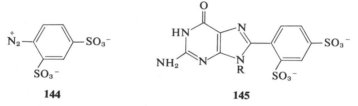

144   145

I. HALOGENATION

Halogenation of nucleosides is dealt with in detail in Chapter 2, Volume I. Here we refer only to those reactions which are sufficiently mild to be applicable to the nucleic acids themselves.

The reaction can be effected in water by bromine [221–224], iodine [225], *N*-bromosuccinimide(NBS), *N*-iodosuccinimide [226], iodide-thallous chloride [227], and iodine chloride (ICl) [228]; solvent dimethylformamide [228,229] has also been used. No detailed mechanistic study has been made in any instance.

Bromination occurs at C-5 in the pyrimidine derivatives and at C-8 in guanine residues; adenine derivatives are remarkably resistant. With uracil derivatives in water near neutrality the following transformations may occur.

Under somewhat acidic conditions the 5-bromouracil (**146**) is formed and this is so too for the nucleoside and nucleotide. Cytosine derivatives behave similarly giving the 5-bromo product. The saturated product (**148**) derived from uridylic (and correspondingly from cytidylic) acid releases the sugar following mild treatment with base, then acid; the ureide (**149**) is a possible intermediate [222].

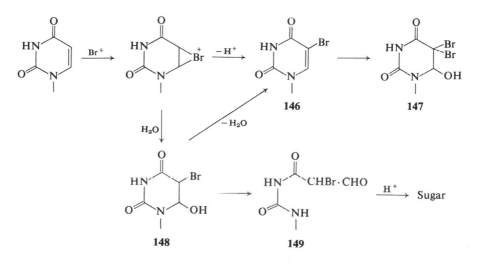

NBS reacts with polynucleotides. In reaction with TMV–RNA, guanine, and cytosine residues are modified, the former best at pH 9, the latter at pH 7 (as found with bromine itself) [230]. Although uridylic acid is readily brominated, very little reaction of uracil residues in TMV–RNA is claimed to be found. NBS has been used for selective bromination of yeast tRNA^{Ala} [231], while bromine itself has been used with tRNA in DMF solution [229], and with poly C [232].

Iodination is not nearly so readily effected. Iodine in potassium iodide solution is ineffective at 37°C. ICl in DMF, has been investigated and the order of reactivity U > C ≫ G found [228]. *N*-Iodosuccinimide is rapidly destroyed by water but it does iodinate uracil and cytosine residues in TMV–RNA, the latter more rapidly [226]. A recently investigated system is iodide in presence of the oxidizing agent thallic trichloride in water. As applied to DNA, essentially the only base modified is cytosine, native DNA reacting at 5% of the rate of heat-denatured material. Large amounts of iodine can be incorporated into the latter and into poly C with little strand breakage [227].

While iodine reacts very poorly with the common bases it rapidly modifies two of the minor nucleosides of tRNA [233,234]. N_6-Isopentenyladenosine (**150**) is converted to the cyclized product (**152**; X = I) [235]. The formation of the alcohol (**153**) and the cyclized product (**152**; X = H) in acid must be mechanistically similar, proceeding through the tertiary carbonium ion (**151**; X = I or H), or related species. This and other aspects of the chemistry of N_6-isopentenyladenosine are discussed in detail elsewhere [233]. 2-Methylthio-N_6-isopentenyladenosine in tRNA is also reactive, though the reaction may be more complex [236].

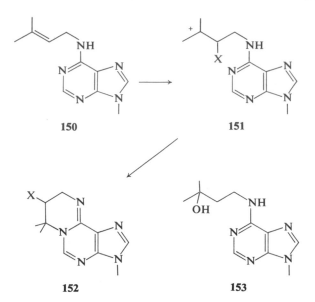

150   151

152   153

4-Thiouridine is oxidized rapidly by iodine to the 4,4-disulfide [237]. This reaction, too, occurs in *E. coli* tRNA[Tyr] in which two thiouridine residues are in contiguous positions and is not a result of intermolecular coupling [238–240]. Even in tRNA's having only one 4-thiouridine residue a reaction occurs, which (like disulfide formation) is reversible by thiosulfate. The iodine reaction with 2-methylthio-N_6-isopentenyladenosine may be reversible by thiosulfate [236].

We may conclude, with respect to tRNA, that iodine is among the most specific of the base-modifying reagents.

IV. Reactions Affecting the Internucleotide Linkage

A. INTRODUCTION

In this section we will discuss reactions which lead, directly or indirectly, to the cleavage of the internucleotide linkage. The cleavage reactions themselves fall into two classes: those that proceed by hydrolysis (P—O cleavage) and those that proceed by an elimination (C—O cleavage). In the former, the very much greater susceptibility of the polyribonucleotides as compared with their 2′-deoxy analogs is well known. In the latter, the presence of a carbonyl (or related) function β to the phosphate residue is essential to activate the elimination process. Thus, for example, glycerol 1-phosphate is highly stable to base, whereas glyceraldehyde 3-phosphate is extremely labile even at pH values near neutrality. In terms of nucleotides, elimination reactions are important in situations where the glycosidic linkage has been broken, that is, when the heterocyclic aglycone residue has been lost, or when a carbonyl function has been introduced into the sugar residue by oxidation.

$$^{2-}O_3POCH_2CH(OH)CHO \longrightarrow PO_4{}^{3-} + CH_3COCHO \xrightarrow{OH^-} CH_3CH(OH)CO_2{}^-$$

It is important in nucleotide sequence work, where enzymic methods are not applicable (or as an adjunct to these) that some means of increasing the specificity of cleavage of a given internucleotide link should be available. Elimination, in contrast to hydrolysis, can provide the necessary rate difference. In principle, too, changing the degree of esterification of the internucleotide phosphoryl residue can achieve this and in this connection we will discuss phosphate alkylation briefly.

A number of hybrid molecules containing ribo- and deoxyribonucleotides are known, for instance, the RNA-linked DNA replicative fragments from *E. coli* and other sources. Their hydrolytic chemistry is exactly as expected from the known chemistry of RNA and DNA [240a].

B. PHOSPHATE DIESTER HYDROLYSIS

Simple dialkyl phosphates, considered as analogs of the internucleotide linkage in DNA, are extremely stable to base and, moreover, hydrolysis, when it does occur, probably does so by alkyl-oxygen cleavage. Acid-catalyzed hydrolysis is also slow. Hydrolysis rates are increased dramatically by neighboring-group participation, a matter which has been studied much in recent years though mainly in compounds other than nucleotides [241]. The 5-membered cyclic phosphates are exceptional when compared with their acyclic analogs. Thus ethylene phosphate (**154**) is hydrolyzed by acid and by base in the region of 10^7–10^8 times faster than dimethyl phosphate [242]; ribonucleoside 2′,3′-phosphates equally are very rapidly hydrolyzed [243]. Under both conditions P—O cleavage alone occurs [244]. Studies of this and related phenomena in recent years have led to a wider understanding of phosphate ester chemistry [245]. Briefly, it appears that ring strain can account for much (but not all) of the observed rate increase. Mechanistically, the strain can be released by going to a trigonal bipyramidal pentacoordinate intermediate (**155**) by addition of water, the ring including one apical and one basal bond, and hence a 90° O—P—O bond angle. The intermediate can then break down by ring cleavage, or, competitively, by loss of an exocyclic substituent, manifesting itself in rapid ^{18}O-exchange in $H_2{}^{18}$O-enriched solvent. The concept of pseudorotation [246] is important in this context; that is the fast exchange of the apical (a) and basal (b) ligands about phosphorus (**156**) as a result of a deformation process (**157**). Groups may enter or leave only at apical positions [245].

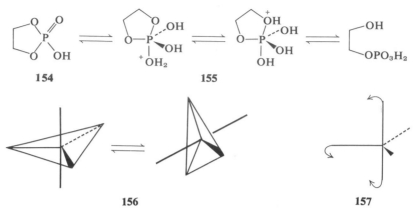

154   155

156   157

These considerations have a bearing on the hydrolysis of ribonucleotide esters (or polyribonucleotides) in which the neighboring 2′-hydroxyl group is involved. Under acid catalysis, a pentacoordinate intermediate (158) is assumed to be formed which can undergo pseudorotation allowing the $C_{2'}O$—P, the $C_{3'}O$—P and the exocyclic RO—P bonds to become apical, whence bond fission leads to starting ester, rearranged ester, and cyclic phosphate, as observed [247]. The possibility of alkylphosphoryl group migration is important in synthetic studies since such rearrangement may occur during removal of acid-labile protecting groups. Choice of suitable protecting groups has, however, essentially obviated this danger. Fortunately there is no evidence that migration occurs in polynucleotides under conditions in which hydrolysis is inappreciable.

In polynucleotide structural studies base-catalyzed hydrolysis is more generally applicable because, under these conditions glycosidic linkages are stable. It is not clear whether pentacoordinate intermediates (159) are important in the hydrolysis of nucleoside 2′- or 3′-phosphate esters (e.g., 160). Evidence that no migration of the alkyl phosphoryl group (2′ ⇌ 3′) occurs has been given so that an "in-line" displacement mechanism probably pertains, through the $S_N2(P)$ transition state (161) [247,248]. Alternatively the pentacoordinate intermediate (159) is formed but cannot pseudorotate since this would put the electropositive —O⁻ ligands into an apical position, energetically an unfavorable state [249]. It is interesting that ribonuclease catalysis of both dinucleoside phosphate and cyclic phosphate hydrolysis proceed exclusively by "in-line" mechanisms [250,251]. Another interesting sidelight is that a 3′,5′-linked polynucleotide in a helical state is in a conformation which precludes hydrolysis by an "in-line" mechanism (see also Section V,C, 2). This is not so for a 2′,5′-linked polymer. It is therefore conjectured that the greater stability achieved, as a result, may account for the prebiotic establishment of the now universal 3′,5′-linkage [252]. By the principle of

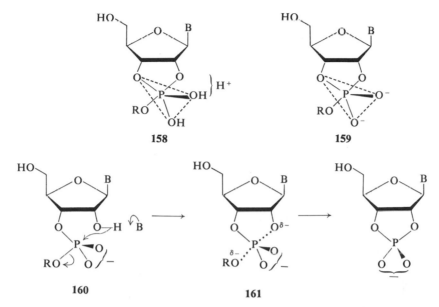

microscopic reversibility it accounts, too, for the observation [253] that chemical generation of oligonucleotides from nucleoside 2',3'-phosphates oriented on complementary polynucleotides leads almost exclusively to products with the 2',5'-linkage. Polymerization using nucleoside 5'-poly-phosphates chemically or, presumably, enzymically does not, of course, suffer from this mechanistic restriction.

Macromolecular DNA in solution undergoes a slow reduction in molecular weight with time [254]. We shall discuss the question in more detail later, but meanwhile one conceivable explanation is that distortion of an inter-nucleotide linkage to a state analogous to that in 5-membered cyclic phos-phates increases the rate of hydrolysis sufficiently at that position to account for the observed rate of molecular weight decrease. Single-strand scission of *E. coli* (or λ duplex) DNA in 0.3 *N* NaOH occurs at a rate of one break/10^6 daltons at 25°C [254a]. Double-strand cleavage results from the application of hydrodynamic stress [255,256]. The mechanism of the shearing process is unknown and no work has been reported on the nature of the end groups in such sheared molecules. The process may be hydrolytic as above, but alterna-tively sufficient strain could be localized in a bond to allow homolytic (radical) cleavage and calculation suggests that this is so [256].

C. POLYRIBONUCLEOTIDE HYDROLYSIS

In the previous section the mechanism of base-catalyzed hydrolysis of ribonucleotide esters was discussed briefly. It is important to consider the factors which may apply in determining the rate of the process. The *cis*-stereochemistry of the neighboring 2'-hydroxyl group in the furanose system is probably optimal. A *trans*-2'-OH group as in *arabino*-nucleotide esters is ineffective as a neighboring group [257].

The acidity of the neighboring function in **162** determines the concentration of **163** and hence the rate of hydrolysis under a given set of conditions. In other series, rate measurements are consistent with the view that $k_{obs} = Kk$ [258]. The factors influencing the rate are difficult to assess because the acidity of the neighboring OH group and the nucleophilicity of the derived alkoxide anion are not simply related. It is also observed that the rate increases with the increased stability of the leaving group RO^- and that a plot of log k_{obs} vs pK_a of ROH is linear over a wide range of pK_a [248,158]. Consistently a very low rate of hydrolysis of **162** (R = Pr^i) is found compared with **162** (R = Me or nucleoside-5') [259]. Metal ion catalysis is another important factor which is little understood in detail (see below).

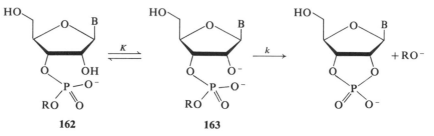

All of these factors must play a part in determining the rate of hydrolysis of the internucleotide linkages in the 16 possible dinucleoside phosphates (and in the corresponding linkages in polynucleotides). One further factor may be the stereochemical restraint imposed on polynucleotides by stacking interactions—these interactions being maximal in aqueous solution [260]. The nature of the base and its state of ionization will be important in this connection. It may be assumed that the nucleophilic activity of the neighboring hydroxyl group and the leaving capacity of the 5'-linked nucleoside residue should, as a first approximation, be taken as constant and variations in rate of hydrolysis would depend on stacking interactions, the latter possibly being disrupted by ionization, particularly of guanine and uracil residues [2].

The evidence available (Table II) shows that dinucleotides of the form PupNp undergo hydrolysis more slowly than those of the form PypNp and that the 3'-linked component has less effect on the rate than the 5'-linked one. There is, thus, about one order of magnitude difference between the fastest and slowest dinucleotide, the latter being ApAp [261,262]. Dinucleoside phosphate [263] and RNA [264–266] hydrolyses give comparable results. These observations are in general agreement with the views expressed above.

TABLE II

RELATIVE RATES FOR THE HYDROLYSIS OF SOME DINUCLEOTIDES
IN 0.86 M KOH AT 26°C[a]

UpUp	1.0	ApUp	0.25
CpC3p	0.62	ApGp	0.21
CpC2P	0.38	GpAp	0.18
GpCp	0.25	ApAp	0.10

[a] From Lane and Butler [261].

It is worth recalling here that alkaline hydrolysis of RNA can be accompanied by modification of some bases. For example, 1-methyladenine rearranges to N_6-methyladenine [87], 3-methyluracil is degraded [267], 3-methylcytosine [268], and, more slowly, cytosine [269] are deaminated, while 5,6-dihydrouracil [270], and 7-methylpurines [271] are ring-opened. 2'-O-Methylnucleosides appear in hydrolysates as alkali-stable dinucleotides [272,273].

In general it seems that the difference in rate of hydrolysis between one internucleotide linkage and another by acid- or base-catalyzed hydrolysis is not sufficient to have any useful application in structural work. It is also true that the pH conditions, being at the extremes of the scale, are not such as to allow the retention of secondary structure. Catalyzed hydrolysis which occurs around neutrality is much more likely to be useful and this can be effected by means of heavy metal ions.

Metal ions may act as electrophilic catalysts in the hydrolysis of phosphate esters. In no case is the catalysis understood in detail, and in many cases it is not possible to obtain rate data as the reactions do not always occur in homogeneous solution. Nevertheless, the rate enhancements can be very large. Phosphate monoesters have been studied intensively. La^{3+} and Ce^{3+} are very effective [274,275]. It is often noted, as it is with diester hydrolyses, that the reactions are sharply pH-dependent and that the optimum pH may vary with the concentration both of the metal ion and of the substrate. A neighboring oxygen function has an important effect on the rate and may be involved, in some cases, in metal binding. For example, the lanthanum ion catalyzed hydrolysis of 2-methoxyethyl phosphate at pH 8.5 has been thought of as proceeding through a complex such as **164** [276].

Turning to polynucleotides, it appears that in general, the polyribonucleotides are much more susceptible to metal ion-catalyzed hydrolysis than polydeoxyribonucleotides [274]. Products range from oligonucleotides through nucleoside-2',3' phosphates and nucleotide-2' (and -3') phosphates to nucleosides [277–281]. Many metal ions, including Ca^{2+}, Zn^{2+} and Ba^{2+},

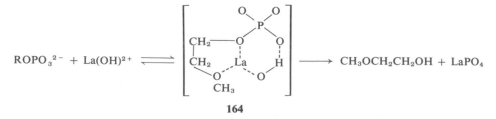

$$ROPO_3^{2-} + La(OH)^{2+} \;\rightleftharpoons\; \mathbf{164} \;\longrightarrow\; CH_3OCH_2CH_2OH + LaPO_4$$

164

act catalytically, the effect being noted *inter alia* by loss in biological activity (e.g., of TMV–RNA) or molecular weight diminution if not by product detection [282]. The Pb^{2+} ion has the largest catalytic activity so far observed [274,280,281,283]. Thus at 1.1 mM concentration and pH 7.5 rapid depolymerization of polyribonucleotides occurs. With homopolynucleotides the observed rate order is poly A $\simeq$ poly U $\gg$ poly I > poly C. Secondary structure protects against hydrolysis so that poly U·poly A is hydrolyzed at one-third the rate of either component and poly I·poly C is stable. Transfer RNA is in part hydrolyzed, the rate being reduced by added Mg^{2+}, while, surprisingly, CpC is unaffected. It is evident that behind these observations must lie much interesting and possibly useful chemistry [280,283].

D. PHOSPHATE ELIMINATION REACTIONS

In this section we deal with phosphate elimination reactions. Their importance to nucleic acid chemistry is due to the fact that conditions may be found which lead to internucleotide bond cleavage under conditions, both acidic and basic, much milder than those that pertain in hydrolysis. In practice a nucleoside residue in the polynucleotide is altered so as to activate the system for elimination. If the terminal residue is so altered, a method for stepwise degradation is available. If a nonterminal residue is suitably modified chain scission at that site can be effected. Normally in nucleic acid chemistry the activating function is a carbonyl group or derived function and this may be introduced into a nucleoside residue either by oxidation or by removal of the base.

1. *Stepwise Degradation of Polynucleotides*

A rational process for the stepwise degradation of a polyribonucleotide (**165**; R = remainder of chain) is based on periodate oxidation of the 3′-terminal residue to give the "dialdehyde" (**166**). Under mildly basic conditions this undergoes a β-elimination reaction to give **167** from which the terminal 3′-phosphate residue may be removed enzymically and the whole process then repeated [284–286]. This process has been extensively developed to increase the rate and the extent of reaction under milder conditions and to liberate the terminal base as such. The latter is achieved *inter alia*, by

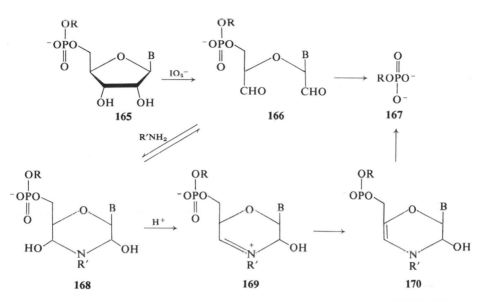

overoxidation with periodate in an amine-catalyzed reaction [287]. The perfection of the elimination reaction has come about by the introduction of primary amine catalysis at pH's around neutrality [288–292]. The amine, it appears [293,294], performs two functions, since high concentrations are efficacious; (1) the formation of an intermediate, and (2) acid–base catalysis. The reaction may be envisaged as proceeding through **168, 169,** and **170** or closely related structures with the latter undergoing the elimination. Evidence for a cyclic intermediate is available, the stability of the latter being dependent on the nature of R′ (in R′NH$_2$). With methylamine the intermediate (**168**) is stable at pH 8–9 showing no sign of elimination, as would be expected [290,293]. Lowering of the pH to 5–7 leads to fast and essentially quantitative elimination in model experiments using adenosine-5′ phosphate. On the other hand reduction of the "dialdehyde" (**166**) with borohydride gives a stable system and, if [^{3}H]BH$_4^-$ is used, a labeled 5′-terminus. Reduction of an amine adduct gives a stable morpholine system which likewise cannot undergo elimination [293]. In this way an RNA can be derivated at its 5′-end, allowing for attachment to other molecules. Using this principle a ferritin–tRNA adjunct has been made [294a], and a nucleic acid–protein crosslink has been generated in 30 S ribosomes [294b].

The point to note is that, so far as eliminations in general are concerned, the order of effectiveness for activation of the α-hydrogen by the carbonyl or derived function is

$$\mathrm{C{=}N{-}} \ll \mathrm{C{=}O} < \mathrm{C{=}N^+}$$

With very careful attention to detail [291] it has been possible to analyze several 3'-terminal RNA sequences, e.g., of TMV–RNA using aniline [295, 296], of phage f2 using cyclohexylamine [297–299], and of *E. coli* tRNA[Phe] using lysine as base [300]. In the latter work 26 cycles were performed, the first 19 nucleotides being sequenced unequivocally.

It is interesting that the terminal dialdehyde (**166**) and acyl hydrazides, such as thiosemicarbazide and isonicotinic hydrazide (isoniazid), react to give a product (**168**; R = NH·CO·C₅H₄N) essentially stable at neutrality, which may be used to label, and hence to identify the terminal sequence [301–303]. Dimedone reacts in the same way [304]. An extension of this is its use in attaching a large organic residue for various purposes such as purification of specific transfer RNA's [305]. Correspondingly functionalized cellulose columns may be used for isolation of periodate-oxidized nucleic acids or terminal oligonucleotides [306,307]. It may be that the efficiency of these latter processes could be increased. With careful pH control, a periodate-oxidized oligonucleotide may be attached in stable linkage to an aminoethyl-cellulose support and then be released (by β-elimination at lower pH) [298]. This is the basis of an automatic sequenator.

Stepwise degradation of polydeoxyribonucleotides has not yet been developed to a practically useful state. Oxidation of thymidine 3'-phosphate by platinum–hydrogen peroxide or –oxygen gives rise to the carboxylic acid (**171**). This residue (as the anion) is of course incapable of promoting a base–catalyzed elimination, although a decarboxylative elimination in this and the correspondingly oxidized dithymidine phosphate can be effected [308]. Derivatization (in two stages) to the propylamide (**172**) gives a weakly activated system. Quantitative elimination can be achieved under rather vigorous alkaline conditions [309].

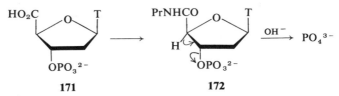

Oxidation of the 3'-hydroxyl by dimethyl sulfoxide–acetic anhydride to the ketone (**173**) followed by treatment with alkali leads to the elimination of a

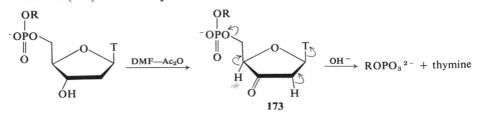

5′-phosphoryl residue as well as the base residue, but the conditions, though milder than those used above, are still not satisfactory with polynucleotides [310]. Amine catalysis was apparently not investigated.

2. *Glycosidic Bond Hydrolysis*

So far as methods developed up to the present are concerned, chain scission has depended on the initial removal of a base residue. This is therefore a convenient place to discuss the hydrolysis of the glycosidic linkage.

As is well known, ribonucleosides are more stable than deoxyribonucleosides and, in each class, the purine bases are lost much more rapidly than the pyrimidines. Recently rate measurements made over a wide pH range have deepened our understanding of the process [311–317]. Figure 1 shows the pH-rate profiles for the hydrolysis of the natural deoxynucleosides. Individually, they show interesting variations in behavior. Thus thymidine shows both spontaneous and acid-catalyzed modes shared by deoxyuridine and the 5-bromo and iodo analogs [313,315,317]. The former is not observed with

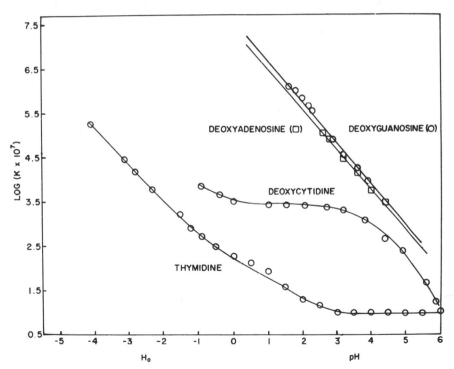

Fig. 1. Plot of the logarithms of the rate constants (sec^{-1}) at 95°C *vs* pH (above 0) and H$_0$ (below 0) for the hydrolysis of the four common naturally occurring deoxyribonucleosides [317].

deoxycytidine [317] but hydrolysis of the monocation (plateau at pH 0–3) and the dication occurs, as it does with deoxyguanosine and deoxyadenosine [316]. The two latter compounds have closely similar reactivities to acid hydrolysis and are more than 500 times more rapidly hydrolyzed than guanosine [316].

The observations fall into place if it is assumed that a preequilibrium protonation of the heterocyclic ring is followed by a rate-limiting ionization of the glycosidic bond, the deoxyuridine derivatives showing an additional water hydrolysis of the neutral species. Mechanistically hydrolysis of deoxyuridine (**174**) is visualized as follows:

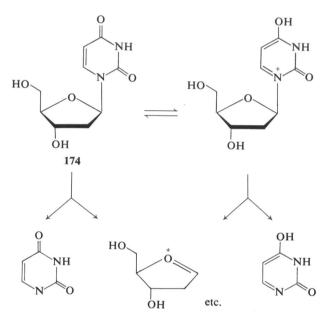

The effect of substituents at C-5 is in accord with this, i.e., electron-withdrawing halogens accelerate hydrolysis [311,315,317]. In the case of the purine derivatives preequilibrium protonation at C-7, the most basic site, is followed by a unimolecular cleavage of the N-9 to C-1' bond [316]. Much evidence has been adduced in favor of this view [317,318] rather than the earlier one [319] that protonation of the sugar ring-oxygen is followed by opening of the lactol ring and generation of a Schiff's base which then undergoes hydrolysis. (For references to glycosylamine chemistry, see Capon [318].) Decisive evidence comes from a study of 7-methyldeoxyguanosine (**175**) and of 1,7-dimethyguanosine [316]. The former shows a pH-independent plateau in the pH-rate profile corresponding to the species (**176**). The latter is more than 25,000 times as reactive as guanosine. 7-Alkylation

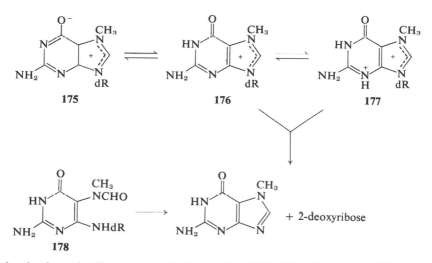

also leads to lability in the alkaline region [320,321]. Here reversible ring-opening of the imidazolium ring occurs. Though the glycosylamine bond in **178** is stable to base it is labile to acid, hydrolysis to 7-methylguanine competing with ring closure to **175**.

In addition to the above, 3-methyldeoxyadenosine has been shown to be more labile than 7-methyldeoxyguanosine in mild acid [120]. Deoxy-inosine and -xanthosine are said to be more labile than the corresponding adenine and guanine nucleosides from which they are derived by deamination.

In considering the stability of DNA, before and after alkylation and deamination, the new data on nucleoside hydrolysis are clearly very important.

3. *Eliminative Polynucleotide Chain Scission*

As mentioned above, removal of a base residue may be the first step in polynucleotide chain cleavage. Elimination is then activated by the unmasked sugar aldehyde group (**179**) or a derived function, and proceeds under acidic (**180**), basic (**181**), including amine (**182**) catalysis.

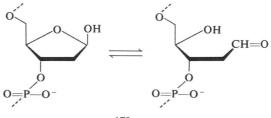

179

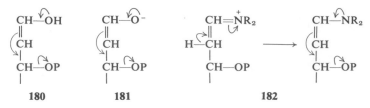

180   **181**   **182**

In this section we will first consider work with DNA. A variety of methods are available for removing base residues. In practice the purine residues are readily removed from DNA by mild acid treatment, for example, by HCl or formic acid [322,323]. The product, apurinic acid (**183**), which itself has suffered chain scission, may then be degraded further by acid [324] or, more cleanly, by an aromatic amine in aqueous formic acid. With the latter, purine removal and degradation of the formed apurinic acid is conveniently carried out in one step [325,326]. The products are of the form Py_nP_{n+1} (**184**). Using diphenylamine in aqueous formic acid at room temperature the products are

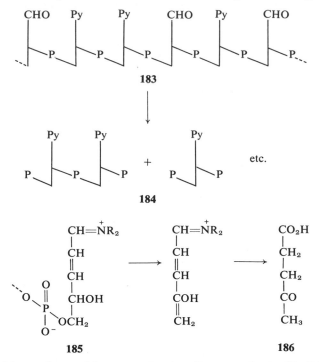

formed with no detectable contamination from products (Py_nP_n) resulting from the loss of the monoesterified phosphate or from the (slow) loss of pyrimidine residues. It is particularly interesting that the method leads to complete elimination of the free sugar residue, i.e., both 3'- and 5'-phosphoryl groups are eliminated; phosphatase treatment in consequence gives products solely of the form Py_nP_{n-1}. Presumably the first intermediate (**185**) can undergo a second elimination efficiently from the vinylogous β-position, which in the acid-catalyzed reaction finally yields levulinic acid (**186**). The products of the diphenylamine/formic acid reaction can be separated into fractions of equal chain length (isostichs or isopliths) [327,328], DNA from ϕX174 gave pyrimidine sequences up to 11 nucleotides long [327], that from fl up to 19 nucleotides long [329], and in the case of fd DNA a unique stretch of 20 nucleotides was formed and sequenced [330].

Degradation of apurinic acid with alkali gives chain cleavage but is less satisfactory since, as is well known, aldehydes, including free sugars, undergo a variety of reactions in base. In the reaction with apurinic acid some of the products apparently terminate in the residue (**189**; R = O) [331]. This result can be rationalized in terms of C_3—OP elimination to **187** followed by base-catalyzed prototropic shifts to give **188**, whence cyclization gives the cyclopentenone. Products terminating in (**189**; R = O) which is an enol phosphate, should be readily converted to the true elimination product by mild acid, though this has not been shown.

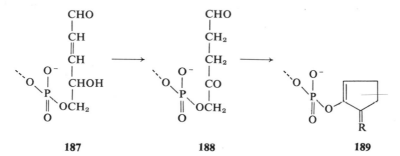

In attempts to bring about chain cleavage in apurinic acid under even milder conditions amine catalysis has been studied. Alphatic amines were essentially ineffective and hydroxylamine gave the stable oxime. Phenylhydrazine and semicarbazide were apparently effective but again the products were pyrimidine polynucleotides terminated by an organic residue, probably **189** (R = N·NHPh or N·NH·CO·NH$_2$) [332,333]. It is not quite clear why phenyl hydrazine should lead to elimination; it and *unsym* dimethylhydrazine are explicably excellent catalysts for α-elimination, e.g., from glycolaldehyde phosphate [334,335], viz:

$$^-HO_3POCH_2\cdot CHO \longrightarrow \ ^-HO_3PO{-}CH_2{-}CH{=}N{-}NHPh \longrightarrow PO_4H^{2-}$$

When DNA is treated with hydrazine the pyrimidine bases are degraded (see Section II,B), and the sugar residue remains almost certainly as the hydrazine derivative [336,337]. There are conflicting reports as to the stability of this system to base [337,338]. Acid-catalyzed degradation is precluded by the lability of the purine glycosidic bonds. However, pretreatment with benz-aldehyde to displace the hydrazine [339], then base, certainly allows chain cleavage to occur. The whole process has been subjected to careful scrutiny and it is evident that the reaction does not go entirely quantitatively [340]. Purine polynucleotides (Pu_nP_{n+1}) are formed and can be separated into their isopliths [341]. There is a relatively good correlation between the polypurine and polypyrimidine isopliths as there should be if both are obtained by degradation of the same complementary double-stranded DNA [337,341]. The disadvantages of strong base-catalyzed elimination are obvious. More recently it has been claimed that aromatic amines, aniline, or *p*-anisidine, under mild (pH 5) conditions give chain rupture. With apyrimidinic acid, isopliths ($Pu_nP_{n=1}$) of $n = 1–8$ can be recognized but these too, it seems, may carry a terminal non-nucleotidic organic residue [342]. This may, in certain applications, be no disadvantage since, for instance, it may allow radioactive labeling with, e.g., ^{14}C-aniline.

Any chemical reaction which degrades a base residue is, in principle, capable of application to DNA chain cleavage. Those above have been studied most extensively. Other methods, particularly those involving base oxidation, have been given some consideration. Osmium tetroxide oxidizes thymine residues much faster than the other bases [9,343]. As applied to DNA, the oxidized residue can be removed by base, whence diphenylamine–formic acid treatment affords groups of cytosine nucleotides, C_nP_{n+1} [344]. Sequences of adenine nucleotides, as yet not sufficiently characterized result from permanganate oxidation, followed by alkali [345]. The problem associated with these reagents is that, as in hydrazinolysis, a nitrogenous residue may be left attached glycosidically and this has to be removed completely before the chain cleavage reaction can be initiated. As yet, it would seem, solutions to this problem are only partial.

Turning to RNA we find that much less work has been devoted to the question of chain cleavage. This is not surprising as ribonucleases of high specificity abound so that the need to devise chemical cleavage reactions is less pressing. The same kind of approach as has been taken with DNA is possible, provided always that it is recognized that in RNA internucleotide linkages are much less stable and the glycosidic linkages are much more stable than those in DNA. This places considerable restrictions on the methods available. The base and acid lability of the internucleotide linkage precludes the preparation of riboapurinic and riboapyrimidinic acid. Removal of uracil from RNA by hydroxylaminolysis at pH 10 [59,346] gives a product which, after mild acid treatment (to remove the oximino residue), can then be cleaved by aromatic amine-catalyzed elimination [347]. Again a terminal non-nucleotide residue is retained, and the hydroxylaminolysis step leaves much to be desired [348].

Probably the best example of chain cleavage as applied to RNA depends upon the very high degree of acid lability of the base Y (**190**) in yeast tRNA[Phe] [349]. The tRNA contains 76 bases and base Y which is in position 37 is

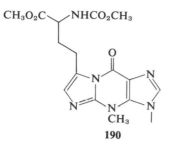

190

removed by 0.1 N HCl at room temperature [350]. The aniline-catalyzed elimination then cleaves the RNA chain with, apparently, no other damage to the system. Indeed also present in this tRNA is a 7-methylguanine residue at position 50. Somewhat stronger acid treatment removes it as well and elimination of the intervening oligonucleotide can be effected [351].

4. Thermal Degradation of Polynucleotides

When a biologically active nucleic acid is heated in solution (or sometimes in the dry state) the rate of inactivation is greater at higher temperatures and lower pH values. The process is irreversible and is to be distinguished from denaturation at the melting temperature in double-stranded DNA. Many studies of subcritical heating have been made [352]. The process in DNA is considered to depend on loss of purine bases, guanine rather faster than adenine, and as a consequence, chain scission. Both can be measured [352–355]. In circular single-stranded ϕX174 DNA the rate of depurination and of inactivation are similar and are considered to be distinguishable from the subsequent scission step [356,357]. Poly dAT undergoes slow chain scission ($\sim$ pH 4) which is associated with loss of adenine and appearance of terminal 3'-phosphate residues [358]. A considerable increase in rate is observed at temperatures between 60° and 70°C, i.e., around the T_m value, so that secondary structure protects against depurination. The pH-rate profiles for the acid-catalyzed hydrolysis of purine deoxyribonucleosides (Section IV,D,2) are consistent with slow glycosidic bond hydrolysis at, or somewhat below, neutrality. Both this step and the subsequent elimination reaction leading to chain scission may involve general species catalysis.

Alkylation at N-7 in guanine and N-3 in adenine residues, as we noted, leads to much-enhanced rates of hydrolysis and in DNA a slow release of alkylated purines occurs [120,359,360]. Inactivation results from alkylation. Indeed following alkylation by EMS, bacteriophage T4 survival studies show a primary and a delayed inactivation which are reasonably ascribed, respectively, to the alkylation and the subsequent chain cleavage [361]. Alkylation mutagenesis presents a confused picture [188,362,363]. In addition to the question of the site of alkylation, explanations require consideration, *inter alia*, of replication and of repair error before and after depurination. Since both transitions and transversions occur it is probable that more than one mechanism is involved [188].

In contrast to DNA, thermal degradation of RNA appears to proceed by direct hydrolysis of internucleotide bonds [254]. The pH-inactivation rate profile for TMV–RNA differs markedly from that for ϕX174 DNA [357] and evidence for the generation of terminal 3'- or 2',3'-phosphate end groups has been given [364].

E. ALKYLATION ON PHOSPHATE

We have seen that removal of a base from a nucleic acid leaves a sugar residue which is then a point of lability in the system. Alkylation of guanine at N-7 or of adenine at N-3, as occurs in DNA, can lead to this result. Alkylation of the internucleotidic phosphoryl group can, in principle, also lead to chain cleavage, but for a different reason. In the latter lability is introduced because of the much greater ease of hydrolysis of phosphate triesters than of diesters.

The stability of an alkylated internucleotide linkag will depend very much on whether we are dealing with the deoxyribo- or the ribo- series. In the former, the system will be a normal trialkyl phosphate and thus will be susceptible to neutral or base-catalyzed hydrolysis. This may be quite slow (for some rates, see Cox and Ramsay [365]) and it is likely that the introduced alkyl group will often be the more easily displaced. The ease of base-catalyzed displacement, broadly, is in the order 1° > 2° > 3°; more accurately it reflects the stability of the alkoxide ion formed or, in other words, the acidity of the corresponding alcohol. Thus of the simple alkoxy groups, methoxy is the most easily displaced. It follows that chain cleavage is not a necessary consequence of *P*-alkylation since the newly introduced group may be more easily lost in the triester. It is interesting that the neutral hydrolysis of trimethyl phosphate proceeds by alkyl–oxygen cleavage, in effect, acting as an alkylating agent. This is particularly the case in alkylation by *N*- and *S*-mustards and imines, since the reaction is reversible, 2-aminoethyl

$$\text{>NCH}_2\text{CH}_2\text{Cl} \rightleftharpoons \text{>}\overset{+}{\text{N}}\!\!\begin{array}{c}\text{CH}_2\\|\\\text{CH}_2\end{array} \rightleftharpoons \text{>NCH}_2\text{CH}_2\text{OPO(OR)}_2$$

dialkyl phosphates, themselves being alkylating agents [88,366]. *P*-Alkylation by a Michael reaction, for instance by acrylonitrile, should in principle occur reversibly, though the question does not seem to have been studied in connection with polynucleotides. Information is rapidly accumulating on the stability of internucleotide triesters as a result of increased interest in intermediates carrying such groupings in the chemical synthesis of oligonucleotides (see Chapter by Letsinger in Volume III).

Alkylation of the internucleotide linkage in a polyribonucleotide must give an exceedingly labile system. Model experiments show that a nucleoside 2′ (or 3′)-dialkylphosphate (**191**) is unstable over the entire pH range [367]. The internucleotide linkage could be broken at two stages; (1) during the formation of the cyclic intermediate (**192**) and (2) during the hydrolysis of the

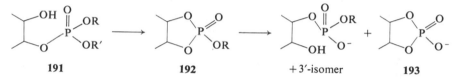

| **191** | **192** | +3′-isomer | **193** |

latter. Opening of 5-membered cyclic triesters is exceptionally fast and concomitant hydrolytic cleavage of the exocyclic P—O bond to give **193** is also largely accelerated [245].

Having discussed the expected results of forming phosphotriester groups in polynucleotides, we should now look at the evidence as to whether it does or does not occur. It is evident from model experiments that alkylating proceeds very much more readily on base residues than on phosphate residues. The nucleophilicities of dialkyl phosphate anions (a model for the internucleotide linkage) are relatively low. Table III gives a basis for comparison with other nucleophiles [368]. Monoalkyl phosphates (models for terminal residues) are more reactive. Their reactivity is pH-dependent, the dianion, naturally, being more nucleophilic than the (protonated) monoanion.

TABLE III

COMPETITION FACTORS OF SOME PHOSPHATES FOR
N,*N*-DI-2′-CHLOROETHYL-2-NAPHTHYLAMINE[a]

HPO_4^{2-}	108	$HOCH_2CH(OH)CH_2OPO_3^{2-}$	66
$H_2P_2O_7^{2-}$	45	$(C_2H_5O)_2PO_2^-$	5.3
$C_2H_5OPO_3^{2-}$	153	$[HOCH(CH_3)CH_2O]_2PO_2^-$	3.0
$HOCH_2CH_2OPO_3^{2-}$	100	CH_3COO^-	100
$HOCH(CH_3)CH_2OPO_3^{2-}$	134		

[a] From Davis and Ross [368].

These considerations apply to Me_2SO_4, alkyl sulfonates and the like. On the other hand, diazoalkanes react, in effect, with the free acid. Since the secondary phosphoryl dissociation of phosphomonoesters is around 6, at neutrality there will be about 10% of the monoprotonated form present. Adenosine-5' phosphate at pH 7 gives essentially only the monomethyl ester with diazomethane, while with Me_2SO_4 (especially at pH 5) N_1-methylation is very much favored. With cytidylic acid *P*-alkylation alone occurs with diazomethane; with Me_2SO_4 alkylation is mainly at N-3 [108].

Work with dinucleoside phosphates, both in the ribo- and the deoxyribo-series make it clear that methylation by both Me_2SO_4 and diazomethane occurs selectively on the base residue and that there is no *P*-alkylation. There is in consequence, no internucleotide bond scission. This view has been contested in regard to diribonucleoside phosphate alkylation but a careful reexamination has confirmed the original findings [369]. Thus at least from the point of view of preparative chemistry the internucleotide linkage in solution at neutral pH is extremely unreactive; ApU for instance gives ApMeU quantitatively with diazomethane and MeApU with dimethyl sulfate. Homopolynucleotides are also readily alkylated. In no case has any direct evidence of internucleotide *P*-alkylation been found and so the process is preparatively useful. The terminal 5'-phosphate in poly U is alkylated by sulfur mustard [370] and, one may presume, by other alkylating agents. Some workers have carried out alkylations with a variety of reagents without, however, commenting on any changes in sedimentation constant [111,371, 372]. Others have noted that diazomethane methylation of poly U but not poly A [108] and of rRNA and tRNA leads to some degradation [373]. Ethyl methane sulfonate ethylation of poly A gives rise to some chain scission [114].

In the case of diazomethane, chain scission may be dependent on *P*-alkylation; on the other hand, the interesting suggestion has been made that the diazomethane is acting as a base as in **194** [108].

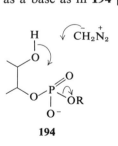

194

Chain scission of DNA as a result of alkylation certainly occurs. Whether this is due to *P*-alkylation in addition to events following base modifications is unclear [374]. But it should be reiterated that, on chemical grounds alkylation on the internucleotide linkages of ribo- and deoxyribopolynucleotides must occur; the nucleophilicity of dialkyl phosphates is low but finite. It was a matter only whether a sensitive enough direct test could be devised. It has recently been shown that at a level of alkylation of DNA by EMS and MMS, introducing 26–28 groups/10^6 daltons, 10–15% of the alkyl groups are accounted for as alkyl phosphate after acid hydrolysis. The corresponding alcohol can be released from the latter by phosphatase treatment [374a].

Alkylation on the terminal mono, pyro, or triphosphate group of poly-nucleotides *in vivo* is also a very probable event and has not been sufficiently considered as a cytotoxic process.

V. Influence of Conformation on Reactivity in Polynucleotides

A. METHODS OF STUDY

The extent to which a polynucleotide is modified in a reaction will largely determine the way in which the process is studied. At one extreme, a muta-genic response provides strong evidence for reaction. Further genetic analysis of a mutant may allow classification into one or other of the several types of lesion and this, in turn, with the help of knowledge of the chemical specificity of the reagent, may lead to conclusions as to the nature of the primary event. Such conclusions must inevitably be tentative since the reaction, per se, cannot at present be studied in a direct way. Likewise changes in the hydro-dynamic, or other physical behavior of nucleic acid solutions are very sensitive to structural alterations. Such methods give evidence of chain cleavage, but the nature of the alteration is necessarily obscure. For example, unwinding of supercoiled DNA as a result of one single-strand break is readily observed. Bisulfite-Pb^{2+} causes such a strand cleavage [375]. A possible hypothesis that this is due to radical oxidation of a sugar residue, may be inferred from the known chemistry of bisulfite but this remains a conjecture. What matters is that incredible sensitivity is available in this and a variety of other potential test systems which involve chain breakage. Interstrand crosslinking at very low levels can be demonstrated by the resultant rapid reannealing [191,352].

At the next level, small extents of reaction of bases can be detected and quantitated by using either radioactively labeled reagent or nucleic acid. Labeled reagent is much used, for example, in alkylation studies and in this instance product ratios can be accurately measured. Using unlabeled reagent and ^{32}P-nucleic acid is convenient and more economical since the reagent is very often in large molar excess over the substrate and in general is not recovered. Quantitation is conveniently carried out at the mono- and oligo-nucleotide level [376].

Larger extents of reaction can be followed by spectroscopic methods or changes in melting behavior, and normal analytical methods can be used.

A variety of reagents have been applied to biosynthetic polynucleotides with several purposes in mind, for example, to provide modified templates for enzymic studies, to study potential mutagenic (or carcinogenic) reactions, and, importantly, to relate reaction rates to higher structure. In this chapter we are solely concerned with the reactions and not with the interest the products provide. In this context polynucleotides are to be considered as test objects to give information applicable to the nucleic acids themselves. It is not always clear that, as simple test objects, they are very informative. For this reason their reactions will be discussed along with, and not prior to, those of the nucleic acids.

B. DNA

1. *Effect of Secondary Structure*

When duplex DNA undergoes reaction in aqueous solution the most frequently encountered observation is that the process is extremely slow. Indeed for osmium tetroxide [377] it is said that no reaction occurs. This, of course, is merely to say that in the time for complete reaction of a mononucleotide, no observable change has taken place with DNA. The lack of reactivity is to be ascribed, in the first place, to double-strandedness. Thus highly structured single-stranded systems react at rates much nearer to those of the monomers. The single-stranded circular phage S13 DNA can be inactivated by hydroxylamine under conditions in which the replicative form is unaffected [378]. The relative rates of the hydroxylamine–cytosine reaction at pH 6.5 and 37°C for 5′-CMP, poly C and DNA are approximately 1500:300:1 [54,379]. In the same way iodination [227], deamination [182], and reaction with the carbodiimide (**119**) [168] of native DNA is extremely slow (Fig. 2). So too is perphthalic acid oxidation [380] and the reaction with acylhydrazides (e.g., semicarbazide) [381]. Work with homopolynucleotides lends support to the view that it is the duplex character of the DNA that is responsible for the rate decrease. In the carbodiimide [168] and hydroxylamine [382] reactions, rates for uridylic acid and the essentially unstructured poly U are similar but are sharply diminished in poly A·poly U. Deamination of poly d(A–T) is slow and increases markedly at the melting temperature (Fig. 3) [180]. Some values for extent of modification for perphthalic acid oxidation are given in Table IV.

We may ask whether interstrand hydrogen-bonding or base-stacking is more important in the rate diminution. Of course these factors are mutually dependent in a duplex structure, but the evidence from the hydroxylamine–

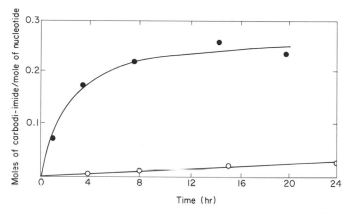

Fig. 2. The reaction of DNA with the carbodiimide (**119**) in borate buffer, (pH 8.0) at 30°C. (○) Native DNA. (●) Denatured DNA prepared by heating in borate buffer, pH 8.0 at 100°C then fast cooling [168].

poly C reaction shows that base-stacking does reduce the rate, but by a small factor. The effect is more marked in alkylation of poly C [112]. Two more pertinent questions may be asked. Why do the bases in native DNA react at all, and what importance attaches to the stereochemistry of the reaction in question? The former question has been raised in connection with proton exchange studies. All five protons in the G–C base pair exchange slowly and 2-3 exchange slowly in the A–U pair [383]. It is inferred that for any exchange to occur, local structural distortion must take place, though this breathed-open state is to be distinguished from a simple pair of locally

TABLE IV

REACTION OF PERPHTHALIC ACID WITH SOME POLYNUCLEOTIDES[a,b]

Polynucleotide	Nucleotide modified	Percent modified
Poly A	A	84
Poly A·poly U	A	2
Poly C	C	73
Poly C·poly G	C	16
	G	15
Poly C·poly I		Undetected
DNA, native	A	5
DNA, heat-denatured	A	15
tRNA$_{yeast}^{Phe}$	A	27

[a] From von der Haar *et al.* [200].

[b] The conditions for the reaction are perphthalic acid (10 moles excess), pH 7.0, 20°C for 7.5 hr.

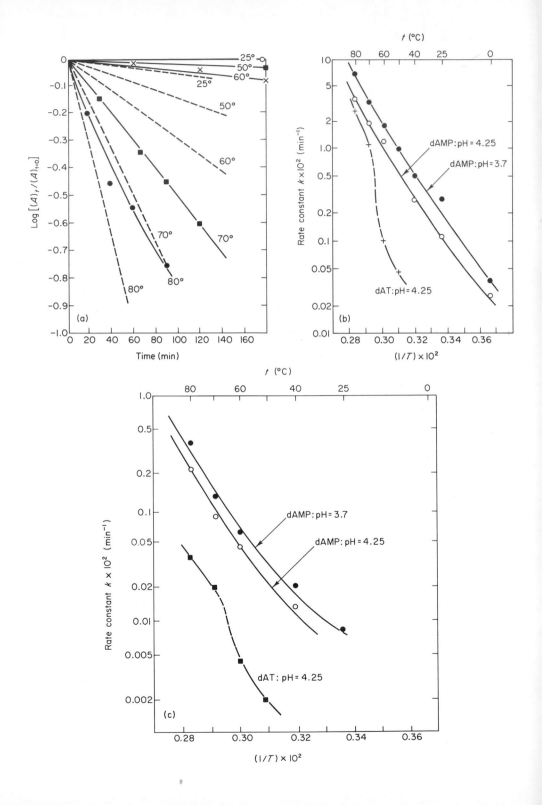

single-stranded stacks. Of course the exchange process itself is subject to general acid–base catalysis [384] but the stereochemistry and steric requirements are unknown.

The stereochemistry of other reactions are known or more predictable. *N*-oxidation at N-1 in adenine and N-3 in cytosine presumably has its reaction coordinate in the plane of the ring. In-plane stereochemistry of the transition state for alkylation at N-1, N-3, and N-7 in the purines and N-3 in the pyrimidines is also certain. In the addition reactions of the 5,6- double bonds of the pyrimidines trigonal carbon atoms become tetrahedral. The attack at C-6 by hydroxylamine (and presumably by bisulfite) is from above (or below) the plane of the ring as in **40**. Models suggest that for a small reagent this is possible in a double helical system, with some steric hindrance, without, however, taking into account solvent or catalytic species. C-6 is a prochiral center and in the asymmetric environment of a right-handed helix one of the two configurations of the new asymmetric center should be generated in excess or even exclusively; this has not yet been investigated but it could cast light on the degree to which the helical system is unfolded during reaction, e.g., in photochemical hydration and reduction and in addition reactions.

Displacement of the amino group at C-4 of cytosine by water or by hydroxylamine should have a transition-state stereochemistry approximating that in aromatic substitution, i.e., **195**. The same displacement, but consequent on a prior 5,6-double bond addition presumably involves a transition entity

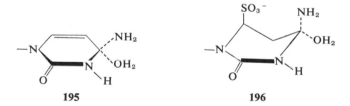

195   **196**

not unlike **196**. So far, then, as addition at the 5,6- double bond of the pyrimidines is concerned distortion of the helical system in the transition state is probably necessary. The product, whose stereochemistry approximates to a half-chair, **40**, **56**, will seriously disturb (or interrupt) base-stacking.

For reaction to occur at all at the hydrogen-bonding centers it is likely that an opened structure such as has been discussed in connection with proton

Fig. 3. Kinetics of deamination and deamination-dependent depurination of dAMP and poly d(A–T) in 1 *M* NaNO₂, 0.2 *M* sodium acetate buffer, pH 3.7 and 4.25, respectively. (a) Extent of reaction *vs* time at different temperatures (solid lines and experimental points, poly d(A–T); dashed lines dAMP). (b) Dependence of reaction rate on temperature, plotted as *k* (on a logarithmic scale) *vs* $1/T$. (c) Rates of depurination at different temperatures during reaction with HNO₂ [180].

exchange will be necessary. It may therefore be that the rates of a variety of reactions can be discussed in terms of an equilibrium constant for the "breathing" process together with a rate constant for the unhindered reaction itself. The formaldehyde reaction (Section III,D,1), which has been studied in much detail, should be amenable to such an analysis. Formaldehyde reacts with denatured DNA much faster than it does with the native material and, for example, it has been used for holding locally melted regions of the molecule in the denatured state for electron microscopy [385]. The reaction in principle resembles proton exchange in that hydroxymethylation occurs on N-3 of uracil, thymine, and hypoxanthine base residues and on the amino groups of adenine and guanine, the former substitutions being very fast in the forward and reverse directions. Although at the concentrations employed formaldehyde can hardly be a denaturant in the usual sense of the word, it reduces the T_m value of DNA if present during heating. Moreover, if added after heating at temperatures in the transition range, the optical density (hyperchromism) is raised to the value that would have been attained had the formaldehyde been present during heating [386,387]. This could come about as a result of fast reaction stabilizing the "breathed-open" state, at temperatures where the latter begins to merge with melting per se. A recent kinetic analysis of the reaction of formaldehyde with T7 DNA, poly d (A–T) and poly dA·poly dT is consistent with a model in which hydroxymethylation of a base pair occurs as a result of the "breathing" process. This in turn induces spontaneous denaturation of adjacent base pairs and the induction effect increases with temperature [388]. We should note that the helical conformation of *single-stranded* poly C is not destabilized by reaction with formaldehyde [388a]. Whatever the reaction coordinate for the formation of the hydroxymethyl derivative, the product once formed will have the hydroxymethyl group in the ring plane. Since it can lie in the helical groove, it need not destabilize the helix and it may even stabilize it by increasing the hydrophobic character of the system. On the other hand, in *double-stranded* helical systems, reaction at ring nitrogens (N–1 in guanine, N–3 in the pyrimidines) will effectively prevent reestablishment of the base-pair. Hydroxymethylation of the amino groups, with consequent reduction in basicity, may also diminish their hydrogen-bonding potentialities. Nitrous acid deamination (reaction with NO^+) should have similar stereochemical requirements to hydroxymethylation but will of course differ in that the final step is irreversible.

When a compound can react with more than one base, it is of interest to ask whether product ratios are altered when the bases are present in DNA. Alkylating agents come into this category. In the first place, for the simple alkylating agents such as sulfate and sulfonate esters reaction rates are depressed by secondary structure. For example, poly C [112] and double helical systems such as poly A·poly U become almost completely unreactive

[389]. In DNA the bases compete with each other and with water for the reagent. The product ratios are changed. The major alkylation positions in mononucleotides are N-1 of adenine and N-7 of guanine. In native DNA, adenine N-1 alkylation is depressed and the N-3 position is favored (Section III,B) [90,390,391]. A model of double-stranded DNA shows that approach by a reagent in the base pair plane at the N-3 and N-7 positions is sterically acceptable; N-3 and N-7 lie in the large groove. Thus for these two positions alkylation of the double-helical system can occur and it may be reasonable to presume that the adenine N-3 alkylation is a measure of this process. The minor products, viz., those resulting from alkylation of N-1 of adenine, of N-3 of cytosine and thymine, and O-6 of guanine must arise from reactions occurring in an unfolded state of the DNA molecule. There is some evidence that increasing S_N1 character in the reaction gives a corresponding increase in the O-6 alkylation product [392].

Some evidence has been adduced, more recently, that alkylation of guanine by the ethylene iminium ion derived from $(C_2H_5)_2N \cdot CH_2 \cdot CH_2Cl$ proceeds *faster* in DNA than it does with monomeric species, as judged by the competition factor (vs water) [94]. The suggestion is made that complexing of the positive iminium ion with the polyanionic nucleic acid is responsible for the rate increment. By extension this should hold for other mustard-type reagents. No evidence has been given that this is so and the whole matter may need reinvestigation. The question is of some importance since it is well known that polyamines complex with DNA and it should be possible to use this feature to design reagents (alkylating or otherwise) of defined specificity. It does, in any case, seem clear that the mustards lead to considerably greater extents of reaction than the sulfonates under comparable conditions [393].

The bifunctional alkylating agents, e.g., the *S*- and *N*-mustards, show one further feature in their reactions with DNA, which is not noticeable with the monomeric species. Crosslinking occurs exclusively between guanine residues; N-7 is again alkylated giving, for example, **197**. The degree of crosslinking is remarkably high and depends on the GC content of the DNA [90]. About 25% of *S*-mustard residues covalently linked to DNA are isolated as **197**; the rest are singly linked as in **198** [360,394]. It is presumed [393] that the first intermediate (**199**) is formed and that this reacts rapidly with a contiguous guanine residue or with water. Models suggest that for intrastrand reaction a

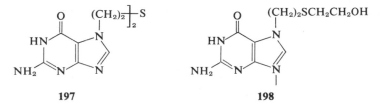

197   **198**

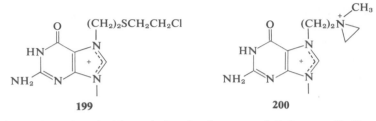

| 199 | 200 |

GpG sequence is suitable and that for interstrand linkage a GpC sequence (and its complement) provide the required base–base distance. Even so, considerable distortion of the DNA structure is necessary. Evidence has accumulated which makes it clear that both inter- and intrastrand cross-linking occurs, but the base sequence requirements suggested above have not been demonstrated. The most recent work shows that, of the crosslinks with *S*-mustard, one in three is interstrand [360]. One may expect quantitative differences in the amounts of inter-, intra-, and single-arm alkylations for different bifunctional alkylating agents. For instance the second (cross-linking) step during alkylation with HN2, $[CH_3N(CH_2CH_2Cl)_2]$, is apparently slow compared with the first step; presumably the iminium ion (**200**) is less reactive than the episulfonium ion derived from the chloroethyl inter-mediate (**199**). The biological outcome of treatment with various bifunction-al reagents may accordingly show variations.

2. *Denatured DNA*

When DNA is denatured by heating and then fast cooling the reactivity of the system to external reagents increases. This is not surprising since the original secondary structure has been lost. It is observed, however, with several reagents that, although the initial rate is increased, the extent of reaction is nevertheless small and the suggestion is made that many hydrogen-bonded pairs are present. The reactions with perphthalic acid [380], the carbodiimide (**119**) [168] semicarbazide [381], and formaldehyde [386] are cases in point. Figure 4 shows the rate profile for the reaction with perphthalic acid, in which a considerable rate diminution in the second phase of the reaction is seen. This suggests that the higher structure present in denatured DNA at low temperatures is not in a fluxional state and that its stability can be investigated by reactions of the type we are discussing. It would be of interest to compare the reaction profiles for several reagents with the same denatured material.

This leads on to the more important question of the reactivity of DNA when it is in states that may be described as denatured but which at present are poorly defined in structural terms. Included would be complexes of DNA with polylysine and with histones [395,396]. Of particular interest is the state

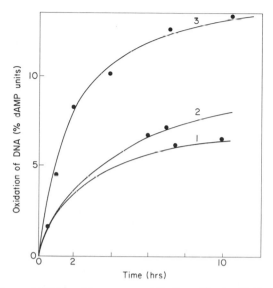

Fig. 4. Oxidation of DNA with monoperphthalic acid at pH 7 and 20°C. Initial concentrations; DNA 7.75 μmoles/ml; perphthalic acid, 23.10 μmoles/ml. (1) native DNA; (2) DNA denatured with $HgCl_2$; (3) DNA denatured by heating and fast cooling [380].

of DNA in concentrated polymer solutions. Physical evidence, including electron microscopy, shows that a spontaneous rearrangement of double-stranded DNA occurs to give an ordered tertiary structure [397–399]. This state approaches the compactness of the contents of bacteriophage heads. No chemical studies of this material have been made, but it is well known that the mutability of T4 phage by hydroxylamine is much greater than that of native *Bacillus subtilis* transforming DNA; the mutability of the latter is markedly increased by treatment at higher temperatures or in presence of high concentrations of ethyleneglycol [400]. Other studies on the reaction between methoxyamine and free and intraphage DNA from phage S_d indicate that in the latter some segments of the DNA do not react, while others react as fast as the monomer. Some 16–17% of the cytosines are so modified and the experiments are said to lend support to earlier evidence that some at least of the DNA in bacteriophages is not double-helical [401]. It would be of considerable interest if further work could show whether or not reactivity was confined to definite regions of the intraphage genome. It is of some interest that nonrandom mutagenesis of the *Escherischia coli* genome by *N*-methyl-*N*'nitro-*N*-nitrosoguanidine (NG) can be shown in synchronized cultures and this is clearly associated with increased reactivity of the replication point [402,403].

3. *Intercalating Reagents*

There are a number of compounds that react with nucleic acids in an apparently anomalous way. One may assume that a relatively strong, oriented complex is first formed and that within the complex covalent binding occurs. Complex formation by intercalation has been shown for some reagents; others are assumed to associate in this way, though subsequent work may correct this. In this group we include polycyclic hydrocarbons, derivatives of acridine dyestuffs and a number of antibiotics [404].

The carcinogenic hydrocarbons form intercalation complexes with DNA which, on activation, by light or by oxidation, give covalently linked products [405]. The chemistry is still far from clear due essentially to the fact that, though the intercalative binding energy is large, solubility is very low, and so the extent of reaction of, say, 3,4-benzpyrene is correspondingly low [405]. Iodine or hydrogen peroxide/ferrous ion oxidation lead to reaction with guanine residues, predominantly, as evinced by homopolynucleotide studies. It is conjectured that relatively long-lived radical cation and neutral radical intermediates, respectively, are involved [405]. It is not clear what bearing these experiments have on the oxidative activation of polycyclic hydrocarbons *in vivo* [406]. Polycyclic hydrocarbon epoxides may be the proximal carcinogens. The K-region epoxide benz[*a*]anthracene 5,6-oxide (**201**) is not only carcinogenic; it is also mutagenic and in *Salmonella typhimurium* leads almost exclusively to frame-shift mutants [407]. The chemistry of binding has not yet been elucidated.

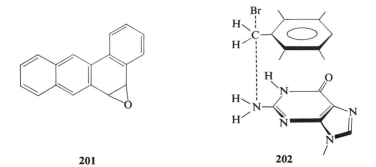

201   **202**

7-Bromomethylbenz[*a*]anthracene (**84**) is carcinogenic. It is very reactive. It is hydrolyzed rapidly by water and is peculiar in that it reacts faster with DNA than with mononucleotides [408]. Some relative rates are given in Table V [409]. Evidently rapid complex formation prior to reaction must account for this behavior. A further peculiarity is that alkylation occurs on the amino group of guanine and of adenine [410]. If the initial complex were intercalative in character reaction on the amino group might be expected as

TABLE V

EXTENT OF REACTION OF SOME POLYNUCLEOTIDES WITH
7-BROMOMETHYLBENZ[4] ANTHRACENE[a]

Polynucleotide	Substituted nucleotides number/1000 nucleotides
Poly rG	25
Poly rG·poly rC	10
Poly dG·poly dC	100
Poly dG	30
Poly rA	10
Poly dA	15
Poly r(A–U)	≤1
Poly d(A–T)	10
Poly rA·poly rU	≤1
Poly rU, poly rC, poly rI	≤1
Poly Me$_2{}^2$G	25
M. luteus DNA	200

[a] From Pochon and Michelson [409].

the transition state for the reaction (202) should have approximately the correct geometry. This is not so for the other basic centers (other than O-6) where the lone pair is in the base-pair plane (Section V,B,1). (It would not be surprising if the polycyclic epoxides followed the same reaction course.) It has also been found that deoxyguanosine reacts with the bromomethyl compound in the same way, i.e., at N-2, when in aqueous solution, but that in dimethyl-sulfoxide the more conventional 7-substitution occurs [410]. A variety of other related compounds, some of which are bifunctional, are biologically active [411].

Another group of frame-shift mutagens are based on the acridine ring system with, in addition, a basic side chain carrying a mono- or bifunctional nitrogen mustard residue [412,412a]. Even without an alkylating side chain some activity is shown, but the bifunctional compounds are very active with a spectrum of activity wider than that of proflavine itself. There can be little doubt that the compounds, e.g., ICR 191 (203) intercalate and then alkylate a neighboring base. This would have the effect of increasing the residence time of the intercalation complex. No evidence about the chemistry has as yet been given. The irreversible binding of the intensely fluorescent (204) has been put to good use. Differential reactivity of the heterochromatic regions along the chromosomes of *Vicia faba* is found, and observed by fluorescence microscopy [413]. The Y chromosome of humans is also stained specifically [414]. It is clear from this one example that it should be possible to

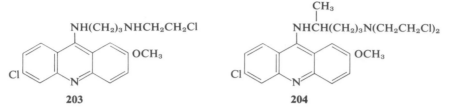

203   204

cast light on the state of DNA in the chromosomes by making use of reagents with specificities for higher structure [415], and to relate reactivity to microscopic features. The presence or otherwise of "globular" DNA [416] should be demonstrable.

The biological activity of mitomycin C (**205**), or the related porfiromycin, depends on initial reduction, presumably of the quinone system [417]. The product is then an active alkylating agent for native or denatured DNA and RNA [418]. It is bifunctional, crosslinks DNA, and shows specificity for reaction with guanine residues [419]. Like the mustards one in four or five alkylations are crosslinking. It has been suggested that the reduction product eliminates methanol to generate an indole system (**206**) and that its alkylating ability is dependent on the aziridine system and the benzylic reactivity of the —$CH_2 \cdot O \cdot CO \cdot NH_2$ residue [404,417,420]. The chemistry of the alkylation process has not been elucidated, but an interesting method has shown that

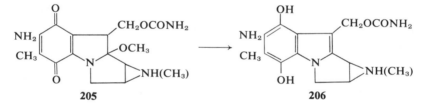

205   206

neither N-7 nor C-8 of guanine can be involved. It is well known that the proton at C-2 in thiazolium and imidazolium (**207**) systems exchanges rapidly with solvent protons and that this is due to stability of the zwitterion (**208**). The 7-methylguanosinium ion likewise exchanges [421]. When applied

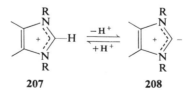

207   208

to guanine-8-[^{3}H] DNA reduced mitomycin causes no loss of tritium, though alkylation with a nitrogen mustard causes stoichiometric release [422]. This test reaction should have wide applicability. If mitomycin C intercalates prior to reaction, alkylation at O-6 [420] or, more likely, at the 2-amino group of guanine is possible. The stereochemistry of the alkylation reaction should be not unlike that for **84** and **201**.

C. RNA

1. *Introduction.*

The reactivity of bases in RNA, often yeast ribosomal RNA, has been the subject of much study. Comparisons with DNA and with homopolymers are to be found in papers already alluded to, but, broadly, the reactivity of RNA has been found to be less than that of mononucleotides and to correspond to that expected for a loosely structured single-stranded system. TMV–RNA has received much attention [423]. More recently the success in deducing the nucleotide sequences of many RNAs has, at once, raised new problems and provided the means of investigating them. The comparative reactivity of individual bases can be established by sequencing the nucleic acid after modification.

Without entering into details, sequencing has shown, in many cases, that base complementarity between contiguous regions in the same chain exists, leading to loops and helical stems. The transfer RNAs show these features *par excellence*, which are generalized in Fig. 5 [424]. There are strong physico-

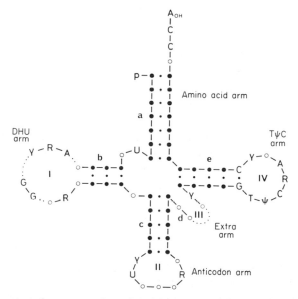

Fig. 5. The general representation of the clover-leaf model of tRNA. I, II, III, IV are loops and a, b, c, d, and e are double-helical regions. Capital letters indicate nucleotides common to most tRNA sequences and p is 5′-terminal phosphate. R; Y: purine; pyrimidine nucleotides (after Cramer [424]).

chemical grounds for believing that tertiary structure also exists [424,425], the nature of which is far from clear. It is generally assumed that base-pairing and base-stacking interactions occur between the loops and it may be, too, that nucleotide residues are "buried" in the folded structure. The reactivities of bases in double- and single-stranded helical systems have already been discussed. There are, however, no guidelines for bases in looped or buried regions or for those involved in maintaining tertiary structure. The tRNA's, as a group, have widely differing sequences but very similar dimensions. It is normally assumed that all are closely related, conformationally, and that in consequence chemical modification studies on individual species should be capable of interpretation, largely, in terms of a general model.

2. Transfer RNA

When unfractionated tRNA, in presence of magnesium ions is treated with a reagent specific for one or other of the common bases it is invariably found that the reaction is biphasic. Thus, typically, methoxyamine reacts rapidly with about 20% of the cytosines in mixed E. coli tRNA [426] and then the rate drops to 1/200th of the initial rate. The first condition for differentiating structural features, however crudely, is therefore met. Using conditions which allow reaction to proceed to the plateau region, with purified tRNA's, and then sequencing the modified material permits the reactive bases to be identified. In so doing one point on the reaction curve for each base is obtained. It is found that a number of bases have undergone complete reaction; others have only been partially modified. The clover leaf representations of a number of tRNA's are shown in Figs. 6a–i and the positions of the bases modified by a variety of specific reagents are indicated. It is obvious that a kinetic analysis for each modifiable base, would be more revealing. This, though experimentally tedious, has been done in the case of the carbodiimide reaction on the E. coli tyrosine suppressor $tRNA_{Su_{III}+}^{Tyr}$ [427] (Fig. 7) and $tRNA^{fMet}$ (Fig. 8) [428] and the bisulfite-catalyzed cytosine to uracil conversion on $tRNA^{Glu}$ [429].

Any study of macromolecular structure in which chemical modification of a structural unit is involved is open to criticism on the grounds that one alteration may, lead to conformational changes in turn reflected in changes in the rates of modification of further sites. False conclusions concerning the original structure may follow. No single solution to this fundamental problem is available. Consistency in the results obtained with a number of specific reagents on one nucleic acid should be indicative of structural integrity. Initial rates of reaction at individual sites, as judged by experiments using small extents of reaction has also been advocated [430]. If the plateau level of reaction increases with increasing concentration of reagent, particularly for reactions which are intrinsically reversible, modification must be assumed to be effecting conformational changes. Kethoxal [158] and the carbodiimide [431], but not methoxyamine, show this phenomenon.

On the other hand, the conditions of the reaction can be altered in a controlled way with a consequent change in the degree of modification. Thus two additional adenine residues in yeast tRNA^{Ser} are oxidized by perphthalic acid at 40°C, as compared with 20°C (Fig. 6f), corresponding to the first plateau in the melting profile [432].

What conclusions can be drawn from the data in Figs. 6a–i? In the first place, the helical stems are essentially devoid of reactivity and this, in itself, provides support for the clover-leaf structure. There are exceptions. A nitrogen mustard-type reagent alkylates certain stem guanine residues in yeast tRNA^{Val} (Fig. 6a) [430]; as noted earlier (Section V,B,1), N-7 is sterically available for reaction in double-helical systems. Photooxidation of *E. coli* tRNA^{fMet} affects stem guanines [433]. This result could be accounted for if intercalation of the photosensitizer, methylene blue, leads to more efficient energy transfer. Photohydration of uracil and cytosine residues occurs in loop and stem regions [434]. In consequence alkylation and photocatalyzed reactions are unlikely to be of value in defining structured regions.

The nucleotides in loop IV are also exceptionally unreactive compared with those in loops I–III and in the CCA$_{OH}$ terminus. Thus the loop shows virtually no reactivity to the carbodiimide [427,428,435]. The pseudouridine residue in the constant TψC region reacts with acrylonitrile only under denaturing conditions [436–438]. The neighboring cytosine is quite unaffected by methoxyamine [426,439], nitrous acid [440], or bisulfite [441–443] under normal conditions. Is the lack of reactivity in this loop an intrinsic property of the loop conformation, or does it reflect an involvement of the bases in tertiary structure maintenance? The question, though capable of solution, has not been investigated. It raises the point, however, that the modification results can only be discussed adequately by reference to detailed molecular models.

Although a variety of models have been suggested [424,425] only two describe tertiary base-pairing and -stacking interactions in sufficient detail. These are designated A [432] and B [444] in Fig. 9. A detailed model of the anticodon loop has been described in which the (predominantly purine) bases on the 5'-side and the anticodon are stacked, while the two pyrimidine residues which complete the loop project from the structure (Fig. 10) [445]. Models A and B differ considerably in their tertiary base-pairing connections but both include the anticodon loop essentially as in Fig. 10. Inspection of Fig. 9 shows that the degree of pairing interactions of bases in the loops, in both models, accounts in a general way for the reactivity that these regions exhibit.

Specific points can be tested. In model B G15 is hydrogen-bonded to C57 in the *E. coli* suppressor tRNA (Fig. 6b). A mutant tRNA with G15 replaced

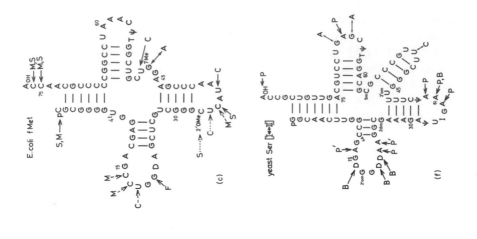

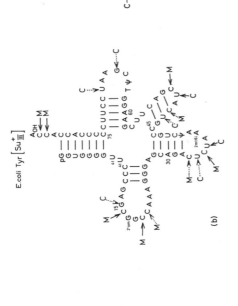

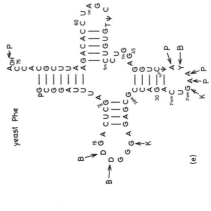

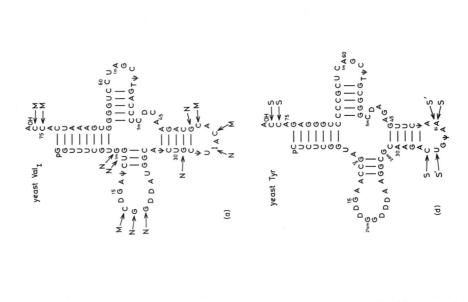

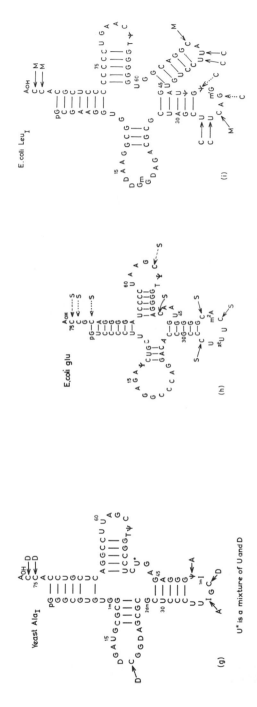

Fig. 6. Clover-leaf representations of a number of tRNA sequences. Positions modified by reagents are indicated by arrows. A, acrylonitrile; B, borohydride; C, the carbodiimide (119); D, nitrous acid; F, *N*-acetoxyfluorenyl-2-acetamide; K, Kethoxal; *M*, methoxyamine; N, a nitrogen mustard; P, perphthalic acid (at 20°C; P' additional modification at 40°C); S, bisulfite (pH 6); S', bisulfite (pH 7). Dashed and dotted arrows indicate positions of medium and low reactivity, respectively. References, giving the reaction conditions are in the text. All reaction solutions contain Mg²⁺ except for the bisulfite reaction with *E. coli* tRNA^Glu (see *h*).

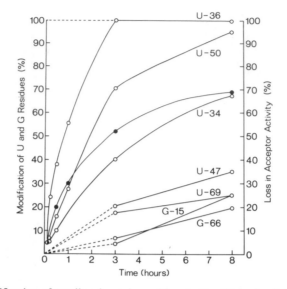

Fig. 7. Modification of uracil and guanine residues in *E. coli* tyrosine Su^+_{III} tRNA by the carbodiimide (**119**) in borate buffer, pH 8.0 with 10 m*M* MgCl₂ at 37°C for varying times (○). Loss in amino acid acceptor activity is also plotted (●) [427].

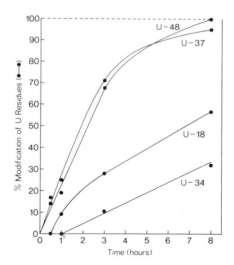

Fig. 8. Modification of uracil residues in *E. coli* formylmethionine tRNA with the carbodiimide as in Fig. 7 [443].

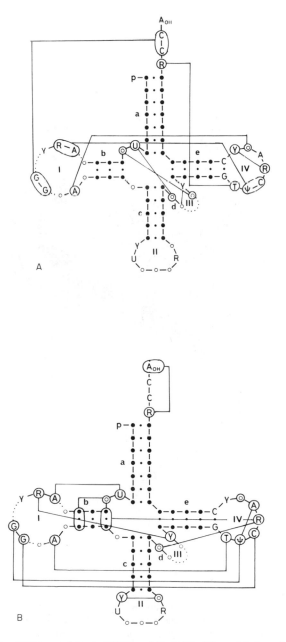

Fig. 9. Transfer RNA models A [432] and B [444]. Lines joining circled residues indicate proposed additional base-pairing in tertiary structure. Other symbols are as in Fig. 5.

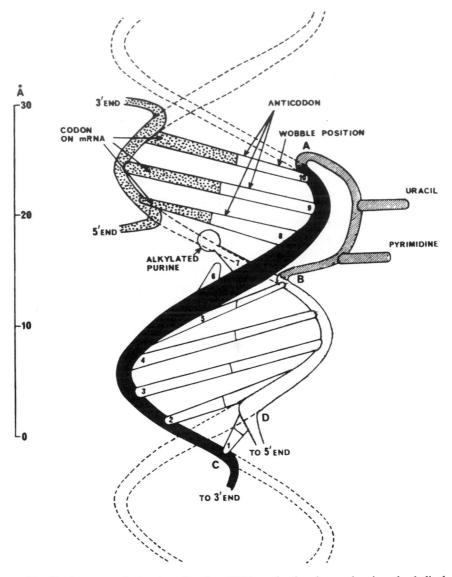

Fig. 10. A proposed structure for the tRNA anticodon loop, showing the helical character of the structure and its relation to the mRNA codon. The bases 1 to 10 (on the 5′- side of the loop) are stacked on one another. The dotted lines indicate the generic helix from which the structure can be imagined to be derived [445].

by A can be isolated. Comparison of the two tRNAs shows that in the former C57 is inert to methoxyamine and in the latter it reacts completely [446]. This evidence favors model B, since model A involves a G15–C65 interaction; C65 is inert in both tRNAs. G15 is unreactive to the carbodiimide reagent in tRNA^{fMet} and marginally in the tyrosine tRNA (Figs. 6b,c) [427,428].

If it is assumed that unpaired nucleotides, stacked or not, undergo reaction then model A accounts satisfactorily for the positions of perphthalic acid oxidation [432], borohydride reduction [85], and Kethoxal addition [157] in yeast tRNA^{Phe} and tRNA$^{Ser}_{I,II}$ (Figs. 6e,f). Where tested, Kethoxal reacts at G20 (and with a G in the anticodon, where present) [447,448]. An unpaired G20 is therefore favored. However, G20 also reacts preferentially with *N*-acetoxyAAF (**138**) in tRNA^{fMet} and others [449]. This favors a *syn* conformation for G20, allowing reaction at C-8 [450]. In model B G20 shows this feature but is also hydrogen-bonded. Conformational instability in this region could reconcile these two sets of observations.

The reactivity of the anticodon loop warrants some discussion. The availability of the bases on the 3'-side, and those constituting the anticodon, for reaction with acrylonitrile [437], perphthalic acid [432], Kethoxal [447], methoxyamine [439], bisulfite [429], and the carbodiimide [425] argues in favor of a structure such as that in Fig. 10. The pyrimidines on this structure should also show high reactivity, but this is not observed in practice. Figures 6b and c show a gradation of reactivity toward the 5'- end of the loop with methoxyamine [426,428], bisulfite [443], and the carbodiimide [427,428]. The reaction of bisulfite with C34 in yeast tRNA^{Tyr} [442] is probably to be ascribed to conformational disturbance due to reversible reaction at U35; certainly the corresponding cytosines in *E. coli* tRNA$^{Tyr}_{Su_{III}^+}$ and tRNA^{fMet} are virtually inert. We may add to this the fact that the low reactivity toward methoxyamine of C33 compared with C35 in the tyrosine tRNA (Fig. 6b) is also found in a T1-nuclease digest fragment which includes the anticodon loop and stem. The reactivity order is therefore an intrinsic feature of the loop in this case [451]. The *E. Coli* leucine tRNA (Fig. 6i) is different. The pyrimidines on the the 5'-side of the loop are all very reactive and the pseudo-uridine on the 3'-side much less so [451a]. The chemical results suggest that the structure in Fig. 10 cannot be generally correct. It may hold for tRNA^{Leu} (Fig. 6i) but a loop in which the 5'-side is stacked would be more in accord with the data on tRNA$^{Tyr}_{Su_{III}^+}$ and tRNA^{fmet} (Figs. 6b and 6c), and this is supported by other evidence [452]. The low propensity for stacking of uridine residues and the presence of a uridine as the first residue in the loop on the 5'-side may account for the difference between tRNA^{Leu} and those tRNA's which have a cytidine in this position. In any case, differing base sequences in a loop must, in the last analysis, lead to differing conformations. The reactivity of bases in a dynamic structure [453] that flips between the two extremes is difficult to assess. Further refinement of the recent X-ray crystal structural analysis of yeast tRNA[Phe] [453a] should allow a direct comparison between structure and reactivity to be made.

It is not the present purpose to discuss all of the observations shown in Figs. 6a–i, nor to attempt to reach conclusions on tRNA structure. Rather, it is the intention to show how the chemistry, particularly the stereochemistry, of a modification reaction can be brought to bear on the question of the position of a nucleotide in a macrostructure. In summary, the questions that can be asked relate to whether (a) a nucleotide residue is stacked in a double- or single-stranded helical system, (b) a base is hydrogen-bonded, and (c) is present in the *syn* or, more common, *anti* conformation. It is evident that reagents are available that can throw light on these systems. However, probably the most important parameter to be investigated in defining a folded polynucleotide structure is the conformation of the internucleotide linkage [425,454]. So far no attempt has been made to view this as a chemical problem. The stereochemical relation between the phosphate residue and the 2'-OH group during hydrolysis is fairly clearly defined. However, hydrolysis as normally carried out involves pH's inimical to macrostructure integrity. The discussion given in Section IV,B suggests that heavy metal ion or nuclease-catalyzed hydrolysis near neutrality might yield information about internucleotide bonds present in a conformation suitable for intramolecular transesterification. It is of interest in this connection, and perhaps significant, that in a number of tRNAs the most rapidly cleaved linkage is at G20 during T1 nuclease action [455,456]. This position in model B corresponds to a sharp turn in the sugar phosphate backbone with a conformation approaching that required for hydrolysis. The prediction is also supported by the recent observations that the initial cleavage of tRNA[Phe] by Pb^{2+} [455] and high concentrations of Mg^{2+} [456] are at D17 and D18 respectively.

References

1. G. Felsenfeld and H. T. Miles, *Annu. Rev. Biochem.* **36,** 407 (1967)
2. N. K. Kochetkov and E. I. Budowsky, eds., "Organic Chemistry of Nucleic Acids." Plenum, New York, 1972.

3. D. J. Brown, "The Pyrimidines." Wiley (Interscience), New York, 1962 and Suppl. 1, 1970.
4. J. H. Lister, "Fused Pyrimidines," Part II. Purines. Wiley (Interscience), New York, 1971.
4a. A. Albert, *Chem. Biol. Purines, Ciba Found. Symp.*, p. 97 (1957).
5. A. Bendich, *in* "The Nucleic Acids" (E. Chargaff and J. N. Davidson, eds.), Vol. 1, p. 81. Academic Press, New York, 1955.
6. D. M. Brown, C. A. Decker, and A. R. Todd, *J. Chem. Soc., London* p. 2715 (1952).
7. R. Shapiro and R. S. Klein, *Biochemistry* **5**, 2358 (1966).
8. R. Shapiro and R. S. Klein, *Biochemistry* **6**, 3576 (1967).
9. K. Burton, *Biochem. J.* **104**, 686 (1967).
10. E. B. Ziff and J. R. Fresco, *J. Amer. Chem. Soc.* **90**, 7338 (1968); *Biochemistry* **8**, 3242 (1969).
11. M. Saneyoshi and S. Nishimura, *Biochim. Biophys. Acta* **145**, 208 (1967).
12. R. T. Walker and U. L. RajBhandary, *Biochem. Biophys. Res. Commun.* **38**, 907 (1970).
13. M. Green and S. S. Cohen, *J. Biol. Chem.* **225**, 397 (1957).
14. J. T. Madison, *in* "Methods in Enzymology" (L. Grossman and K. Moldave, eds.), Vol. 12, Part A, p. 141. Academic Press, New York, 1967.
15. D. Shugar and J. J. Fox, *Biochim. Biophys. Acta* **9**, 199 (1952).
16. D. M. Brown and M. J. E. Hewlins, *J. Chem. Soc., C* p. 2050 (1968).
17. W. E. Cohn and D. G. Doherty, *J. Amer. Chem. Soc.* **78**, 2963 (1956).
18. A. R. Hanze, *J. Amer. Chem. Soc.* **89**, 6720 (1967).
19. S. Altman, *Nature (London) New Biol.* **229**, 19 (1971).
20. D. M. Brown, *in* "Methods in Enzymology" (L. Grossman and K. Moldave, eds.), Vol. 12, Part A, p. 31. Academic Press, New York, 1967.
21. D. H. Hayes and F. Hayes-Baron, *J. Chem. Soc., C* p. 1528 (1967).
22. F. Lingens, *Angew. Chem., Int. Ed. Engl.* **3**, 379 (1964).
23. D. M. Brown, A. D. McNaught, and P. Schell, *Biochem. Biophys. Res. Commun.* **24**, 967 (1966).
24. F. Lingens and H. Schneider-Bernlöhr, *Justus Liebigs Ann. Chem.* **686**, 134 (1965).
25. E. I. Budowsky, J. A. Haines, and N. K. Kochetkov, *Dokl. Akad. Nauk. SSSR* **158**, 379 (1964).
26. H. Hayatsu and T. Ukita, *Biochem. Biophys. Res. Commun.* **14**, 198 (1964).
27. K. Kikugawa, H. Hayatsu, and T. Ukita, *Biochim. Biophys. Acta* **134**, 221 (1967).
28. H. Hayatsu, K. Taeisha, and T. Ukita, *Biochim. Biophys. Acta* **123**, 445 (1966).
29. L. Gal-Or, J. E. Mellema, E. N. Moudrianakis, and M. Beer, *Biochemistry* **6**, 1909 (1967).
30. K. Kikugawa, H. Hayatsu, and T. Ukita, *Chem.-Biol. Interact.* **1**, 247 (1969-1970).
31. K. Kikugawa, A. Muto, H. Hayatsu, K. Miura, and T. Ukita, *Biochim. Biophys. Acta* **134**, 232 (1967).
32. W. P. Jencks, "Catalysis in Chemistry and Enzymology." Magraw-Hill, New York, 1969.
33. H-J. Rhaese and E. Freese, *Biochim. Biophys. Acta* **155**, 476 (1968).
34. H.-J. Rhaese, E. Freese, and M. S. Melzer, *Biochim Biophys. Acta* **155**, 491 (1968).
35. J. H. Phillips and D. M. Brown, *Progr. Nucl. Acid Res. Mol. Biol.* **7**, 349 (1967).
36. E. I. Budowsky, E. D. Sverdlov, and G. S. Monastyrskaya, *Biochim. Biophys. Acta* **246**, 320 (1971).
37. D. M. Brown and M. R. Osborne, *Biochim. Biophys. Acta* **247**, 514 (1971).
38. H. Schuster, *J. Mol. Biol.* **3**, 447 (1961).

39. D. M. Brown and P. Schell, *J. Chem. Soc., London* p. 208 (1965).
40. P. D. Lawley, *J. Mol. Biol.* **24**, 75 (1967).
41. E. I. Budowsky, E. D. Sverdlov, R. P. Shibaeva, and G. S. Monastyrskaya, *Mol. Biol.* **2**, 329 (1968).
42. D. M. Brown and M. J. E. Hewlins, *J. Chem. Soc., C* p. 1922 (1968).
43. D. M. Brown, M. J. E. Hewlins, and P. Schell, *J. Chem. Soc., C* p. 1925 (1968).
44. N. K. Kochetkov, E. I. Budowsky, and N. A. Simukova, *Dok. Akad. Nauk SSSR* **153**, 597 (1963).
45. H. E. Johns, J. C. Leblanc, and K. B. Freeman, *J. Mol. Biol.* **13**, 849 (1965).
46. G. D. Smith and M. P. Gordon, *J. Mol. Biol.* **34**, 281 (1968).
47. D. Verwoerd, H. Kohlhage, and W. Zillig, *Nature (London)* **192**, 1038 (1961).
48. L. Grossman and E. Rodgers, *Biochim. Biophys. Acta* **33**, 975 (1968).
49. D. M. Brown, P. F. Coe, and D. P. L. Green, *J. Chem. Soc., C* p. 867 (1971).
50. E. I. Budowsky, E. D. Sverdlov, R. P. Shibaeva, G. S. Monastyrskaya, and N. K. Kochetkov, *Biochim. Biophys. Acta* **246**, 300 (1971).
51. E. I. Budowsky, E. D. Sverdlow, R. P. Shibaeva, and G. S. Monastyrskaya, *Mol. Biol.* **2**, 321 (1968).
52. D. M. Brown and P. F. Coe, *Chem. Commun.* p. 568 (1970).
53. D. M. Brown and P. F. Coe, in press.
54. J. H. Phillips and D. M. Brown, *J. Mol. Biol.* **11**, 663 (1965).
55. K. L. Wierzchowski and D. Shugar, *Acta Biochim. Polon.* **8**, 219 (1961).
56. G. DeBoer, O. Klinghoffer, and H. E. Johns, *Biochim. Biophys. Acta* **213**, 253 (1970).
57. D. M. Brown and C. M. Taylor, *J. Chem. Soc., Perkin Trans.* **1**, 2385 (1972); unpublished observations.
58. C. Janion and D. Shugar, *Acta Biochim. Pol.* **12**, 337 (1965).
59. N. K. Kochetkov and E. I. Budowsky, *Progr. Nucl. Acid Res. Mol. Biol.* **9**, 403 (1969).
60. W. J. Wechter and R. C. Kelly, *Abstr. I.U.P.A.C. Meet. Chem. Nucl. Acid Components* N16 (1967).
61. R. W. Chambers and V. Kirov, *J. Amer. Chem. Soc.* **58**, 2160 (1963).
62. B. Bannister and F. Kagan, *J. Amer. Chem. Soc.* **82**, 3363 (1960).
63. D. V. Santi and C. F. Brewer, *Biochemistry*, **12**, 2416 (1973).
64. D. M. Brown and P. F. Coe, *Chem. Commun.* p. 864 (1970).
65. R. Shapiro, R. E. Servis, and M. Welsher, *J. Amer. Chem. Soc.* **92**, 422 (1970).
66. H. Hayatsu, Y. Wataya, and K. Kai, *J. Amer. Chem. Soc.* **92**, 724 (1970).
67. H. Hayatsu, Y. Wataya, K. Kai, and S. Iida, *Biochemistry* **9**, 2858 (1970).
68. M. Sono, Y. Wataya, and H. Hayatsu, *J. Amer. Chem. Soc.* **95**, 4745 (1973).
69. R. Shapiro, B. I. Cohen, and R. E. Servis, *Nature (London)* **227**, 1047 (1970).
70. H. Hayatsu and A. Miura, *Biochim. Biophys. Res. Commun.* **39**, 156 (1970).
71. F. Mukai, I. Hawryluk, and R. Shapiro, *Biochem. Biophys. Res. Commun.* **39**, 983 (1970).
72. R. Shapiro and J. M. Weigras, *Biochem. Biophys. Res. Commun.* **40**, 839 (1970).
73. D. M. Brown and J. Goodchild, unpublished observations.
74. A. H. Schein and F. T. Schein, *in* "Methods in Enzymology" (L. Grossman and K. Moldave, eds.), Vol. 12, Part A, p. 38. Academic Press, New York, 1967.
75. R. W. Chambers, *Progr. Nucl. Acid Res. Mol. Biol.* **5**, 349 (1966).
76. P. Cerutti, *in* "Methods in Enzymology" (L. Grossman and K. Moldave, eds.), Vol. 12, Part B, p. 461. Academic Press, New York, 1968.
77. P. Cerutti, K. Ikeda, and B. Witkop, *J. Amer. Chem. Soc.* **87**, 2505 (1965).
78. G. Ballé, P. Cerutti, and B. Witkop, *J. Amer. Chem. Soc.* **88**, 3946 (1966).

79. P. Cerutti, J. W. Holt, and N. Miller, *J. Mol. Biol.* **34**, 505 (1968).
80. N. Miller and P. Cerutti, *J. Amer. Chem. Soc.* **89**, 2767 (1967).
81. P. Cerutti, N. Miller, M. G. Pleiss, J. F. Remsden, and W. J. Ramsay, *Proc. Nat. Acad. Sci. U.S.* **64**, 731 (1969).
82. J. F. Remsden, N. Miller, and P. A. Cerutti, *Proc. Nat. Acad. Sci. U.S.* **65**, 460 (1970).
83. P. Cerutti and N. Miller, *J. Mol. Biol.* **26**, 55 (1967).
84. R. Adman and P. Doty, *Biochem. Biophys. Res. Commun.* **27**, 579 (1967).
85. T. Igo-Kemenes and H. G. Zachau, *Eur. J. Biochem.* **10**, 549 (1969).
86. W. Wintermeyer and H. G. Zachau, *FEBS Lett.* **11**, 160 (1970).
87. J. B. Macon and R. V. Wolfenden, *Biochemistry* **7**, 3453 (1968).
88. W. C. J. Ross, "Biological Alkylating Agents." Butterworth, London, 1962.
89. A. Loveless, "Genetic and Allied Effects of Alkylating Agents." Butterworth, London, 1966.
90. P. D. Lawley, *Progr. Nucl. Acid Res.* **5**, 89 (1968).
91. R. Shapiro, *Ann. N.Y. Acad. Sci.* **163**, 624 (1969).
92. T. Ueda and J. J. Fox, *Advan. Carbohyd. Chem.* **22**, 307 (1967).
92a. C. B. Reese, *in* "Protective Groups in Organic Chemistry" (J. F. W. McOmie, ed.). p. 95. Plenum, New York, 1973.
93. S. Walles and L. Ehrenberg, *Acta Chem. Scand.* **23**, 1080 (1969).
94. S. Osterman-Golkar, L. Ehrenberg, and C. A. Wachtmeister, *Radiat. Bot.* **10**, 303 (1970).
95. C. C. Price, G. M. Gaucher, P. Koneru, R. Shibakawa, J. R. Sowa, and M. Yamaguchi, *Biochim. Biophys. Acta* **166**, 327 (1968).
96. M. H. Benn, P. Kazmaier, C. Watanatada, and L. M. Owen, *Chem. Commun.* p. 1685 (1970).
97. D. R. McCalla, *Biochim. Biophys. Acta* **155**, 114 (1968).
98. U. Schulz and D. R. McCalla, *Can. J. Chem.* **47**, 2021 (1969).
99. P. D. Lawley and C. J. Thatcher, *Biochem. J.* **116**, 693 (1970).
100. E. Cerdá-Olmedo and P. C. Hanawalt, *Mol. Gen. Genet.* **101**, 191 (1968).
101. R. Süssmuth, R. Haerlin, and F. Lingens, *Biochim. Biophys. Acta* **269**, 276 (1972).
102. P. N. Magee and E. Farber, *Biochem. J.* **83**, 114 (1962).
103. W. Lijinsky, J. Loo, and A. E. Ross, *Nature (London)* **218**, 1174 (1968).
104. A. E. Ross, L. Keefer, and W. Lijinsky, *J. Nat. Cancer Inst.* **47**, 789 (1971).
105. P. Brookes and P. D. Lawley, *J. Chem. Soc., London* p. 3923 (1961).
106. J. J. Roberts and G. P. Warwick, *Biochem. Pharmacol.* **12**, 1441 (1963).
107. J. A Haines, C. B. Reese, and Sir A. Todd, *J. Chem. Soc., London* p. 1406 (1964).
108. R. L. C. Brimacombe, B. E. Griffin, J. A. Haines, W. J. Haslam, and C. B. Reese, *Biochemistry* **4**, 2452 (1965).
109. B. E. Griffin, *in* "Methods in Enzymology" (L. Grossman and K. Moldave, eds.), Vol. 12, Part A, p. 141. Academic Press, New York, 1967.
110. S. Hendler, E. Fürer, and P. R. Srinivasan, *Biochemistry* **9**, 4141 (1970).
111. F. Pochon and A. M. Michelson, *Biochim. Biophys. Acta* **149**, 99 (1967).
112. D. B. Ludlum, *Biochim. Biophys. Acta* **247**, 412 (1971).
113. B. E. Griffin and C. B. Reese, *Biochim. Biophys. Acta* **68**, 185 (1963).
114. D. B. Ludlum, *Biochim. Biophys. Acta* **174**, 773 (1969).
115. A. Dipple, P. Brookes, D. S. Mackintosh, and M. P. Rayman, *Biochemistry* **10**, 4323 (1971).
116. A. D. Broom, L. B. Townsend, J. W. Jones, and R. K. Robins, *Biochemistry* **3**, 494 (1964).

117. B. E. Griffin and C. B. Reese, *Biochim. Biophys. Acta* **68**, 185 (1963).
118. H. G. Windmueller and N. O. Kaplan, *J. Biol. Chem.* **236**, 2716 (1961).
119. J. A. Haines, C. B. Reese, and Sir A. Todd, *J. Chem. Soc., London* p. 5281 (1962).
120. A. K. Kriek and P. Emmelot, *Biochim. Biophys. Acta* **91**, 59 (1964).
121. Y. Kanaoka, E. Sato, M. Aira, S. Yonemitsu, and Y. Mizuno, *Tetrahedron Lett*, p. 3361 (1969).
122. H. T. Miles, *J. Org. Chem.* **26**, 4761 (1961).
123. A. Psoda and D. Shugar, *Biochim. Biophys. Acta* **247**, 507 (1971).
124. R. V. Wolfenden, *J. Mol. Biol.* **40**, 307 (1969).
125. O. M. Friedman, G. N. Mahapatra, B. Dash, and R. Stevenson, *Biochim. Biophys. Acta* **103**, 286 (1965).
126. A. Loveless, *Nature (London)* **223**, 206 (1969).
127. P. D. Lawley *in* "Topics in Chemical Carcinogenesis" (W. Nakahara, S. Takayama, T. Sugimura, and S. Odashima, eds.) p. 237. Univ. Tokyo Press, Tokyo, 1972.
128. A. Loveless and C. L. Hampton, *Mutat. Res.* **7**, 1 (1969).
129. A. Holy and K. H. Scheit, *Biochim. Biophys. Acta* **123**, 430 (1966).
130. B. R. Reid, *Biochemistry* **9**, 2852 (1970).
131. E. D. Bergmann, D. Ginsburg, and R. Pappo, *Org. React.* **10**, 179 (1959).
132. R. W. Chambers, *Biochemistry* **4**, 219 (1965).
133. J. Ofengand, *Biochem. Biophys. Res. Commun.* **18**, 192 (1965).
134. J. Ofengand, *J. Biol. Chem.* **242**, 5034 (1967).
135. M. Yoshida and T. Ukita, *Biochim. Biophys. Acta* **157**, 455 and 466 (1968).
136. G. E. Means and R. E. Feeney, "Chemical Modification of Proteins." Holden-Day, San Francisco, California, 1971.
137. J. A. Carbon and H. David, *Biochemistry*. **7**, 3851 (1968).
137a. M. J. Feldman, *Prog. Nucl. Acid Res. Mol. Biol.* **13**, 1 (1973).
138. R. P. Bell, "The Proton in Chemistry." Cornell Univ. Press, Ithaca, New York, 1959.
139. H. Fraenkel-Conrat, *Biochim. Biophys. Acta* **15**, 308 (1954).
140. M. Staehelin, *Biochim. Biophys. Acta* **29**, 410 (1958).
141. L. Grossman, S. Levine, and W. S. Allison, *J. Mol. Biol.* **3**, 47 (1961).
142. S. Lewin, *J. Chem. Soc., London* p. 792 (1964).
143. E. J. Eyring and J. Ofengand, *Biochemistry* **6**, 2500 (1967).
144. R. Hazelkorn and P. Doty, *J. Biol. Chem.* **236**, 2738 (1961).
145. J. Penniston and P. Doty, *Biopolymers* **1**, 145 and 209 (1963).
146. M. J. Feldman, *Biochim. Biophys. Acta.* **149**, 20 (1967).
147. D. Freifelder and P. F. Davison, *Biophys. J.* **2**, 249 (1962).
148. V. D. Axelrod, M. Ya. Feldman, I. I. Chugaev, and A. A. Bayev, *Biochim. Biophys. Acta* **186**, 33 (1969).
149. A. H. Alegria, *Biochim. Biophys. Acta* **149**, 317 (1967).
150. T. L. V. Ulbricht, *Progr. Nucl. Acid Res. Mol. Biol.* **4**, 189 (1965).
151. R. E. Cline, R. M. Fink, and K. Fink, *J. Amer. Chem. Soc.* **81**, 2521 (1959).
152. C. Heidelberger, *Progr. Nucl. Acid Res. Mol. Biol.* **4**, 1 (1965).
153. M. Staehelin, *Biochim. Biophys. Acta* **31**, 448 (1959).
154. R. Shapiro, B. I. Cohen, S-J. Shiuey, and H. Maurer, *Biochemistry* **8**, 238 (1969).
155. R. Shapiro and J. Hachmann, *Biochemistry* **5**, 2799 (1966).
156. R. B. Inman, *J. Mol. Biol.* **18**, 464 (1966).
157. M. Litt, *Biochemistry* **8**, 3249 (1969).
158. L. L. Kisselev, personal communication.
159. M. Litt and V. Hancock, *Biochemistry* **6**, 1848 (1967).

160. N. K. Kochetkov, E. I. Budowsky, M. E. Broude, and L. M. Klebanova, *Biochim. Biophys. Acta* **134**, 492 (1967).
161. R. Shapiro and S. C. Agarwal, *J. Amer. Chem. Soc.* **90**, 474 (1968).
162. B. M. Goldschmidt, T. P. Blazej, and B. L. Van Duuren, *Tetrahedron Lett.* p. 1583 (1968).
163. N. K. Kochetkov, V. W. Shibaev, and A. A. Kost, *Tetrahedron Lett.* p. 1993 (1971).
164. J. R. Barrio, J. A. Secrist, III, and N. J. Leonard, *Biochem. Biophys. Res. Commun* **46**, 597 (1972).
165. R. K. Ralph, R. J. Young, and H. G. Khorana, *J. Amer. Chem. Soc.* **85**, 2002 (1963).
166. P. T. Gilham, *J. Amer. Chem. Soc.* **84**, 687 (1962).
167. N. W. Y. Ho and P. T. Gilham, *Biochemistry* **6**, 3632 (1967).
168. D. H. Metz and G. L. Brown, *Biochemistry* **8**, 2312 (1969).
169. S. W. Brostoff and V. M. Ingram, *Biochemistry* **9**, 2372 (1970).
170. R. Naylor, N. W. Y. Ho, and P. T. Gilham, *J. Amer. Chem. Soc.* **87**, 4209 (1965).
171. N. W. Y. Ho and P. T. Gilham, *Biochemistry* **10**, 3651 (1971).
172. J. C. Lee, N. W. Y. Ho, and P. T. Gilham, *Biochim. Biophys. Acta* **95**, 503 (1965).
173. G. G. Brownlee, F. Sanger, and B. G. Barrell, *J. Mol. Biol.* **34**, 379 (1968).
174. G. Augusti-Tocco and G. L. Brown, *Nature (London)* **206**, 683 (1965).
175. O. I. Ivanova, D. G. Knorre, and E. G. Malygin, *Mol. Biol.* **1**, 335 (1967).
176. D. H. Metz and G. L. Brown, *Biochemistry* **8**, 2329 (1969).
177. A. Gierer and K.-W. Mundry, *Nature (London)* **182**, 1457 (1958).
178. H. Schuster and W. Vielmetter, *J. Chim. Phys.* **58**, 1005 (1961).
179. H. Fraenkel-Conrat, "The Molecular Basis of Virology." Van Nostrand-Reingold, Princeton, New Jersey, 1968.
180. T. Kotaka and R. L. Baldwin, *J. Mol. Biol.* **9**, 323 (1964).
181. H. Schuster and G. Schramm, *Z. Naturforsch. B* **13**, 697 (1958).
182. H. Schuster, *Z. Naturforsch. B* **15**, 298 (1960).
183. H. Schuster and R. C. Wilhelm, *Biochim. Biophys. Acta* **68**, 554 (1963).
184. R. Shapiro and S. H. Pohl, *Biochemistry* **7**, 448 (1968).
185. J. A. Carbon, *Biochim. Biophys. Acta* **95**, 550 (1965).
186. J. A. Carbon, *Proc. Nat. Acad. Sci. U.S.* **59**, 467 (1968).
187. M. S. May and R. W. Holley, *J. Mol. Biol.* **52**, 19 (1970).
188. J. W. Drake, "The Molecular Basis of Mutation." Holden-Day, San Francisco, California, 1970.
189. I. Tessman, *J. Mol. Biol.* **5**, 442 (1962).
190. E. P. Geiduschek, *Proc. Nat. Acad. Sci. U.S.* **47**, 950 (1961).
191. E. F. Becker, B. K. Zimmerman, and E. P. Geiduschek, *J. Mol. Biol.* **8**, 377 (1964).
192. R. Litman, *J. Chim. Phys.* **58**, 997 (1961).
193. W. Lijinsky and M. Greenblatt, *Nature (London), New Biol.* **236**, 177 (1972).
194. B. C. Challis and M. R. Osborne, *Chem. Commun.* p. 5810 (1972).
195. A. S. Vanderbilt and I. Tessman, *Genetics* **66**, 1 (1970).
196. A. R. Katritzky and J. M. Lagowski, "Chemistry of Heterocyclic N-Oxides." Academic Press, New York, 1971.
197. G. B. Brown, *Progr. Nucl. Acid. Res. Mol. Biol.* **8**, 209 (1968).
198. F. Cramer and H. Seidel, *Biochim. Biophys. Acta* **72**, 157 (1963).
199. H. Seidel and F. Cramer, *Biochim. Biophys. Acta* **108**, 367 (1965).
200. F. von der Haar, E. Schlimme, V. A. Erdmann, and F. Cramer, *Bioorganic Chem.* **1**, 282 (1971).
201. T. J. Delia, M. J. Olsen, and G. B. Brown, *J. Org. Chem.* **30**, 2766 (1965).

202. L. R. Subbaraman, J. Subbaraman, and E. J. Behrman, *Biochemistry* **8**, 3059 (1969).
203. H. Seidel, *Biochim. Biophys. Acta* **138**, 98 (1967).
204. N. J. M. Birdsall, J. C. Parham, U. Wolcke, and G. B. Brown, *Tetrahedron* **28**, 3 (1972).
205. N. J. M. Birdsall, U. Wolcke, T.-C. Lee, and G. B. Brown, *Tetrahedron* **27**, 5969 (1971).
206. J. A. Miller, *Cancer Res.* **30**, 559 (1970).
207. C. C. Irving and L. T. Russell, *Biochemistry* **9**, 2471 (1970).
208. C. M. King and B. Phillips, *Science* **159**, 135 (1968); *J. Biol. Chem.* **244**, 6208 (1969).
209. J. R. DeBraun, E. C. Miller, and J. A. Miller, *Cancer Res.* **30**, 577 (1970).
210. Sir C. Ingold, "Structure and Mechanism in Organic Chemistry," p. 621. Cornell Univ. Press, Ithaca, New York, 1953.
211. E. Boyland and R. Nery, *J. Chem. Soc., London* p. 5217 (1962).
212. E. Kriek, *Biochim. Biophys. Acta* **161**, 273 (1968).
213. E. Kriek, J. A. Miller, U. Juhl, and E. C. Miller, *Biochemistry* **6**, 177 (1967).
214. L. M. Fink, S. Nishimura, and I. B. Weinstein, *Biochemistry* **9**, 496 (1970).
215. A. M. Kapuler and A. M. Michelson, *Biochim. Biophys. Acta* **232**, 436 (1971).
216. T. H. Corbett, C. Heidelberger, and W. F. Dove, *Mol. Pharmacol.* **6**, 667 (1970).
217. V. M. Maher, E. C. Miller, J. A. Miller, and W. Szybalski, *Mol. Pharmacol.* **4**, 411 (1968).
218. M. K. Agarwal and I. B. Weinstein, *Biochemistry* **9**, 503 (1970).
219. H. Erickson and M. Beer, *Biochemistry* **6**, 2694 (1967).
220. H. D. Hoffman and W. Müller, *Biochim. Biophys. Acta* **123**, 421 (1966).
221. S. Y. Wang, *Nature (London)* **180**, 91 (1957); *J. Org. Chem.* **24**, 11 (1959).
222. W. E. Cohn, *Biochem. J.* **64**, 28P (1956).
223. A. M. Moore and S. M. Anderson, *Can. J. Chem.* **37**, 590 (1959).
224. C.-T. Yu and P. C. Zamecnik, *Biochem. Biophys. Res. Commun.* **12**, 457 (1963).
225. U-C. Hsu, *Nature (London)* **203**, 152 (1964).
226. K. W. Brammer, *Biochim. Biophys. Acta* **72**, 217 (1963).
227. S. L. Commerford, *Biochemistry* **10**, 1993 (1971).
228. H. Yoshida, J. Duval, and J.-P. Ebel, *Biochim. Biophys. Acta* **161**, 13 (1968).
229. J. H. Weil, *Bull. Soc. Chim. Biol.* **47**, 1303 (1965).
230. A. Tsugita and H. Fraenkel-Conrat, *J. Mol. Biol.* **4**, 73 (1962).
231. J. A. Nelson, S. C. Ristow, and R. W. Holley, *Biochim. Biophys. Acta* **149**, 590 (1967).
232. G. E. Means and H. Fraenkel-Conrat, *Biochim. Biophys. Acta* **247**, 441 (1971).
233. R. H. Hall, *Progr. Nucl. Acid Res. Mol. Biol.* **10**, 57 (1970).
234. R. H. Hall, "The Modified Nucleosides in Nucleic Acids." Columbia Univ. Press, New York, 1971.
235. M. J. Robins, R. H. Hall, and R. Thedford, *Biochemistry* **6**, 1837 (1967).
236. R. D. Faulkner and M. Uziel, *Biochim. Biophys. Acta* **238**, 464 (1967).
237. W. H. Miller, R. O. Roblin, and E. B. Eastwood, *J. Amer. Chem. Soc.* **67**, 2201 (1945).
238. M. N. Lipsett and B. P. Doctor, *J. Biol. Chem.* **242**, 4072 (1967).
239. M. N. Lipsett, *J. Biol. Chem.* **242**, 4067 (1967).
240. J. A. Carbon and H. David, *Biochemistry* **7**, 3851 (1968).
240a. S. Hirose, R. Okazaki and F. Tamanoi, *J. Mol. Biol.* **77**, 501 (1973).
241. T. C. Bruice and S. Benkovic, "Bioorganic Mechanisms," Vol. II. Benjamin, New York, 1966.
242. F. H. Westheimer, *Chem. Soc., Spec. Publ.* **8**, 1 (1957).

243. D. M. Brown, D. I. Magrath, and A. R. Todd, *J. Chem. Soc., London* p. 2708 (1952).
244. P. C. Haake and F. H. Westheimer, *J. Amer. Chem. Soc.* **83**, 1102 (1961).
245. F. H. Westheimer, *Accounts Chem. Res.* **1**, 70 (1968).
246. I. Ugi, D. Marquardring, H. Klusacek, P. Gillespie, and F. Ramirez, *Accounts Chem. Res.* **4**, 288 (1971).
247. D. M. Brown, D. I. Magrath, A. H. Neilson, and A. R. Todd, *Nature (London)* **177**, 1124 (1956).
248. D. M. Brown and D. A. Usher, *J. Chem. Soc., London* p. 6547 (1965).
249. D. S. Frank and D. A. Usher, *J. Amer. Chem. Soc.* **89**, 6360 (1967).
250. D. A. Usher, D. I. Richardson, and F. Eckstein, *Nature (London)* **228**, 663 (1970).
251. D. A. Usher, E. S. Evenrich, and F. Eckstein, *Proc. Nat. Acad. Sci. U.S.* **69**, 115 (1971).
252. D. A. Usher, *Nature (London), New Biol.* **235**, 207 (1972).
253. M. Renz, R. Lohrmann, and L. E. Orgel, *Biochim. Biophys. Acta* **240**, 463 (1971).
254. J. Eigner, H. Boedtker, and G. Michaels, *Biochim. Biophys. Acta* **51**, 165 (1961).
254a. W. E. Hill and W. L. Fangman, *Biochemistry* **12**, 1772 (1973).
255. A. D. Hershey and E. Burgi, *J. Mol. Biol.* **2**, 143 (1960).
256. C. Levinthal and P. F. Davison, *J. Mol. Biol.* **3**, 674 (1961).
257. W. J. Wechter, *J. Med. Chem.* **10**, 762 (1967).
258. D. M. Brown and D. A. Usher, *J. Chem. Soc., London* p. 6558 (1965).
259. G. R. Barker, M. D. Montague, R. J. Moss, and M. A. Parsons, *J. Chem. Soc., London* p. 3786 (1957).
260. J. Bartolomé and F. Orrego, *Biochemistry* **9**, 4509 (1970).
261. B. G. Lane and G. C. Butler, *Can. J. Biochem. Physiol.* **37**, 1329 (1959).
262. B. G. Lane and F. W. Allen, *Can. J. Biochem. Physiol.* **39**, 721 (1961).
263. H. Witzel, *Justus Liebigs Ann. Chem.* **635**, 182 (1960).
264. R. M. Bock, *in* "Methods in Enzymology" (L. Grossman and K. Moldave, eds.), Vol. 12, Part A, p. 218. Academic Press, New York, 1967.
265. B. G. Lane and G. C. Butler, *Biochim. Biophys. Acta* **33**, 218 (1959).
266. H. B. Kaltreider and J. F. Scott, *Biochim. Biophys. Acta* **55**, 379 (1962).
267. W. Szer and D. Shugar, *Acta Biochim. Pol.* **7**, 491 (1960).
268. P. Brookes and P. D. Lawley, *J. Chem. Soc., London* p. 1348 (1962).
269. D. H. Marrian, V. L. Spicer, M. E. Balis, and G. B. Brown, *J. Biol. Chem.* **189**, 533 (1951).
270. D. I. Magrath and D. C. Shaw, *Biochem. Biophys. Res. Commun.* **26**, 32 (1967).
271. R. H. Hall, *in* "Methods in Enzymology" (L. Grossman and K. Moldave, eds.), Vol. 12, Part A, p. 305. Academic Press, New York, 1967.
272. B. G. Lane, *Biochemistry* **4**, 212 (1965).
273. L. Hudson, M. W. Gray, and B. G. Lane, *Biochemistry* **4**, 2009 (1965).
274. G. L. Eichhorn and J. J. Butzow, *J. Biopolym.* **3**, 79 (1965).
275. E. Bamann, F. Fischler, and H. Trapmann, *Biochem. Z.* **325**, 413 (1951), and papers cited therein.
276. W. W. Butcher and F. H. Westheimer, *J. Amer. Chem. Soc.* **77**, 2420 (1955).
277. K. Dimroth, H. Witzel, W. Hülsen, and H. Mirbach, *Justus Liebigs Ann. Chem.* **620**, 94 (1959).
278. K. Dimroth and H. Witzel, *Justus Liebigs Ann. Chem.* **620**, 109 (1959).
279. E. Bamann, H. Trapmann, and F. Fischler, *Biochem. Z.* **325**, 89 (1954).
280. G. L. Eichhorn, *in* "Inorganic Biochemistry" (G. L. Eichhorn, ed.), Vol. 2, p. 1210. Elsevier, Amsterdam, 1973.
281. R. Britten, *C. R. Trav. Lab. Carlsberg* **32**, 371 (1962).

282. J. W. Huff, K. S. Shastry, M. P. Gordon, and W. E. C. Wacker, *Biochemistry* **3**, 501 (1964).
283. W. R. Farkas, *Biochim. Biophys. Acta* **155**, 401 (1968).
284. D. M. Brown, M. Fried, and A. R. Todd, *Chem. Ind. (London)*, p. 352 (1953); *J. Chem. Soc. London* p. 2206 (1955).
285. P. R. Whitfield and R. Markham, *Nature (London)* **171** 1151 (1953).
286. P. R. Whitfeld, *Biochem. J.* **58**, 390 (1954).
287. D. H. Rammler, *Biochemistry* **10**, 4699 (1971).
288. C.-T.Yu and P. C. Zamecnik, *Biochim. Biophys. Acta* **45**, 148 (1960).
289. M. Ogur and J. D. Small, *J. Biol. Chem.* **235**, PC60 (1960).
290. W. E. Cohn and J. X. Khym, *in* "Acides ribonucléiques et polyphosphates. Structure, synthese et fonctions," p. 217. CNRS, Paris, 1962.
291. H. C. Neu and L. A. Heppel, *J. Biol. Chem.* **239**, 2927 (1964).
292. J. X. Khym, *Biochemistry* **2**, 343 (1963).
293. D. M. Brown and A. P. Read, *J. Chem. Soc, C* p. 5072 (1965), also unpublished work.
294. A. Steinschneider, *Isr. J. Chem.* **9**, 597 (1971).
294a. M. Wu and N. Davidson, *J. Mol. Biol.* **78**, 1 (1973).
294b. R. A. Kenner, *Biochem. Biophys. Res. Commun.* **51**, 932 (1973).
295. P. R. Whitfeld, *Biochim. Biophys. Acta* **108**, 202 (1965).
296. A. Steinschneider and H. Fraenkel-Conrat, *Biochemistry* **5**, 2735 (1966).
297. H. L. Weith and P. T. Gilham, *J. Amer. Chem. Soc.* **89**, 5473 (1967).
298. J. C. Lee, H. L. Weith, and P. T. Gilham, *Biochemistry* **9**, 113 (1970).
299. H. L. Weith and P. T. Gilham, *Science* **166**, 1004 (1969).
300. M. Uziel and J. X. Khym, *Biochemistry* **8**, 3254 (1969).
301. R. Dulbecco and J. D. Smith, *Biochim. Biophys. Acta* **39**, 358 (1960).
302. J. A. Hunt, *Biochem. J.* **95**, 541 (1965).
303. J. A. Hunt, *in* "Methods in Enzymology" (L. Grossman and K. Moldave, eds.), Vol. 12, Part B, p. 240. Academic Press, New York, 1968.
304. D. G. Glitz and D. S. Sigman, *Biochemistry* **9**, 3433 (1970); **10**, 3810 (1971).
305. P. C. Zamecnik, M. L. Stephenson, and J. F. Scott, *Proc. Nat. Acad. Sci. U.S.* **46** 811 (1960).
306. H. von Portatius, P. Doty, and M. L. Stephenson, *J. Amer. Chem. Soc.* **83**, 3351 (1961).
307. R. S. Yolles, *Biochim. Biophys. Acta* **87**, 583 (1964).
308. G. P. Moss, C. B. Reese, K. Schofield, R. Shapiro, and Lord Todd, *J. Chem. Soc., London* p. 1149 (1963).
309. J. P. Vizsolyi and G. Tener, *Chem. Ind. (London)* p. 263 (1962).
310. T. Gabriel, W. Y. Chen, and A. L. Nussbaum, *J. Amer. Chem. Soc.*, **90**, 6833 (1968).
311. A. Wacker and L. Träger, *Z. Naturforsch. B* **18**, 13 (1963).
312. K. E. Pfitzner and J. G. Moffatt, *J. Org. Chem.* **29**, 1508 (1964).
313. E. R. Garrett, T. Suzuki, and D. J. Weber, *J. Amer. Chem. Soc.* **86**, 4460 (1964).
314. H. Venner, *Hoppe-Seyler's Z. Physiol. Chem.* **344**, 189 (1966).
315. R. Shapiro and S. Kang, *Biochemistry* **8**, 1806 (1969).
316. J. A. Zoltewicz, D. F. Clark, T. W. Sharpless, and J. Grahe, *J. Amer. Chem. Soc.* **92**, 1741 (1970).
317. R. Shapiro and M. Danzig, *Biochemistry* **11**, 23 (1972).
318. B. Capon, *Chem. Rev.* **69**, 407 (1969).
319. C. A. Dekker and L. Goodman, *in* "The Carbohydrates" (W. Pigman and D. Horton, eds.), 2nd ed., p. 1, Vol. 2A. Academic Press, New York, 1970 and references cited therein.

320. J. A. Haines, C. B. Reese, and A. R. Todd, *J. Chem. Soc., London* p. 5281 (1962).
321. P. D. Lawley and P. Brookes, *Biochem. J.* **89**, 127 (1963).
322. G. B. Petersen and K. Burton, *Biochem. J.* **92**, 666 (1964).
323. H. S. Shapiro and E. Chargaff, *Biochim. Biophys. Acta* **26**, 608 (1957).
324. H. S. Shapiro, *in* "Methods in Enzymology" (L. Grossman and K. Moldave, eds.), Vol. 12, Part A, p. 205. Academic Press, New York, 1967.
325. K. Burton and G. B. Petersen, *Biochem. J.* **75**, 17 (1960).
326. K. Burton, *in* "Methods in Enzymology" (L. Grossman and K. Moldave, eds.), Vol. 12, Part A, p. 222. Academic Press, New York, 1967.
327. J. B. Hall and R. L. Sinsheimer, *J. Mol. Biol.* **6**, 115 (1963).
328. R. Cerny, E. Cherná, and J. H. Spencer, *J. Mol. Biol.* **46**, 145 (1969).
329. G. B. Petersen and J. M. Reeves, *Biochim. Biophys. Acta* **179**, 510 (1969).
330. V. Ling, *J. Mol. Biol.* **64**, 87 (1972).
331. A. S. Jones, A. M. Mian, and R. T. Walker, *J. Chem. Soc., C* p. 2042 (1968).
332. D. C. Livingston, *Biochim. Biophys. Acta* **87**, 538 (1964).
333. M. M. Coombs and D. C. Livingston, *Biochim. Biophys. Acta* **174**, 161 (1969).
334. P. Fleury, J. Courtois, and A. Desjobert, *Bull. Soc. Chim. Fr.* **15**, 694 (1948).
335. D. M. Brown and J. C. Stewart, *J. Chem. Soc., London* p. 5362 (1964).
336. A. Temperli, H. Türler, P. Rüst, A. Danon, and E. Chargaff, *Biochim. Biophys. Acta* **91**, 462 (1964).
337. H. S. Shapiro, *in* "Methods in Enzymology" (L. Grossman and K. Moldave, eds.), Vol. 12, Part A, p. 212. Academic Press, New York, 1967.
338. V. Habermann, *Collect. Czech. Chem. Commun.* **26**, 3547 (1961).
339. F. Baron and D. M. Brown, *J. Chem. Soc., London* p. 6263 (1955).
340. A. R. Cashmore and G. B. Petersen, *Biochim. Biophys. Acta* **174**, 591 (1969).
341. J. Sedat and R. L. Sinsheimer, *J. Mol. Biol.* **9**, 489 (1964).
342. B. Vaniushin and J. Burianov, *Biokhimiya* **32**, 218 (1969).
343. K. Burton and W. T. Riley, *Biochem. J.* **98**, 70 (1966).
344. K. Burton, *Essays Biochem.* **1**, 57 (1965).
345. G. K. Darby, A. S. Jones, J. R. Tittensor, and R. T. Walker, *Nature (London)* **216**, 793 (1967).
346. N. K. Kochetkov, E. I. Budowsky, M. F. Turchinsky, N. A. Simukova, and E. D. Sverdlov, *Biochim. Biophys. Acta* **142**, 35 (1967).
347. M. F. Turchinsky, L. I. Guskova, I. Hazai, E. I. Budowsky, and N. K. Kochetkov, *Biochim. Biophys. Acta* **254**, 366 (1971).
348. E. I. Budowsky, N. A. Simukova, and L. I. Guskova, *Biochim, Biophys. Acta* **166**, 754, (1968).
349. K. Nakanishi, N. Furutachi, M. Funamizu, D. Grunberger, and I. B. Weinstein, *J. Amer. Chem. Soc.* **92**, 7617 (1970).
350. P. Philippsen, R. Thiebe, W. Wintermeyer, and H. J. Zachau, *Biochem. Biophys. Res. Commun.* **33**, 922 (1968).
351. H. J. Zachau, personal communication.
352. W. Szybalski, *in* "Thermobiology" (A. H. Rose, ed.), p. 73, and references cited therein. Academic Press, New York, 1967.
353. S. Zamenhof, *Progr. Nucl. Acid Res. Mol. Biol.* **6**, 1 (1967).
354. S. Greer and S. Zamenhof, *J. Mol. Biol.* **4**, 123 (1962).
355. S. Zamenhof and S. Arikawa, *Mol. Pharmacol.* **2**, 570 (1966).
356. W. Fiers and R. L. Sinsheimer, *J. Mol. Biol.* **5**, 420 (1962).
357. W. Ginoza, C. J. Hoelle, K. B. Vessey, and C. Carmack, *Nature (London)* **203**, 606 (1964).
358. R. Kotaka and R. L. Baldwin, *J. Mol. Biol.* **9**, 323 (1964).

359. P. Brookes and P. D. Lawley, *Biochem. J.* **89**, 138 (1963).
360. P. D. Lawley, J. H. Lethbridge, P. A. Edwards, and K. V. Shooter, *J. Mol. Biol.* **39**, 181 (1969).
361. A. Ronen, *Biochem. Biophys, Res. Commun.* **33**, 190 (1968).
362. D. R. Krieg, *Progr. Nucl. Acid Res. Mol. Biol.* **2**, 125 (1963).
363. B. S. Strauss, *Nature (London)* **191**, 730 (1961).
364. M. P. Gordon and J. W. Huff, *Biochemistry* **1**, 481 (1962).
365. J. R. Cox and O. B. Ramsay, *Chem. Rev.* **64**, 317 (1964).
366. D. M. Brown and J. O. Osborne, *J. Chem. Soc., London* p. 2590 (1957).
367. D. M. Brown, D. I. Magrath, and Sir A. R. Todd, *J. Chem. Soc., London* p. 4396 (1955).
368. W. Davis and W. C. J. Ross, *J. Chem. Soc., London* p. 4297 (1952).
369. B. E. Griffin, J. A. Haines, and C. B. Reese, *Biochim. Biophys. Acta* **142**, 536 (1967).
370. C. W. Abell, L. A. Rosini, and M. R. Ramseur, *Biochemistry* **54**, 608 (1965).
371. F. Pochon and A. M. Michelson, *Proc. Nat. Acad. Sci. U.S.* **53**, 1425 (1965).
372. A. M. Michelson and F. Pochon, *Biochim. Biophys. Acta* **114**, 469 (1966).
373. A. K. Kriek and P. Emmelot, *Biochemistry* **2**, 733 (1963).
374. A. O. Olson and K. M. Baird, *Biochim. Biophys. Acta* **179**, 513 (1969).
374a. P. Bannon and W. Verly, *Eur. J. Biochem.* **31**, 103 (1972).
375. J. Vinograd, A. Lebowitz, R. Radloff, R. Watson, and P. Laipis, *Proc. Nat. Acad. Sci. U.S.* **53**, 1104 (1965).
376. G. G. Brownlee, "Laboratory Techniques in Biochemistry and Molecular Biology." North-Holland Publ., Amsterdam, 1971.
377. M. Beer, S. Stern, D. Carmalt, and K. H. Mohlenrich, *Biochemistry* **5**, 2283 (1966).
378. E. S. Tessman, *J. Mol. Biol.* **17**, 218 (1966).
379. D. M. Brown and M. R. Osborne, unpublished observations.
380. F. Cramer and H. Seidel, *Biochim. Biophys. Acta* **91**, 14 (1964).
381. H. Hayatsu and T. Ukita, *Biochim. Biophys. Acta* **123**, 458 (1966).
382. N. K. Kochetkov, E. I. Budowsky, and V. Demuskin, *Mol. Biol.* **1**, 583 (1967).
383. J. J. Englander, N. R. Kallenbach, and S. W. Englander, *J. Mol. Biol.* **63**, 153 (1972).
384. B. McConnell and P. H. von Hippel, *J. Mol. Biol.* **50**, 297 (1970).
385. R. B. Inman, *J. Mol. Biol.* **18**, 464 (1966).
386. D. Stollar and L. Grossman, *J. Mol. Biol.* **4**, 31 (1962).
387. L. Grossman, *in* "Methods in Enzymology" (L. Grossman and K. Moldave, eds.), Vol. 12, Part B, p. 467. Academic Press, New York, 1968.
388. H. Utiyama and P. Doty, *Biochemistry* **10**, 1254 (1971).
388a. G. D. Fasman, C. Lindblow and L. Grossman, *Biochemistry* **3**, 1015 (1964).
389. D. B. Ludlum, *Biochim. Biophys. Acta* **95**, 674 (1965).
390. P. D. Lawley and P. Brookes, *Biochem. J.* **92**, 194 (1964).
391. P. Brookes and P. D. Lawley, *Biochem. J.* **89**, 138 (1963).
392. P. D. Lawley, personal communication.
393. P. Brookes and P. D. Lawley, *Biochem. J.* **80**, 496 (1961).
394. K. W. Cohn and C. L. Spears, *Biochim. Biophys. Acta* **145**, 734 (1967).
395. J. T. Shapiro, M. Lang, and G. Felsenfeld, *Biochemistry* **8**, 3219 (1969).
396. T. Y. Shih and G. D. Fasman, *Biochemistry* **10**, 1675 (1971).
397. L. S. Lerman, *Proc. Nat. Acad. Sci. U.S.* **68**, 1886 (1971).
398. C. F. Jordan, L. S. Lerman, and J. H. Venable, *Nature (London) New Biol.* **236**, 67 (1972).
399. Yu. M. Evdokimov, A. L. Platonov, A. S. Tikhonenko, and Ya. M. Varshavsky, *FEBS Lett.* **23**, 180 (1972).

400. E. Freese and H. B. Strack, *Proc. Nat. Acad. Sci. U.S.* **48**, 1796 (1962).
401. V. R. Skyladneva, N. P. Kiseleva, E. I. Budowsky, and T. I. Tikchonenko, *Mol. Biol.* **4**, 110 (1970).
402. E. Cerdá-Olmedo, P. C. Hanawalt, and N. Guerola, *J. Mol. Biol.* **33**, 705 (1968).
403. D. Botstein and E. W. Jones, *J. Bacteriol.* **98**, 847 (1969).
404. M. J. Waring, "The Molecular Basis of Antibiotic Action" (E. F. Gale, E. Cundliffe, P. E. Reynolds, M. H. Richmond, and M. J. Waring, authors), Chapter 5, p. 173. Wiley, New York, 1972.
405. S. A. Lesko, H. D. Hoffman, P. O. P. Ts'O, and V. M. Maher, *Progr. Mol. Subcell. Biol.* **7**, 434 (1971).
406. M. J. Cookson, P. Sims, and P. L. Grover, *Nature (London), New Biol.* **234**, 186 (1971).
407. B. N. Ames, P. Sims, and P. L. Grover, *Science* **176**, 47 (1972).
408. F. Pochon, P. Brookes, and A. M. Michelson, *Eur. J. Biochem.* **21**, 154 (1971).
409. F. Pochon and A. M. Michelson, *Eur. J. Biochem.* **21**, 144 (1971).
410. P. Brookes, personal communication.
411. R. M. Peck and A. P. O'Connell, *J. Med. Chem.* **15**, 68 (1972).
412. B. N. Ames and H. J. Whitfield, Jr., *Cold Spring Harbor Symp. Quant. Biol.* **31**, 221 (1966).
412a. B. N. Ames, F. D. Lee and W. E. Durston, *Proc. Nat. Acad. Sci. U.S.* **70**, 782 (1973).
413. T. Caspersson, L. Zech, E. J. Modest, G. E. Foley, U. Wagh, and E. Simmonsson, *Exp. Cell Res.* **58**, 128 and 141 (1969).
414. K. P. George, *Nature (London)* **226**, 80 (1970).
415. L. S. Lerman, *Progr. Mol. Subcell. Biol.* **2**, 382 (1971).
416. F. H. C. Crick, *Nature (London)* **234**, 25 (1971).
417. W. Szybalski and V. N. Iyer, *in* "Antibiotics" (D. Gottlieb and P. D. Shaw, eds.), Vol. I, Mechanisms of Action, p. 211. Springer-Verlag, Berlin and New York, 1967.
418. A. Weissbach and A. Lisio, *Biochemistry* **4**, 196 (1965).
419. M. N. Lipsett and A. Weissbach, *Biochemistry* **4**, 206 (1965).
420. V. N. Iyer and W. Szybalski, *Science* **145**, 55 (1964).
421. M. Tomasz, *Biochim. Biophys. Acta* **199**, 18 (1970).
422. M. Tomasz, *Biochim. Biophys. Acta* **213**, 288 (1970).
423. B. Singer and H. Fraenkel-Conrat, *Progr. Nucl. Acid Res.* **9**, 1 (1969).
424. F. Cramer, *Progr. Nucl. Acid Res.* **11**, 391 (1971).
425. S. Arnott, *Progr. Biophys. Mol. Biol.* **22**, 181 (1971), and references cited therein.
426. A. R. Cashmore, D. M. Brown, and J. D. Smith, *J. Mol. Biol.* **59**, 359 (1971).
427. S. E. Chang, A. R. Cashmore, and D. M. Brown, *J. Mol. Biol.* **68**, 455 (1972).
428. S. E. Chang, *J. Mol. Biol.* **75**, 533 (1973).
429. R. P. Singhal, *J. Biol. Chem.* **246**, 5848 (1971); also personal communication.
430. V. V. Vlasov, N. I. Grineva, and D. G. Knorre, *FEBS Lett.* **20**, 66 (1972).
431. A. S. Girshovich, M. A. Grachev, and L. V. Obukhova, *Mol. Biol.* **2**, 351 (1968).
432. F. Cramer, H. Doepner, F. von der Haar, E. Schlimme, and H. Seidel, *Proc. Nat. Acad. Sci. U.S.* **61**, 1384 (1968).
433. L. H. Schulman, *J. Mol. Biol.* **58**, 117 (1971).
434. L. H. Schulman, *Proc. Nat. Acad. Sci. U.S.* **66**, 507 (1970).
435. S. W. Brostoff and V. M. Ingram, *Science* **158**, 666 (1967).
436. A. V. Lake and G. M. Tener, *Biochemistry* **5**, 3992 (1966).
437. M. Yoshida, Y. Kaziro, and T. Ukita, *Biochim. Biophys. Acta* **166**, 646 (1968).
438. M. A. Q. Siddiqui and J. Ofengand, *J. Biol. Chem.* **245**, 4409 (1970).
439. T. L. Jilyaeva and L. L. Kisselev, *FEBS Lett.* **10**, 229 (1971).

440. J. A. Nelson, S. C. Ristow, and R. W. Holley, *Biochim. Biophys. Acta* **149**, 590 (1967).
441. Y. Furuichi, Y. Wataya, H. Hayatsu, and T. Ukita, *Biochem. Biophys. Res. Commun.* **41**, 1185 (1970).
442. Z. Kućan, K. A. Freude, I. Kućan, and R. W. Chambers, *Nature (London) New Biol.* **232**, 177 (1971).
443. L. H. Schulman and J. P. Goddard, *J. Biol. Chem.* **247**, 3864 (1972); also personal communication.
444. M. Levitt, *Nature (London)* **224**, 759 (1969).
445. W. Fuller and A. Hodgson, *Nature (London)* **215**, 817 (1967).
446. A. R. Cashmore, *Nature (London), New Biol.* **230**, 236 (1971).
447. M. Litt, *Biochemistry* **10**, 2223 (1971).
448. M. Litt and C. M. Greenspan, *Biochemistry* **11**, 1437 (1972).
449. D. Grunberger, I. B. Weinstein, and S. Fujimura, personal communication.
450. D. Grunberger, I. B. Weinstein, L. M. Fink, J. H. Nelson, and C. R. Cantor, *Progr. Mol. Subcell. Biol.* **2**, 371 (1971).
451. A. R. Cashmore, unpublished observations; see Cashmore *et al.* [426].
451a. S. E. Chang and D. Ish-Horowicz, *J. Mol. Biol.* (1974) (in press).
452. O. C. Uhlenbeck, J. Baller, and P. Doty, *Nature (London)* **225**, 508 (1970).
453. C. Woese, *Nature (London)*, **226**, 817 (1971).
453a. S. H. Kim, G. J. Quigley, F. L. Suddath, A. McPherson, D. Sneden, J. J. Kim, J. Weinzierl and A. Rich, *Science* **179**, 285 (1973).
454. J. Rubin, T. Brennan, and M. Sundaralingam, *Biochemistry* **11**, 3112 (1972).
455. H. G. Zachau, D. Duffing, H. Feldmann, F. Melchers, and W. Karau, *Cold Spring Harbor Symp. Quant. Biol.* **31**, 417 (1966).
456. G. Samuelson and E. B. Keller, *Biochemistry* **11**, 30 (1972).
457. G. Dirheimer, J.-P. Abel, J. Bonnet, J. Gangloff, G. Keith, B. Krebs, B. Kuntzel, A. Roy, J. Weissenbach, and C. Werner, *Biochimie*, **54**, 127 (1972).
458. W. Wintermeyer and H. G. Zachau, *Biochim. Biophys. Acta* **299**, 82 (1973).

2

ULTRAVIOLET SPECTROSCOPY, CIRCULAR DICHROISM, AND OPTICAL ROTATORY DISPERSION

C. Allen Bush

I. Theoretical Introduction

A. UNITS

1. *Units for Optical Absorption*

The light intensity in an absorbing sample decays exponentially according to

$$I = I_0 e^{-kl} \tag{1}$$

The usual way of reporting absorbence is in terms of the log I_0/I, which is known as the absorbence (A) or optical density (OD). If the absorbing molecules follow Beer's law, their optical density is proportional to the concentration, and a molecular quantity, ϵ, the molar decadic extinction coefficient may be defined

$$\epsilon = OD/cl \tag{2}$$

The concentration c is in moles per liter and the path length, l, is in centimeters. ϵ is a molecular quantity and may be related to quantities calculated from quantum-mechanical models of the molecules. Since the absorption bands of many molecules are broadened by vibronic effects and solvent effects, we may integrate the area under the absorption band to derive molecular parameters characteristic of the entire electronic absorption band. These parameters may then be compared with parameters computed from molecular theories. In order to integrate areas under absorption bands, the spectrum must be broken down into regions or bands which may be distinguished and identified. This process requires considerable judgment and knowledge of the chromophores involved in the molecule under investigation.

One such parameter which characterizes the intensity of a given absorption band is the oscillator strength [1],

$$f_a = 4.32 \times 10^{-9} \int_a \epsilon_a(\nu) \, d\nu \tag{3}$$

We have indicated an integration over the frequency, ν. However, one may equivalently integrate over the wavelengths. The dipole strength also measures the intensity of the band and it may be derived from the square of the electric transition dipole moment as computed by molecular theory. It is given experimentally by [1]

$$D_{0a} = |\mu_{0a}|^2 = 0.92 \times 10^{-38} \int_a \frac{\epsilon_a(\nu) \, d\nu}{\nu} \tag{4}$$

2. Units of Optical Rotation

Optical rotation is defined as the angle of rotation of plane polarized light by a sample and is given the symbol α. For optically active molecules in isotropic solution the rotation is proportional to the path length of the sample and the concentration. Hence we may define the specific rotation,

$$[\alpha] = \alpha/l'c' \tag{5}$$

For historical reasons, the angles are measured in degrees, the path length in decimeters, and the concentration in grams per milliliter. More often we will discuss the molar rotation which is given by

$$[\phi] = \alpha M/100 = 100\alpha/cl \tag{6}$$

In this formula M is the molecular weight of the solute, c the concentration of sample in moles per liter, and l is the path length in centimeters. In polymeric systems, in order to facilitate comparison among oligomers and polymers of differing chain length, one often reports mean rotation per residue. Thus for polymers, M is the average molecular weight per residue and c is the molar concentration of residues.

A factor correcting for the index of refraction of the solvent is sometimes included in published reports of the rotation. The unit of rotation, $[m']$ is equal to our $[\phi] \times 3/(n^2 + 2)$. Although this factor might be expected to improve agreement between molecular calculations performed on a molecule in a vacuum and experimental results in solution, this has not been found to be the case. Hence this correction has not been used by all workers. When comparing experimental results from different laboratories, one should make certain which units of rotation are being used. The differences in these units, while minor, are generally beyond experimental error.

3. Definition of ORD

The optical rotatory dispersion (ORD) is simply the dependence of (ϕ) on wavelength of light, λ. There has been considerable interest in the general functional form that this dependence takes on. Experimental investigations have led to various proposed formulas to represent the dispersion quantitatively. In 1905, Drude proposed a theory of optical activity based on a medium consisting of electrons oscillating in helical paths [2]. The formula for the ORD which he derived is valid in nonabsorbing samples.

$$[\phi(\lambda)] = \sum_{m} \frac{k_m}{(\lambda^2 - \lambda^2_m)} \tag{7}$$

The sum extends over the oscillators, m, of various rotatory powers, k_m, and characteristic wavelengths, λ_m. A large number of compounds are found to follow a single-term Drude equation in regions in the spectrum where they do not absorb light. Substances departing widely from this formula are recognized as anomalous.

We may get a picture of what an ORD curve looks like by plotting a single-term Drude equation. In Fig. 1, the solid line represents the equation in the wavelength region where it is valid. At $\lambda = \lambda_0$, the formula gives infinite values of rotation, but as we go to wavelengths shorter than λ_0, the rotation reappears with opposite sign. We will give a more quantitative discussion of ORD curve shape below, but as a plausible construction, let us draw a smooth

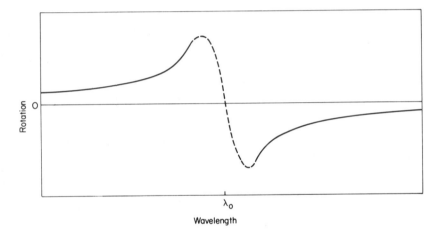

Fig. 1. ORD curve; Drude equation in the region of transparency (———). A plausible construction joining the two branches of the Drude curve (– – –).

dotted line connecting the two branches of the curve. This dotted line illustrates the behavior of the ORD in the region of λ_0. This is the absorption region for the molecule, where the Drude equation does not apply. This reversal in the sign of the rotation at $\lambda = \lambda_0$ is called a Cotton effect and we will discuss it much more extensively below. We leave the subject ORD for now, with the reminder that it is just this Cotton effect that gives us most of the information on molecular structure.

B. Circular Polarization

1. *Definition*

Let us now take a second way of looking at optical rotation. This alternative point of view will allow us to introduce the concept of circular dichroism

(CD) as well as provide a simple qualitative way of illustrating the origin of optical activity in molecules.

We have been discussing plane polarized light qualitatively and we now wish to represent it mathematically. We introduce a cartesian coordinate system in which the light beam is directed along the z axis (in the direction of unit vector **k**). In this coordinate system we may represent the electric vector of a light beam traveling in the **k** direction and polarized with the electric vector along the x axis (unit vector **i**) as

$$E_p(r, t) = E_0 \mathbf{i} \cos \omega(t - z/c) \tag{8}$$

This beam has an angular frequency ω and is polarized in the **i** direction. This plane polarized beam may also be viewed as a sum of two circularly polarized components, $\mathbf{E}_l + \mathbf{E}_r$.

$$\mathbf{E}_{l,r} = \tfrac{1}{2}E_0[\mathbf{i} \cos \omega(t - z/c) \pm \mathbf{j} \sin \omega(t - z/c)] \tag{9}$$

We see immediately that the sum of $\mathbf{E}_l + \mathbf{E}_r = \mathbf{E}_p$, the plane polarized beam. For a physical picture of the circularly polarized beams, take $\mathbf{E}_l$ as an example. If we pick a position on the z axis and watch the beam go by, we see that the plane of polarization rotates a full circle in every period, $\tau = 2\pi/\omega$. Likewise, if we pick a fixed time and move along the z axis in the direction of the light beam, we see that the plane of polarization rotates in a clockwise direction once in every wavelength, $\lambda = 2\pi c/\omega$. For $\mathbf{E}_r$, the plane of polarization rotates in a counterclockwise sense when viewed along the direction of the light beam. Thus, the electric vectors, $\mathbf{E}_l$ and $\mathbf{E}_r$ describe helices having opposite handedness. The sum of these two oppositely polarized beams is a plane polarized beam.

2. *Circular Birefringence*

Continuing to consider our plane polarized beam as a sum of $\mathbf{E}_l + \mathbf{E}_r$, imagine that the indices of refraction of the sample for $\mathbf{E}_l$ and $\mathbf{E}_r$ differ; that is $n_l \neq n_r$. The resulting phase shift between $\mathbf{E}_l$ and $\mathbf{E}_r$ as they emerge from the sample causes a rotation of the plane of polarized light. The resulting optical rotation was shown by Fresnel is 1825 to be

$$\alpha = (\pi l/\lambda)(n_l - n_r) \tag{10}$$

where α is in radians [2]. This description of optical rotation, which is completely equivalent to the one based on plane polarization, suggests the name circular birefrigence as a synonym for optical rotation; that is, the two circularly polarized beams are differently refracted.

If we consider the optical rotation as circular birefrigence, we may view optically active enantiomers as helices having opposite handedness. If our optically active sample is a solution of right-handed helices, the right-handed beam, E_r, will see something different in these molecules than does the left handed beam, E_l. Since these two beams are treated differently by the right-handed helical molecule, it is not surprising that $n_l \neq n_r$ and we observe optical rotation.

3. *Circular Dichroism*

Now that we have been introduced to circular polarization and circular birefringence, it is useful to consider the case of absorbing molecules. Since E_l and E_r see different indices of refraction, we expect them to see different absorption coefficients also. The difference between the extinction coefficients for left- and right-handed circularly polarized light is called circular dichroism (CD).

$$\Delta\epsilon = \epsilon_l - \epsilon_r \tag{11}$$

The CD is generally measured by passing alternately right- and left-handed circularly polarized light through the sample and measuring the difference in their optical transmission. In the literature we find CD reported in units of $\Delta\epsilon$, and also in units of molar ellipticity, $[\theta]$. In order to understand the relationship between these two measures of CD, let us return to consideration of the rotation of plane polarized light by an optically active sample which is also absorbing light. Plane-polarized light enters the sample; the plane of polarization is rotated and, in addition, its two circularly polarized components are differently absorbed due to circular dichroism. The light which emerges is no longer truly plane-polarized due to the unequal amplitudes of the two emerging circular components. Light of this type is intermediate between plane and circular polarization and is called elliptically polarized light. The ellipticity of this light, θ, is the ratio of the minor to the major axis of the ellipse. It is related to the difference in extinction coefficients of the right- and left-handed beams by [1]

$$\theta \text{ (radians)} = (2.303c'l'/4)\Delta\epsilon \tag{12}$$

where the factor 2.303 accounts for the fact that ϵ is measured in a logarithmic base 10 rather than base e (see Section I, A,1 above). As in the case of optical rotation, the units of path length are decimeters. Circular dichroism is often reported in units of molar ellipicity in degrees per decimeter [1], where

$$[\theta] = \theta(\text{radians}) \left(\frac{180}{\pi}\right) 10/cl \tag{13}$$

the concentration, c, is in moles per liter and l is now the path length in centimeters. In the literature we find circular dichroism reported either in units of molar ellipticity or units of $\Delta\epsilon$. They are simply proportional, as may be seen from the above considerations.

$$[\theta] = 3300\Delta\epsilon \tag{14}$$

For polymers we follow the same convention as in the rotation [Eq. (6)], reporting the mean ellipticity per residue.

C. PROPERTIES OF CD AND ORD CURVES

1. *CD Curve Shape*

The dependence of $[\theta]$ on wavelength of light is best understood by comparison with the ordinary bulk absorption curve shape. The CD curve having the simplest appearance is that of a single Cotton effect. An example which looks much like an isolated band is seen in Fig. 2. The CD falls to zero outside the absorption band and may be approximately represented by a Gaussian function. Of course, ϵ_r may be greater than ϵ_l, leading to CD bands of negative sign. In fact, one often finds a negative band adjacent to a positive one, giving rise to a double Cotton effect such as the one pictured in Fig. 2b. More

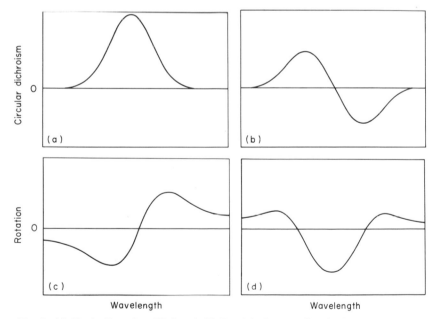

Fig. 2. (a) Single Gaussian CD band. (b) Double Cotton effect in CD, the sum of two Gaussian bands. (c) ORD corresponding to a single Gaussian CD band. (d) ORD of a double Cotton effect.

complicated combinations of bands may be found, as will be discussed below. In general, they can be understood as a combination of the above cases [3].

Since CD bands can be either positive or negative, the CD curve resulting from addition of positive and negative bands has maxima and minima which occur at wavelengths which are not the true locations of the absorption bands. Hence, one should use some caution in attempting to assign CD maxima or minima to a band originating at that wavelength. For example, ApA (Fig. 3) [4,5] has CD bands of opposite sign near the 260 nm absorption band. The

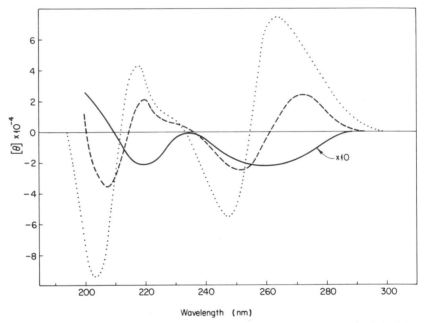

Fig. 3. CD of adenylic acids. AMP in water, pH 7, × 10 (——) [4]. ApA in 0.01 *M* tris, pH 7.4, 23°C (– – –) [5] Poly A in 0.01 *M* tris, pH 7.4, 23°C (· · ·) [5].

observed CD has a maximum at 265 nm and a minimum at 255 nm. Changes of CD band maxima and minima in a conformational transition may not reflect true shifts in absorption bands, but simply changes in the relative magnitudes of Cotton effects. Such difficulties are all the more important in ORD curves, where the curve shapes are complex and overlapping of bands is more common.

2. *Rotational Strength*

We have defined above the experimental as well as the molecular parameters for absorption and we have also defined the experimental parameters of optical activity. We now discuss the rotational strength which is a measure of the rotatory power of an absorption band, and which may be calculated from molecular theories as well as from experimental measurements. As in the case of the absorption spectrum, the circular dichroism spectrum must be decomposed into bands, whose rotational strength may then be calculated. The rotational strength of an electronic transition is given from experiment by [1]

$$\mathbf{R}_{0a} = \frac{3hc}{8\pi^3 N} \int_a \frac{[\theta(\lambda)] \, d\lambda}{\lambda} \tag{15}$$

N is the number of absorbing molecules per milliliter. The integration is carried out over a given CD band, a, leading to an evaluation of the rotational strength of electronic transition 0 to a.

This parameter, the rotational strength of an electronic transition, may be calculated from molecular transition moments [6],

$$R_{0a} = \text{Im} \, (\mathbf{m}_{a0} \cdot \mathbf{\mu}_{0a}) \tag{16}$$

We will discuss in detail how this formula may be evaluated from the molecular values of the electric transition dipole moment, $\mathbf{\mu}_{0a}$, and the magnetic dipole transition moment, $\mathbf{m}_{0a}$. A number of techniques for evaluating theoretically the rotational strength have been developed for various molecules including polynucleotides.

3. Relationship of ORD to CD

The relationship of the anomaly or change in sign of the ORD curve and the appearance of a CD band in optically active compounds was described by Aimé Cotton in 1895 [2]. In honor of this discovery, we now describe a maximum in the CD at $\lambda = \lambda_0$ or a reversal in the sign of the rotation at $\lambda = \lambda_0$ as a Cotton effect at λ_0.

It is well known that ORD and CD curves, extended over all wavelengths, contain identical information about molecular structure [7]. Moreover, the relationship between these two experimental quantities (the Kronig–Kramers transformation) holds, with certain limitations, even when the CD and ORD curves are known over only a portion of the electromagnetic spectrum. Although relatively little use has been made of this relationship in the field of bipolymers, in fact it is quite straightforward to use this relationship to compare CD and ORD curves. The conversion from ORD to CD provides a convenient method for analysis of ORD curves when the CD is not available. Knowledge of the CD curves, in cases of multiple Cotton effects, allows us to make assignments of transitions more easily and to compute rotational strengths explicitly.

Some studies in the literature utilizing the Kronig–Kramers transform, have been discussed especially for single Cotton effects [1]. In polypeptides, Carver *et al.* [8] extended the Moscowitz method to include several Gaussian-shaped CD bands. They then used a nonlinear least squares fit of experimental ORD curves to Kronig–Kramers transforms of Gaussian CD bands. Holzwarth and Doty [9] used a direct numerical method to compute the ORD curves corresponding to the CD curves which they reported for polypeptides. Bush and Tinoco [10] used a Gaussian curve method to calculate ORD curves for dinucleoside phosphates. They first obtained theoretical CD curves by regarding each CD band as a sum of Gaussian components chosen to simulate the absorption band shape; the rotational strength was computed directly from quantum-mechanical theory. They then obtained ORD curves by summing the Kronig–Kramers transforms of the Gaussian CD bands.

A convenient method for comparing CD and ORD measurements uses a direct numerical evaluation of the Kronig–Kramers integrals. This method involves no assumptions about the position, band shape, or number of Cotton effects, and works well for transformations in either direction. Recently, Thiery [11] developed a method which has been applied to polynucleotides. Also Bush [12] has made available a program based on a similar method of Emeis *et al.* [13].

4. *A Method For Calculation of Kronig–Kramers Integrals*

We write the Kronig–Kramers transform in terms of wavelength, λ, since most CD and ORD curves are reported in this manner. For the transformation of CD to ORD we have [1]

$$[\phi(\lambda)] = (2/\pi) \int_0^\infty \frac{[\theta(\lambda')]\lambda'\, d\lambda'}{\lambda^2 - \lambda'^2} \tag{17}$$

where λ' is the variable of integration. This integral is evaluated numerically by the method of Emeis *et al.* [13]. The integral is expressed as the sum of two integrals; one of them has a singularity and the other is integrated numerically by Simpson's rule. The integral with the singularity is evaluated up to the pole, and from the pole to infinity, also by Simpson's rule; in the region of the singularity, the $[\theta(\lambda')]$ curve is expanded in a Taylor series and then integrated [12].

For the transformation of ORD to CD, we use a formula similar to that given by Moscowitz [1], and equivalent to it,

$$[\theta(\lambda)] = \frac{-2\lambda}{\pi} \int_0^\infty \frac{[\phi(\lambda')]\, d\lambda'}{\lambda^2 - \lambda'^2} \tag{18}$$

This integral is evaluated in the same way as Eq. (17) for the region of absorption. However, unlike $[\theta(\lambda)]$, the function $[\phi(\lambda)]$ does not approach zero outside the region of absorption; hence, the integral in Eq. (18) receives contributions from wavelengths beyond the range generally considered in ORD measurements. If ORD data is not available outside the region of absorption, we may represent $[\phi(\lambda)]$ by a Drude equation,

$$[\phi(\lambda')] = A/(\lambda'^2 - \lambda_0^2) \tag{19}$$

where A and λ_0 are the usual Drude constants. This function, when inserted into Eq. (18), may be integrated in closed form [13].

This simple form of the Kronig–Kramers transform works rather well with polynucleotides since there are several Cotton effects in an easily accessible range of the ultraviolet spectrum. These Cotton effects dominate both the CD and ORD curves in the range of 200–300 nm. In cases where the only Cotton effects are nearer the limit of wavelength of the instrument (near 190 nm) the Kronig–Kramers transforms become considerably more difficult to calculate. In the case of polynucleotides the reliability of the Kronig–Kramers transforms is about 2–5%. This has been observed by Thiery [11] as well as in our laboratory where a similar method of calculation is used.

The foregoing discussion is meant to make clear that there is no real difference in the molecular information obtained in a measurement of a CD or ORD curve. Both experiments are measuring the same property, even though the experiments may seem rather different in principle. Each measurement has its place and we will outline is some detail below situations in which one may be preferred over the other. But, in general, there is little profit in presenting both measurements side by side for a single sample without comment. Computation of the Kronig–Kramers transforms is not difficult and one should at least compare the two measurements to test for self-consistency.

5. *ORD Curve Shape*

A qualitative indication of the appearance of an ORD curve in the Cotton effect region has been given in Fig. 1. A more rigorous discussion has been given by Moscowitz [1]. He begins with a representation of the CD curve as a Gaussian function centered at $\lambda = \lambda_a$.

$$[\theta(\lambda)] = [\theta_a^\circ] \exp \{-(\lambda - \lambda_a)^2/\Delta_a^2\} \tag{20}$$

He then shows how the Gaussian function may be integrated using Eq. (17) to give the ORD curve. The resulting ORD curve for a single Cotton effect is shown in Fig. 2c and that for a double Cotton effect in Fig. 2d. No simple functional form has been found to represent the ORD curve as well as the Gaussian curve represents the CD or absorption curve shapes.

D. ABSORPTION CURVES

1. *Chromophores*

We have emphasized in the foregoing development that the optical activity arises from the rotational strength of a given absorption band, and the Cotton effect associated with it. The emphasis in classical organic chemistry on asymmetric carbon atoms is somewhat misleading in this regard. In fact the easiest way to understand the origin of optical activity is in terms of the asymmetric placement of chromophores. Through analysis of various chromophores, we will get the maximum information about molecules in general and about polynucleotides in specific.

Since chromophores usually involve π-electron systems having planar symmetry, the isolated chromophores themselves are rarely optically active. In most of the biological systems we discuss, the chromophores are made optically active by their asymmetric environment. The basis for understanding how an asymmetric environment induces optical activity in a chromophore was given by Kirkwood [14]. This approach is based on perturbation theory and has been extended by Tinoco [6]. He has shown that a number of effects in optical activity as well as the absorption spectra of polymers may be understood within the perturbation theory approach. We will discuss some of these methods in detail below.

2. *Hypochromism and Hypochromicity*

The fact that DNA in the native, or Watson–Crick double helical form, absorbs light less strongly than either the denatured form or the mononucleotides has been recognized for some time. Since the increase in absorption coefficient at 260 nm on melting of DNA is about 40%, it is easy to follow spectrophotometrically and is the most common technique for following DNA melting. The native to denatured transition in DNA takes place over a range of only a few degrees. The sharpness of the transition in thermal profile results from the cooperative statistical nature of the transition.

The underlying reason for the lower absorption of the native form compared to that of the denatured or random form has been the center of some discussion in the literature, but the nature of the effect seems to be rather well established at this time. The earliest successful theoretical attempt at explaining hypochromism was that of Tinoco [15]. Despite a minor error in the original formulation, this theory is now recognized as essentially correct. Rhodes [16] has also developed a theory which in fact is equivalent to that of Tinoco. The Tinoco–Rhodes theory is based on the assumption that there is no electron overlap or exchange among the residues of the polymer. This assumption appears to be substantially correct for the electrons of the aromatic nucleotide base which is the chromophore responsible for the 260 nm absorption. In the absence of electron exchange among the bases, one considers the influence of the electric field of the electrons on one base on the electronic transition moments of the absorbing base.

We will discuss some of the details illustrative of this theory below, but a few simple conclusions can be readily summarized. The hypochromism of the 260-nm band is associated with an increase in the ₍absorption intensity of bands at shorter wavelengths. The hypochromism is common to systems which are stacked with the chromophores one on top of another like a deck of cards, while systems in which the chromophores are in an end-to-end aggregate are generally predicted to be hyperchromic. The stacked array in native DNA is rather rigid, and hence it shows a strong hypochromic effect. Other polynucleotides may also show lesser amounts of stacking but also some hypochromism. RNA, synthetic polynucleotides and even dinucleoside phosphates show this hypochromic effect as a result of base-stacking. The variation in hypochromism is interpreted as a variation in the degree of base stacking as we will discuss below [17].

In order to better understand the nature of some of the controversy in the literature, we should make clear the distinction between hypochromism and hypochromicity. The former is based on the oscillator strengths which are determined by the integrated absorption intensity [18].

$$H = 1 - f_{polymer}/f_{monomer} \tag{21}$$

Hypochromicity, on the other hand, is more often reported because it is easier to measure,

$$h = 1 - \epsilon_{polymer}/\epsilon_{monomer} \tag{22}$$

In fact, the hypochromism and hypochromicity are quite similar in DNA, RNA, and polynucleotides, since the absorption band shape has approximately the same wavelength maximum and shape in both the native and denatured forms as well as in the mononucleotides. Hence, any theory used to explain hypochromicity in polynucleotides should in fact also predict true hypochromism of the integrated absorption intensity.

Following the publication of the Tinoco and Rhodes theory of hypochromism, several semiclassical coupled oscillator theories were introduced [19,20]. These theories predict hypochromicity to result from the interaction of two identical absorption bands on different bases. This type of interaction is also known as exciton interaction and is similar to the effect widely discussed in molecular crystals. In contrast to the theory of Bolton and Weiss [19] no hypochromism arises in the theories of Tinoco and Rhodes from systems having only one absorption band.

It has been shown by Thiery [21] and by Rhodes and Chase [22] that exciton-type theories do not predict true hypochromism as is seen in DNA and polynucleotides. The fact that the absorption band shapes in the monomers is similar to that of the polymers requires that the oscillator strengths be diminished in the polymer in order to give the hypochromic effect we see in DNA and polynucleotides. The sum rule on oscillator strengths [21] requires that hypochromism in the 260-nm band be accompanied by increased absorption at some shorter wavelength. Since this sum rule is obeyed quite generally, the theories of Bolton and Weiss [19] and of Nesbet [20] clearly cannot predict hypochromism since they treat only a single absorption band. These classical theories do predict hypochromicity but no hypochromism and hence are apparently incorrect for polynucleotides.

DeVoe [23,24] has developed a classical theory which does in fact give true hypochromism. However, it requires oscillators at different frequencies to exchange intensity, so it is in practice equivalent to the theories of Rhodes [16] and Tinoco [15] which predict hypochromism for stacked chromophores as a result of intensity lending. As Rhodes and Chase [22] have pointed out, the original theory of Rhodes and Tinoco is essentially a classical one. This fact results from the neglect of exchange of electrons among the chromophores of the polymer. The fact that neglect of electron exchange allows one to understand both the hypochromism and the optical activity of polynucleotides has been taken as evidence for the correctness of this assumption.

The strong absorption bands in polynucleotides arise mainly from the aromatic $\pi-\pi^*$ transitions of the bases. However, there are also weak $n-\pi^*$ transitions polarized parallel to the helix axis. Rhodes [16] has discussed this transition and shows that it should be hyperchromic in the stacked polymer. Experimental verification of this hyperchromic effect has not been found [25]. Due to their weak absorption intensity, these transitions are difficult to observe in the presence of the strong $\pi-\pi^*$ transitions which apparently are quite near in wavelength. The influence of the $n-\pi^*$ transitions is better recognized in the CD as will be pointed out below [26].

E. EXCITON AND HYPOCHROMIC INTERACTION IN A DIMER

1. *Exciton Optical Activity*

Two chromophores, when they are brought into proximity, will interact by their electric fields. In polynucleotides and other polymers as well, the monomer and polymer absorption band shapes are similar. If there were electron exchange (charge-transfer bands) we would expect these bands to look quite different. Since we do not observe large changes in band shape or band position, we will ignore the possibility of electron overlap and consider only the interaction of the electrons by their electric fields.

In this approximation the wavefunction of a dimer in the ground state may be given by [6]

$$\psi_0 = \phi_{10}\phi_{20} \qquad (23)$$

We will assume the wavefunctions of the monomer ϕ_1 and ϕ_2 are known. For the excited state wavefunction we take the symmetric and antisymmetric combination as directed by degenerate perturbation theory.

$$\psi_a\pm = (1/\sqrt{2})(\phi_{10}\phi_{2a} \pm \phi_{1a}\phi_{20}) \qquad (24)$$

The energies of these two states are

$$E\pm = E_a \pm V_{12} \qquad (25)$$

where

$$V_{12} = \langle\phi_{1a}\phi_{20}|V_{12}|\phi_{10}\phi_{2a}\rangle \qquad (26)$$

We can see the heart of this calculation is the interaction energy V_{12}. There are several functional forms that have been used for V_{12} in doing practical calculations. A point monopole approximation has been used by Bush and Tinoco [10], for calculations on dinucleoside phosphates. A point monopole-distributed dipole approximation has been used by Bush [26], for $n-\pi^*$ transitions. The dipole–dipole approximation for V_{12}, while perhaps not the most accurate, is the easiest to understand and gives a good illustration of the find of effects we see in dimers. In this approximation, it is the interaction of the transition dipole moments which leads to the interaction energy [27].

$$V_{12} = (\mu_{10a}\cdot\mu_{20a})/R_{12}{}^3 - (3R_{12}\cdot\mu_{10a}R_{12}\cdot\mu_{20a})/R_{12}{}^5 \qquad (27)$$

R_{12} is the distance between the chromophores and the transition electric dipole moment of the monomer is defined as

$$\mu_{0a} = \langle\phi_0|er|\phi_a\rangle \qquad (28)$$

Its square is proportional to the absorption intensity [see Eq. (4)]. According to Eq. (25) the dimer will have two absorption bands at energies separated by V_{12}. Their transition dipole moments differ depending on the geometry of the dimer. The dipole strengths [see Eq. (4)] of the two bands are [27]

$$D_{0a}\pm = \mu_{0a}{}^2 \pm \mu_{10a}\cdot\mu_{20a} \qquad (29)$$

We see from this formula that there may be a redistribution of the absorption among the two split bands. This results in the appearance of a shift or a shoulder in the dimer absorption. However, no true hypochromism is predicted and the sum of the plus and minus bands is just the sum of the two monomer absorptions.

In most polynucleotide systems which have been studied, the splitting $(2V_{12})$, is considerably smaller than the width of the absorption band of the monomer. Hence, this splitting is not usually observed in the absorption spectra of polynucleotides, and the wavelength of the maximum and the band shape appear quite similar in the monomers and polymers. The effect of this splitting is seen clearly in some polypeptide systems including the alpha helical system which has a $\pi-\pi^*$ absorption band at 200 nm [28].

In polynucleotide systems, the splitting of these two bands is most clearly seen in the CD or ORD of the dinucleoside. In that case, the signs of the CD of the plus and minus bands are opposite, and the two exciton bands appear quite distinctly. Although other effects are operative in addition to this exciton effect, it still may be clearly recognized in the CD of a number of dinucleosides.

In order to compute the optical rotation, we will calculate the rotational strength which is given by Eq. (16). We may assume real wavefunctions for the calculation of this quantity. Appropriate wavefunctions have been given in Eqs. (23) and (24) and we may use these to compute the values of the operators in Eq. (16). For the magnetic moment operator, we must include not only the inherent magnetic moment of the monomer transition, but also the magnetic moment seen at one chromophore as a result of the linear momentum at a distance, $\mathbf{R}$. For the magnetic moment of the dimer we write [6]:

$$\mathbf{m} = \mathbf{m}_1 + \mathbf{m}_2 + (e/2mc)(\mathbf{R}_1 \times \mathbf{P}_1 + \mathbf{R}_2 \times \mathbf{P}_2) \qquad (30)$$

where $\mathbf{P}$ is the linear momentum operator. For an electrically allowed transition, such as a $\pi-\pi^*$ in nucleotides, the second term may be easily arranged to contain the entire magnetic moment simply by choice of origin for the measurement of the distance vector, joining the chromophores, $\mathbf{R}_{12}$. If we follow this procedure, the magnetic moment of the dimer may be written as

$$\mathbf{m}_{a0}\pm = (e/2mc\sqrt{2})(\mathbf{R}_2 \times \mathbf{P}_{2a0} \pm \mathbf{R}_1 \times \mathbf{P}_{1a0}) \qquad (31)$$

We now replace the linear momentum operator by the electric transition dipole operator [29],

$$\mathbf{P}_{a0} = (2\pi i \, m/e)v_a \boldsymbol{\mu}_{0a} \qquad (32)$$

$$v_a = E_a/h$$

We can now compute the rotational strength of the two exciton bands as indicated by Eq. (16) by taking the product of the magnetic moment and the electric moment [27].

$$R_{0a}\pm = \mp (\pi v_a/2c)\mathbf{R}_{12}\cdot\boldsymbol{\mu}_{10a} \times \boldsymbol{\mu}_{20a} \qquad (33)$$

$$\mathbf{R}_{12} = \mathbf{R}_2 - \mathbf{R}_1$$

We may make certain statements about the sign of the Cotton effect of dinucleosides based on this simple formula. If the transition moments (μ_{0a}) in the bases make an angle of less than 90°, the dipole interaction described by Eq. (27) is repulsive and the sign of V_{12} is positive. In this case, which is that of DNA geometry, $R-$ occurs at the long wavelength side of the absorption band [see Eq. (25)]. If the helix is right-handed, $R-$ is a positive number, and at ν^+ there will be a negative CD band. This is the case of ApA, Fig. 3[30]. This situation is reversed for left-handed helical arrangements of the bases. This fact provides a simple way of assigning the skew sense of the bases in a dinucleoside. Conclusions based on this reasoning are most useful in dinucleosides or polynucleotides of identical bases. Its application to dinucleosides of different bases depends on the relative direction of the transition moments, μ_{0a}, in the bases [10]. These directions are difficult to determine with certainty.

2. *First-Order Theory of Hypochromism in a Dimer*

We have stated that the exciton theory is conservative. In optical activity, the rotational strengths are of equal magnitude and opposite sign [see Eq. (33)]. Also, the total oscillator strength of the dimer is the same as that for the two-component monomers [see Eq. (29)]. In fact, polynucleotides also show other effects, such as hypochromism and single Cotton effects. These effects may be understood by including higher excited states in the perturbation theory treatment. We will extend our discussion of dimers to include some of these effects. We do not imply that this treatment is by any means thorough, but simply illustrative of some of the effects which are seen experimentally. In the perturbation expansion we include higher excited states, indexed by b, in our expansion of the wavefunction. For the ground state [31],

$$\psi_0{}^1 = \phi_{10}\phi_{20} + \sum_{b,b'} \phi_{1b}\phi_{2b'} \frac{V_{10b;20b'}}{E_{b'} + E_b} \tag{34}$$

and for the excited state,

$$\psi_a{}^1 \pm = \psi_a \pm + \sum_b \phi_{2b}\phi_{10} \frac{V_{10a;20b}}{E_b - E_a} \tag{35}$$

The coupling energy in Eq. (35), $V_{10a;20b}$, between different states is similar to the exciton coupling energy, $V_{10a;20a}$ in Eq. (25).

$$V_{10a;\,20b} = \langle \phi_{10}\phi_{2b}|\mathbf{V}_{12}|\phi_{1a}\phi_{20}\rangle \tag{36}$$

The exciton effect has been shown not to contribute to the hypochromism [16].

If we sum the dipole strengths for the two exciton levels, we obtain the band hypochromism of the dimer.

$$|\mu_{0a}^{1}|^2 = 2\mu_{0a}^2 + \sum_b \frac{4E_b}{E_b^2 - E_a^2} (\mu_{20b} \cdot \mu_{10a} V_{10a;20b} + \mu_{10b} \cdot \mu_{20a} V_{10b;20a}) \quad (37)$$

This equation shows how intensity may be traded by band *a* to other bands by means of the interband dynamic coupling energy, $V_{10a;20b}$.

The sum rules on optical absorption [11] tell us that interaction of identical bands or nearly identical bands may redistribute the absorption, but lead to no hypochromism. Equation (37) tells us that hypochromism in the 260 band will be accompanied by hyperchromism in some shorter wavelength absorption band. This hyperchromic effect has been looked for in the easily accessible ultraviolet bands in the neighborhood of 200 nm, but such a hyperchromic effect has not been found in DNA. Presumably the hyperchromic bands important in the summation over *b* in Eq. (37) are in the vacuum ultraviolet. Indeed, experiments on the vacuum ultraviolet absorption of crystalline films of nucleotide bases have detected hyperchromicity at wavelengths shorter than 160 nm [32].

3. *Single Cotton Effects*

The exciton interactions give rise to conservative rotation in the optical activity spectra. The interaction with higher energy bands may give rise to single Cotton effects. This nonconservative behavior in the optical activity may be calculated through a polarizability approximation much like that used by Rhodes [16] in the hypochromism calculation. The polarizability approximation was first used in the calculation of optical activity by Kirkwood [14]. A similar procedure was used by Bush and Brahms [33] in showing how nonconservative bands could arise in polynucleotides through interaction with higher energy absorption bands. The mechanism is similar to conservative optical activity, but the single Cotton effect in the near ultraviolet is balanced by other Cotton effects in the far ultraviolet.

The mechanisms that we have been discussing above are for the relatively well-understood π–π^* bands seen in polynucleotides at 260 nm. In the optical activity, there is also the possibility that a forbidden band (n–π^*) contributes. These bands are not prominent in the absorption, as a result of their relatively small transition dipole moments compared to the strong π–π^* transitions. In the circular dichroism of some polynucleotides, however, there is evidence that n–π^* effects may be important in some cases [5]. We have not discussed mechanisms leading to optical activity of weak transitions such as the n–π^*. However, these transitions have been extensively studied as a result of their importance in polypeptides and in numerous organic molecules. Mechanisms leading to the optical activity of n π^* transitions include static field or one-electron Cotton effects [34] and also coupling of $n-\pi^*$ to strong $\pi-\pi^*$ transitions [35]. Both effects have been proposed to be important in polynucleotides where azine $n-\pi^*$ transitions occur in the 280–290 nm range in purine derivatives [26].

4. *Comments on the Nature of the Coupling Energy*

The coupling energy mentioned in the exciton and hypochromicity theory [see Eqs. (25) and (36)], as well as that giving rise to $n-\pi^*$ Cotton effects, is rather short-range in terms of polymer dimensions. Hence, all optical effects depend mainly on neighboring residues in the polymer. Interactions among farther neighbors are relatively unimportant unless they are spatially close. This is in contrast to some of the other physicochemical properties such as sedimentation rate, viscosity, light-scattering dimensions, etc. Hence, hydrodynamic and optical measurements are complementary in the information they give on polymer conformation. Apparent discrepancies between the results of these two kinds of measurements may often be reconciled giving rise to a more complete picture of polymer conformation in solution. The hydrodynamic properties are often sensitive to interactions of relatively far neighbors in the polymer, while optical properties often depend only on nearest-neighbor interactions.

Certain general conclusions about the ultraviolet optical properties of polynucleotides may be drawn from the above discussion. In general, stacking of the base planes in a single, double, or triple helical structure implies hypochromism. Base-pairing in the polynucleotide structure, while it may stabilize the structure, does not contribute directly to hypochromism. The hypochromism is approximately linearly related to the degree of stacking.

Likewise, asymmetric or helical stacking of bases in a polynucleotide leads to Cotton effects. One may observe double Cotton effects due to coupled oscillators of the exciton type and in addition there may be single Cotton effects. The existence of these single and double Cotton effects have some geometric implications as we will discuss below [3]. Although both hypochromism and optical activity arise from stacking of the bases, the experimental observation that the two measurements differ in their geometry dependence is not difficult to explain. Hypochromism and optical activity do not necessarily give the same picture of a given structure. For example, the temperature dependence of the optical activity may be found to differ from the temperature dependence of hypochromism. These two melting curves need not be alike if the melting is not a single transition between two states. For example, a dinucleoside could exist in equilibrium of a right- and a left-handed as well as an unstacked form. The optical activity of the right-

and left-handed helical forms would cancel while their hypochromicities would add. In an attempt to provide a formal model predicting differences in optical activity and hypochromism melting curves, Glaubiger et al [36]. have proposed a model in which the bases behave as a torsional oscillator. Melting in this model is not a simple two-state transition and therefore the model features different circular dichroism and hypochromism melting profiles. While no clear picture has emerged on the nature of the states involved in the melting transition, it is apparently not a simple process.

Moreover, there is some danger in estimating the degree of stacking of a polynucleotide structure from optical activity alone. While the optical activity melting curve is very useful in estimating polynucleotide stacking, it is desirable to have other physical measurements on a given structure. For example, the 2′–5′ linked dinucleosides discussed by Brahms et al. [37] were said to be unstacked. These workers observed small circular dichroism and a weak dependence of the circular dichroic maxima on temperature, implying a small value of the energy of stacking. On the other hand, hypochromism studies [38] and NMR [39], indicate that the 2′–5′ compounds are indeed stacked. The apparent discrepancy between the circular dichroism and other physical measurements has been clarified by Kondo et al. [39]. Kondo et al. [39] have given a detailed scheme of proposed geometry based mainly on NMR and have explained how the optical activity of this fairly rigidly stacked 2′–5′ dinucleoside is observed to be small.

F. FARADAY EFFECT

In 1845 Michael Faraday discovered that in the presence of a magnetic field parallel to the light path, optically inactive material showed rotation. Shortly thereafter, a number of investigators attempted to relate the effect to molecular structure. They drew their inspiration in this effort from the spectacular successes achieved in relating natural optical rotation to molecular dissymmetry. Their lack of success in this effort emphasizes the fact that, despite the apparent similarities, these two effects are quite different.

Natural optical activity has been productive of a number of techniques for investigating chemical structure and polymeric conformation. That such a productivity is less likely from magnetically induced optical activity, may be seen from the following qualitative argument. If one imagines the dinucleoside phosphate ApA as two planar symmetric chromophores held by the ribose phosphate in a skewed conformation, the asymmetric perturbation of each chromophore on its neighbor is held in place as the molecule tumbles in solution. Hence, the sign and magnitude of the natural Cotton effects are quite sensitive to the geometric relationship of the two chromophores. If, however, the perturbation giving rise to the optical activity is a magnet fixed in the instrument, this sensitivity to molecular geometry is not present [40]. Therefore one does not expect the spectacular effects of polymerization which appear in natural CD to be found in Faraday effect. The effects of polymer conformation on magnetic CD are of the same order of magnitude as are the effects on the absorption spectrum [41,42]. Hypochromism is typically 20–40% in polynucleotides and can be easily measured in absorption. On the other hand, measurements of this accuracy are quite difficult in magnetic CD.

When measuring the magnetic ORD or Faraday rotation of a solution, one finds an enormous background effect due to the solvent. Therefore, it is much more convenient to measure the magnetic CD, to which transparent solvents do not contribute. Magnetic CD is related to magnetic ORD in the same way as natural CD is related to ORD [43]. Magnetic CD has been useful in the difficult problem of the electronic assignments in nucleotide bases, a problem which we will discuss below [44]. Little has been published on the magnetic CD of polymers but Maestre *et al.* [44a] have reported MCD of polynucleotides and of bacteriophage which clearly show some interesting effects of polymer conformation.

II. Nucleotide Chromophores

Since the basis for understanding the ultraviolet absorption and optical activity of a molecule is its chromophoric nature, we will now review present knowledge of the chromophores of nucleotide bases. These planar aromatic heterocycles have $\pi-\pi^*$ transitions with large extinction coefficients in the range of 190–300 nm. We chose the near ultraviolet since this is the range in which measurements are most easily performed. Although measurements at wavelengths shorter than 190 nm are possible by use of vacuum ultraviolet techniques, little work on nucleotides in the vacuum ultraviolet has been done. Recent advances in instrumentation make it likely that the short wavelength limit can be extended substantially by the use of vacuum techniques and improved modulators [44b]. Through the use of very short pathlength cells, one should be able to measure CD in water solution into the range of 160–170 nm. Water absorption below this wavelength will probably limit the shorter wavelength region (160–130 nm) to gas phase or dry film work.

The presence of ribose or deoxyribose does not greatly modify the absorption spectrum. Hence experiments on the absorption of nucleotides, bases, and nucleosides and the CD of nucleosides and nucleotides all refer to the same chromophore. In addition to $\pi-\pi^*$ transitions, the nitrogen nonbonding electrons give rise to $n-\pi^*$ transitions, but these have small extinction coefficients and are not prominent in the absorption. They may, however, contribute to the optical activity as we will discuss below.

A. ABSORPTION SPECTRA

The complete ultraviolet spectra of the nucleotides and nucleosides has been reported by Voet et al. [45]. Their results are summarized in Table I. They gave complete accurate curves over the wavelength range of interest, but no attempt at decomposition of the spectra into individual components was made. An assignment of the electronic origin of the transitions was made by Clark and Tinoco [46]. They classified the spectral transitions and attempted to relate them to parent benzene bands. From examination of analogous compounds, they correlated the purine and pyrimidine bands to the B_{1u}, B_{2u} and E_u bands of benzene (See Table I)[45–48]. A similar classification has been used by Pullman and Pullman [49] who have carried out calculations of spectra and attempted to improve our understanding of the electronic nature of the transitions.

We now summarize the attempts to identify and classify the absorption bands in nucleotide bases. The experimental techniques which have been found most useful are absorption spectra, linear dichroism of crystals, CD of nucleosides and derivatives, and magnetic circular dichroism (MCD).

1. Adenine

Adenine exhibits an absorption band of simple curve shape at 260 nm, and additional bands at 206 and 186 nm [46]. In spite of the simple appearance of the curve shape of the 260 and 206 nm bands, Miles et al. [4] have presented CD evidence on adenosine derivatives which indicate that both the 260 and 206 nm bands are composites. In agreement with polarized fluorescence data [50], they assert that the 260 nm band is composed of a main band (B_{2u}) and a weaker π–π* (B_{1u}) band at slightly shorter wavelength (245 nm). MCD work of Voelter et al. [44] supports this assignment as well as the assignment of the 206 nm band as a composite of degenerate E_u bands. Apparently it is these E_u bands which give rise to the CD bands in adenosine at 220 and 200 nm [4]. The polarization of the longest wavelength π–π* band in adenine has been determined from polarized absorption of crystals to be along the short axis of the molecule (in the direction of the C_4—C_5 bond) [51]. In addition to the π–π* bands, Stewart and Davidson [52] found some perpendicularly polarized absorption bands in crystals of adenine which they assigned to n–π* bands. Miles et al. [4] also assign an n–π* band in adenine in the 290 nm region.

While the above-quoted experiments show reasonable internal consistency, the assignments are difficult to interpret in the CD of polyadenylic acid. We will discuss this question in more detail below, but in summary the polymer exhibits evidence of a single absorption band at 260 nm as well as at 206 nm. In addition to a weak CD band at 280 nm, there is a band of unexplained origin at 230 nm in poly A. (See Fig. 3.)

TABLE I

ULTRAVIOLET ABSORPTION BANDS[a] IN WATER AT NEUTRAL pH

Compound	Ref.[b]	Long-wave region (B_{2u}) λ_{max}	$\varepsilon \times 10^{-3}$	Middle region (B_{1u}) λ_{max}	$\varepsilon \times 10^{-3}$	Short-wave region (E_u) λ_{max}	$\varepsilon \times 10^{-3}$	λ_{max}	$\varepsilon \times 10^{-3}$
Adenine	a	260.5	13.4	—	—	207	23.2	185	—
Adenosine	a	259.5	14.9	—	—	206	21.2	190	19.8
Guanine	a	275	8.1	245	10.7	—	—	196	22.1
Guanosine	a	276	9.0	252.5	13.7	—	—	188	26.8
9-Methyl Hypoxanthine	b	—	—	248	11.0	201	21.0	—?	—
Inosine	c	—	—	248	12.2	?	?	?	—
Cytosine	a	267	6.1	230	6.3	—	—	196.5	22.5
Cytidine	a	271	9.1	230	8.2	—	—	198	23.2
Uracil	a	259.5	8.2	—	—	202.5	9.2	—	—
Uridine	a	261.1	10.1	—	—	205	9.8	—	—
Thymine	a	264.5	7.9	—	—	205	9.5	—	—
Thymidine	a	267	9.7	—	—	206.5	9.8	—	—

[a] The bands are inferred from absorption maxima and correlations among analogous compounds, following Ts'o [47].

[b] References: a, Voet et al. [45]; b, Clark and Tinoco [46]; c, Ts'o et al. [48].

2. Thymine and Uracil

In most approximations, these two molecules are taken to have identical spectra, and they are usually discussed together. From CD of uracil nucleosides, Miles *et al.* [53] assign 262 nm bands as the B_{2u} and a weak transition at 240 nm as the B_{1u}. The E_{1u} pair of bands are seen at 216 nm and at 196 nm. They also used solvent effects in an attempt to discover any $n–\pi^*$ bands in thymine and uracil, but fail to see any clear evidence for $n–\pi^*$ transitions.

Stewart and Davidson [52] and more recently Eaton and Lewis [54] have reported polarized crystal absorption spectra of 1-methyl thymine and 1-methyl uracil. The polarization of the longest wavelength $\pi–\pi^*$ transitions in these two molecules are similar and are approximately parallel to the C_5—C_6 bond [54]. Although the CD data of Miles *et al.* [53] did not indicate the presence of any $n–\pi^*$ bands, the data of Eaton and Lewis [54] show evidence for an $n–\pi^*$ band at 264 nm. The relative weakness of these perpendicularly polarized bands makes an unambiguous assignment difficult.

3. Guanine

From CD studies on guanosine derivatives, Miles *et al.* [55] place the B_{2u} transition at 265 nm. They also assign an $n–\pi^*$ band at 270 nm. The absorption band at 248 nm is assigned as the B_{1u} and a band at 210 nm is assigned as the E_u. The MCD studies of Voelter *et al.* [44] indicate bands at 277, 248, and 215 nm. While it is possible that the MCD band at 277 is a composite of the B_{2u} and the $n–\pi^*$ bands of the Miles *et al.* [55] assignment, it seems difficult to make a consistent interpretation of the bands in guanine. The only data available on the polarization of these transitions are from studies on the polarization of the fluorescence which indicate that the two longest wavelength transitions are approximately perpendicular [50].

4. Cytosine

This chromophore demonstrates clearly that our knowledge of the nature of the observed bands in nucleotides is less than perfect. Miles *et al.* [56] assign the band at 271 nm to the B_{2u} and they place the B_{1u} band at 247 nm where they observe a CD band. No absorption band is seen at 247 nm, but one is seen at 235 nm for which there is no corresponding CD band in the nucleoside. Bands at 220 and 195 nm are assigned to the E_{1u} by Miles *et al.* [56]. These workers also used solvent effects to test their earlier assignment of the 235 nm absorption band as $n–\pi^*$ [55]. The results of these solvent tests are somewhat confusing and interpreted by them as due to conformational changes in the nucleoside. The failure of the CD and absorption maxima to coincide indicate to the present author that perhaps assignment of the 247 nm band as $n–\pi^*$ might be more likely [55].

The polarization of the 265 nm and the 230 nm absorption bands of cyto-sine has been studied using polarized absorption of crystals and polarized fluorescence by Callis and Simpson [57]. They find the polarization of the 265 nm band to be similar to that in uracil; i.e., along the N_1—C_4 axis. They also find, somewhat surprisingly, that the 230 nm band is nearly parallel to the longer wavelength absorption band.

5. Summary

Certainly not enough facts are available to unambiguously resolve the spectra of nucleosides into component bands of known absorption intensity, polarization, and electronic nature. The nature of the azine n–π^* transitions, which appear to be important in CD, is probably not simple. The nonbonding electrons were originally thought to be localized at the nitrogen. Recent calculations including all-valence electrons, which we will discuss in detail below, have shown that the highest occupied σ-orbitals are quite delocalized. Also, Chen and Clark [58] find a surprising amount of perpendicularly polarized absorption in purine. Apparently, more experiments, especially on the polarization of the absorption bands, will be needed to fully interpret the absorption spectra.

B. Calculations of Transitions

1. Simple Methods

The initial attempts at calculation of the spectra of nucleotide bases utilized simple Hückel π-electron theory [18]. This simple theory predicts π–π^* transition energies, intensities, and polarizations necessary for hypochromism calculations [18]. While the calculated values are qualitatively useful, the approximations made are too crude to lead to reliable values of spectral polarization and intensity. An additional shortcoming of the simple Hückel theory is that, like other π-electron theories, simple Hückel does not treat σ-electrons. The σ-electrons must be treated separately. The Del Re [59] method has been used in calculation of ground-state dipole moments, but it is not useful in calculations of spectra.

Improved results on the π–π^* transitions have been sought through the use of the semiempirical Pariser–Parr–Pople method. Since this method contains parameters which may be adjusted to give agreement with experimental data, one might hope to be able to predict reliable spectral properties. The strategy in this semiempirical method is to adjust parameters to give good results for experimentally known quantities. We have reasonably good experimental information about the energies and, in some cases, the polarizations of transitions. If we adjust the theory to agree with known quantities, we might be able to predict unknown quantities such as energies and polarizations of far-ultraviolet transitions and charge distributions.

Unfortunately, it is not clear that this procedure is reliable. Indeed, one can do quite well in calculating the correct number and spectral position of absorption bands [60]. Through adjustment of the Pariser–Parr–Pople parameters, Berthod et al. [60] were able to shed some light on the nature of the transitions. In general Berthod et al. [60] were able to rationalize their calculated spectra in terms of the experimental assignments discussed in Section II, A above. The intensities calculated were also in approximate agreement with experiments, but the directions of polarization are more doubtful. As we shall see below, use of the molecular orbitals of Berthod et al. [60] in optical activity calculations works fairly well in some cases but not in all cases [61,62].

2. All-Valence Electron Semiempirical Calculations

With the introduction of bigger and faster computers, certain advances in calculation have made it possible to lift some of the approximations in earlier molecular orbital theories. The CNDO method is an extension of Pariser–Parr–Pople π-electron method which treats all valence electrons. Although still semiempirical, it avoids σ–π separation and also allows calculation of n–π^* as well as π–π^* transitions. The method has been applied to spectra of simple molecules [63,64]. This method should be useful in interpreting the spectra of nucleotides.

Bush [26] has used extended Hückel methods for calculation of n–π^* absorption intensities and a theory of circular dichroism of n–π^* transitions in polynucleotides. Extended Hückel theory is capable of treating n–π^* transitions since it includes all-valence electrons. Reasonable agreement is found for the n–π^* transition in adenine at 280 nm. The extended Hückel theory is very approximate and probably not reliable in detail because of the extreme assumptions made.

3. All Electron Ab Initio Calculations

The rapid growth in the capacity of modern computers has made possible the calculation of nonempirical wavefunctions for molecules of increasing size. There are a few reports in the literature of ground-state calculations on some of the nucleotide bases [65] and a recent report of calculations on some of the excited states of thymine [66].

The advantage of such all electron ab initio calculations is that one has a much better idea of the meaning of his approximations. The semiempirical theories (e.g., extended Hückel, Pariser–Parr–Pople, CNDO, INDO), introduce experimental parameters in such a way as to lead to correct answers for some experimentally known quantities such as transition energies. However, since the use of empirical parameters introduces approximations whose nature is unknown, the reliability of predictions of unknown quantities (e.g., transition charge distribution, magnetic transition moments, and higher excited states) is difficult to evaluate.

The current state of the art in *ab initio* calculation of large molecules such as nucleotide bases, makes a few assumptions even in the ground-state calculation. The atomic basis orbitals are far from complete. The atomic orbitals usually have a Gaussian radial dependence contracted into five atomic basis orbitals for the first-row atoms. These atomic basis functions approximate an atomic $1s$, $2s$ and three $2p$ functions. While such a limitation of the basis appears drastic, it has been extensively tested and appears to be adequate [67,68]. These calculations also ignore electron correlation; in an infinite basis they converge to the Hartree–Fock result. One-electron properties, such as ground-state charge distribution, are thought to be well represented by Hartree–Fock functions. The reliability of calculated energies, excited states, and transition moments is open to some question. It is not known whether the current state of the art *ab initio* calculations will be adequate to accurately describe excited-state properties and transition moments.

Snyder *et al.* [66] have reported calculations for thymine and its anions, which include excited singlet and triplet states using configuration interaction. More calculations on the other nucleotide bases will be needed before a full evaluation of this method can be made, but their calculation does lead to an improved understanding of the charge distribution in the excited states. The polarization of the singlet transition agrees with experiment, but the energies appear unreliable, apparently due to correlation effects. It appears that an improved theoretical understanding of the electronic properties of nucleotides may be available in the next few years.

4. *Summary*

Simple approximate theories of π-electrons lead to a qualitative understanding of the π–π^* transitions. However, approximations inherent in the σ–π separation and semiempirical theory make it doubtful that these calculations are quantitatively exact [65]. Also the σ-orbitals may be necessary to achieve a full interpretation of the spectra. While it is likely that the rapid progress in computation of molecular wavefunctions may lead to a better understanding in the near future, we cannot say that our present knowledge is complete. The meaning of this situation to our understanding of the hypochromicity and optical activity of polynucleotides is that at present we cannot expect theoretical treatments to lead to quantitatively exact results. As we have described in Section I,E above, the monomer wavefunctions are assumed to be known in the theories of polymer hypochromism and optical activity. In fact, these monomer wavefunctions are not really well known. Through the use of experimental data on energies, intensities, and polarization, we may get a qualitative theoretical understanding of the optical properties of polynucleotides. Most of the useful techniques for interpreting optical activity and hypochromism are quite empirical in nature. Theory can guide us in formulating empirical relationships, but we cannot expect the theories to give quantitative interpretation in a wide variety of cases. In our description of useful techniques using optical absorption and optical activity, we will describe a number of the most useful empirical techniques.

III. Experimental Techniques—Ultraviolet Absorption Spectra

A. Base Composition and Concentration

Ultraviolet absorption spectra are routinely used for measuring concentrations of mononucleotides as well as synthetic and natural polynucleotides. The measurement is simple and reasonably accurate and sensitive. In polymeric systems, the hypochromism may be significant and should be taken into account. If the extinction coefficient is known, the concentration may be measured by measuring the optical density at the absorption maximum near 260 nm. The extinction coefficients for most monomers and a large number of oligomers have been tabulated [69,70]. The correct extinction coefficient for an unknown system may be measured by base hydrolysis to the monomers if the base composition is known. Alternatively, the extinction coefficient may be measured by quantitative analysis for phosphate.

The base composition may be approximately determined from hydrolyzed mixtures by exploiting the differences in the absorption curve shape of the different nucleotides [70]. In addition, for oligomers of moderate length, the base composition may be determined without hydrolysis by means of denaturing solvents which unstack the bases, removing hypochromism. Mandeles and Cantor [71] used 7 *M* urea at acid, neutral, and basic pH to determine the base composition of dimers and trimers. It is possible that use of sophisticated treatment of the absorption data might improve the accuracy and sensitivity of their method. Such a procedure would involve computerized data gathering, smoothing, and curve fitting.

B. Hypochromism and Stacking

1. *Oligomers*

Even dinucleoside phosphates may show significant hypochromism at room temperature. The values found in dimers range from 0 to 13% [17]. Warshaw and Tinoco [17] have interpreted their hypochromism data in conjunction with ORD data, to estimate the degree of stacking of the various bases in dimers. U is shown to prefer unstacked arrangements, while A, G, and C show strong preference for stacking. This technique has also been extended to trinucleoside diphosphates by Cantor and Tinoco [72].

The hypochromicity of a series of A oligomers of differing chain length has been analyzed by several workers [73,74]. The results are in reasonable agreement with those of optical activity studies [75,76]. The results indicate a noncooperative or pairwise stacking mechanism for single-stranded oligomers at neutral pH. From thermal melting curves, the ΔH of stacking is estimated to be -6 to -12 kcal per base-stacking interaction. In general, one must correct for thermal expansion of the solution in these melting experiments. The discrepancies among various estimates of ΔH are apparently due to different methods of data treatment [73].

2. Synthetic Polymers

In poly A at neutral pH, the same pairwise noncooperative stacking mechanism predominates, so that the melting curves appear similar to those of the oligomers [73]. In acid solutions, a double-stranded structure is formed whose melting is cooperative. The critical length for the cooperative formation of the double-stranded structure has been estimated from CD data to be six base pairs [75]. The double-stranded structure is more rigidly stacked than the single-stranded one, so additional hypochromism of the double-stranded over the single-stranded structure is seen in poly rA. Poly dA also shows a double stranded structure at acid pH as seen in CD, but its formation does not lead to additional hypochromism [77]. The reason for the appearance of additional hypochromism in acid poly rA double helices and its absence in poly dA remains obscure.

The formation of double-stranded structures of complementary base pairs has been extensively studied using the hypochromicity technique. Both 2- and 3-stranded structures have been documented in a number of cases. The stoichiometry of these multiply stranded helices is often determined by mixing curves (Job plots) using ultraviolet hypochromicity. The extensive literature on this subject has been summarized in the review of Felsenfeld and Miels [78]. Table I of their review lists some 40 known complementary helices of synthetic ribo- and deoxypolynucleotides. The majority of these structures has been characterized by ultraviolet hypochromicity.

3. Mononucleotide Hypochromicity

Mononucleosides are known to bind cooperatively to their complementary synthetic polynucleotides. Adenosine binds to poly U but such complexes occur at concentrations sufficiently high that hypochromicity studies are difficult [79]. The self-stacking interaction of nucleosides in water has been studied by several physical methods [80]. Again the high concentrations indicate that hypochromicity is not the method of choice for studying these systems. However, Sollie and Schellman [81] have observed significant hypochromism in deoxyadenosine solutions by means of very short path length cells.

4. *Hypochromicity vs Wavelength*

The differences in the absorption spectrum of an A–T base pair and a G–C base pair have been exploited in order to study the nature of the regions of a DNA or RNA which are disrupted at a given temperature. Felsenfeld and Sandeen [82] first observed the dispersion of the hyperchromic effect in DNA and showed that A–T base pairs melt at a slightly lower temperature than do G–C base pairs.

This method has also been utilized in RNA, where the looping of the single chain to form base-paired double-helical regions has been studied [83]. In addition to the thermal melting of the double-stranded regions studied by Cox [83], Gould and Simpkins [84] have extended their studies to include the spectrophotometric titration of RNA and obtained detailed estimates of the lengths and base composition of the base-paired regions of ribosomal RNA. While a number of assumptions are implicit in this type of treatment of titration and thermal melting absorption spectra of RNA and DNA, the interpretation is essentially phenomenological, and hence can give some general insights into polynucleotide structure. Thus, the method is independent of any theoretical model bases on a quantum-mechanical mechanism of hypochromism.

5. *Nearest-Neighbor Effects*

The simplest theories, which we have discussed above, assume that all base pairs titrate or melt essentially independently from one another. However, the stacking of oligonucleotides depends on neighbor–neighbor interaction along the chain. Felsenfeld and Hirschman [85] have analyzed the hypochromicity spectrum of native DNA relative to denatured DNA and relative to the component monomers. They developed a theory which takes account of nearest-neighbor interactions and which, for random sequences, predicts a quadratic as well as linear dependence on base composition. Nearest-neighbor effects, as measured by the relative magnitude of the quadratic term, are found to be negligible over most of the spectrum in the native to denatured DNA melting [85]. On the other hand, the denatured coil to monomer hyperchromicity shows significant quadratic dependence on base composition.

This nearest-neighbor effect probably arises mainly in the single-stranded stacks as observed in model compounds [17]. The appearance of nearest-neighbor effect allows one to estimate the degree of randomness of the DNA base sequence. Most DNA's are found to be random, but λ phage DNA has significant departures from randomness. Hirschman and Felsenfeld [86] have also described a simplified method for treating the case of the native to denatured DNA melting. In this case the nearest-neighbor interactions may be neglected. Using their method, one can determine the concentration and base composition of mixtures of partially denatured DNA.

C. LINEAR DICHROISM

In DNA double helices, one would expect strong negative dichroism for the $\pi-\pi^*$ transitions. In fact, if DNA is in a B form, with all bases perpendicular to the helix axis, and if all the transition moments are in plane, there should be no absorption at all for a fully oriented sample when the electric vector of the light is parallel to the helix axis. In practice, the difficulty of obtaining fully oriented helices and measuring their dichroism accurately has rendered conclusive experiments on this effect difficult.

In early experiments on oriented films of polynucleotides, Rich and Kasha [87] found differences in the absorption curve shape for parallel and perpendicularly polarized light. They interpreted their results as evidence for significant perpendicularly polarized absorption bands of the $n-\pi^*$ type. More recent experiments have failed to confirm their results in detail [25]. The nature and degree of orientation of stroked films is difficult to determine, and one might expect some difficulty in reproducing such experiments. Orientation in solution by flow or electric fields, in which one could determine the degree of orientation, might help settle the question of base-tilting and the nature of the perpendicularly polarized transitions.

D. RAMAN HYPOCHROMISM

The advent of laser-excited Raman spectrometers has made it possible to examine the vibrational frequencies of polynucleotides in water in some detail. Most of the work in this field has involved vibrational frequencies and shifts and is thus better treated in a discussion of vibrational spectra (see Chapter 5 in Volume I of this treatise). The intensities of Raman lines are difficult to measure or calculate, but the question of Raman intensity has recently been treated theoretically by Peticolas *et al.* [88].

Moreover, Tomlinson and Peticolas [89] have measured the increase in Raman scattering intensity due to the thermal melting of neutral poly A. Their theory of Raman intensity predicts a dependence on the square of the electronic oscillator strength (f_a) for the resonance Raman effect. However, their experiments do not truly correspond to resonance scattering and they observe that the increase in Raman intensity roughly parallels that of the ultraviolet absorption in thermal melting of poly A. Apparently, in the off-resonance case, which has not been treated theoretically, the effect is linear rather than quadratic in the oscillator strength. In spite of the difficulty of measuring absolute intensities of Raman scattering, this method has some promise. The ring vibration modes of the various bases are well separated in frequency, and it may be possible to distinguish them in the Raman spectrum of polynucleotides. Hence this Raman hypochromism measurement offers the promise of observing the hypochromicity of the different bases separately as they melt individually in a copolymer or natural polynucleotide. See Chapter 5, Volume I for a further discussion of this subject.

IV. Experimental Techniques—Optical Activity

A. OPTICAL ACTIVITY OF NUCLEOSIDES AND NUCLEOTIDES

Since the nucleosides and nucleotides are the simplest monomeric components, they might seem to be the logical starting point for the understanding of polynucleotide optical activity. In fact, this has been shown not be the case. The optical properties of polynucleotides are often found to be more dependent on the interaction among the nucleotides than on the monomers themselves. This fact has led to both experimental and theoretical difficulties in the study of mononucleotides.

The planar chromophoric bases are weakly perturbed by the attached sugars. Hence, many nucleosides have rather small Cotton effects and their experimental determination is difficult. This is especially so for purine nucleosides which have generally smaller Cotton effects than do pyrimidine nucleosides. Recent advances in CD instrumentation have improved the signal-to-noise ratio, making possible good measurements on these weak Cotton effects in many purine nucleosides. Although one may hope for progress on purine nucleosides in the near future, the pyrimidine nucleosides are better understood at present.

The theoretical work on mononucleosides is also less extensive than that in polynucleotides. The mechanism of perturbation of the nonchromophoric sugar is quite different in character from that of the strong coupled oscillators at similar frequencies found in polynucleotides. One would hope that from the optical activity of nucleosides we would be able to gain information on the relative relationship of the sugar to the base. But only recently has a theoretical treatment been proposed which is able to shed some light on this relationship.

1. *Theoretical Treatments*

Although it is impossible to carry out a rigorous theoretical calculation of the CD of nucleosides, the theories give a general semiquantitative indication of behavior. Correlation of the theoretical predictions with experimental results can be extremely useful in the interpretation of data.

The theoretical method that has been most widely used is the coupled oscillator approach. It is based on the division of the nucleoside into electron groups whose interaction is calculated by perturbation theory. In the original formulation due to Kirkwood [14] effects of magnetic moments were ignored. The theory considered only the coupling of strong electronic transitions and is therefore called coupled oscillator theory. Although Tinoco [6] has extended the formalism to include one electron Cotton effects, this extension has not yet been applied to nucleosides.

In a recent paper, Inskeep *et al.* [90] have described a theory which seems to offer a basis for understanding at least some of the effects in the optical activity of nucleosides. This technique allows one to calculate the rotational strength induced in the $\pi-\pi^*$ transitions of the base by the asymmetric perturbation of the sugar. In effect, one calculates the coupled oscillator effect between the $\pi-\pi^*$ transitions of the base and the far ultraviolet transitions of the sugar, which are approximated by polarizabilities. This theory seems to explain many of the Cotton effects in the pyrimidine nucleosides but the neglect of $n-\pi^*$ transitions in this theory could be a serious deficiency.

Miles *et al.* [91] have used their coupled oscillator method to compute the $\pi-\pi^*$ Cotton effects for a large number of pyrimidine nucleosides and derivatives for which they have experimental CD data. They find that neglect of one-electron effects leads to a reasonable understanding of the B_{2u} ($\pi-\pi^*$) Cotton effect in both cytosine and uracil nucleosides. They are able to correlate their theory and experimental work leading to the conclusion that the predominant effect is the interaction of the furanose ring with the base. Hence, the feature of the nucleoside which determines the sign of the Cotton effect is the torsion angle, ϕ_{CN}. This dihedral angle of the glycosyl bond determines the relative orientation of the base and the furanose. (See Chapter 6 of Volume I for a more complete definition of ϕ_{CN}.) The ring puckering of the furanose and the substituents, e.g., ribose vs deoxyribose, are found to have a smaller but significant effect on the magnitude of the Cotton effect. The torsion angle is highly restricted in pyrimidine nucleosides by steric effects. Its value is usually between $-40°$ and $-60°$ in the common β anomeric nucleosides. This leads to positive Cotton effects for the B_{2u} transition in both cytidine and uridine as is seen in Fig. 4. The similarity in behavior of cytidine and uridine in the dependence of rotational strength on ϕ_{CN} results from the similarity in the nature of the B_{2u} transition in these two bases (see Section II,A,2,4).

The relation of the 2'-OH of the ribose to the attachment of the base at C-1' has an influence on the computed rotational strength. The magnitude of the effect of the 2'-OH is not large enough to cause differences in the sign of the Cotton effect for most pyrimidine nucleosides. Hence ribose and deoxyribose nucleosides look qualitatively similar in their CD. However, Miles *et al.* [91] are unable to quantitatively predict the difference in magnitude be-

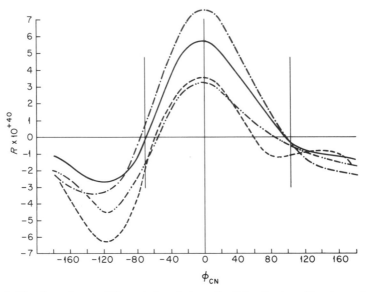

Fig. 4. The dependence of the rotational strength of the B_{2u} transition in pyrimidine nucleosides on the torsion angle, ϕ_{CN}, computed from the theory of Miles *et al.* [91]. The vertical lines at $-75°$ and $105°$ represent the angles of sign change of the rotational strength in the empirical correlation of Rogers and Ulbricht [96]. (——) Uridine 3'-*endo*; (– – –) Uridine 2'-*endo*; (– ·) cytidine 3'-*endo*; (– — ··) cytidine 2'-*endo*.

tween uridine and deoxyuridine. They propose a difference in the furanose ring-puckering for the ribose (3'-*endo*) and deoxy (2'-*endo*) which could explain the experimental observation.

Miles *et al.* [97] have also published calculations and experiments on guanosine derivatives. In this case their theoretical results again support a bisector rule for the longest wavelength π–π* transition. Their experiments on guanosine derivatives substituted in the 8 position so as to force a *syn* conformation agree with this rule. Their rule is also similar to that proposed for purine nucleosides by Rogers and Ulbricht [99]. Miles *et al.* [97] also quote calculated results for xanthosine, but in this case their figure implies a quadrant rule for the behavior of the longest wavelength CD band.

A second method for computing coupled oscillator effects in the $\pi-\pi^*$ CD bands of nucleosides has been introduced by Teng *et al.* [92]. This method assigns the $\pi-\pi^*$ chromophore transition moment to the center of the base. This transition moment interacts with all the bond transition moments of the sugar according to the usual scalar triple product. The perturbation energy in this calculation is computed from a transition monopole assigned to all the atoms of the chromophore. The transition monopoles interact with all the bond transition moments of the sugar. The far ultraviolet transition moments of the bonds of the sugar are approximately calculated from the bond polarizabilities.

The calculation of Teng *et al.* [92] includes Cotton effects of two different $\pi-\pi^*$ transitions on each base. This is an improvement over the calculation of Miles *et al.* [91] which yields the rotational strength of only one $\pi-\pi^*$ transition for each base. The results of Teng *et al.* [92] for pyrimidines are similar to those of Miles *et al.* [91]; a bisector rule is predicted for the longest wavelength $\pi-\pi^*$ transition. On the other hand, Teng *et al.* [92] find at least one case of a predicted quadrant rule for CD bands in pyrimidines and additional examples for the purines. For the case of purines, the complexities of the spectra again make interpretation difficult. Teng *et al.* [92] conclude that their theoretical predictions are not followed in the case of guanosine.

Both these coupled oscillator methods calculate only the effects of $\pi-\pi^*$ transitions. They ignore the effects of $n-\pi^*$ transitions and other one electron effects. Experiments on the polarized absorption spectra of crystals of nucleotide bases have consistently suggested the importance of $n-\pi^*$ transitions (see Section II,A above).

The contribution of $n-\pi^*$ bands to the CD of nucleosides remains to be fully documented. Assignments of $n-\pi^*$ bands have been suggested in cytosine as well as in other nucleosides [55]. The results of Ingwall on adenosine penta-furanosides strongly suggest the presence of $n-\pi^*$ bands in the 225 nm and the 270–290 nm region. In polynucleotides, CD bands of $n-\pi^*$ origin have been assigned in poly dA and in poly G as well as in other cases [5,26].

A recent calculation in our laboratory used a method for calculating the CD of adenosine which was substantially different from the coupled oscillator methods discussed above [94e]. The rotational strengths were calculated directly from the wave functions of the entire nucleoside. Since the extended Hueckel wave functions used in the calculation had both σ and π electrons in them, the rotational strengths contain the effects of coupled oscillators, $n-\pi^*$ bands, and other one electron effects in an unbiased way. The calculations were carried out for α and β anomeric configurations of deoxyadenosine as a function of the sugar–base torsion angle, ϕ_{CN}.

Unfortunately the requirement for the wave function of the entire nucleoside made it impossible to use very sophisticated semiempirical functions. In order to get good agreement between experimental spectra and semiempirical wave functions, it is usually necessary to introduce a configuration interaction calculation. Since a CI calculation was not carried out, the agreement between the calculated transitions and the experimental spectra of adenosine is not especially good [94e]. Nevertheless, we could draw a plausible correlation between the four longest wavelength CD bands in the data of adenosine pentafuranosides [94b] and the four longest wavelength bands in our calculation.

Two qualitative results of this calculation are quite significant. First, we find that $n–\pi^*$ CD bands are calculated to have substantial rotational strengths. They are about as strong as the $\pi–\pi^*$ bands. Second, we find a quadrant rule behavior for the signs of all our CD bands as a function of torsion angle. We have also carried the calculation for guanosine and the results are similar to those for adenosine.

Although our results disagree with the coupled oscillator calculation of Miles et al. [91] on the importance of one electron effects and $n–\pi^*$ bands, we do agree on the major geometric features which determine the sign of CD; the position of the furanose ring relative to the base, as given by ϕ_{CN}, is the main determinant. The furanose pucker and substituents have a lesser effect. Their main role is to determine the torsion angle by steric interaction.

2. Experimental Studies on Pyrimidines

We will not attempt here to give a complete review of all the experimental studies on the optical activity of nucleosides and nucleotides. Rather, we will concentrate on the more recent studies and on those which we feel illustrate the basic principles. For more thorough reviews, see Ts'o [47] and Yang and Samejima [93].

A major use of the optical activity of nucleosides has been in discriminating the anomeric configuration at C-1'. Emerson et al. [94] have proposed that pyrimidine β-nucleosides have positive Cotton effects, while α-pyrimidine nucleosides have negative Cotton effects. The opposite situation holds in the purine nucleoside case, where the β-anomer has a negative Cotton effect and the α-anomers have positive Cotton effects. This empirical rule has been abundantly verified by subsequent experiments [95]. However, its true basis probably rests on the preference of the two anomers for distinct ranges of the torsion angle, ϕ_{CN}. Miles et al. [91] have been able to prepare anomers of pyrimidine nucleosides having sufficient steric constraint to force the torsion angle beyond its normal range. These compounds violate the anomeric rule of Emerson et al. [94] but follow the correlations of rotational strengths vs torsion angle proposed by Miles et al. [91].

Rogers and Ulbricht [96] have recently proposed an empirical rule relating the sign of the long wavelength Cotton effect with the torsion angle in pyrimidine nucleosides. They predict positive Cotton effects for torsion angles, ϕ_{CN} between 0° and −75°, and 0° and 105°. In Fig. 4 we have marked the angles −75° and 105° at which this rule predicts the Cotton effect to change sign. We observe good agreement between this empirical rule and the calculations of Miles *et al.* [91].

Although early studies tended to ignore the slight differences between deoxy and ribose nucleosides, it is clear that differences in magnitude do occur. The differences are small and have no obvious systematic variation [93]. The effect of the 2'-OH as predicted by the theory of Miles *et al.* [91] would be expected to lead to consistently smaller Cotton effects for the β-anomeric ribose compounds at least in the pyrimidines. This prediction is not born out by the observed differences between uridine and deoxyuridine [91].

3. *Anomeric Linkage in Nucleosides*

Although nucleosides in RNA and DNA quite generally have the β,D-ribosyl configuration, recent interest in synthetic nucleoside analogs for pharmocological purposes has stimulated an outburst of creativity on the part of synthetic organic chemists in the area of nucleoside analogs. Since synthetic nucleosides may have either α- or β-glycosyl linkage, the use of CD as an analytical tool for determining this linkage has assumed some importance. An anomeric configuration is not easily determined analytically. Although NMR can be used to determine the linkage, the large sample required for NMR may pose a problem. The CD measurement requires only about 0.1 mg of sample and the measurement is often faster and easier than NMR.

The original proposal for correlating CD and anomeric configuration was proposed by Ulbricht and co-workers [94]. This rule forms a useful starting point for the discussion. It states simply that for pyrimidines, the β,D-nucleoside should have positive Cotton effects in the 260 nm region while the α,D-pyrimidine nucleosides have negative Cotton effects in the 260 nm region. This rule should be applicable to cytosine uracil and thymine nucleosides since they all have similar chromophoric properties (see Section II,A). The rule should apply to nucleosides of other pyrimidine bases so long as the substituents on the ring are not so polarizable as to change substantially the transition energies or the directions of the transition moments. Likewise, this rule should hold for any sugar pentose or hexose. Probably some caution should be exercised in applying the rule to sugars which have chromophoric properties themselves. Groups such as amides or carbon–carbon double bonds could couple to the base in an unusual way causing the rule to break down.

For purine nucleosides there are certain complications but the rule simply stated is that β,D-purine nucleosides have negative Cotton effects in the 260 nm region while the α,D-anomers have the opposite sign [94a]. This rule is apparently valid for adenosine and guanosine derivatives at 260 nm but has not been adequately documented in inosine which has Cotton effects at wavelengths longer than the main absorption band at 248 nm [95].

One of the difficulties in applying simple anomeric rules in purine nucleosides arises from the apparent complexity of the electronic absorption and CD spectra. A careful study of adenine pentafuranosides which covered a wide range of wavelengths reveals the nature of the problem. Ingwall [94b] finds that the 260 nm region does indeed follow the rule proposed by Ulbricht [94a]. On the other hand, Ingwall's results on a series of anomeric pentafuranosides show changes in the CD band shape in the 260 nm region which imply that there are several distinct electronic transitions between 250 and 280 nm. In addition, she finds that the bands in the 225–200 nm region do not always change sign as a result of inversion of the anomeric configuration.

We feel that there is chemical and geometric information to be gleaned from accurate measurements of the CD of nucleosides recorded over an extended range of wavelengths. While such information is not easily extracted, we will outline some possible schemes that one can use to attack this problem. It is instructive to explore the reason underlying the sign change in CD between α and β anomeric pairs of nucleosides. If we ignore the 2'- and 3'-OH and the 4'-CH$_2$OH of the ribose, then the only remaining asymmetry is at C-1', and the α and β anomers become topological enantiomers. The α anomer is the geometric mirror image of the β anomer whose sugar–base torsion angle, ϕ_{CN}, is the opposite sign.*

Recent theoretical work of Miles *et al.* [91] and Bush [94e] has shown that the ribo substituents 2'-OH, 2'-OH and 4'-CH$_2$OH have a minor influence on the CD of nucleosides. Therefore we interpret the deviation from mirror image CD curves between α and β anomeric pairs found by Ingwall [94b] as indicating that the substituents influence the torsion angle; the α and β anomers do not always have exactly opposite sign of the torsion angle. This discussion now brings us to the important question of the dependence of the nucleoside CD bands on the sugar–base torsion angle, ϕ_{CN}.

---

* The sugar–base torsion angle that defines the relative orientation between the ribose and the base was first defined by Donohue and Trueblood (94c). For the purposes of the present discussion, this angle may be thought of as the dihedral angle defined by the C-8, N-9, C-1', O-1' bonds in a purine and by the C-6, N-1, C-1', O-1' bonds in pyrimidine. We retain the sign convention of Donohue and Trueblood [94c] in which left-handed screw rotation of the C-6, N-1 bond into the C-1', O-1' bond implies a positive torsion angle. This definition of ϕ_{CN} is similar to the definition of the torsion angle χ of Sundaralingam [94d] except that the sign convention is opposite.

4. Dependence of Cotton Effects on Torsion Angle

As we have mentioned, considerable effort has been invested in determining conformations of nucleosides. We will not attempt to review this work but will mention the comprehensive crystallographic study of Sundaralingam [94d] and the calculation of nonbonded energies by Wilson and Rahman [94f].

There are two major features that determine the conformation of an isolated nucleoside. First, is the sugar–base torsion angle ϕ_{CN} which we have already discussed. A second feature is the puckering of the ribofuranose. This five membered ring is highly restricted in the number of low energy conformations available to it. The five atoms are rarely found to be co-planar and one generally finds four atoms to be nearly co-planar with one atom of the furanose deviating to one side or the other of this plane. Although more detailed subclassifications exist, most nucleosides can be classified in either one of two groups of furanose puckering conformation. The classes are known as 2'-*endo* and 3'-*endo*. These two classes are defined by four atom planes from which the 2'-carbon, respectively, the C-3' deviates from the average plane of the other four furanose atoms in the direction of C-5'. Such a simplified classification of furanose puckering could hardly satisfy a crystallographer but will serve as a basis for our present discussion.

Although it is clear that other geometric features of nucleosides such as dihedral angles about the 3'-C—O and 5'-C—O bonds become important in consideration of the geometry of polynucleotides, the pucker and the torsion angle are the most significant features of the conformation of isolated nucleosides. Moreover, the non-bonded energy calculations show that these two geometric features are not independent. The plots of conformational energy as a function of ϕ_{ON} given by Wilson and Rahman [94f] differ for a given nucleoside between 3'-*endo* and 2'-*endo* sugar puckering.

Donohue and Trueblood [94c] introduced the simplifying suggestion that the torsion angle is restricted two narrow regions called *syn* and *anti*. They suggested that for ϕ_{CN} other than $-30°$ (*anti*) or $+150°$ (*syn*) the steric interaction energy would be high. While nucleoside conformation is usually discussed in this simple binary notation, we will discover below that the resulting description may be an oversimplification. Nevertheless, we must begin our discussion of ϕ_{CN} and Cotton effects in these terms.

Rogers and Ulbricht [96] proposed a rule intended to determine whether a pyrimidine β nucleoside is in the *syn* or *anti* conformation. This rule is based on empirical correlations with the CD of cyclonucleosides whose conformation was inferred from the chemical structure that presumably holds them rigid. The rule says that *anti* nucleosides have positive Cotton effects in the 260 nm region while *syn* pyrimidine nucleosides have negative Cotton effects. In fact, this rule may be phrased as a bisector rule declaring that pyrimidine nucleosides with ϕ_{CN} ranging from $-75°$ up to $+105°$ have positive Cotton effects while torsion angles outside this range lead to negative Cotton effects (see Fig. 4). The reader will recognize that consistency of this rule with the anomeric rule requires that β pyrimidine nucleosides be in the *anti* conformation. Although this is generally believed to be the case, we will show that it may be an oversimplification and we will present a conflicting point of view below.

Rogers and Ulbricht [99] have also proposed a similar rule to be applied to purine nucleosides. It states that negative values of ϕ_{CN} give negative Cotton effects while positive values of ϕ_{CN} give positive CD in the 260 nm region. This rule, like the pyrimidine rule, is a bisector rule. The complexity of the electronic spectra of purines has already been mentioned. This fact makes the application of the rule for purines ambiguous. The group headed by Miles and Eyring has also published studies that include both experimental work on model compounds and theoretical calculations based on coupled oscillator theory [91]. We have discussed their theoretical work above and the experimental conclusions we discuss here are based on these calculations. Their experimental interpretation is based on the assumption that ordinary nucleosides are in an *anti* conformation. From their theoretical work, the Miles and Eyring group finds a bisector rule for the sign of the 260 nm Cotton effect in pyrimidine nucleosides. Although it is satisfying that this rule is similar to the empirical rule of Rogers and Ulbricht [96], we will point below certain other results which disagree with the bisector rules.

Miles *et al.* [97] have also studied the question of the dependence of Cotton effects on torsion angle for purine nucleosides, but the complexity of the purine spectra make a simple interpretation difficult. For guanosine, they find a simple bisector rule which agrees with the rule of Rogers and Ulbricht [99]. For the case of xanthosine, Miles *et al.* [97] predict a quadrant rule rather than a simple bisector rule. The differences between the prediction of quadrant and bisector rules can be very significant and we will discuss the problem further below.

A somewhat different point of view on nucleoside conformation is presented by the work of Hart and Davis [98]. These workers use an NMR technique known as nuclear Overhauser effect (NOE) which involves a kind of spin decoupling, a complete discussion of which is beyond the scope of the present chapter. Hart and Davis [98,98a] have proposed that the *syn* conformation is found in ordinary pyrimidine nucleosides in water solution with substantial probability. This interpretation is not consistent with the rules of Ulbricht and of Miles and Eyring who assign *anti* conformations to ordinary nucleosides.

Using an infrared technique, Pitha [99a] has been able to detect hydrogen bonding in nucleosides. Using nucleoside derivatives with blocking groups on the sugar, Pitha has been able to restrict the possibilities for intramolecular hydrogen bonding, thus concentrating attention on intramolecular hydrogen bonds which stabilize the nucleoside in a particular conformation. Specifically, Pitha studied the 2',3'-isopropylidene nucleosides in nonpolar solvents. In these experiments only the 5'-OH is free to hydrogen bond and when it forms an intramolecular hydrogen bond to the 0-2 of pyrimidines or to the N-3 of purines, the nucleoside is held in a *syn* conformation. Unfortunately, physical measurements on 2', 3'-ipA and 2',3'-ipU in nonpolar solvents are complicated as a result of low solubility and aggregation. Nevertheless, Pitha has given a convincing argument that 5'-OH intramolecular hydrogen bonds are formed to the base in carbon tetrachloride and chloroform solutions [99a].

These observations suggest that the CD of 2',3'-isopropylidene nucleosides in aprotic solvents should be that of the nucleoside in the *syn* conformation. Therefore, Hart and Davis [98a] measured the CD of 2',3'-ipC and 2',3'-ipU in 1,2-dichloroethane (DCE). They find some aggregation of the blocked nucleoside even at the relatively low concentrations possible in ultraviolet CD experiments. Nevertheless the low concentration CD curves for both ipC and ipU appear to have positive Cotton effects. This result is in disagreement with the bisector rules of Miles and Eyring [91] and Rogers and Ulbricht [96].

The apparent contradictions among the results we have discussed above may in fact result from an attempt to oversimplify the problem. Careful study of nucleoside conformation by nonbonded interaction energies indicates that a simple binary division of nucleoside conformations into *syn* and *anti* may not be adequate. The crystallographic data can be interpreted as an indication that there are at least two ranges of *syn* conformation near $\phi_{CN} = 140°$ and $\phi_{CN} = -120°$ [99b]. Conformational calculations indicate that the *anti* range could extend to ϕ_{CN} as high as $+20°$ [94f]. The *anti* range must extend as low as $\phi_{CN} = -85°$, the value in B form DNA. Moreover, we point out in the survey of theoretical work that a bisector rule for the sign of CD bands may be incorrect. There are some indications that a quadrant rule may in fact be operating.

In order to glean enough information from CD spectra to specify the torsion in greater detail than simply to say it is *syn* or *anti*, we recommend some extensions to the CD theory and experiment. We suggest that advantage be taken of the complexity of the spectra by extending the range of wavelength under study as far as possible to short wavelengths in the ultraviolet. Present instruments work well to 190 nm and the range can certainly be extended to 165 nm by using a CD machine of modern design [44b]. By taking accurate data over a wide range of wavelengths, one can resolve a number of CD bands. This gives one several distinct Cotton effects which can be correlated with the torsion angle. This approach maximizes the information context of the CD spectrum and gives one some hope of being able to determine geometry.

B. OLIGONUCLEOTIDES

1. *Oligomers Containing A Single Base*

Simple stacking interaction, not involving base pairs, can be recognized in dinucleoside phosphates. In ApA (Fig. 3) a double Cotton effect is observed which can be explained by the coupled oscillator effects discussed in Eqs. (23)–(33) [30]. This simple theory which assumes stacked bases in a right-handed helical stack having DNA geometry, approximately agrees with experiment. Longer oligomers of A have a CD curve similar to that of ApA [75]. The monomer lacks any neighbor to couple with and has a weak Cotton effect qualitatively different from that of oligomers. The similarity among oligomers indicates that the relative position of the bases in oligonucleotides is the same as in the dimer. The CD curves of oligo A have been computed from nearest-neighbor formulas by Brahms *et al.* [75].

Thermal melting of the stacked bases has been observed in CD [75], ORD [76], and hypochromicity [73]. The melting curves of the oligomers and the polymer are similar. This implies a noncooperative stacking. The enthalpy of stacking has been estimated to be -6 to -10 kcal per stack. The ORD of the polymer is slightly greater than one would expect on the basis of a non-cooperative nearest-neighbor interaction. Poland *et al.* [76] interpret this result as a thermodynamic cooperativity of small degree. However, it could also result from a non-nearest-neighbor optical interaction. The likelihood of the latter interpretation is illustrated by the slight differences in the CD curve shape between poly A and the dinucleoside phosphate (Fig. 3) [5].

The dinucleoside phosphate, CpC (Fig. 5), has a single Cotton effect in the near-ultraviolet. The absence of a paired negative band can be understood by the addition of an interaction with far-ultraviolet bands to the coupled oscillator effect [33]. The single Cotton effect probably reflects differences in the transition moment directions rather than differences in the dinucleoside geometry between ApA and CpC. Recent calculations of Johnson and Tinoco [62] on dinucleosides including both single and double Cotton effects successfully interpret the CD of CpC.

The thermal melting behavior of oligomers of cytidylic acid have been studied by Brahms *et al.* [100]. The situation seems to be quite similar to that in oligomers of adenylic acid. Mainly nearest-neighbor optical interactions and a noncooperative melting behavior are observed.

2. *Dinucleosides of Differing Bases*

The most complete theory of dinucleoside circular dichroism is that of Johnson and Tinoco [62] which calculates the $\pi–\pi^*$ Cotton effects of single nonconservative bands as well as that of conservative bands or double

Cotton effects. Their results are in reasonable agreement with experiment for most cases but show discrepancies with certain chromophores. These discrepancies could result either from an inadequate knowledge of the $\pi-\pi^*$ transitions or the neglect of $n-\pi^*$ transitions.

ORD has been used in conjunction with hypochromism to estimate the degree of base-stacking in various dinucleoside phosphates [17,101]. Warshaw and Tinoco [17] have measured the ORD and hypochromism of all 16 dinucleosides composed of the normal RNA bases. They have classified their

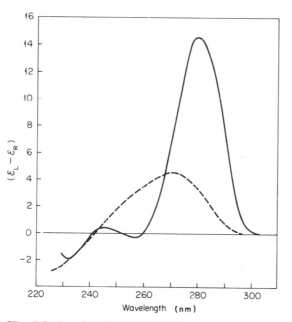

Fig. 5. The CD of CpC and CMP in 4.7 *M* KF at neutral pH, −17°C. CpC (———); and CMP (– – –) [33].

results according to the tendency of the bases to form stacked structures. In their classification, at neutral pH A, G, and C stack with their neighbors, while U does not. Although this classification is rather arbitrary, it has been useful in understanding polynucleotide behavior. Since the method is based on the magnitudes of Cotton effects and hypochromism at room temperature, it has the weakness that it includes any differences in Cotton effects or hypochromism among the different fully stacked dinucleosides.

In an attempt to reach conclusions about the stacking of dinucleosides which would be fully thermodynamic in nature, Brahms *et al.* [102] have measured the CD of a number of dinucleosides as a function of temperature.

By assuming a simple two-state melting, they calculated van't Hoff ΔH for the stacking. This method assumes a knowledge of the CD of the fully stacked compounds. These investigators used measurements at low temperatures ($-20°$C) in 4.7 M KF solutions to determine the latter parameters. This method of classification leads to the conclusion that G tends not to stack [103]. This method is also open to some criticism due to the assumptions of two-state melting and fully stacked circular dichroism. In view of the probable complexity of the system, it is not too surprising that the methods of Brahms *et al.* [103] and Warshaw and Tinoco [17] lead to different conclusions.

In order to answer the important question of the optical properties of the fully stacked dinucleoside, measurements have been carried out at low temperatures in solutions of rather high salt concentration. It has been claimed that the effect of added salts on the optical properties is negligible. Bush and Brahms [33] showed that the CD of poly C in 4.7 M KF is similar to that in modest salt strength solutions at neutral pH. Davis and Tinoco [104] have compared CpC and ApA in low salt and in 6 M LiCl and found the differences to be negligible. On the other hand, it is clear that the CD of GpC in 4.7 M KF is quite different in shape from that in lower salt strength solutions [105]. This fact illustrates the difficulty of making general statements about dinucleosides based on experiments on a few compounds.

Warshaw and Tinoco [17] have also investigated the effect of pH on the ORD of dinucleosides. Of course, ionization of a chromophoric base should change the dimer optical properties as a result of the ionization of the chromophore. The ionized base is a different chromophore from the neutral species. On the other hand, they were able to classify their data reasonably well on the assumption that the predominant effect of ionizing both bases is to cause unstacking of the dinucleoside by charge repulsion. This hypothesis explains the data on most dimers, but certain apparently doubly charged dimers show large optical activity [17].

3. *Longer Oligonucleotides of Differing Bases*

The fact that the thermodynamic interaction between stacked bases is mainly independent and pairwise, while the optical interaction is mainly nearest-neighbor, implies that one need only know the ORD of dimers in order to predict the ORD of longer single-stranded oligomers. This hypothesis was first tested on a series of trinucleoside diphosphates by Cantor and Tinoco [72]. Their empirical equation assumes that a trimer rotation receives contributions from its component monomers plus a contribution from each of its two nearest-neighbor interactions as evidenced in the ORD curves of dinucleoside phosphates. For the trinucleoside IpJpK, the rotation per residue is

$$[\Phi_{IJK}(\lambda)] = \frac{2[\Phi_{ij}(\lambda)] + 2[\Phi_{jK}(\lambda)] - [\Phi_{j}(\lambda)]}{3} \tag{38}$$

In this equation $[\Phi_{ij}(\lambda)]$ is the molar rotation per residue of the dinucleoside phosphate, IpJ, and $[\Phi_{j}(\lambda)]$ is that of the mononucleoside J. Within this approximation, Cantor and Tinoco [72] were able to represent the rotation of a trinucleoside diphosphate rather well, especially in the longer wavelength region of the Cotton effect spectrum. Their results are illustrated in Fig. 6. Their success indicates that in order to interpret the ORD of a single-stranded ribonucleotide, one need know only the 16 ORD curves of the component dinucleoside phosphates and that of the 4 mononucleosides. These curves have been catalogued by Warshaw [106]. While this method was originally intended for ORD, it works equally well in CD as we will see below [107–109].

All the dinucleoside and trinucleoside data available indicate considerable sequence dependence in both CD and ORD. This sequence dependence is not

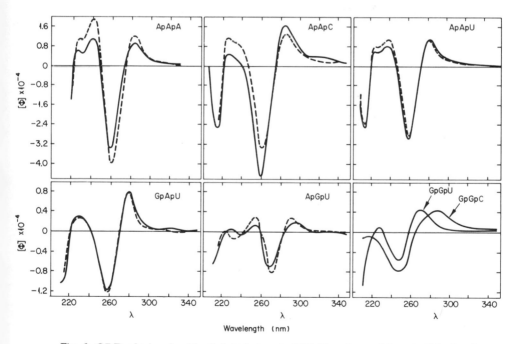

Fig. 6. ORD of trinucleoside diphosphates at pH 7. Experimental (——). Calculated from Eq. (38) (– – –) [72].

evident in the absorption properties. Cantor and Tinoco [72,110] have discussed the promises and problems inherent in using ORD to determine the sequence of oligomers. The measurements are relatively rapid and are easily amenable to automation. The ORD or CD measurement requires approximately 0.03 mg of sample, but it can be recovered undegraded after the experiment. There are a large number of trimers which can be successfully sequenced by a single ORD measurement. In certain favorable cases, longer oligomer sequences can also be identified [110].

The difference between a dinucleoside phosphate and a dinucleotide is just a terminal phosphate. The effect of this terminal phosphate on the optical activity seems to be variable. The ORD of the trinucleotide ApApCp [111] is quite similar to that of the trinucleoside diphosphate measured by Cantor and Tinoco [72] (Fig. 6). Cantor *et al.* [107] tested the effect of terminal phosphates in both the 3'- and 5'-position in oligodeoxythymidylic acids. Their results indicated that the 3'-phosphate has very little influence, and in their system the 5'-phosphate had little influence also. On the other hand, Inoue *et al.* [112] found some disagreement between their ORD data on trinucleotides and nearest-neighbor calculations. They ascribed these discrepancies to effects of the 3' terminal phosphate. Moreover, it is clear that the CD of ApAp is weaker than that of ApA [5]. Also the presence of the 5'-phosphate in pdApdA has an ORD different from that of dApdA [113]. These facts illustrate the characteristic difficulty of generalizing results found on a few oligonucleotides to all oligonucleotides.

4. *Thermal Melting*

We have discussed above the experiments of Brahms and co-workers [102,103]. They studied the thermal melting of various oligomers, fitting their results to a transition between stacked and unstacked states for the nearest neighbors. This model would indicate that the thermal melting in hypochromicity should be identical to the melting curve in optical activity. Davis and Tinoco [104] point out that these two experiments do not always give identical melting curves. These data as well as simple consideration of models indicate that the two-state model for melting is probably an oversimplification. Glaubiger *et al.* [36] have suggested a model in which the bases oscillate on a torsional spring. This model, while not able to match all the data of Davis and Tinoco [104], does have the feature of differing hyperchromicity and ORD melting curves. This feature results from the fact that the sign of the ORD differs for left- and right-handed stacks, while both have equal hypochromism. Any model which mixes right- and left-handed helical arrangements can give different hypochromism and optical activity melting curves.

5. *Double Strands or Aggregates*

In general, short oligonucleotides are good models for single-stranded polymers. Most do not aggregate under neutral solution conditions. However, certain guanine containing trinucleotides aggregate even at low concentrations such as those used in ultraviolet optical studies. Jaskunas *et al.* [114] attempted to find interaction between Watson–Crick complementary triplets of trinucleoside diphosphates. Their results were essentially negative, but they did find specific aggregation between the complementary trinucleoside diphosphates GpGpC and GpCpC. From mixing plots they determined that the complex is most likely three-stranded. Moreover, the process is complicated by self-aggregation of GGC as well as GCC. Likewise, Brahms *et al.* [103] also identified self-aggregation in GpGpC as well as in GpGpU. Both research groups find that the aggregation is promoted by low-temperature, high-salt concentration, and high-oligomer concentrations.

C. POLYNUCLEOTIDES

1. *Single Strands*

Most of the single-stranded polynucleotides which have been studied are homopolymers. Such a single strand may be considered as simply an extension of an oligomer. In many cases, the geometric relationship of adjacent bases is the same in polymers as that in dinucleosides and no cooperative effects are seen.

Thus, Cantor *et al.* [115] were able to approximately calculate the ORD of poly C, poly U, and poly A neutral single strands from the ORD of CpC, UpU, and ApA. If only nearest-neighbor interactions are considered, the polymer optical activity should be just twice that of the dinucleoside phosphate. Referring to Fig. 3 we see that the CD of poly A is approximately twice that of ApA. The magnitude of the 260 nm double Cotton effect in the polymer is slightly larger than twice that in the dimer. Also, there is a slight red shift of the double Cotton effect due to an exciton effect [72,116]. Also there are slight discrepancies in the 230 nm region and also in the 278 nm region where a shoulder appears in the polymer but not in the dimer. This latter effect has been discussed by Bush and Scheraga [5] and assigned to an $n–\pi^*$ transition. In both poly C and poly U, Cantor *et al.* [115] found somewhat better agreement between the ORD measured and that calculated from the ORD of dinucleosides.

The approach of Cantor *et al.* [115] has been criticized by Inoue *et al.* [108] who used dimer ORD curves extracted from the ORD curves of longer oligomers. They correctly claim that the molar rotation of the tetranucleotide AAAC minus the molar rotation of the dinucleotide ApCp leads to a better

ORD curve for ApAp for model purposes. This difference curve is closely related to the ORD curve of poly A. Their improved agreement results from the presence of some exciton and farther neighbor interactions in the tetra-nucleotide AAAC which mimics those interactions in poly A. In fact, Gray and Tinoco [109] have formalized this "polymer" approach to nearest-neighbor interactions. They compare the ORD or CD of the polymer to that of nearest-neighbor interactions in model polynucleotides having defined sequences. The reason that Cantor and Tinoco's [72] dimer approach is most widely used is that all the dinucleoside data are available, while there are not sufficient optical data available on longer oligonucleotides or sequenced polynucleotides to serve as a basis for the theory of Gray and Tinoco[109].

Any failure of the nearest-neighbor theory to correctly predict polymer CD or ORD could be ascribed either to interactions among farther neighbors or to geometric differences between the polymer and dimer. The importance of far-neighbor interactions is probably not greater than 10–20% in the magnitude of a Cotton effect [26,117]. The discrepancies in curve shape between poly A and ApA (Fig. 3) are not large and it is possible that they may result from non-nearest-neighbor effects of unknown origin. However, we will see below some examples of much stronger deviation from nearest-neighbor behavior which are clearly indicative of specific geometric differences between the polymer and its corresponding dinucleosides.

2. *Multiple Strands of Homopolymers*

There are several examples of multistranded polymer structures composed of identical homopolymers which have been extensively studied by optical acitivity. The relevance of these structures to biological systems remains unclear, especially since the structures which have been most extensively studied occur in acidic solutions.

Polyadenylic acid forms a stable double helix in acidic solutions from which fibers may be prepared. A structure involving parallel strands held by salt bridges between the phosphate and the protonated adenine ring has been proposed on the basis of x-ray scattering from fibers [118]. Subsequent studies on the ultraviolet absorption and optical activity of acid poly A have revealed at least two distinctly different helix geometries. These forms of acid poly A can be interconverted by changes in salt strength and pH [77,119]. These two structures for acid poly A can be easily distinguished by optical activity (Fig. 7),[120] but we have no detailed information on the exact nature of the geometric difference detected by optical activity. The structure of acid poly A determined by x-ray scattering has an appreciable tilt of the base planes from the helix axis [118]. It has been suggested that the structures in solution may differ in the amount of base-tilting [119].

The acid poly A structure is formed cooperatively as a result of its double-stranded nature. Hence, melting curves in hypochromicity or optical activity show a sharp melting point. Also, there is a critical cooperative length necessary for the formation of the structure. In studies of the CD of oligomers of varying chain length, Brahms *et al.* [75] observed that only oligomers of length 7 residues or more were able to form the acid double helix.

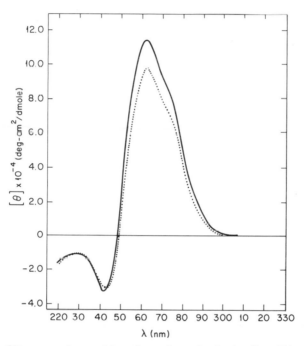

Fig. 7. The CD curves of two acid conformations of polyadenylic acid in similar states of protonation at 27°C. $R = [\theta]_{262}/[\theta]_{290}$. (——) 0.15 M KCl, pH 4.1, $R = 9.8$; ($\cdots$) 0.001 M KCl, pH 4.0, $R = 13.8$. The extent of protonation of adenine residues, determined by titration of these solutions is 94% in the former solvent, 92% in the latter [120].

Poly C may also be protonated to form a double helix. The structure of fibers has been determined by x-ray scattering and the solution structure is assumed to be similar to it [121]. The CD of the neutral single strand and the acid double strand differ appreciably (see Fig. 8). It should be noted that the chromophoric properties of a protonated base differ from those of the neutral chromophore. Hence, differences between the CD of the acid double-stranded helix and the neutral single-stranded form arise both from geometric and chromophoric differences. The acid poly C helix, like the poly A double helix is cooperative and thus shows a sharp melting transition. In addition, Brahms

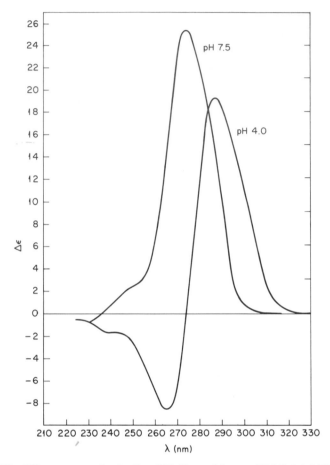

Fig. 8. The CD spectrum of poly C at 0°C. Neutral form, pH 7.5, 0.1 *M* KF, 0.01 *M* tris. Acid form, pH 4.0, 0.1 *M* NaCl, 0.05 *M* acetate [100].

et al. [100] show that a critical length of 7 residues is necessary to form the double helix between cytidylic acid oligomers.

Poly U is generally considered to be unstacked at room temperature; its hypochromism and optical activity are relatively small. However, at lower temperatures, it shows a rather large Cotton effect as may be seen in the ORD (Fig. 9). There is a sharp melting point in the ORD near 10°C implying that the structure is probably multistranded [122]. The low stability of this structure has discouraged further study of its geometric nature.

Guanine residues are particularly prone to aggregation even in short oligonucleotides and the mononucleotide [123]. Poly G is apparently known

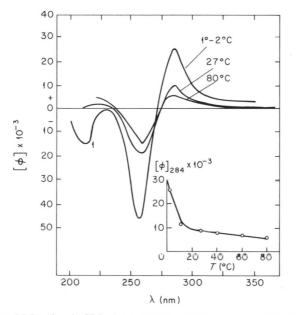

Fig. 9. The ORD of poly U in 0.15 *M* KF, pH 7.5 at several temperatures. Inset: dependence of the mean residue rotation at 284 nm on temperature [122].

only in aggregates in solution. These extremely stable aggregates are not disrupted either by ethyleneglycol or by slightly acidic pH [124,125]. We do not know either the geometry or the number of strands in this complex so it is difficult to interpret the CD (Fig. 10). The shoulder in the 295 nm region does not correspond to any strong absorption band and probably results from an *n–π** transition [26,124].

Although inosine does not occur naturally in most RNA's, poly I shows

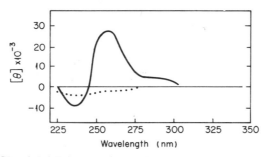

Fig. 10. The CD of GMP (· · ·) and poly G (——) in 0.05 *M* KF, 0.001 *M* EDTA pH 5.5 at 27°C [124].

some especially puzzling properties in the optical activity. There is apparently a transition between a multistranded and a single-stranded helix stimulated by changes in salt strength [126]. In Fig. 11 we compare the ORD of poly I in low-salt concentration (0.1 M) with that in high-salt solutions (1 M). The melting curve for the low-salt form is broad between 25° and 85°C, while the high-salt form melts sharply at 42°C. The ORD curves for the two structures are similar and they both differ in a basic way from that of most other polynucleotides. An ORD curve such as that in Fig. 11a represents a double Cotton effect having negative CD at long wavelength (compare Fig. 11a to Figs. 7–10). Theoretical considerations imply that for π–π* transitions such a pattern should be seen for left-handed helices of single-stranded polymers of identical chromophores (see Section I,E,1) [10]. All of the well-known polynucleotide structures are right-handed, but there remains the tantilizing possibility that some left-handed helices may occur in certain cases.

3. *Multiple Strands of Complementary Polynucleotides*

The study of such complementary double strands as poly A plus poly U and poly G plus poly C has been quite extensive because of their obvious relevance to Watson–Crick base-pairing in RNA. However, the situation is not straightforward as may be seen from the study of poly A plus poly U. The ORD [122] and CD [127] of poly A · poly U as well as the triple-stranded poly A · 2 poly U show surprising similarity between the double-stranded and triple-stranded helices (see Fig. 12). Such a situation indicates that optical activity cannot be used to distinguish the number of strands in a complex. Thermal melting and mixing curves are usually used for these purposes. Brahms [127] has also reported the CD of a random copolymer, poly AU. Its CD (Fig. 12) is similar in shape to that of the homopolymer complex.

Poly G · poly C forms an extremely stable complex over a range of temperature and pH. The ORD [126] and CD [124] have been reported. It is probable that this complex is a true double-stranded polymer which should act as a model for GC pairing in RNA. The geometry of the helix is not known. The calculation by Johnson and Tinoco [62] of the circular dichroism of this polymer does not agree with experiment for either DNA double-strand geometry or RNA geometry. It is possible that this failure could result from some theoretical shortcoming such as incorrect assignment of the π–π* transitions in their calculation.

Inosine often serves as a model for guanosine in polymer systems and poly I · poly C apparently forms a stable double helix similar to poly G · poly C. Its ORD shows a double Cotton effect, whose longer wavelength component has positive rotational strength [126]. This result is in contrast to poly I which shows negative rotation at longer wavelength (cf. Fig. 11).

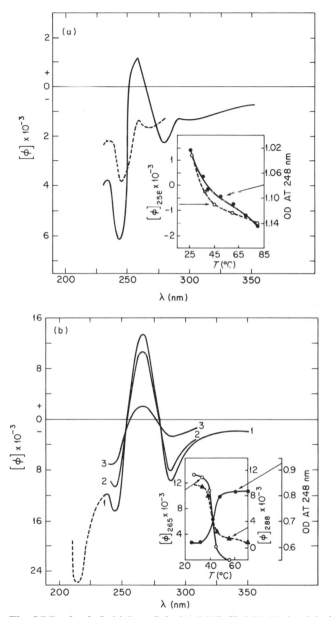

Fig. 11. The ORD of poly I. (a) Low Salt, 0.1 *M* NaCl, 0.01 *M* glycylglycine, pH 7.4 at 27°C (———) and at 80°C (– – –). Inset: dependence of mean residue rotation at 258 nm and the absorbence at 248 nm on temperature. (b) High-salt, 1 *M* NaCl, 0.01 *M* EDTA, pH 7.0. Curve 1, 27°C; curve 2, 39°C; curve 3, 45°C. Poly I in 1 *M* NaClO₄, at 27°C (– – –). Inset: dependence of mean residue rotation at 265 nm peak and at 288 nm trough, and the absorbence at 248 nm on temperature [126].

143

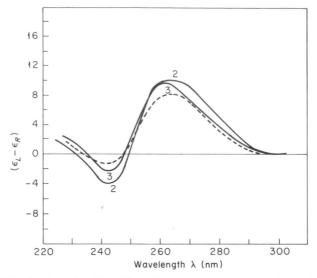

Fig. 12. CD of polynucleotides of A and U. Curve 1 (– – –), poly A·2 poly U in 0.02 M NaCl, 0.01 M MgCl$_2$, 0.01 M tris, pH 7.4 at 10°C. Curve 2, poly A·poly U in 0.005 M NaCl, 0.01 M tris, pH 7.4 at 15°C. Curve 3, poly AU random copolymer in 0.1 M NaCl, 0.1 M tris, pH 7.4 at 15°C [127].

4. *Generalizations about Polynucleotides*

In the formation of complementary double-stranded helices of the Watson–Crick type, we see blue shifts in the Cotton effects. Poly A·poly U as well as poly G·poly C when compared to their single-stranded components, exhibit blue shifts of several nanometers in both CD and ORD. Such a shift is not seen in the absorption bands [93].

Most of the Cotton effects in polynucleotides arise from $\pi-\pi^*$ transitions. The theory of Johnson and Tinoco [62] is able to interpret the Cotton effects in many polynucleotides and able to make accurate generalizations about them. Some of the discrepancies between their calculated curves and the experimental CD curves could be the result of incorrect assumptions about the polymer geometry or transition moments. On the other hand, certain Cotton effects, especially those at long wavelength, are difficult to explain by $\pi-\pi^*$ transitions. Cotton effects in poly A and poly dA at 278 nm and in poly G at 295 nm are difficult to reconcile with $\pi-\pi^*$ assignments. The nearest strong absorption bands to these CD bands are several nanometers to the blue. Bush and Scheraga [5] have demonstrated that bands resulting from exciton splitting of strong transitions in polynucleotides should lead to blue shifts; they do not lead to red shifts. Thus some workers have favored an $n-\pi^*$ assignment for these long wavelength Cotton effects [26].

D. Polydeoxynucleotides

Optical activity is generally quite capable of distinguishing polydeoxynucleotides from polyribonucleotides. In fact, DNA always shows CD curves which differ in a characteristic way from those of RNA [128]. This characteristic difference is also easily perceived in ORD curves and has aroused considerable interest [129]. The chromophoric nature of the DNA and RNA bases is essentially the same; T and U have very similar electronic spectra. These characteristic differences must be due to characteristic geometric differences arising from the influence of the 2'-H and 2'-OH. The optical activity of synthetic polynucleotides has been extensively studied in an attempt to illuminate the biologically significant question of the structural differences between DNA and RNA.

1. *Deoxydinucleosides*

The earliest studies on deoxydinucleosides were interpreted as indicating a weaker stacking interaction than in the ribo analogs. The main CD bands in dCpdC and dApdA in the 260 nm region are similar in shape but lower in magnitude than those of CpC and ApA [5,130]. The thermal melting also indicates a smaller value of ΔH for the deoxy than for the ribo dimers [38,131].

It appears, however, that the CpC and ApA cases may be exceptional. Cantor and Warshaw [38] have addressed themselves to the question of whether the difference in stacking between ribo- and deoxydinucleosides was one of geometric nature or simply one of degree. They have studied a wide variety of ribo- and deoxydinucleoside phosphates and concluded that there must be differences in geometry for a large number of analogous pairs. Their case is abundantly illustrated in Fig. 13 which compares the CD of the two ribo and two deoxy sequence isomeric dinucleosides of guanine and adenine. Whatever the detailed nature of the interaction between these chromophores, it appears to be quite sensitive to small geometric differences. The differences in relative magnitudes of the Cotton effects in Fig. 13 show that the geometry of stacking must differ in the ribo and deoxy series [38]. Cantor *et al.* [107] report that all the 16 dinucleosides show evidence of base–base interaction in their CD.

2. *Oligodeoxynucleotides*

Using the CD curves they measured for deoxydinucleosides, Cantor *et al.* [107] have attempted to construct CD curves for oligomers using the nearest-neighbor method found to be successful in ribonucleotides. They find that the method works in some but not all of the deoxytrinucleotides they tested.

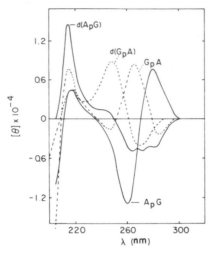

Fig. 13. CD of dinucleoside phosphates of adenine and guanine in 0.1 *M* NaClO₄, 0.01 *M* phosphate, pH 7.2 at 26°C [38].

They compared experimental CD curves for deoxytrinucleoside diphosphates and trinucleotides with those calculated from the CD of deoxydinucleosides by Eq. (38). The nearest-neighbor technique is clearly not as general for treating oligodeoxynucleotides as for oligoribonucleotides. This point is clear from the example of dApdA (Fig. 14) which has a CD curve qualitatively different in shape from that of deoxyoligomers. It is probable that there are

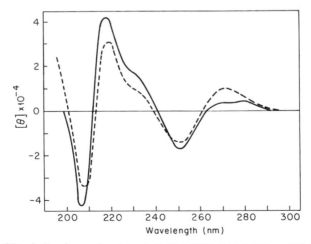

Fig. 14. CD of oligodeoxyadenylates in 0.05 *M* KF, 0.01 *M* tris, pH 7.4 at 23°C [15]. (——) (pdA)₆; (– – –) pdApdA.

significant differences in the geometry of dApdA and that of adjacent bases in poly dA.

3. *Polydeoxynucleotides*

The CD and ORD curves of poly dC are similar in curve shape to those of poly rC, dCpdC, and CpC (see Fig. 15). The deoxy compounds have bands of appreciably smaller magnitude [130]. One might conclude that the stacking interaction in the deoxy series is simply weaker than in the ribo series. This

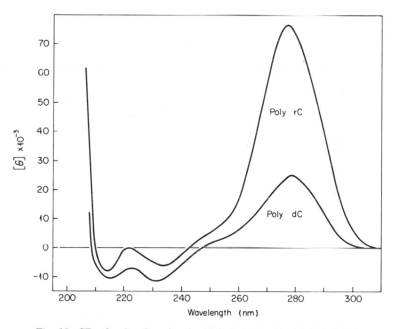

Fig. 15. CD of poly rC and poly dC in 2 *M* NaCl, pH 10.0 [130].

statement cannot be general, however, as the CD curve for poly dA is qualitatively different from that for poly rA (Fig. 16). In general, geometric differences in the stacking of the ribo and deoxy polymers must be invoked to explain the differences in circular dichroism.

Wells *et al.* [132] have studied the CD of polydeoxynucleotides of repeating defined sequences which form Watson–Crick complementary double strands. In specific, they have compared the CD of polymers having all pyrimidines on one strand and all purines on the other strand with the CD of polymers having alternating purine and pyrimidine sequences. Certainly, one would expect the

CD of the double-strand poly dA · poly dT to differ from that of the alternating copolymer double-strand poly d(A–T); the nearest neighbors on the individual strands differ. In fact, Wells *et al.* [132] find differences in the CD of poly d(A–T–C) · poly d(G–A–T) and poly d(T–A–C) · poly d(G–T–A). These differences can be explained by differences in the sequence-dependent nearest-neighbor interactions along the single strands.

In addition to this sequence dependence, Wells *et al.* [132] have been able to demonstrate significant geometric differences between polymers having all

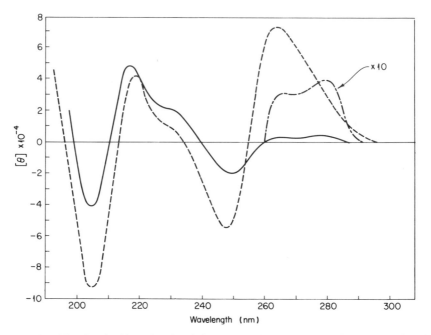

Fig. 16. CD of poly dA and poly rA in 0.01 *M* tris, pH 7.4 at 23°C [5]. (——) poly dA; (– ·) poly dA × 10; (– – –) poly rA.

pyrimidines on one strand and all purines on the other and ordinary DNA-type polymers having both pyrimidines and purines on each strand. They can compare nearest-neighbor interactions of identical type and sequence among their polymers using the method of Gray and Tinoco [109]. This method is similar to the nearest-neighbor method of Cantor *et al.* [115]. But, instead of using dinucleoside phosphates to represent the nearest-neighbor interactions, one uses nearest-neighbor interaction in sequenced polymers. This method is extremely useful in double-stranded polymers since the base-paired interactions present in the polymers cannot be observed in dinucleosides [114].

Using Gray and Tinoco's [109] method to calculate the CD of poly dA·poly dT from the circular dichroism of model polymers which have both purines and pyrimidines in each strand, they obtain a calculated curve which does not agree with the experimental CD of poly dA·poly dT [132]. Their results are given in Fig. 17 which compares the experimental CD of poly dA·poly dT with that calculated for this polymer from the CD of poly d(A–A–C)·poly d(G–T–T) and poly d(A–C)·poly d(G–T). The curves in Fig. 17 differ appreciably, having dissimilar shapes. This difference is probably too large to

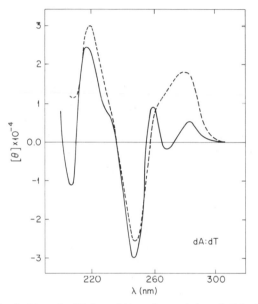

Fig. 17. CD of poly dA·poly dT. (——) Experimental. (– – –) Calculated from copolymers having purines and pyrimidines on each strand. [132].

be explained by non-nearest-neighbor interactions and has been interpreted as an indication that poly dA·poly dT has a geometry different from that of ordinary DNA [132].

4. *Comparison of Ribo- and Deoxynucleotides*

We have argued above that the difference in optical activity of the ribo- and deoxypolynucleotides must be geometric in its origin. This fact has been recognized for some time and a number of proposals have arisen to explain the role of the 2′-OH in the conformation and optical activity of polynucleotides.

One might imagine that in the ribo series, the 2'-OH could form hydrogen bonds which influence the conformation. Ts'o *et al.* [113] have argued in favor of a hydrogen bond from the 2'-OH to the base. This hypothesis would imply that both sugars of a dinucleoside would be equally important in determining its conformation. Maurizot *et al.* [131,133] have shown that only the 3'-linked sugar is important in determining dinucleoside CD. They argue in favor of a 2'-OH bond to the phosphate.

In fact, several recent experiments imply strongly that no hydrogen bond formed by the 2'-OH influences the CD of ribonucleotides. Adler *et al.* [77,130] find no salt effects on the CD of poly C or poly A. In addition, the CD of poly 2'-*O*-methyl A is very similar to that of poly A, and not at all like that of poly dA [134]. In a comparison of the ORD spectra of 2'-O methyl dinucleoside phosphates with the ORD of ordinary dinucleosides, Singh, and Hillier [134a] have concluded that the 2'-O methyl compounds behave much more like the ribo dimers than like their deoxyribo analogs. It is becoming increasingly apparent that hydrogen bonding does not play a major role in the differences between ribo and deoxy polymers and that the primary effect of the 2' substituent is steric. The 2'-OH or 2'-O methyl groups interact with neighboring residues, restricting the ribo-type polymers to base tilted geometries having the sugar in a 3'-*endo* conformation.

5. *Polydeoxyadenylic Acid*

This polymer has a CD rather different not only from its ribo analog, but also quite different from its deoxydinucleoside, dApdA (Figs. 14 and 16). The ORD melting curve of a series of deoxyoligomers shows the unusual property that the longer oligomers have a smaller rotation per residue than do the shorter oligomers and the dimer. This result has been interpreted as an anti-cooperative phenomenon [135]. Vournakis *et al.* [135] assume that the contribution per stacking interaction to the ORD is constant, and that the origin of the decreased rotation in longer oligomers is thermodynamic in nature. Hence, they infer a repulsive interaction between next-nearest-neighbors in the polymer.

A much more straightforward interpretation would result from the hypothesis that interior residues in the longer oligomers of dA have a contribution to ORD that differs from end residues or from dimers. The curve shape of the ORD and CD differs greatly for dimers and longer oligomers as may be seen in Fig. 14. Bush and Scheraga [5] have suggested that the difference between interior and exterior residues is one of solvation, but the possibility of a change in deoxyribose puckering would be an attractive hypothesis. Such a hypothesis would suggest that ApA, dApdA, and poly A all have identical ribose-puckering. But in poly dA, a cooperative change to different puckering is seen, leading to modified polymer geometry and the CD curve shape seen for poly dA in Fig. 16.

E. RNA

1. *Single-Stranded Stacked RNA*

By single-stranded stacked RNA, we mean RNA that is free of base-pairing. Single-stranded RNA forms a substantial number of base pairs in neutral solutions of moderate ionic strength. These base pairs can be disrupted by repulsion between negatively charged phosphates if all the positive ions are removed by dialysis. Thus, Cantor *et al.* [115] were able, after exhaustive dialysis to remove salts, to obtain ORD curves of several single-stranded RNA's which were apparently free from base-pairing. The absence of salt does not disrupt base-stacking and this salt-free RNA still shows large Cotton effects. Since the stacking in these single-stranded stacked RNA's is similar to that in oligomers, Cantor *et al.* [115] were able to apply the nearest-neighbor method used in short oligonucleotides [72]. They showed that in order to interpret the ORD of a single-stranded stacked RNA, one need know only the nearest-neighbor frequencies and the ORD of the 16 dinucleoside phosphates.

Since nearest-neighbor frequencies are not known for many RNA's, they proposed a further assumption that the sequence is random. This assumption is supported for a number of RNA's by hydrolysis and nearest-neighbor frequency data. Using the random sequence assumption, Cantor *et al.* [115] computed the ORD of TMV–RNA from that of dinucleoside phosphates using the formula,

$$[\phi_{RNA}(\lambda)] = \sum_i X_i \{2 \sum_j X_j [\phi_{i,j}(\lambda)]\} - [\phi_i(\lambda)] \tag{39}$$

X_i and X_J are the fraction of bases I and J in the RNA. The rotations are the same as in Eq. (38) above. They showed that the ORD curve calculated on these assumptions agreed reasonably well with that measured for salt-free TMV–RNA as shown in Fig. 18. Cantor *et al.* [115] were also able to interpret the ORD of R-17, MS-2 as well as mixed transfer RNA's in salt-free solution with the random sequence approximation. Subsequently, this method has been successfully applied to *E. coli* ribosomal RNA in salt-free solution [136].

There are conditions other than salt-free solution which can be used for observation of single-stranded stacked RNA. It is likely that the effect of reacting the amino groups of mixed transfer RNA's with formaldehyde prevents base-pairing in a way similar to salt-free solution. The ORD curve reported by Fasman *et al.* [137] for transfer RNA reacted with formaldehyde is similar to that found for salt-free transfer RNA by Cantor *et al.* [115].

2. Effect of Base-Pairing

The formation of base-pairing in poly A·poly U and in poly G·poly C causes blue shifts in the Cotton effects [93]. This same shift is seen in TMV–RNA where a comparison of the ORD in 0.1 M KCl with that in salt-free solution reveals a blue shift (Fig. 18). Cantor *et al.* [115] could quantitatively interpret this blue shift by adding a curve representing the effect of base-pairing to the ORD curve of single-stranded stacked RNA. A curve representing the effect of A–U base pairs was determined from the difference in ORD curves of poly A·poly U and poly A and poly U. The effect of G–C base pairs was determined similarly from the difference in the ORD curve of poly G·poly C and poly G and poly C. Single-stranded poly G is not known, but the ORD curve may be derived from the ORD of GpG.

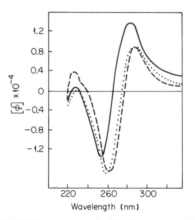

Fig. 18. ORD of TMV–RNA at neutral pH. Experimental in 0.15 M salt (——) Experimental in absence of salt (· · ·). Calculated from Eq. (39), (– – –) [115].

The utility of the method of Cantor *et al.* [115] may be seen in its application to 5 S ribosomal RNA of *E. coli.* Little is known about the function of this 120 residue RNA, but the sequence has been determined [138]. Cantor [139] calculated the ORD for the model containing 23 base pairs proposed by Brownlee *et al.* [138] and also that for a model having 49 base pairs. He found that the latter model agreed much better with his ORD experiments than did the former. Such data, in conjunction with some other optical data, indicate that models proposed for 5 S RNA should contain approximately 43 base pairs [139].

A method known as matrix rank analysis has been applied to the ORD of TMV–RNA [140]. This method seeks the smallest number of linearly independent ORD curves needed to synthesize the ORD of an RNA over a wide range of solution conditions. McMullen *et al.* [140] found that only two independent curves were needed and that these curves approximate the single stranded stacked and the base-paired double helical ORD curves for TMV–RNA (See Fig. 18).

3. *Complementary Double-Stranded RNA*

True complementary double-stranded RNA has optical activity curves very similar to those of the base-paired regions of single-stranded RNA containing base pairs. A comparison of the single-stranded, base-paired form of MS-2 RNA with that of the complementary double-stranded replicative intermediate of MS-2 showed great similarity [141]. The latter curve showed a slight blue shift in the Cotton effect, indicating perhaps a greater extent of base-pairing in the complementary double-stranded RNA. The complementary double-stranded RNA from rice dwarf virus has a CD which is also similar to that of single-stranded base-paired RNA [142]. The wavelength of the maximum is about 5 nm to the blue of that for TMV–RNA whose base composition is similar.

An approach to nearest-neighbor empirical theory of polynucleotide optical properties based on sequenced polymers has been proposed by Gray and Tinoco [109]. This method is similar to the nearest-neighbor method based on dimers of Cantor *et al.* [115]. This method should be especially useful for base-paired double-stranded polynucleotides. Unfortunately, a lack of optical activity curves for sequenced polynucleotides has prevented its use up to the present. The method has been applied to a limited set of sequenced polydeoxynucleotides as we have discussed in Section IV,D,3.

4. *CD at Longer Wavelengths*

Small Cotton effects which are near in wavelength to large ones are not easily recognized in ORD curves. Thus, the small negative Cotton effect reported in spectra of yeast s RNA and *E. coli* ribosomal RNA was seen readily only when CD instruments became available [143]. In Fig. 19 we show examples of this effect in both single-stranded and complementary double-stranded RNA's. This Cotton effect has been observed in rice dwarf virus RNA [142], 5 S RNA and double-stranded F-2 RNA [110] and *E. coli* transfer RNA [144]. This Cotton effect is strongly overlapped by the strong positive band at 265 nm, and hence, its magnitude is very sensitive to slight shifts in the 265 nm band. This overlapping also makes it difficult to determine the true magnitude and wavelength position of this band and its origin is unknown. An intriguing possibility is that it is related to the long wavelength bands seen in poly dA and in poly G which are probably of $n-\pi^*$ origin [26].

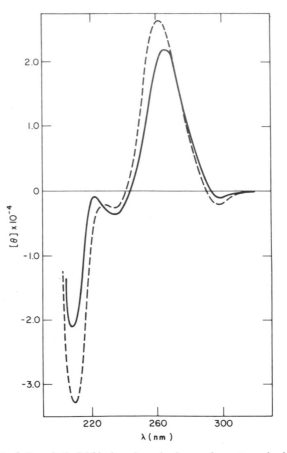

Fig. 19. CD of *E. coli* 5 sRNA (———) and of complementary double-stranded F2 RNA (– – –) at neutral pH and room temperature [110].

The negative band at 295 nm is too large to be the result of some undetected minor chromophore in RNA. But the long wavelength CD and absorption of 4-thiouracil has been studied in several RNA's. Scott and Schofield [144] find that the denaturation of transfer RNA by organic solvents causes changes in the CD due to 4-thio U at 336 nm which parallel those at 265 nm. They also report some interesting differences in this wavelength region among the behavior of several different purified transfer RNA's. Cotter and Gratzer [145] have reported CD bands in *E. coli* ribosomal RNA at 340 nm which can be assigned to 4-thio U. They find that the thermal denaturation at 340 nm parallels that at 260 nm indicating that the 4-thio U chromophores are in base-paired regions of the RNA. Although these Cotton effects due to minor components are small, their CD per chromophore may be quite large as is the case with 4-thio U. Although these special chromophores have not been extensively investigated, they offer some promising uses as probes of specific regions of the RNA.

F. DNA

1. *Single Strands*

The nearest-neighbor method analysis has been recently applied to single-stranded DNA by Cantor *et al.* [107]. The calculation of the CD of oligo-deoxynucleotides from that of deoxydinucleotides was found to be less reliable than was the case for the ribo series. Thus, it is surprising that Cantor *et al.* [107] had reasonable success in their calculation of the CD of single-stranded Fl DNA in salt-free solution. The reason for their success is not clear, but may reside in fortuitous cancellation of errors. There are not enough optical activity data on single-stranded DNA available to critically judge the method, but certainly further study is warranted. There is not a large number of known single-stranded natural DNA's. The only other curve for a single-stranded DNA known to the present author is the ORD of ΦX174 DNA [146]. This curve does not differ remarkably from the typical DNA ORD. The separated single strands of a double-stranded DNA might be amenable to study by these methods.

2. *Double-Stranded DNA*

Most of the study on DNA has involved the more common complementary double-stranded DNA's. The ORD of a number of bacterial, viral, and mammalian DNA's has been reported by Samejima and Yang [147]. The curves have been found to be similar and characteristically different from the ORD of RNA's. There are small differences in the position of peaks and troughs which seem to reflect differences in base composition [147].

The similarity among the optical activity curves for a number of DNA's has also been noted in the CD studies of Brahms and Mommaerts [128]. In the CD, the characteristic double Cotton effect of DNA and its contrast to the characteristically larger single Cotton effect in RNA is easily recognized as is seen in Fig. 20 [148].

The CD and ORD of several mitochondrial DNA's has been reported by Bernardi and Timasheff [149]. These curves, although they retain the double Cotton effect characteristic of DNA, show some unusual properties. There is a number of distinct shoulders in the long wavelength positive CD band of various yeast mitochondrial DNA's. Such shoulders are not seen in other

DNA's and these shoulders differ among various yeast strains which were examined. These mitochondrial DNA's also show differences from nuclear DNA's in a number of physical properties. The origin of these differences is not yet understood [149].

In principle, the method of Gray and Tinoco [109] ought to be a good semi-empirical technique for interpreting the CD and ORD of double-stranded DNA. Their method compares the CD of DNA to that of a group of poly-nucleotides of known sequence. Unfortunately, we do not have CD data on enough sequenced polydeoxynucleotides to fully test this method in DNA, but a simplified version of the Gray and Tinoco [109] method, which requires

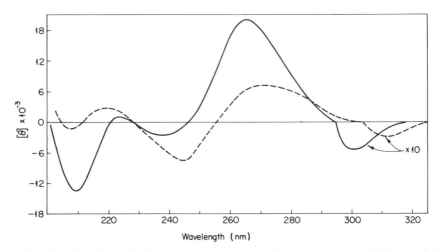

Fig. 20. CD of *E. coli* ribosomal RNA (——) and of salmon sperm DNA (– – –) in 0.1 *M* KCl, 0.0005 *M* EDTA, 0.005 *M* tris. The small negative Cotton effects at 290–310 nm are × 0 [148].

additional assumptions but eliminates the need for certain unavailable model polymers, has been applied with good success to double-stranded DNA by Wells *et al.* [132].

Allen *et al.* [149a] have presented an application of the Gray and Tinoco method in which one compares the CD of a number of DNA's of varying base composition and hence of nearest-neighbor frequency. By choosing DNAs of a wide range of nearest-neighbor frequency, they were able to calculate the CD curves of a number of hypothetical polynucleotides of various sequences all in DNA geometry. The results were very interesting and confirm the notion that DNA's having all purines on one strand and all pyrimidines on the other must have a geometry different from that of ordinary B form DNA.

3. Differences between RNA and DNA

All the RNA's which have been investigated to date have optical activity curves dominated by a single Cotton effect near the absorption maximum. This generalization applies to single-stranded stacked RNA, to single-stranded RNA containing base-paired hairpin loops and to complementary double-stranded RNA (see Figs. 18–20). DNA, on the other hand, generally shows a weaker double Cotton effect (see Fig. 20). This generalization applies to all the double-stranded DNA's which have been studied and also to single-stranded DNA. Even mitochondrial DNA, which demonstrates some unusual properties, fits this general picture. These generalizations do not extend to synthetic polymers of regular repeating sequence (see Figs. 15 and 16).

A generalization about the helical structure of DNA and RNA can be discovered in the analysis of x-ray scattering from fibers. RNA's are always found to have a similar structure, independent of the method of preparation [150,151]. This RNA structure involves a substantial tilt of the base planes with respect to the helix axis. DNA, on the other hand, generally appears in the B form, which has the base planes perpendicular to the helix axis. Depending on the conditions of preparation, DNA may also be found in helices which have various degrees of tilting of the base planes [151]. Yang and Samejima [129] have proposed that base-tilting correlates with the appearance of optical activity curves of the RNA type; that is, a large single Cotton effect. Recent theoretical calculations indicate that this attractive hypothesis is correct for polynucleotides not having a repeating sequence [3,62].

The appearance of a single Cotton effect was recognized by Bush and Brahms [33] to be the result of the interaction of the 260 nm absorption bands with far-ultraviolet transitions. However, they incorrectly assumed that this interaction, which they computed by a Kirkwood polarizability approximation, would not depend strongly on tilting of the bases. Subsequently, Tinoco [3] and Johnson and Tinoco [62] explicitly calculated the magnitudes of the single and double Cotton effects for different degrees of base-tilting, and showed that one could explain the single Cotton effect in RNA as the result of base-tilting. In a DNA of random sequence, having base planes perpendicular to the helix axis, the single Cotton effects cancel out, leading to the small double Cotton effect observed in experiment [62]. In synthetic polymers the cancellation effects may be absent, leading, for example, to the single Cotton effect in poly dC (Fig. 15), a case considered exceptional by Yang and Samejima [129]. The generalization that RNA shows large single Cotton effects, while DNA shows small double Cotton effects extends to both single and double strands. The arguments of Johnson and Tinoco [62] are applicable to either system.

4. Distorted DNA

In our discussion of x-ray scattering of RNA and DNA, we pointed out that, although RNA is always found in helices having bases tilted, DNA is found either in the B form or in other forms which have some base-tilting. It appears that the ability of DNA to occur in helices of differing geometry is also evidenced in solution and that optical activity curves can be used to detect these differing forms.

DNA in 80% ethanol (Fig. 21) was shown to be only very slightly hyperchromic and to have a positive Cotton effect in the CD which was not unlike that of RNA. Brahms and Mommaerts [128] proposed that the DNA re-

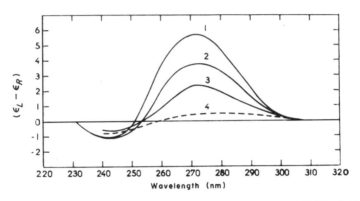

Fig. 21. CD of DNA in 80% ethanol, 20% aqueous 0.0001 *M* EDTA, 0.01 *M* tris, pH 7.4 at various temperatures. Curve 1, 22°C; curve 2, 40°C; curve 3, 60°C; curve 4, 80°C [128].

sembles the A form geometry which has tilted bases. In view of recent developments outlined in the preceding section, we would conclude that their interpretation was correct. Further, the tilt must be in the same direction as that for RNA, which is toward the positive helix axis [3].

Ethyleneglycol, like ethanol, does not induce large hyperchromism but does induce a Cotton effect in the CD which is negative [152] (see Fig. 22). Likewise, 6 *M* LiCl or CsCl were found to induce similar negative Cotton effects in DNA [153]. Subsequently, Green and Mahler [124] attempted to find effects of ethyleneglycol, similar to those seen in DNA, in studies on synthetic polynucleotides. They were unable to find such effects, apparently because they studied polyribonucleotides which do not have the ability to assume different geometries.

In an attempt to determine the nature of the DNA geometry responsible for the negative Cotton effects observed in high salt strength solutions, Tunis and

Hearst [153] studied the ORD of RNA–DNA hybrids. These double helices are known to have geometry similar to the A form of DNA, having base planes tilted toward the positive helix axis. Unlike the DNA in high salt strength solutions which show a negative Cotton effect, these hybrids showed a positive Cotton effect. Tunis and Hearst [153] concluded that DNA in high-salt solutions has a conformation which is different from that of ordinary DNA but that it was not that of the A form of DNA. In view of the recent theoretical work of Tinoco [3], we would conclude that DNA in ethanol (Fig. 21) and RNA–DNA hybrids have a positive tilt such as that in RNA. On the other hand, DNA in high-salt solutions and in ethyleneglycol (Fig. 22)

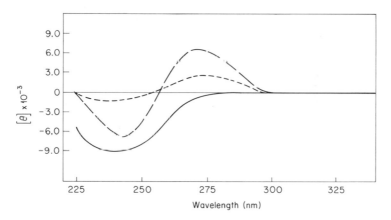

Fig. 22. CD of calf thymus DNA (— —) in aqueous 0.05 *M* KF, 0.001 *M* EDTA, at 25°C; (——) in ethyleneglycol containing the same salts; (– – –) in ethyleneglycol at 50°C [152].

is seen to have a negative Cotton effect and thus has a tilt opposite to that of the A form of DNA. This conclusion further implies that in high-salt solutions, such as are used in buoyant density centrifugation, DNA is not in the B form but is in a form having the bases in a different geometric relationship.

Recent experiments on the CD of films of DNA under various conditions of salt and relative humidity allow one to make a connection between the x-ray studies on fibers and CD in solution. Tunis-Schneider and Maestre [153a] found that DNA films under conditions of high relative humidity show CD patterns nearly identical with those of DNA in solution (e.g., Fig. 20). Since such preparation conditions are similar to those of fibers of the B form of DNA, these results confirm the idea that DNA in solution is in the B form with bases perpendicular to the helix axis. Sodium DNA at lower relative humidity shows single positive Cotton effects like those of DNA in ethanol

(Fig. 21), or of RNA (Fig. 20) [153a]. Such preparation conditions are those of fibers of the A form of DNA. Therefore, one may conclude that single positive Cotton effects are characteristic of the A form of DNA. Lithium DNA films at lower relative humidity exhibit CD curves having single negative Cotton effects similar to those seen in DNA in ethylene glycol (Fig. 22) or 4 to 6 M LiCl or CsCl solutions. Since lithium DNA at low relative humidity yields the C form in x-ray fibre studies, Tunis-Schneider and Maestre [153a] conclude that these single negative Cotton effects are characteristic of the C form of DNA.

In contrast to these interesting modifications of the optical activity of DNA are the effects in RNA. Ethanol causes either precipitation or denaturation of RNA [128]. Ethylene glycol and dimethylsulfoxide cause denaturation of RNA without other structural modification [124,154]. High salt concentration has no effect on the optical activity of RNA up to the concentrations that cause it to precipitate [153].

We discussed above the interesting case of poly dA which seems to have a different geometric relationship between nearest neighbors than does the deoxydinucleoside dApdA [5]. In Section IV,D,5, we hypothesized that the dinucleoside has the "RNA-like" conformation, while the single stranded polymer may have a "DNA-like" geometry (see Figs. 3, 14, and 16). Another example of conformational mobility in polydeoxynucleotides in solution has been given in Section IV,D,3 and Fig. 17 [132]. We might hope to find further examples of conformational changes in polydeoxynucleotide models which could help us to explain the changes we see in DNA. The biological implications for control mechanisms of cooperative conformational changes in DNA could be quite far reaching.

In summary, we may make an approximate correlation for sugar pucker, base tilting, and optical activity. The A form of DNA and all RNA's have 3′-*endo* sugars, positive tilting, and single positive Cotton effects. DNA may also be found in the B form, having 2′-*endo* sugars, base planes perpendicular to the helix axis, and small double Cotton effects. The C form of DNA also has 2′-*endo* sugars, bases tilted slightly toward the negative helix axis, and negative Cotton effects.

G. NUCLEOPROTEIN COMPLEXES

One useful feature of spectroscopic techniques, shared by ORD and CD, is found in examination of impure systems. Under favorable circumstances, we may study RNA and DNA as they exist in nucleoprotein complexes and cell organelles. A general requirement is that substances in the sample other than RNA and DNA do not absorb too much light in the wavelength region of interest near 260 nm. Also, if there are other substances such as protein contributing to the optical activity, we must have some way of separating their contribution from that of RNA or DNA.

We will discuss several examples from viruses, ribosomes, and nucleohistones below. The fact that the major bands in the optical activity of proteins lie in the 190 to 230 nm region makes it possible to separate their contributions from that of polynucleotides in the 240 to 290 nm region. Although ORD has been used for this purpose in the past, CD is probably more appropriate since the contributions to the CD are localized in wavelength near the absorption band responsible for the Cotton effects [105].

1. *Light Scattering*

The fact that cell organelles and such particles as viruses are large causes certain problems associated with light scattering in the interpretation of optical activity spectra. Unfortunately, these difficulties have not always been recognized, and only recently has a serious effort been made to understand the ORD and CD of particulate suspensions. It is not yet possible to give a definitive description of the effects to be expected in optical activity measurements on particulate systems. However, it seems certain that such a quantitative description will be available in the near future. Particulate suspensions do give ORD and CD curves rife with artifacts, but a good theoretical understanding of these effects should allow the experimentalist to correct for them, allowing extension of the techniques outlined in this chapter to particulate systems. The most obvious problem associated with light scattering might seem to be differential scattering of right- and left-handed circularly polarized light. Recent calculations show this effect to be important in some cases [154a]. Another important effect in the CD of particulate suspensions is "concentration obscuring" which decreases the apparent magnitude of the measured CD. Although the wavelengths of the crossings are not changed by these effects, the depression of measured CD values depends on wavelength so that shifts of peaks and troughs may be observed) [156]. In polypeptides, the shifts of the troughs in particulate systems are to be red. These red shifts result from the relatively weak absorption at the wavelength of the $n-\pi^*$ Cotton effect at 225 nm compared to that of the $\pi-\pi^*$ at 190 nm. In polynucleotides, the primary source of CD is the $\pi-\pi^*$ transitions, and the CD curves follow the absorption curves more closely than in the polypeptide case. Hence, we expect shifts to be less pronounced in polynucleotides. Of course, the depression in the magnitude of the ellipticity will be observed.

2. *Ribosomes*

The first optical study of the structure of ribosomal RNA in ribosomes was a measurement of hyperchromism due to thermal melting [157]. The increase in absorption on melting parallels that in free RNA, suggesting that the RNA structure is similar in ribosomes to that in solution. On the other hand, dye binding studies imply that the RNA is single-stranded stacked in ribosomes as compared to a largely base-paired structure in solution [158].

The ORD assay for base-pairing which we have discussed above may also be used in this nucleoprotein complex. Bush and Scheraga [136] investigated the ORD of ribosomes, their subunits, and ribosomal RNA under salt-free as well as 0.1 M salt conditions. The ORD of the ribosomal protein was not large in the 260 nm region of the RNA Cotton effect (see Fig. 23). From

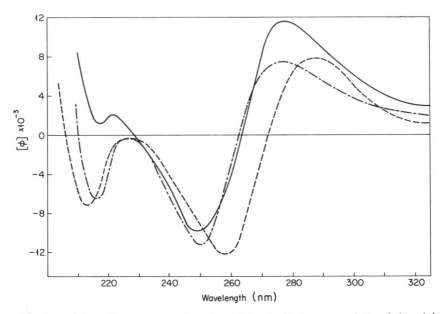

Fig. 23. Molar rotation per residue of RNA in: 70 S ribosomes of *E. coli* (——) in 0.01 M magnesium acetate, 0.01 M tris, pH 7.4 at 8°C. Free RNA in 0.0001 M EDTA, pH 6.8, 24°C (– – –). Free RNA in 0.1 M KCl, 0.01 M tris, pH 7.4 at 6°C (·–) [136].

comparison of the RNA Cotton effect in whole ribosomes with the Cotton effects of free RNA in the presence and absence of salt, they were able to conclude that RNA in ribosomes is extensively base-paired, as if found in RNA in 0.1 M salt solutions. Their ORD curve for salt-free RNA could be calculated from the base composition and dinucleoside ORD curves by the method of Cantor *et al.* [115] [Eq. (39)]. Similar experiments reaching similar conclusions have also been reported by Cotter *et al.* [159] who were able to rationalize the erroneous findings from dye-binding studies of Furano *et al.* [158].

3. *RNA Virus*

A similar approach was used to interpret the ORD of TMV [136]. This structure is known from x-ray scattering to hold the RNA in a large helix of 23 Å pitch. In this conformation, base-pairing would not be expected. The virus is 95% protein so that the protein contribution dominates even the RNA Cotton effect region at 260 nm. However, RNA-free ghosts are easily prepared by aggregation of the TMV protein under appropriate conditions. Subtraction of the protein ghost ORD curve from that of the native TMV leads to an ORD curve of the RNA as it is found in the virus. This curve is then compared to the salt free RNA–ORD curve and that calculated by Cantor *et al.* [115] from dinucleoside phosphates. These curves are all found to be in reasonable agreement, indicating the absence of base-pairing in TMV–RNA in the virus [136].

The ORD of mengo virus, an RNA virus from mouse L cells has also been reported (Fig. 24) [160]. In contrast to TMV, this virus contains 21% RNA and is spherical in shape. One does not have protein ghosts for this case to facilitate comparison between the RNA in the virus and in solution. However, it does not appear that there is a significant wavelength shift in the ORD curve between the virus and the free RNA. Although the ORD peaks and troughs occur at rather long wavelength for both the RNA and the virus, presumably the RNA is extensively base-paired in solution, and hence is

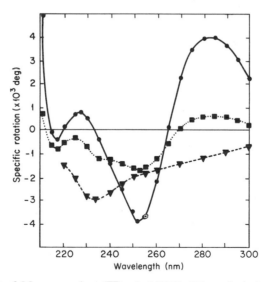

Fig. 24. ORD of M-mengo virus (■), viral RNA (●), and viral protein (▼). The solvent is 0.1 *M* KCl, 0.001 *M* MgCl₂, 0.001 *M* EDTA, 0.01 *M* tris, pH 7.2 [160].

base-paired in the virus. These conclusions must be regarded as tentative, but they present an interesting contrast to the TMV case.

4. DNA Virus

Many bacteriophages have modest protein content and their optical activity in the 260 nm region is dominated by their DNA. The ORD curves for a number of phages have been reported by Maestre and Tinoco [146,161]. For the even-numbered T phages, they find essentially single negative Cotton effects not unlike those for DNA in ethyleneglycol and 6 M salt solution (see Section IV,F,4). Other phages show mixtures of the single negative Cotton effect and ordinary DNA double Cotton effect curves.

The tight packing required to store the large viral DNA inside the virus protein coat apparently causes changes in the geometry of the DNA helix. Maestre and Tinoco [161] were able to correlate the changes in ORD among a series of phages with the DNA content per unit volume of the phage. They interpreted this result as an indication of differing amounts of distortion required to pack the DNA inside the protein coat. The modifications in the DNA Cotton effects seen in high-salt concentrations by Tunis and Hearst [153], are similar to those seen in bacteriophage by Maestre and Tinoco [146]. If the changes in DNA optical activity caused by high-salt concentrations result from changes in the tilting of the bases relative to the helix axis a similar distortion might be occurring in the phage.

5. Nucleohistone

The study of DNA conformation in chromatin is one that has only recently been attempted by optical activity measurements. The CD of calf thymus nucleohistone has been studied by Permogorov et al. [162]. They find a slight difference in the CD of the nucleohistone from that of free DNA. The change is similar to that induced in free DNA by 2 M salt. We have discussed the effect of high-salt concentrations in Section IV,F,4, and have interpreted these effects as the addition of a negative Cotton effect at 245 nm. Sponar et al. [163] have measured the ORD of native, partially disassociated and reconstituted calf thymus nucleohistone. They report curves which resemble those of free DNA in 2 M salt and also find that these curves are retained even in nucleohistone with a substantial part of protein removed.

Fasman et al. [164] have approached histone–DNA interaction from the opposite viewpoint, studying the aggregation of purified Fl histone with calf DNA. They find that negative Cotton effects are induced in DNA by complexation with this basic protein. Results similar to those of Fasman et al. [164] were reported by Wilhelm et al. [165]. They also report that the positive CD band of DNA is depressed on the addition of various histone fractions.

Tunis-Schneider and Maestre [153a] suggest that single negative Cotton effects are characteristic of the C form of DNA. The appearance of such negative Cotton effects in chromatin and DNA–histone complexes implies that perhaps the C form of DNA plays some role in the control of transcription or replication.

The CD curves of DNA in nucleohistone complexes seem to depend strongly on the method of preparation. Isenberg *et al.* [166] report only a small effect of histone on the CD of DNA. In studies on whole chromatin, Wagner and Spelsberg [167] find that the DNA has a positive Cotton effect similar to that in RNA. The recent outburst of publications of CD curves for various native, partially denatured and reconstituted DNA–nucleohistone complexes present some discrepancies on matters of experimental fact. It is clear that these differences as well as questions of light scattering must be answered before a consistent interpretation can be given.

Acknowledgment

We wish to thank Dr. S. Hanlon for making available some unpublished results. We also thank Dr. C. R. Cantor for providing manuscripts prior to publication.

References

1. A. Moscowitz, *in* "Optical Rotatory Dispersion" (C. Djerassi, ed.), p. 150. McGraw-Hill, New York, 1960.
2. T. M. Lowery, "Optical Rotatory Power." Longmans-Green, New York, 1935 (reprinted, Dover, New York, 1964).
3. I. Tinoco, Jr., *J. Chim. Phys.* **65**, 91 (1968).
4. D. W. Miles, M. J. Robins, R. K. Robins, and H. Eyring, *Proc. Nat. Acad. Sci. U.S.* **62**, 22 (1969).
5. C. A. Bush and H. A. Scheraga, *Biopolymers* **7**, 395 (1969).
6. I. Tinoco, Jr., *Advan. Chem. Phys.* **4**, 113 (1962).
7. W. Moffit and A. Moscowitz, *J. Chem. Phys.* **30**, 648 (1959).
8. J. P. Carver, E. Shechter, and E. R. Blout, *J. Amer. Chem. Soc.* **88**, 2550 (1966).
9. G. Holzwarth and P. Doty, *J. Amer. Chem. Soc.* **87**, 218 (1965).
10. C. A. Bush and I. Tinoco, Jr., *J. Mol. Biol.* **25**, 601 (1967).
11. J. Thiery, *J. Chim. Phys.* **65**, 98 (1968).
12. C. A. Bush, Program ORPD-02. Perkin-Elmer Program Exchange Library, Perkin-Elmer Corp., Norwalk, Connecticut, 1969.
13. C. A. Emeis, L. J. Oosterhof, and G. deVries, *Proc. Roy. Soc., Ser. A* **297**, 54 (1967).
14. J. G. Kirkwood, *J. Chem. Phys.* **5**, 479 (1937).
15. I. Tinoco, Jr., *J. Amer. Chem. Soc.* **82** 4785 (1958); **83**, 5047 (1961).
16. W. Rhodes, *J. Amer. Chem. Soc.* **83**, 3609 (1961).
17. M. M. Warshaw and I. Tinoco, Jr., *J. Mol. Biol.* **20**, 29 (1966).

18. H. DeVoe and I. Tinoco, Jr., *J. Mol. Biol.* **4**, 518 (1962).
19. H. C. Bolton and J. J. Weiss, *Nature (London)* **195**, 666 (1962).
20. R. K. Nesbet, *Biopolym. Symp.* **1**, 129 (1964).
21. J. Thiery, *J. Chem. Phys.* **43**, 553 (1965).
22. W. Rhodes and M. Chase, *Rev. Mod. Phys.* **39**, 349 (1967).
23. H. DeVoe, *Biopolym. Symp.* **1**, 251 (1964).
24. H. DeVoe, *J. Chem. Phys.* **41**, 393 (1965).
25. J. Brahms, J. Pilet, H. Damary, and V. Chandrasekharan, *Proc. Nat. Acad. Sci. U.S.* **60**, 1130 (1968).
26. C. A. Bush, *J. Chem. Phys.* **53**, 3522 (1970).
27. I. Tinoco, Jr., *Radiat. Res.* **20**, 133 (1963).
28. I. Tinoco, Jr., A. Halpern, and W. T. Simpson, *in* "Polyamino Acids, Polypeptides and Proteins" (M. Stahlman, ed.), p. 147. Univ. of Wisconsin Press, Madison, 1962.
29. H. Eyring, J. Walter, and G. E. Kimball, "Quantum Chemistry," p. 111. Wiley, New York, 1944.
30. M. M. Warshaw, C. A. Bush, and I. Tinoco, Jr., *Biochem. Biophys. Res. Commun.* **18**, 633 (1965).
31. I. Tinoco, Jr., *J. Chem. Phys.* **33**, 1332 (1960); **34**, 1067 (1961).
32. T. Yamada and H. Fukutome, *Biopolymers* **6**, 43 (1968).
33. C. A. Bush and J. Brahms, *J. Chem. Phys.* **46**, 79 (1967).
34. D. Stigter and J. A. Schellman, *J. Chem. Phys.* **51** 3397 (1969).
35. R. W. Woody and I. Tinoco, Jr., *J. Chem. Phys.* **46**, 4927 (1967).
36. D. Glaubiger, D. A. Lloyd, and I. Tinoco, Jr., *Biopolymers* **6**, 409 (1968).
37. J. Brahms, J. C. Maurizot, and A. M. Michelson, *J. Mol. Biol.* **25**, 481 (1967).
38. M. M. Warshaw and C. R. Cantor, *Biopolymers* **9**, 1079 (1970).
39. N. S. Kondo, H. M. Holmes, L. M. Stempel, and P. O. P. Ts'o, *Biochemistry* **9**, 3479 (1970).
40. A. Moscowitz, *Proc. Roy. Soc., Ser. A* **297**, 16 (1967).
41. R. A. Harris, *J. Chem. Phys.* **47**, 4481 (1967).
42. I. Tinoco, Jr. and C. A. Bush, *Biopolym. Symp.* **1**, 235 (1964).
43. B. Vriat and C. Djerassi, *Nature (London)* **217**, 918 (1968).
44. W. Voelter, R. Records, E. Bunnenberg, and C. Djerassi, *J. Amer. Chem. Soc.* **90**, 6163 (1969).
44a. M. F. Maestre, D. M. Gray, and R. B. Cook, *Biopolymers* **10**, 2537 (1971).
44b. W. C. Johnson, *Rev. Sci. Instr.* **42**, 1283 (1971).
45. D. Voet, W. B. Gratzer, R. A. Cox, and P. Doty, *Biopolymers* **1**, 193 (1963).
46. L. B. Clark and I. Tinoco, Jr., *J. Amer. Chem. Soc.* **87**, 11 (1965).
47. P. O. P. Ts'o, *in* "Biological Macromolecules" (G. D. Fasman and S. N. Timasheff, eds.), Vol. IV. Dekker, New York, 1970.
48. P. O. P. Ts'o, N. S. Kondo, R. K. Robins, and A. D. Broom, *J. Amer. Chem. Soc.* **91**, 5625 (1969).
49. A. Pullman and B. Pullman, *Advan. Quantum Chem.* **4**, 267 (1968).
50. P. R. Callis, E. J. Rosa, and W. T. Simpson, *J. Amer. Chem. Soc.* **86**, 2292 (1964).
51. R. F. Stewart and L. H. Jensen, *J. Chem. Phys.* **40**, 2071 (1964).
52. R. F. Stewart and N. Davidson, *J. Chem. Phys.* **39**, 255 (1963).
53. D. W. Miles, M. J. Robins, R. K. Robins, M. W. Winkley, and H. Eyring, *J. Amer. Chem. Soc.* **91**, 824 (1969).
54. W. Eaton and T. P. Lewis, *J. Chem. Phys.* **53**, 2164 (1970).
55. D. W. Miles, R. K. Robins, and H. Eyring, *Proc. Nat. Acad. Sci. U.S.* **57**, 1138 (1967); *J. Phys. Chem.* **71**, 3931 (1967).

56. D. W. Miles, M. J. Robins, R. K. Robins, M. W. Winkley, and H. Eyring, *J. Amer. Chem. Soc.* **91**, 831 (1969).
57. P. R. Callis and W. T. Simpson, *J. Amer. Chem. Soc.* **92**, 3593 (1970).
58. H. H. Chen and L. B. Clark, *J. Chem. Phys.* **51**, 1862 (1969).
59. G. Del Re, *Theor. Chim. Acta* **1**, 188 (1963).
60. H. Berthod, C. Giessner-Prettre, and A. Pullman, *Int. J. Quantum Chem.* **1**, 123 (1967).
61. W. C. Johnson and I. Tinoco, Jr., *Biopolymers* **8**, 701 (1969).
62. W. C. Johnson and I. Tinoco, Jr., *Biopolymers* **7**, 727 (1969).
63. J. Del Bene and H. H. Jaffee, *J. Chem. Phys.* **48**, 4050 (1968).
64. C. Giessner-Pretre and A. Pullman, *Theor. Chim. Acta* **13**, 265 (1969).
65. E. Clementi, J. M. Andre, M. C. Andre, D. Klint, and D. Hahn, *Acta Phys.* **27**, 493 (1969).
66. L. C. Snyder, R. G. Shulman, and D. B. Newman, *J. Chem. Phys.* **53**, 256 (1970).
67. B. Mely and A. Pullman, *Theor. Chim. Acta* **13**, 278 (1969).
68. E. Clementi, H. Clementi, and D. R. Davis, *J. Chem. Phys.* **46**, 4725 (1967).
69. J. N. Toal, *in* "Handbook of Biochemistry" (H. A. Sober, ed.), p. G-93. Chem. Rubber Publ. Co., Cleveland, Ohio, 1968.
70. A. Pratt, J. N. Toal, G. W. Rushizky, and H. A. Sober, *Biochemistry* **3**, 1831 (1964).
71. S. Mandeles and C. R. Cantor, *Biopolymers* **4**, 759 (1966).
72. C. R. Cantor and I. Tinoco, Jr., *J. Mol. Biol.* **13**, 65 (1965).
73. M. Leng and G. Felsenfeld, *J. Mol. Biol.* **15**, 455 (1966).
74. J. Applequist and V. Damle, *J. Amer. Chem. Soc.* **88**, 3895 (1966).
75. J. Brahms, A. M. Michelson, and K. E. Van Holde, *J. Mol. Biol.* **15**, 467 (1966).
76. D. Poland, J. N. Vournakis, and H. A. Scheraga, *Biopolymers* **4**, 223 (1966).
77. A. Adler, L. Grossman, and G. D. Fasman, *Biochemistry* **8**, 3846 (1969).
78. G. Felsenfeld and H. T. Miles, *Annu. Rev. Biochem.* **36**, 407 (1967).
79. W. M. Huang and P. O. P. Ts'o, *J. Mol. Biol.* **16**, 523 (1966).
80. A. D. Broom, M. P. Schweizer, and P. O. P. Ts'o, *J. Amer. Chem. Soc.* **89**, 3612 (1967).
81. T. N. Sollie and J. A. Schellman, *J. Mol. Biol.* **33**, 61 (1968).
82. G. Felsenfeld and G. Sandeen, *J. Mol. Biol.* **5**, 587 (1962).
83. R. A. Cox, *Biochem. J.* **98**, 841 (1966).
84. H. J. Gould and H. Simpkins, *Biopolymers* **7**, 223 (1969).
85. G. Felsenfeld and S. Z. Hirschman, *J. Mol Biol.* **13**, 407 (1965).
86. S. Z. Hirschman and G. Felsenfeld, *J. Mol. Biol.* **16**, 347 (1966).
87. A. Rich and M. Kasha, *J. Amer. Chem. Soc.* **82**, 6197 (1960).
88. W. L. Peticolas, L. Nafie, P. Stein, and B. Franconi, *J. Chem. Phys.* **52**, 1576 (1970).
89. B. L. Tomlinson and W. L. Peticolas, *J. Chem. Phys.* **52**, 2154 (1970).
90. W. H. Inskeep, D. W. Miles, and H. Eyring, *J. Amer. Chem. Soc.* **92**, 3866 (1970).
91. D. W. Miles, W. H. Inskeep, M. J. Robins, M. W. Winkley, R. K. Robins, and H. Eyring, *J. Amer. Chem. Soc.* **92**, 3872 (1970).
92. N. H. Teng, M. S. Itzkowitz, and I. Tinoco, Jr., *J. Amer. Chem. Soc.* **93**, 6257 (1971).
93. J. T. Yang and T. Samejima, *Prog. Nucl. Acid Res. Mol. Biol.* **9**, 224 (1968).
94. T. R. Emerson, R. J. Swan, and T. L. V. Ulbricht, *Biochemistry* **6**, 843 (1967).
94a. T. R. Emerson, R. J. Swan, and T. L. V. Ulbricht, *Biochem. Biophys. Res. Commun.* **22**, 505 (1966).
94b. J. S. Ingwall, *J. Amer. Chem. Soc.* **94**, 5487 (1972).
94c. J. Donohue and K. N. Trueblood, *J. Mol. Biol.* **2**, 353 (1960).
94d. M. Sundaralingam, *Biopolymers* **7**, 821 (1969).

94e. C. A. Bush, *J. Amer. Chem. Soc.* **95**, 214 (1973).

94f. H. R. Wilson and A. Rahman, *J. Mol. Biol.* **56**, 129 (1971).

95. T. Nishimura, B. Shimizu, and I. Iwai, *Biochim. Biophys. Acta* **157**, 221 (1968).

96. G. T. Rogers and T. L. V. Ulbricht, *Biochem. Biophys. Res. Commun.* **39**, 414 (1970).

97. D. W. Miles, L. B. Townsend, M. J. Robins, R. K. Robins, W. H. Inskeep, and H. Eyring, *J. Amer. Chem. Soc.* **93**, 1600 (1971).

98. P. A. Hart and J. P. Davis, *J. Amer. Chem. Soc.* **94**, 2572 (1972).

98a. P. A. Hart and J. P. Davis, *J. Amer. Chem. Soc.* **93**, 753 (1971).

99. G. T. Rogers and T. L. V. Ulbricht, *Biochem. Biophys. Res. Commun.* **39**, 419 (1970).

99a. J. Pitha, *Biochemistry* **9**, 3678 (1970).

99b. S. T. Rao and M. Sundaralingam, *J. Amer. Chem. Soc.* **92**, 4963 (1970).

100. J. Brahms, J. C. Maurizot, and A. M. Michelson, *J. Mol. Biol.* **25**, 465 (1967).

101. M. M. Warshaw and I. Tinoco, Jr., *J. Mol. Biol.* **13**, 54 (1965).

102. J. Brahms, J. C. Maurizot, and A. M. Michelson, *J. Mol. Biol.* **25** 481 (1967).

103. J. Brahms, A. M. Aubertin, G. Dirheimer, and M. Grunberg-Manago, *Biochemistry* **8**, 3269 (1969).

104. R. C. Davis and I. Tinoco, Jr., *Biopolymers* **6**, 223 (1968).

105. C. A. Bush, *in* "Physical Techniques in Biological Research" (G. Oster, ed.), 2nd ed., Vol. 1, Part A, p. 347. Academic Press, New York, 1971.

106. M. M. Warshaw, *in* "Handbook of Biochemistry" (H. A. Sober, ed.), p. G-96. Chem. Rubber Publ. Co., Cleveland, Ohio, 1968.

107. C. R. Cantor, M. M. Warshaw, and H. Shapiro, *Biopolymers* **9**, 1059 (1970).

108. Y. Inoue, M. Masuda, and S. Aoyagi, *Biochem. Biophys. Res. Commun.* **31**, 577 (1968).

109. D. M. Gray and I. Tinoco, Jr., *Biopolymers* **9**, 223 (1970).

110. I. Tinoco, Jr. and C. R. Cantor, *Methods Biochem. Anal.* **18**, 81 (1970).

111. J. N. Vournakis, H. A. Scheraga, G. W. Rushizky, and H. A. Sober, *Biopolymers* **4**, 33 (1966).

112. Y. Inoue, S. Aoyagi, and K. Nakanishi, *J. Amer. Chem. Soc.* **89**, 5701 (1967).

113. P. O. P. Ts'o, S. A. Rapaport, and F. J. Bollum, *Biochemistry* **5**, 4153 (1966).

114. S. R. Jaskunas, C. R. Cantor, and I. Tinoco, Jr., *Biochemistry* **7**, 3164 (1968).

115. C. R. Cantor, S. R. Jaskunas, and I. Tinoco, Jr., *J. Mol. Biol.* **20**, 39 (1966).

116. D. F. Bradley, I. Tinoco, Jr., and R. W. Woody, *Biopolymers* **1**, 239 (1963).

117. A. S. Schneider and R. A. Harris, *J. Chem. Phys.* **50**, 5205 (1969).

118. A. Rich, D. R. Davies, F. H. C. Crick, and J. D. Watson, *J. Mol. Biol.* **3**, 71 (1961).

119. S. Hanlon and E. O. Major, *Biochemistry* **7**, 4350 (1968).

120. S. Hanlon, unpublished results (1970).

121. R. Langridge and A. Rich, *Nature (London)* **198**, 725 (1968).

122. P. K. Sarkar and J. T. Yang, *J. Biol. Chem.* **240**, 2088 (1965).

123. J. Brahms and C. Sadron, *Nature (London)* **212**, 1309 (1966).

124. G. Green and H. R. Mahler, *Biochemistry* **9**, 368 (1970).

125. F. H. Wolfe, K. Oikawa, and C. M. Kay, *Can. J. Biochem.* **47**, 637 (1969).

126. P. K. Sarkar and J. T. Yang, *Biochemistry* **4**, 1238 (1965).

127. J. Brahms, *J. Mol. Biol.* **11**, 785 (1965).

128. J. Brahms and W. H. F. M. Mommaerts, *J. Mol. Biol.* **10**, 73 (1964).

129. J. T. Yang and T. Samejima, *Biochem. Biophys. Res. Commun.* **33**, 793 (1968).

130. A. Adler, L. Grossman, and G. D. Fasman, *Biochemistry* **7**, 3836 (1968).

131. J. C. Maurizot, J. Brahms, and F. Eckstein, *Nature (London)* **222**, 559 (1969).

132. R. D. Wells, J. E. Larson, R. C. Grant, B. E. Shortle, and C. R. Cantor, *J. Mol. Biol.* (in press).

133. J. C. Maurizot, W. J. Wechter, J. Brahms, and C. Sadron, *Nature (London)* **219**, 377 (1968).
134. A. M. Bobst, F. Rottman, and P. A. Cerruti, *J. Amer. Chem. Soc.* **91**, 4603 (1969).
134a. H. Singh and B. Hillier, *Biopolymers* **10**, 2445 (1971).
135. J. N. Vournakis, D. Poland, and H. A. Scheraga, *Biopolymers* **5**, 403 (1967).
136. C. A. Bush and H. A. Scheraga, *Biochemistry* **6**, 3036 (1967).
137. G. D. Fasman, C. Lindblow, and E. Seaman, *J. Mol. Biol.* **12**, 630 (1965).
138. G. G. Brownlee, F. Sanger, and B. G. Barrell, *Nature (London)* **215**, 735 (1967).
139. C. R. Cantor, *Proc. Nat. Acad. Sci. U.S.* **59**, 478 (1968).
140. D. W. McMullen, S. R. Jaskunas, and I. Tinoco, Jr., *Biopolymers* **5**, 589 (1967).
141. P. J. Oriel and J. Widler, *Nature (London)* **214**, 702 (1967).
142. T. Samejima, H. Hashizumi, K. Imahori, I. Fujii, and K. Miura, *J. Mol. Biol.* **34**, 39 (1968).
143. P. K. Sarkar, B. Wells, and J. T. Yang, *J. Mol. Biol.* **25**, 563 (1967).
144. J. F. Scott and P. Schofield, *Proc. Nat. Acad. Sci. U.S.* **64**, 931 (1969).
145. R. I. Cotter and W. B. Gratzer, *Biochem. Biophys. Res. Commun.* **39**, 766 (1970).
146. M. Maestre and I. Tinoco, Jr., *J. Mol. Biol.* **23**, 323 (1967).
147. T. Samejima and J. T. Yang, *J. Biol. Chem.* **240**, 2094 (1965).
148. P. K. Sarkar and J. T. Yang, *in* "Conformations of Biopolymers" (G. N. Ramachandran, ed.), Vol. 1, p. 197. Academic Press, New York, 1967.
149. G. Bernardi and S. N. Timasheff, *J. Mol. Biol.* **48**, 43 (1970).
149a. F. S. Allen, D. M. Gray, G. P. Roberts, and I. Tinoco, Jr., *Biopolymers* **11**, 853 (1972).
150. S. Arnott, M. H. F. Wilkins, and W. Fuller, *J. Mol. Biol.* **27**, 549 (1967).
151. K. Tomita and A. Rich, *Nature* (London) **201**, 1160 (1964).
152. G. Green and H. R. Mahler, *Biopolymers* **6**, 1509 (1968).
153. M.-J. Tunis and J. E. Hearst, *Biopolymers* **6**, 1218 (1968).
153a. M.-J. Tunis-Schneider and M. F. Maestre, *J. Mol. Biol.* **52**, 521 (1970.)
154. J. H. Strauss, R. B. Kelley, and R. L. Sinsheimer, *Biopolymers* **6**, 793 (1968).
154a. D. J. Gordon, *Biochemistry* **11**, 413 (1972).
155. D. W. Urry, T. A. Hinners, and L. Masotti, *Arch. Biochem. Biophys.* **137**, 214 (1970).
156. D. W. Urry and T. H. Ji, *Arch. Biochem. Biophys.* **128**, 802 (1968).
157. D. Schlessinger, *J. Mol. Biol.* **2**, 92 (1960).
158. A. V. Furano, D. F. Bradley, and L. G. Childers, *Biochemistry* **5**, 3044 (1966).
159. R. I. Cotter, P. McPhie, and W. B. Gratzer, *Nature (London)* **216**, 864 (1967).
160. D. G. Scriba, C. M. Kay, and J. S. Colter, *J. Mol. Biol.* **26**, 67 (1967).
161. M. Maestre and I. Tinoco, Jr., *J. Mol. Biol.* **12**, 287 (1965).
162. V. I. Permogorov, V. G. Debabov, I. A. Shadkova, and B. A. Rebentish, *Biochim. Biophys. Acta* **199**, 556 (1970).
163. J. Sponar, M. Boublik, and I. Fric, *Biochim. Biophys. Acta* **209**, 532 (1970).
164. G. D. Fasman, B. Schaffhausen, J. Goldsmith, and A. Adler, *Biochemistry* **9**, 2814 (1970).
164a. T. Y. Shih and G. D. Fasman, *Biochemistry*, **11**, 398 (1972).
165. F. X. Wilhelm, M. H. Champagne, and M. Daune, *Eur. J. Biochem.* **15**, 321 (1970).
166. I. Isenberg, H. J. Ji, and W. C. Johnson, *Biochemistry* **10**, 2587 (1971).
167. T. Wagner and T. C. Spelsberg, *Biochemistry* **10**, 2599 (1971).

3

HYDRODYNAMIC AND THERMODYNAMIC STUDIES

Henryk Eisenberg

LIST OF SYMBOLS

a, b	Virtual bonds from P to $C_{5'}$ and $C_{5'}$ to P, respectively (Fig. 1).
a_n, a_s	Exponents of M in Eqs. (4) and (6)
a	Persistence length in wormlike chains, $a = A/2$ [Eq. (21)]
b	Force constant in bending of chain (Section II,A,3)
b	Distance between charged groups (Section II,D,4)
c, c_J	Concentration (of component J), grams per milliliter of solution [grams per deciliter in Eqs. (3) and (7)]
f, f_J	Translation frictional coefficient of single particle ($f = P\eta^0 < r^2 > ^{1/2}$)
h	Scattering parameter ($4\pi n/\lambda$) sin ($\theta/2$)
l	Length of bond in chain backbone (Section II,A,3)
m_J	Molality (moles n_J of component J per kilogram of component 1)
m_u	Equivalent molality, nucleotides per kilogram of component 1
n	Number of bonds in chain backbone (Section II,A,3)
n, n^0	Refractive index of solution and of solvent
n	Number of scattering elements (Section II,A,3)

r	Distance to center of rotation in ultracentrifuge
r_b	Center of band in equilibrium sedimentation in a density gradient (Section II,D,4)
$\langle r^2 \rangle, \langle r^2 \rangle_0$	Mean square, and unperturbed mean square end-to-end distance of chains [Eq. (21)]
s, s^0	Sedimentation constant at finite and at vanishing macromolecular concentration, Svedberg units
$\bar{v}, \bar{v}^0$	Partial specific volumes at finite and at vanishing component concentration
w_J	Molality (grams of component J per gram of component 1)
x	Segments [Eq. (13)] or virtual bonds [Eq. (28)] in chain
z	Excluded volume variable [Eq. (13)]
A	Length of Kuhn statistical element ($A = 2a$) [Eqs. (1) and (2)]
A_2	Second virial coefficient (mole $\times$ ml/gm^2)
$[B1], [B3], [B4]$	Binding (grams per gram) of components 1, 3, and 4 to macromolecular species (Section II,B,4)
C_u	Concentration, equivalents of phosphor (or nucleotides) per liter of solution
C_J	Concentration, moles of component J, per liter of solution
D_2	Diffusion coefficient [Eq. (59)]
$[E3]$	Donnan exclusion (grams per gram) of component 3 by macromolecular species (Section II,B,4)
H	Light scattering constant [Eq. (11)]; equal to $(2\pi^2/N_{Av})(n^2/\lambda^4)$ $(\partial n/\partial c_2)^2_{P,\mu_3}$ for unpolarized incident light
J_2	Flow of component 2 [Eq. (58)]
K_η, K_s	Proportionality constants in Eqs. (4) and (6)
L	Contour length of chain
L_{22}, L_{23}	Onsager phenomenological coefficients [Eq. (58)]
M	Molecular weight
M_n, M_w	Number and weight average molecular weight
M_L	Mass per unit length (dalton per angstrom)
N_{Av}	Avogadro's number
N	Number of statistical elements in chain [Eqs. (1) and (2)]
P	Coefficient in equation for frictional coefficient f
$P_z(h)$	z-average of particle scattering factor [Eq. (19)]
$\langle R^2 \rangle, \langle R^2 \rangle_0$	Mean square radius (of gyration) of chains and of Gaussian chains
$\Delta R(h), \Delta R(0)$	Reduced scattering intensity
$\langle R^2 \rangle_z$	z-Average of mean square radius [Eq. (20)]
T_p, T_c	Phase separation and critical temperatures [Eq. (18)]
V_m	Volume (in milliliters) of solution containing 1 kg of component 1 [Eq. (29)]
X	Concentration ratio C_u/C_3 of polyelectrolyte to added simple salt (Section II,C,3)
X	$h^2\langle R^2 \rangle$ or $h^2\langle r^2 \rangle/6$, for Gaussian chains [Eq. (24)]
X	$L/2a$, Kuhn elements in chain [Eq. (25)]
Z	Number of nucleotides per nucleic acid, or number of charges per macromolecule

α	Cosine of bond angle complement θ (Section II,A,3,a)
α, α_η	Chain expansion parameters [Eq. (9)]
α	Ratio of weights of ^{15}N and ^{14}N DNA species (Section II,D,4)
β	Flory–Mandelkern constant, Eq. (7)
β	Mutually excluded volume for pair of segments [Eq. (13)]
$\gamma_\pm, \gamma_i, \gamma_1, \gamma_{1p}$	Activity coefficients (Section II,C,3)
$\delta = r - r_b$	Distance from center of band in equilibrium sedimentation in a density gradient (Section II,D,4)
ε	Excluded volume parameter [Eq. (8)]
η, η^0	Viscosity of solution and of solvent
$[\eta]$	Intrinsic viscosity, $\lim_{c \to 0} (\eta_{sp}/c)$
θ	Scattering angle
θ	Bond angle complement (Section II,A,3)
θ	Effective charge (Section II,D,3)
κ	Characteristic Debye radius (Section II,C,2)
λ	Wavelength *in vacuo*
μ	Chemical potential
ξ_3, ξ_4, ξ'_1	Preferential interaction parameters on weight molality basis (Section II,B,2)
ξ	Manning ion-cloud condensation parameter (Section II,C,3)
π	Osmotic pressure
ρ, ρ^0	Density of solution and of solvent
ρ	Curvature of chain (Section II,A,3)
ρ_b	Density at center of band (Section II,D,4)
σ	Standard deviation of band [Eq. (56)]
ϕ, ϕ_p	Osmotic coefficients (Section II,C,3)
ϕ'	Apparent specific volume $1 - \phi'\rho \equiv (\partial\rho/\partial c_2)_\mu$ (Section II,D,3)
χ	Gradient of $(\partial\rho/\partial c_2)_\mu$ at r_b [Eq. (54)]
ψ	Electrostatic potential
ψ', ψ''	Bond rotation angles (Fig. 1)
ω	Angular velocity
ω', ω''	Bond rotation angles (Fig. 1)
Γ_3	Membrane distribution parameter in three-component system (Section II,B,2)
Γ'	Part of Γ_3 due to Donnan exclusion [Eq. (41)]
Θ	Flory (Θ) temperature
Φ	Flory universal constant [Eq. (3)]

I. Introduction

In recent years a number of excellent reviews have been published on hydrodynamic studies, and to a somewhat lesser extent, on thermodynamic aspects in nucleic acids research. An extensive recent survey of hydrodynamic properties of DNA is due to Bloomfield [1]; within the framework of a symposium Sadron [2] has discussed physicochemical properties of the DNA macromolecule. The rheology of DNA solutions has been dealt with by

Robins [3], and there are two fairly recent survey articles on macromolecular structure and properties of DNA by Josse and Eigner [4], and Edwards and Shooter [5]. The most recent comprehensive review on polyelectrolytes, in which the reader will find information both with respect to theoretical aspects and examples referring to the polyelectrolyte nature of DNA, is the article by Armstrong and Strauss [6]. Extensive references to earlier papers, review articles, and books will be found in these articles.

A review article on a subject of topical interest can be written in one of two ways. It is possible to give a broad survey and mention every possible aspect of the study. Such an approach risks being shallow and often precludes a study in greater depth. It is possible, on the other hand, to pick a smaller number of topics, necessarily within the specific range of interest of the writer himself, and to attempt a more critical discussion, hopefully in greater depth. In view of the fact that the encyclopedic approach is adequately covered in annual reviews, progress reports, and archives volumes, the second alternative will be attempted in this chapter. In particular the writer hopes that he will adequately discuss conflicting viewpoints on controversial problems, but despite this avowed intention he carries the conviction of his own opinion and a complete lack of bias should therefore not be expected. The discussion must perforce follow a pattern closely related to his own past and present research interests.

When discussing *basic theoretical aspects* of nucleic acids in solution (Section II) one may stress their macromolecular nature or their polyelectrolyte character. These long-chain molecules may assume in solution open, linear, or circular conformations, as nucleic acids often do; alternatively, as has for instance been found in the case of transfer RNA [7–9] and of supercoiled circular DNA [10,11], they may fold into specific tertiary structures, almost recalling globular proteins. In *macromolecular aspects* (Section II,A) we shall discuss some basic features embodied in the polyribo- and polydeoxyribonucleotide backbone chains; we shall briefly refer to the possibility of rotation around backbone bonds in the "random" single-stranded form and shall discuss short- and long-range interactions in idealized and real chains.

The presence of charged groups on the nucleic acid chains, as well as counterions required for electroneutrality and added low-molecular weight simple electrolytes, create problems to be discussed within the scope of *thermodynamic aspects* (Section II,B) such as thermodynamics of multicomponent systems, osmotic pressure, membrane distribution, binding of small molecules and of polycyclic dyes. *Polyelectrolyte aspects* (Section II,C) also play an important, if not dominant, role in the unraveling of nucleic acid behavior. We shall briefly review basic polyelectrolyte theory and its application to flexible and rigid particles, in particular the rigid-rod model. The molecular weight, size, and shape of large molecules, and interactions between them are best determined by *equilibrium properties* (Section II,D) such as light- and small-angle x-ray scattering, equilibrium sedimentation, and the use of buoyant densities in a density gradient. *Transport phenomena* (Section II,E), such as intrinsic viscosity, flow birefringence, and dichroism and sedimentation velocity may also be used, albeit with less reliability than the better founded equilibrium methods. The usefulness of the transport methods often resides in the relative ease with which viscosity or sedimentation velocity experiments may be performed, although precise theoretical interpretation may not always be possible.

Finally, we would like to focus on some *specific topics* in our concluding discussion (Section III). These include the size and shape of intact nucleic acids and of nucleic acid fragments, the obtainment of correct molecular weights from equilibrium sedimentation in a density gradient, the inherent flexibility and folding of DNA, problems related to hydration, "preferential" solvation, Donnan effects and the partial specific volumes in DNA solutions, the binding to DNA of metallic ions, of polycyclic dyes, and some properties of synthetic polynucleotides.

Among the many important topics which will not be discussed in this chapter we mention the problem of nucleic acid–protein interactions [12], the study of "compact" forms of DNA [13–15], or the unusual physicochemical properties of mitochondrial DNA [16,17] or RNA [18] for instance.

II. Basic Theoretical Aspects

A. MACROMOLECULAR ASPECTS

1. *The Gaussian Chain*

Ever since Kuhn [19] and Guth and Mark [20] pioneered the use of the statistical approach in the study of the conformation of macromolecules, it was generally accepted that conceptual and computational means then at the disposal of the investigators did not permit evaluation of properties of "real" molecules; the latter are characterized by chemical bonds of known structure, hindered rotation around these bonds, and interactions between close and distant groups along the molecular chains. To describe the properties of chain molecules in solutions, it was necessary to substitute "ideal" models for the "real" chains, to compute (with variable degree of success) measurable properties of the systems. A well-known and useful idea is the random flight concept, in which the real chain is replaced [19] by a number of so-called statistical chain elements, whose orientations in space are assumed to be mutually independent. The random flight problem yields, for a chain of contour length L composed of N statistical elements of length $A (L = NA)$

$$W(\mathbf{r}, N) \, d\mathbf{r} = (2\pi NA^2/3)^{-3/2} \exp \left(-3r^2/2NA^2\right) d\mathbf{r} \tag{1}$$

$$\langle r^2 \rangle_0 = NA^2 \tag{2a}$$

$$= LA \tag{2b}$$

where $W(\mathbf{r}, N) \, d\mathbf{r}$ is the probability that one end of the chain is at position $\mathbf{r}$ with respect to the other end and $\langle r^2 \rangle_0$ is the unperturbed mean square end-to-end distance of this so-called Gaussian chain ($\langle r^2 \rangle_0$ may change to $\langle r^2 \rangle$ for chains with excluded volume perturbation, as will be shown below). The ideal chain has the same values of L and $\langle r^2 \rangle_0$ as the real chain and any properties which depend on these quantities are adequately represented. On the other hand, quantities which depend on the detailed arrangements of segments in the chain (such as, for instance, some optical, dielectric, and other properties) cannot be satisfactorily interpreted by this idealized model, in which the local structure of the chain is not taken into account.

Experimental quantities characteristic of the overall dimensions of the macromolecules, for instance, the mean square radius $\langle R^2 \rangle_0$ (equal to $\langle r^2 \rangle_0/6$ for Gaussian chains), are derived from light-scattering experiments [21] or from the intrinsic viscosity $[\eta]$ and the sedimentation coefficient s [22]. Generally for high molecular weight chains

$$[\eta] = \Phi \langle r^2 \rangle^{3/2}/M \tag{3}$$

where Φ is a universal constant [23,24] and M is the molecular weight; $[\eta] = \lim_{c \to 0}(\eta_{sp}/c)$, where η_{sp} is the specific viscosity of the solution and c is the concentration. For Gaussian chains $\langle r^2 \rangle_0$ is proportional to N, and N is proportional to M; substitution into (3) yields the familiar result

$$[\eta] = K_\eta M^{a_\eta} \tag{4}$$

where K_η is a constant characteristic of a given polymer solvent system and $a_\eta = 0.5$, for the Gaussian chain.

The sedimentation coefficient s^0, for $c \to 0$, for a two-component polymer–solvent system is given by

$$s^0 = \frac{M(1 - \bar{v}^0 \rho^0)}{N_{Av} f} \tag{5}$$

where $\bar{v}^0$ is the limiting partial specific volume of the macromolecule, ρ^0 is the density of the solvent, N_{Av} is Avogadro's number and f is the translational frictional coefficient. The latter is, for Gaussian chains, proportional to $\langle r^2 \rangle_0^{1/2}$, and it therefore follows that

$$s^0 = K_s M^{a_s} \tag{6}$$

where the constant K_s is again characteristic of a given polymer solvent system and $a_s = 0.5$ for Gaussian chains.

A useful relationship is derived [25] on the consideration that, whereas $[\eta]$ is proportional to the cube of a characteristic linear dimension, s is inversely proportional to the linear dimensions itself, multiplied by the solvent viscosity η^0. Combination of Eqs. (5) and (3) with $f = P\eta^0\langle r^2\rangle^{1/2}$ yields

$$\frac{s^0\eta^0[\eta]^{1/3}N_{Av}}{M^{2/3}(1 - \bar{v}^0\rho^0)} = \beta \tag{7}$$

In the limiting case of nondraining random coils, the dimensionless constant β in Eq. (7) has been derived by Mandelkern and Flory [25] from the Kirkwood–Riseman theory, and is equal to $P^{-1}\Phi^{1/3}$; theoretical values of P and Φ for Gaussian coils are 5.2 and 2.66×10^{21} [26] respectively, with $[\eta]$ given in the conventional units, dl/gm. This leads to a value of about 2.7×10^6 for β.

2. *The Excluded Volume Problem*

Gaussian chains have no excluded volume [24,27]. Random flight statistics which lead to Eqs. (1) and (2) do not take into account the fact that real chains occupy space and many configurations available to idealized chains are excluded from the real chain. This contingency cannot be overcome by adjusting the size of the statistical element; $\langle r^2\rangle/N$, for the chain with excluded volume is not independent of N, for large N. It is customary to write [28] instead of Eq. (2a)

$$\langle r^2\rangle = N^{1+\varepsilon}A^2 \tag{8}$$

where ε may range from a slightly negative value to around 0.2; the excluded volume refers to actual volume exclusion (which leads to positive values ε) and repulsive or attractive interaction potentials, which raise or lower ε, respectively. In a good solvent ε may increase to its highest value; on the other hand, a poor solvent (or solvent mixtures) can be found in which attraction between chain segments predominates and ε vanishes (further decrease of ε to negative values may lead to phase separation and precipitation of the macromolecules). One of the definitions of the Flory, or Θ temperature, in polymer solutions, is the temperature at which, in a given polymer solvent system $\varepsilon = 0$ for real chains. Under these circumstances long range volume exclusion is compensated by attractive forces and Gaussian coil statistics are believed to apply [24].

For chains with volume exclusion a_η, in Eq. (4), equals $(1 + 3\varepsilon)/2$ and a_s in Eq. (6) equals $(1 - \varepsilon)/2$; the two coefficients are related by [24] $a_\eta - 0.5 = 3(0.5 - a_s)$. The factor α by which the macromolecular chain is expanded beyond random flight dimensions by the excluded volume effect may be defined by

$$\alpha^2 \equiv \langle R^2\rangle/\langle R^2\rangle_0 \qquad (9a)$$

where $\langle R^2\rangle$ is the mean square radius of the real chain and $\langle R^2\rangle_0$ refers to the unperturbed conditions. An expansion coefficient α_η often used is defined by

$$\alpha_\eta \equiv ([\eta]/[\eta]_\Theta)^{1/3} \qquad (9b)$$

where the subscript Θ refers to the value of $[\eta]$ unperturbed by long-range interactions, i.e., at Θ conditions.

Chains with no intramolecular excluded volume are believed to have zero intermolecular excluded volume as well [24]. This means that at Θ conditions certain experimental quantities such as π/RTc and $Hc/\Delta R(0)$ equal M^{-1} (for monodisperse macromolecules) at all reasonably low concentrations (π is the osmotic pressure, $\Delta R(0)$ is the reduced scattering intensity, extrapolated to zero scattering angle θ, and corrected for solvent scattering; H is a constant [29]. Also, at Θ conditions the slope $dlnc/dr^2$ in equilibrium sedimentation equals $(\omega^2/RT)M(1 - \bar{v}^0\rho^0)$, independent of concentration (r is the distance to the center of rotation, ω is the angular velocity). In the more usual case these quantities do depend on concentration and we have, for real chains, at low concentrations c (in grams per millimeter)

$$\frac{\pi}{cRT} = \frac{1}{M} + A_2c + \cdots \qquad (10)$$

$$\frac{Hc}{\Delta R(0)} = \frac{1}{M} + 2A_2c + \cdots \qquad (11)$$

and

$$(1 - \bar{v}^0\rho^0)\frac{\omega^2}{2RT}\left(\frac{dlnc}{dr^2}\right)^{-1} = \frac{1}{M} + 2A_2c + \cdots \qquad (12)$$

where A_2 is the second virial coefficient, and constitutes a thermodynamic measure of the intermolecular excluded volume. When $A_2 = 0$ the excluded volume vanishes. The situation is reminiscent of the behavior of the van't Hoff gas at the so-called Boyle temperature in imperfect gas kinetics; the second virial coefficient in the expansion of the pressure in powers of the concentration vanishes, although higher virial coefficients, important at higher concentrations only, may stay finite.

The excluded volume constitutes one of the major incompletely solved problems in the physical chemistry of polymer solutions [30]. In recent years it has been customary, in the study of the excluded volume in polymer solutions, to use [27,31] a dimensionless variable z

$$z \equiv (2\pi\langle R^2\rangle_0)^{-3/2}\beta x^2 \tag{13}$$

where β is the mutually excluded volume for a pair of segments of which there are x in the chain (we shall not define β and x more precisely but note that $\beta x^2/2$ is the total volume excluded pairwise by one macromolecule). The intramolecular expansion coefficient α defined by Eq. (9a) may be expressed in terms of a slowly converging series in powers of z; an application of the McMillan–Mayer theory to polymer solutions yields

$$\alpha^2 = 1 + (134/105)z + \cdots \tag{14}$$

and of the higher terms, only the term in z^2 is known [32]. Usefulness of the series expression is limited to a region close to the characteristic temperature Θ, at which β vanishes; approximate closed expressions have given [23,30] but their discussion is beyond the scope of this study.

The second virial coefficient is given by a single contact term (invariant with molecular weight), multiplied by a dimensionless function $F(z)$ of z

$$A_2 = (N_{Av}/M^2)(\beta x^2/2)\, F(z) \tag{15}$$

$F(z)$ is an imperfectly known function which reduces to unity when $z \to 0$. Substituting the definition of z, Eq. (13), yields

$$A_2 = 4\pi^{3/2}N_{Av}(\langle R^2\rangle_0/M)^{3/2}M^{-1/2}z\, F(z) \tag{16}$$

The influence of the excluded volume in affecting properties of DNA solutions has been often debated [33]. Detailed information on properties of DNA solutions at the Θ temperature is not known. One way of taking excluded volume into account is by determining a suitable value for ε in Eq. (8) from the study of the functional relationship in a given solvent system between either $[\eta]$ or s and M [28]. A more direct way [33] makes use of the fact that A_2 in DNA solutions can now be determined in more reliable fashion [34]. Combination of Eqs. (9a), (14), and (16) yields (with the assumption that for small values of z, $F(z) \sim 1$)

$$\langle R^2\rangle/\langle R^2\rangle_0 = 1 + 0.0572A_2M^2/N_{Av}\langle R^2\rangle_0^{3/2} \tag{17}$$

The influence of the excluded volume in expanding chain dimensions can thus be estimated from Eq. (17). In view of the approximations inherent in its derivation it will be useful only if the excluded volume is small. Solvent systems can sometimes be found which approximate such conditions, but very few studies of polynucleotide systems have been undertaken at Θ conditions proper [35].

Another characteristic feature of polymer solutions is that precipitation occurs close to the Θ temperature [24,36]. Critical temperatures T_c for phase separation may be either upper or lower critical solute temperatures [37] depending on whether T_c is a maximum or a minimum, respectively, in a plot of phase separation temperature, T_p, against polymer concentration. In the first case phase separation occurs by lowering, in the second case by raising the temperature. Both cases may be found in the same polynucleotide system [35]. In the vicinity of the Θ temperature it is possible to derive [24]

$$T_c{}^{-1} = \Theta^{-1} + bM^{-1/2} \tag{18}$$

where b is a constant. The agreement between the value of Θ obtained from the vanishing of A_2 in the virial expansion of the osmotic pressure, (by either light scattering or equilibrium sedimentation measurements) and the values derived from phase separation studies, provides strong support for the internal consistency of the correlation of properties of polymer solutions in terms of their excluded volume properties.

We have hitherto assumed that the macromolecules in solution are homogeneous and all characterized by the same value of M. While this may be true for some intact virus DNA particles or some well-characterized transfer RNA molecules, for instance, it is not necessarily true for the major large class of DNA fragments, synthetic polynucleotides, and others. In all such cases problems in the analysis arising because of polydispersity or density heterogeneity have to be considered and will be discussed in the proper context.

3. *The Persistent Chain*

a. *General Considerations.* From the angular dependence of scattered light [29] it is possible to derive information on the mass and distribution of mass of macromolecular particles. For Eq. (11) we substitute [38]

$$\frac{Hc}{\Delta R(h)} = \frac{1}{M_w P_z(h)} + 2A_2 c + \cdots \tag{19}$$

where $\Delta R(h)$ is the reduced scattering intensity (corrected for solvent contribution); $h = (4\pi n/\lambda) \sin (\theta/2)$, n is the refractive index of the solution, λ is the incident wavelength *in vacuo*, and θ is the scattering angle; M_w is the weight average molecular weight ($M_w = \sum c_i M_i / \sum c_i$, the summation is over all species i) in polydisperse systems, $P_z(h)$ is the z-average of the particle scattering factor, normalized to $P(0) = 1$ (independent of particle size distribution). For small values of h it is possible to expand the reciprocal particle scattering factor $P_z{}^{-1}(h)$ in powers of h

$$P_z(h) = 1 + \tfrac{1}{3}h^2 \langle R^2 \rangle_z + \cdots \tag{20}$$

where $\langle R^2 \rangle_z$ is the z-average ($\sum c_i M_i \langle R^2 \rangle_i / \sum c_i M_i$) of the mean square radius $\langle R^2 \rangle$ of the particles. The leading term in the expansion in Eq. (20) is independent of any physical model for the chain and the experimental points are on the limiting slope for values $h^2 \langle R^2 \rangle_z < 0.3$. More detailed information may be obtained if the complete angular dependence curve is examined.

Early results indicated that known theoretical curves for Gaussian coils [21] and rigid rods [22] were not in agreement with the experimental results from DNA solutions [39]. A different kind of model was required. Intuitively we would expect that over short distances along the rigid Watson–Crick double-helix rodlike behavior is indicated, whereas over larger distances bending may occur due to the large size of the chains and the natural flexibility of the structure. The type of behavior encountered may depend on the type of the probe used. Small angle x-ray scattering is analogous to scattering with visible light [40] but is more sensitive to the rodlike nature of the DNA structure over short distances. From the angular dependence of the scattered light, on the other hand, information may be derived on the bending of DNA molecules over larger, macromolecular distances.

In 1949, in an attempt to interpret the behavior of stiff macromolecules, the persistent chain model [41,42] was proposed by Kratky and Porod. It envisages a stiff chain which, at short lengths of the chains approximates rodlike behavior. With increasing chain length coiled configurations come into play, and at very long chain lengths the model asymptotically approximates the Gaussian chain. The persistent chain, more commonly known as the "wormlike" chain [43], is characterized by a continuous curvature of the chain skeleton, the direction of curvature at any point of the trajectory being random. Notwithstanding the fact that the persistent chain qualitatively successfully mimics DNA behavior over a very wide range of molecular weights, quantitative agreement with the "real" chains, in particular for intermediate values of the molecular weight, is still lacking.

For evaluation of the basic features of the persistent chain, we start [44] with a hypothetical freely rotating chain constructed of n bonds of length l each, joined at fixed bond angles $\pi - \theta$; the average projection of the kth bond on the direction of the first bond is $l\alpha^{k-1}$, where $\alpha = \cos \theta$. The persistence length, a, is defined by the sum of all these projections, for an infinitely long chain ($n \to \infty$), $a = l/(1 - \alpha)$. Next the finite freely rotating chain, with fixed bond length and angle, is transformed into a continuously curving chain (the wormlike chain) by letting both l and θ go to zero, while keeping the total contour length $L = \alpha n l$ of the chain constant. One derives for $\langle r^2 \rangle_0$

$$\langle r^2 \rangle_0 = 2aL[1 - (a/L)(1 - e^{-L/a})] \tag{21}$$

For very short stiff chains ($L/a \ll 1$) the exponential in Eq. (21) may be expanded to yield the limiting value $\langle r^2 \rangle_0 = L^2$, which is the relationship applicable for a rigid rod; for very long chains ($L/a \gg 1$) Eq. (21) simplifies to $\langle r^2 \rangle_0 = 2aL$, which is the relationship applicable to Gaussian coils [Eq. (2a)] when the persistence length a is just one-half the length of the Kuhn statistical element A. One may ask whether the persistence chain model correctly represents DNA of intermediate length which is neither completely rodlike nor coiled and whether the simple model correctly describes scattering and frictional properties experimentally derived.

A derivation of Eq. (21) on a totally different basis [45] is based on the elastic properties of a curving chain and provides a physical basis for the rather abstract persistence length a, in terms of the energy changes involved in fluctuations (in a Brownian field) in the curvature of long-chain molecules. It is assumed that the chain is originally straight and undergoes weak bending in the sense that the curvature ρ is small at every point. The free energy of the elastically bent molecule is then given as a linear function in the square of the curvature vector. Statistical considerations lead to

$$\langle \cos \theta(l) \rangle = \exp(-lkT/b) \tag{22}$$

where $\theta(l)$ denotes the angle between two tangents to the curve separated by a distance l along the chain, and b is the force constant in the relationship $\Delta F = 0.5\, b\rho^2$. Equation (22) shows that for large values of l the mean values of $\langle \cos \theta(l) \rangle$ approach zero, that is, distant sections along the macromolecule are statistically independent. (In this treatment torsion is neglected and the average cosine of the angle between the planes $(\mathbf{t}_a, \mathbf{t}_b)$ and $(\mathbf{t}_b, \mathbf{t}_c)$ of successive tangents to the chain equals zero.) In the final step of the calculation a result identical to Eq. (21) is obtained for $\langle r^2 \rangle_0$ with the persistence length a in the previous derivation equal to $a = b/kT$. Thus we expect a to decrease with increasing temperatures; at low temperatures, on the other hand, a more rodlike behavior is expected.

The above derivations assumed either free rotation of backbone valence bonds or absence of torsional hindrance in the chain. Recently Bugl and Fujita [46] have calculated some dynamical aspects of a long polymer backbone with both bending and torsional hindrance. They find the total configurational elastic energy of a thin wire to be given by

$$\tfrac{1}{8}\pi R^4 \int_0^L (E\rho^2 + 2\mu\tau^2)\, dl \tag{23}$$

where E is Young's modulus, μ is the modulus of rigidity, R the radius of cross section, ρ the curvature, and τ the torsion of the space curve characterizing the unstrained wire of minimum energy; $\langle r^2 \rangle_0$ for this model has not been evaluated. In an extension of the Kratky–Porod chain model, Kirste [47] introduces torsional properties into the model by considering a hypothetical chain with hindered, rather than with free rotation. By forming a limiting process which includes not only l and θ, but also the angle of internal hindered rotation θ, Kirste derives equations for wormlike chains with parameters for both curvature and torsion. Applications to DNA have not yet been discussed but may shed interesting results with respect to the intermediate range of DNA chain lengths.

The effort expended on the wormlike chain model over more than 20 years derives from the fact that a distribution function is to date not available for the ends of the chain, or for distances along the chains, such as given by Eq. (1) for the Gaussian chain. To calculate the scattering from such molecules, or hydrodynamic interaction parameters for the evaluation of $[\eta]$ and s, various averages (moments of the distribution function) are required. These moments are not easily obtained if a distribution function is not known. We face the twin difficulty of (1) the correspondence of the idealized model to the real chain, and (2) mathematical and other difficulties in the evaluation of the model itself.

b. *Scattering of Light and X-Rays.* The angular dependence of scattering with visible light and x-rays (at small angles) yield essentially identical results, within the understanding that increasingly finer details of macromolecules are scanned with increasing value of the parameter h [29]. An idealized chain or calculated scattering curve may be quite adequate for light scattering (which is sensitive to larger distances between scattering centers) but not sufficient for small-angle x-ray scattering, in which interference between close scattering centers on the same molecule is more important. Kratky and Porod [42] have approximately evaluated the scattering for very long persistent chains; their early calculation was later improved by Porod [48]. According to Debye [21] the particle scattering factor $P(h)$ for n scattering elements connected in a rigid structure (normalization to $P(0) = 1$ allows atomic shape factors to be taken equal to unity) is

$$\frac{1}{n^2} \sum_i \sum_j \frac{\sin hr_{ij}}{hr_{ij}}$$

where the summation extends over all distances r_{ij} between scattering elements. For chains which are capable of assuming a large number of configurations $\sin hr / hr$ has to be replaced by an average $\langle \sin hr/hr \rangle$ which in turn is replaced by an integral

$$\int_0^\infty H(n) \, dr \, \sin hr/hr$$

where $H(n)$ is the distribution for the endpoints of a chain containing n elements. For very long Gaussian chains $\langle \sin hr/hr \rangle \rightarrow \exp\left(-h^2\langle r^2\rangle_n/6\right)$ from which Debye calculated

$$P(h) = (2/X^2)[\exp(-X) - (1 - X)] \tag{24}$$

where $X = h^2 \langle r^2 \rangle_n / 6$. For $X \to 0$ Eq. (24) yields the correct limiting behavior $1 - X/3$, in good agreement with the law of Guinier [49], or the well-known result of Zimm [38] [Eqs. (19) and (20) above]; the latter states that the intercept of a double extrapolation of $Hc/\Delta R(h)$ to $c \to 0$ and against $h^2/3$ yields M_w^{-1}, whereas the ratio of limiting slope over intercept equals $\langle R^2 \rangle_z$. However, M_w^{-1} and $\langle R^2 \rangle_z$ are obtained for chains or particles of any kind, independent of the shape of the particles. It is only necessary that low enough values of $X = h^2 \langle R^2 \rangle_z$ be experimentally accessible, a condition unfortunately rarely applicable in the case of high molecular weight DNA.

The expansion of $P(h)$ of Eq. (24) at large values of X indicates that asymptotic convergence is to be expected for $h^2 P(h)$. This is not the expected behavior for stiff chains in which an asymptotic limit for $hP(h)$ is predicted (this is the scattering behavior expected from infinite rods). In the original calculation of the scattering function of long persistent chains, the Gaussian approximation $\exp(-h^2 \langle r^2 \rangle_n / 6)$ for $\langle \sin hr/hr \rangle$ was maintained but for $\langle r^2 \rangle_n$ the values derived from Eq. (21) were used [42]. Some improvement was achieved by Porod [48] who also calculated $\langle r^4 \rangle_n$ and achieved a somewhat better characterization of the width of the distance distribution function, and a better numerical approximation to the scattering function; the inner branch of the scattering curve varied with h^{-2}, the outer branch with h^{-1}, as expected from stiff rods. A plot of $\Delta R(h)h^2$ with h exhibits an initial flat portion, followed by a curve linearly increasing with h. From the transition region between the two curves the persistence length could be estimated.

Hermans and Hermans [50] developed an exact treatment for the scattering of zigzag chains, a model of rods connected by universal joints. This model was thought [2] to have some relevance to DNA in terms of actual structure. Somewhat later Luzzati and Benoit [51] calculated the asymptotic form of the scattering function for chainlike particles, for various configurations, including chains with continuous curvature and the zigzag chain model. They do not find a term in h^{-2} for the continuously curving chain, contiguous to the region in which a dependence with h^{-1} is predicted.

Hermans and Ullmann [43] set up a differential equation for the distribution of positions and orientations in the wormlike chain (from which the higher even moments of the distribution of ends could be derived) for the improved calculation of scattering functions. These calculations conformed to an approximate distribution function devised by Daniels [52]. Peterlin performed similar calculations [53,54] and appears to be the first worker to interpret early results of scattering from DNA solutions by the model of the persistent chain. Benoit and Doty [55] exactly calculated $\langle R^2 \rangle$ of chains with free and hindered rotation and of the wormlike chain. The latter is given by

$$12\langle R^2 \rangle/L^2 = (3/2X^2)[4X/3 - 2 + (2/X) - (1 - \exp\{-2X\}/X^2] \quad (25)$$

where $X = L/2a$; Eq. (25) correctly yields the rigid rod relation $\langle R^2 \rangle = L^2$ in the limit $X \to 0$ and the Gaussian coil relation $6\langle R^2 \rangle = 2aL$ at large values of X. Peterlin [54] has discussed the use of Monte Carlo stochastic methods and the effect of excluded volume and polydispersity on the scattering curves as well as the inadequacy of the use of the Gaussian approximation for stiff chains (in particular for small-angle x-ray scattering). The radius of gyration of stiff chain molecules as a function of chain length and interaction with solvent was recently evaluated numerically by Kirste [56] by a Monte Carlo procedure.

More work has recently been devoted to the calculation of scattering curves for wormlike chains with excluded volume. Sharp and Bloomfield [57] use the Daniels distribution modified for excluded volume effects to compute averages for scattering elements separated by chain distances large with respect to the persistence length of the chain. For closer scattering elements they use an expansion in terms of the exactly known moments of the distribution for the wormlike coil without excluded volume. For the limiting case of the Daniels distribution with the excluded volume parameter ε [Eq. (8)] equal to zero, their result is significantly different from the result of Peterlin [53,54] who used a Gaussian average for $\langle \sin hr_{ij}/hr_j \rangle$ and then substituted values of r_{ij} derived from the wormlike chain in the resulting exponential. Use has also been made [58] of a distribution function $P_n(r)$ for end-to-end distances r in a polymer chain of n segments with excluded volume, derived from lattice considerations and consistent with computer calculations [59]

$$P_n(r) = C_n r^\delta \exp\left(-r/\sigma\right)^\delta$$

Here C_n is a normalizing constant, σ is related to the standard deviation of r, and $\delta = 2/(1 - \varepsilon)$ [60]. Similar work based on the related distribution function of Mazur [61] has been reported. Hays *et al.* [33] criticize the current procedure of deriving the excluded volume parameter ε from hydrodynamic studies. They prefer to estimate the excluded volume more directly from A_2, and conclude that, even for very high molecular weight DNA, the role of the excluded volume in the evaluation properties has been much overestimated.

Theoretical papers on the wormlike chain and related models, as well as attempts to derive complete distribution functions are still forthcoming. Kumbar [62] used the Rouse–Bueche model to study the mean square radius of ring and chain molecules with increasing local chain stiffness and excluded volume. Sanchez and von Frankenberg [63] discuss the "correlated" chain and Tagami [64] discusses a model for which it is claimed that all the higher moments of the distribution function may be derived. Horta [65] discusses the validity of the Gaussian approximation in simple chain models for which complete distribution functions may be derived. Yeh and Isihara

[66] devise a segment distribution function bearing on light scattering and hydrodynamic properties of chains with excluded volume. It does not seem that a definitive solution of this problem has been obtained. It is a melancholy reflection that so much theoretical work has been devoted to an idealized model which represents only a pale facsimile of the real chain.

c. *Hydrodynamic Properties.* The sedimentation constant of broken chains and wormlike coils without excluded volume was evaluated in graphical form by Peterlin [67] and in analytical form by Hearst and Stockmayer [68] using the hydrodynamic method of Kirkwood. This method considers the translational (or rotational) diffusion constant for an array of spheres (which replace the real chain) in terms of the mean reciprocal distances $\langle l^{-1} \rangle$ between friction elements; these are derived from the statistical model. Hearst and Stockmayer [68] considered the broken chain (the zigzag model of straight rigid segments of equal length connected by universal joints) and the wormlike chain; for high values of the stiffness parameter and short molecular lengths, the latter was aptly labeled the weakly bending rod. The Daniels distribution was used with a suitable correction for small values l/A. The final equations reveal a dependence of s upon the square root of n, the number of elements, or the molecular weight M. In contrast to the Gaussian coil [Eq. (6)] this plot has a nonzero intercept, which is characteristic of the parameters of the chain. The slope of the plot of s (in Svedberg units) vs $M^{1/2}$ obtained by substituting the value of the frictional coefficient $f = P\eta^0(LA)^{1/2}$ into Eq. (5) is given by

$$3.24 \times 10^{-4}(1 - \bar{v}\rho)(M_L/A)^{1/2}/\eta^0 \tag{26}$$

Here M_L is the mass per unit length (dalton per angstrom) of the chain and A is in angstroms. The intercept for wormlike chains is given by

$$1.76 \times 10^{-4}(1 - \bar{v}\rho)(M_L/\eta^0)[b/d + n(A/b) - 2.431 + 0(b/A)] \tag{27}$$

where d is the effective Stokes diameter of a frictional element and b is the spacing between them along the contour of the chain. (A similar expression pertains to the broken zigzag chain.) The authors assume that the chain can be represented by touching spheres (pearl necklace) and therefore $d = b$. This eliminates one parameter and allows analysis of the dependence of s on $M^{1/2}$ in terms of two parameters, a and b. The values obtained are sensitive to the magnitude of the numerical coefficients in Eqs. (26) and (27); these cannot be exactly evaluated because of approximations inherent in the solution of the hydrodynamic problems. Kuhn *et al.* [69] proceeded in different fashion; they determined the friction factors of wire models in very viscous liquids and applied hydrodynamical similarity theorems toward the interpretation of their results on a molecular level. For large values of M they obtain a similar functional dependence as the results given above, although the constants which they derive do not agree with the values as given by Eqs. (26) and (27).

Crothers and Zimm [70] have empirically extended the above result to chains with excluded volume. They found that a plot of log $(s - 2.7)$ vs log M yields a straight line over a range of intermediate values of M; the slope of this plot equaled 0.445 and the deviation from 0.5 was attributed to excluded volume effects. The important observation, also documented by Eigner and Doty [71], is that both s and $[\eta]$ are unique and monotonously increasing functions of M, independent (within the rather large errors of the experiments) of DNA origin and composition. (The values of both M and s have undergone considerable modification since 1965, and in particular it is important to consider well-characterized DNA samples only.) A theory for the sedimentation coefficients comparable to that of Hearst and Stockmayer is due to Ptitsyn and Eizner [72] who also calculated [73] the intrinsic viscosity of wormlike coils.

Bloomfield and Zimm [28] calculated the hydrodynamic properties of straight chain and ring-chain polymers, whose dimensions are expanded over those of similar Gaussian coils, by a perturbation method based on the normal coordinate treatment [74] with hydrodynamic interaction; the expansion due to excluded volume was characterized by the parameter ε [Eq. (8)]. Wormlike coil behavior was incorporated into these calculations by Gray *et al.* [75]. Their equations for the dependence of s on M are similar to those of Hearst and Stockmayer [68] for the case $\varepsilon = 0$. In another series of papers hydrodynamic properties were discussed by Hearst *et al.* [76] on the basis of the theory of Harris and Hearst [77], which introduces stiffness into the dynamic model of Rouse [78] and of Zimm [74]. Ullman has recently [79,80] discussed the limitations of the Kirkwood classical procedures and derives improved sedimentation and viscosity calculations for random chains, wormlike coils, and rods without excluded volume. The translational frictional coefficients using the intrinsic viscosity of chains with excluded volume was also treated by Sharp and Bloomfield [58] in their paper on light scattering and hydrodynamic properties of polymer chains with excluded volume effects.

A critical evaluation of the continuing stream of theoretical papers both with respect to scattering and to hydrodynamic properties is beyond the scope of this work. For the benefit of readers who may desire to engage in a more thorough study of the theoretical aspects of the study of stiff chains we have in Table I, assembled indexed references of pertinent papers in this field. We shall next proceed to discuss the status of our knowledge with respect to the real polynucleotide chains.

TABLE I

PARTIAL LITERATURE SURVEY OF NON-GAUSSIAN CHAIN PROPERTIES

	Excluded volume	
	Without	With
General theory, statistics	Porod, 1949 (41) Daniels, 1952 (52) Benoit and Doty, 1953 (55) Hermans and Hermans, 1958 (50) Harris and Hearst, 1966 (77) Bugl and Fujita, 1969 (46) Kirste, 1967 (81) Kirste, 1971 (47)	Domb et al., 1965 (59) Mazur, 1965 (61) Fisher, 1966 (60) Kirste, 1970 (56) Yeh and Isihara, 1970, 1971 (66)
Scattering function, $P(h)$	Kratky and Porod, 1949 (42) Hermans and Ullman, 1952 (43) Porod, 1953 (48) Peterlin, 1953 (53) Luzzati and Benoit, 1961 (51) Peterlin, 1963 (54) Sanchez and von Frankenberg, 1969 (63) Tagami, 1969 (64) Horta, 1970 (65)	Sharp and Bloomfield, 1968 (57) Sharp and Bloomfield, 1968 (58)[a] McIntyre et al., 1968 (61)
Sedimentation coefficient, s	Peterlin, 1952 (67) Kuhn et al., 1954 (69) Ptitsyn and Eizner, 1961 (72) Hearst and Stockmayer, 1962 (68) Ullman, 1968, 1970 (79)	Bloomfield and Zimm, 1966 (28) Gray et al., 1967 (75)[a]
Intrinsic viscosity, $[\eta]$	Kuhn et al., 1954 (69) Eizner and Ptitsyn, 1962 (73) Ullman, 1969, 1972 (80)	Bloomfield and Zimm, 1966 (28)[a] Hearst, Beals, and Harris, 1968 (76) Sharp and Bloomfield, 1968 (58)[a]

[a] These papers also discuss the properties of cyclic chains, of interest in the study of circular DNA.

4. The Real Polynucleotide Chain

Structural details of the polynucleotide chain are not taken into account in the treatment of idealized models. Kirste [81] has shown by numerical calculations of isotactic and syndiotactic polymethyl methacrylate (with fixed bond angle, hindered rotation, and helical bias in the case of the syndiotactic chain), that the local structure is of paramount importance in determining the significantly dissimilar small-angle x-ray scattering curves of these two materials. Flory and Jernigan [82] discussed Rayleigh scattering by real chain polymethylene molecules and problems deriving from scattering anisotropy [44]. The earlier work of Birstein and Ptitsyn [83] on conformation of macromolecules should also be consulted. Whereas in vinyl chains the repeating skeletal unit is composed of two carbon atoms only, in the case of the polynucleotide chain three carbon, two oxygen, and one phosphorus atoms characterize the six backbone bonds of the repeating unit. A tremendous increase in complexity is therefore generated; simplifying features (akin to the assumption [44] of independent rotations of amino acid units in a polypeptide chain, because of the double bond character of the peptide bond) based on the rigid structure of the pentose ring make the problem more tractable in the approximation of independent units [84].

A number of polynucleotide structures can be visualized. A rather uncommon occurrence in the case of nucleic acids is the random chain in which hindered rotation, steric interference, and electrostatic repulsion between phosphate charges may lead to extended configurations even in the absence of specific structures. The single polynucleotide chain in solution may assume stable helical structures as a result of stacking of purine or pyrimidine bases [85]. The contribution of the 2'-hydroxyl toward the stabilization of these structures is not well known. As a result of folding back upon itself hydrogen bonds may form between stacked bases under suitable circumstances, mimicking double-strand structures characteristic of DNA; the dAT hairpin structure [86] comes to mind as well as the now well-investigated clover-leaf-like structure of transfer RNA [7–9]. Double-helix structures are, of course, well known and occasionally triple helices may form [87]. Problems related to such structures will be dealt with elsewhere in this book. We shall restrict ourselves to the discussion of some recent results [88] relating to the conformational statistics of a single polynucleotide chain.

Sundaralingam and colleagues [89] have analyzed the stereochemistry of nucleic acids and their constituents from known crystal structures of the latter. From an analysis of the torsional angles about the sugar–phosphate bonds in the x-ray structures of nucleosides, nucleotides, phosphodiesters, nucleic acids, and related compounds, and from a consideration of molecular models, he concluded that the possible conformations for the backbone of helical nucleic acids are strikingly limited. Preferred conformations of the nucleotide unit in polynucleotides and nucleic acids are similar to those found for the nucleotide in the crystal structure; base stacking appears to be a consequence of restricted backbone conformation. Already Sasisekharan *et al.* [90,91] noted (in their calculations of allowed conformations in the polynucleotide chains by the hard-sphere model approach based on stereochemical criteria for nonbonded distances) that the backbone rotation in polynucleotides is highly restricted. The number of possible conformations for the nucleotide unit is very likely limited to 3 conformations, or at most 6, out of the 300 it might take up according to the usual ideas about restricted rotation about single bonds.

The quantitative study of Olson and Flory [88] treats the single polyriboadenylic acid (poly A) chain (Fig. 1) and other polynucleotides with

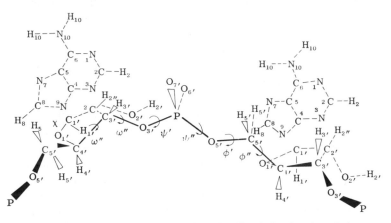

Fig. 1. A section of the extended polyriboadenylic chain showing chain atoms and rotation angles. From Olson and Flory [88].

assumed zero excluded volume (under theta conditions) as a one-dimensional cooperative system influenced only by close-range intramolecular interactions. Continuous bond rotational potentials have been replaced by discrete rotational states [92]. The treatment reduces to the analysis of a collection of independent units [44], from that of a sequence of six interdependent chemical bonds per repeat unit to that of two independent virtual bonds per residue. Rotations about terminal bonds in adjoining units are assumed to be independent, in view of the rigid structure of the pentose ring. Assignment of unique values to the backbone rotation angles ω' and ω'' (Fig. 1) fixes the distance that separates $C_{5'}$ and P. Two virtual bonds a (from P to $C_{5'}$) and b (from $C_{5'}$ to P) can then be drawn; the length of bond a is 2.64 Å, but the length of bond b varies between 4.25 and 4.92 Å, depending on the values assigned to ω' and ω'' and on the puckering of the pentose ring. For the polynucleotide chain comprising x repeat units or virtual bonds of both types a and b, the dimensionless ratio $C_x = \langle r^2 \rangle_0 / xL^2$, characteristic of chain dimensions [44], is expressed as the scalar product

$$C_x = \left[\sum_{i=1}^{x} (a_i + b_i) \sum_{j=1}^{x} (a_j + b_j) \right] / xL^2 \qquad (28)$$

where L^2 is equal to the sum of the squares of bond lengths l in the repeat unit, and a_i and b_j represent the vectors spanning the ith virtual bond a and the jth virtual bond b in the chain sequence. Calculation of the characteristic ratio C_x was carried out using suitable rotational isomeric states and statistical weight matrices. An asymptotic value of about 3 for C_x was determined for the $C_{3'}$-*endo* and about 4.5 for the $C_{2'}$-*endo* configuration of the pentose ring; the values are substantially lower (see Section III,D) than those derived from the experimental viscosity and light scattering results. Only when the angles ψ' and ψ'' (Fig. 1) are fixed at strictly specified values is it possible to obtain calculated values for C_x conforming to the experimental studies.

In an extension of the treatment based on steric restrictions only, Olson and Flory [88] introduce the effect of conformational energies (electrostatic interactions, London dispersion interactions, bond torsional potentials) on the average dimension of the polyribonucleotide chains. Computed values of the characteristic ratio now range between 9 and 24, in much better agreement with the observed dimensions. Only small differences are found between the dimensions of analogous polyribonucleotide and polydeoxyribonucleotide chains. The nature of the purine or pyrimidine base does not affect the computed dimensions of randomly coiled chains. The nature of the heterocyclic base does affect base-stacking. Incorporation of stacked sequences into the random coil chain markedly increases chain dimensions, in confirmation (see Section III,D) of some experimental results.

Although unequivocal quantitative agreement between theory and experiment has not yet been achieved in the calculation of polynucleotide conformations, the above discussion clearly shows that our basic understanding of the phenomena has been significantly advanced.

B. Thermodynamic Aspects

1. *Multicomponent Systems*

Most of the work relating to solutions of nucleic acids deals with systems in which, in addition to the nucleic acid in aqueous solution, additional low molecular weight components (usually uniunivalent symmetrical electrolytes) are present. The low molecular weight electrolytes form the so-called buffer system with the principal solvent water. It is of course possible to study the properties of polynucleotides and DNA in pure water [93]; quite apart from the fact that such systems are of much more limited interest, great care must be exercised to prevent denaturation of the nucleic acid [94] at temperatures above 5°C and at low DNA concentrations. Some recent unexpected results [95] since disproven [93] illustrate this point rather poignantly. In the more usual practice additional electrolytes are present. In the simplest case of a multicomponent system, the three-component system, we designate the principal solvent (water) as component 1, the nucleic acid (PX_z) component 2, and the low molecular weight electrolyte (XY) (diffusible through a semipermeable membrane) component 3. To keep the system at this very simple level, we ensure that the counterion X to the DNA (sodium or cesium, for instance) be common with the cation X of component 3 (NaCl or CsCl, for instance).

We emphasize that, whereas transition from the three-component system to a system comprising a larger number of components does not require the introduction of additional basic concepts, the transition from the two-component to the three-component system profoundly affects the nature of the DNA solution. In the two-component solution (DNA and water only) a large electrostatic potential resides on the ionized DNA macroions (cf. next section); the charges in solution are not screened, the electrostatic forces extend to large distances. As a result, properties are derived which are characteristic of the whole system and not of individual macromolecules. The reduced specific viscosity, η_{sp}/c, of such a system, for instance, increases without limit with decreasing DNA concentration. (This is also so for simple electrolytes and $\eta_{sp}/\sqrt{c}$, rather than η_{sp}/c, is the proper quantity to consider.) However, additions of the simple electrolyte, component 3, produces a major change in the properties of the system. Shielding of the fixed charges (the negative charges, for instance, of the phosphate groups of the polynucleotide chains at neutral pH) by the added simple electrolyte reduces the range of electrostatic forces by the introduction of a screened potential. At constant simple electrolyte concentration, a concentration is reached with dilution of the nucleic acid component at which the distance between macromolecules, for all practical purposes, exceeds the range of the electrostatic forces. We may now speak of individual macromolecules and meaningfully extrapolate to zero nucleic acid concentration. Addition of further low molecular weight components (to form a more complex buffer system) leads to a somewhat more complex analysis (in algebraic terms) but does not produce a fundamental change in the properties of the solution. The first step, from the two-component to the three-component system, is responsible for the fundamental change.

2. Concentration Units and Interaction Parameters

Concentration may be expressed in different units. Molalities $m_J \equiv 10^3 n_J/n_1 M_1$, are usually expressed as moles n_J of component J per kilogram of components 1 (molalities can be based on any other component, should this be found convenient in the interpretation of an experimental design); weight molalities $w_J = m_J M_J/1000$ are in grams of component J per gram of component 1. Units on a volume basis are the molarity, C_J, in moles of component J per liter of solution, and $c_J = C_J M_J/1000$, in grams of component J per milliliter. The two schemes are related by $C_J = 1000\, m_J/V_m$ or $c_J = 1000\, w_J/V_m$. For a three-component system, V_m, the volume (in milliliters) of solution containing 1 kg of component 1, is

$$V_m = (1000 + m_2 M_2 + m_3 M_3)/\rho = 1000(\bar{v}_1 + w_2 \bar{v}_2 + w_3 \bar{v}_3) \qquad (29)$$

where ρ is the density of the solution and the $\bar{v}_J$ are partial specific volumes, $\bar{v}_J \equiv M_J^{-1}(\partial V_m/\partial m_J)_{P,T,m}$; the subscript m in the derivative signifies the restriction that all molalities are constant, except the one with respect to which differentiation is performed.

Interaction between the components in a three-component DNA system occurs and may be measured under various restrictions, depending upon the nature of the experiment [96]. A dialysis equilibrium with component 2 confined to the inner compartment and components 1 and 3 free to diffuse across a semipermeable membrane, the temperature T as well as μ_1 and μ_3 (the chemical potentials of the diffusible components) assume constant values irrespective of the concentration of component 2. However, m_3 (equal to m_Y, the molality of the coion Y) will not be equal to m_3^* (equal to m_Y^*), where starred quantities designate concentrations in the outer, nucleic acid free, compartment. Also, the pressure P will not be equal on both sides of the membrane. Equilibrium distribution of the concentration of Y is expressed in terms of a parameter $\Gamma_3 \equiv (m_Y^* - m_Y)/m_u = (m_3^* - m_3)/m_u$; m_u, which is equal to Zm_2, is the equivalent molality (nucleotides per kilogram of component 1) of the nucleic acid, and Z is the number of charges per macromolecule. At low m_u, Γ_3 is constant and may be expressed in differential form $\Gamma_3 = -(\partial m_3/\partial m_u)_\mu$, where subscript μ signifies constancy of the chemical potentials of the components diffusible through the semipermeable membrane (the subscript T will be omitted here and henceforth as all equations in this chapter refer to constant temperature conditions). On a weight basis we define $\xi_3 \equiv (\partial w_3/\partial w_2)_\mu = -(M_3/M_2)\Gamma_3$ which is the weight, in grams of component 3, to be added to the nucleic acid solution, per gram of DNA, to maintain with a slight change in pressure, constancy of μ_1 and μ_3. Interaction parameters are sometimes based on concentrations defined in terms of the molarity scale; at low nucleic acid concentrations one has [97] the following relation:

$$(\partial c_3/\partial c_2)_\mu = \xi_3 - c_3(\bar{v}_2{}^0 + \xi_3 \bar{v}_3) \qquad (30)$$

The difference between $(\partial c_3/\partial c_2)_\mu$ and ξ_3 becomes increasingly significant with increasing concentration c_3 [96,97].

The above interaction parameters are phenomenological and do not necessarily relate to any well-defined physical binding processes. They are sometimes called preferential or net interaction parameters. These are poor names in the sense that, for some workers, they imply more than is actually meant. As long as their purely operational thermodynamic character is clearly understood no harm may occur. If component 3, for instance, rather than component 1, is taken as the basis of the molality m'_i (or w'_j) then one can define interaction parameters, $\Gamma'_1 = (\partial m'_1/\partial m'_2)_\mu$ and $\xi'_1 = (\partial w'_1/\partial w'_2)_\mu$ $= (M_1/M_2)\Gamma'_1$ in accordance with this frame of reference. At low nucleic acid concentration the connection $\xi'_1/c_1 = -\xi_3/c_3$ (or $\xi'_1 = -\xi_3/w_3$) holds [97–99]. Molecular implications will be discussed in Section II,B,4.

A convenient alternative method to chemical analysis (of compartment contents at dialysis equilibrium) for deriving interaction parameters, is the method of isopiestic distillation, applied to DNA by Hearst [100]. By this method, we may derive $(\partial w_3/\partial w_2)_{P,\mu_1}$ which is rather close to ξ_3 in most cases of practical interest [97]. From electromotive force measurements in cells without liquid junction [101] one may derive a parameter $(\partial w_3/\partial w_2)_{P,\mu_3}$ which is also rather close to ξ_3. In a different context we have examined [102] practical correlations pertaining to four- rather than three-component systems. In particular, consideration of an additional nonionic component may be of some relevance for future nucleic acid studies, aimed at characterizing physical binding to the nucleic acid moiety.

3. *Density and Refractive Index Increments*

Density and refractive index increments are important measurements for the study of DNA volume changes and interaction parameters; they are also essential in the correct evaluation of information from ultracentrifugation, light and small-angle x-ray scattering. For a three-component system at dialysis equilibrium we have, to a very good approximation [97], at low nucleic acid concentration

$$(\partial\rho/\partial c_2)_\mu{}^0 = (1 - \bar{v}_2{}^0\rho^0) + \xi_3{}^0(1 - \bar{v}_3\rho^0) \tag{31a}$$

$$= (1 - \xi_3{}^0) - \rho^0(\bar{v}_2{}^0 + \xi^0\bar{v}_3) \tag{31b}$$

or the completely equivalent form

$$(\partial\rho/\partial c_2)_\mu{}^0 = (1 - \bar{v}_2{}^0\rho^0) + \xi_1{}^{0\prime}(1 - \bar{v}_1\rho^0) \tag{32a}$$

$$= (1 + \xi_1{}^{0\prime}) - \rho^0(\bar{v}_2{}^0 + \xi_1{}^{0\prime}v_1) \tag{32b}$$

(superscript zero refers to properties at the limit of zero polymer concentration).

The density increment $(\partial\rho/\partial c_2)^0_{P,m}$ at constant pressure and composition equals $(1 - \bar{v}_2{}^0\rho^0)$; its determination is required for the evaluation of $\bar{v}_2{}^0$ [97].

The refractive index increment $(\partial n/\partial c_2)_\mu{}^0$ is given by

$$(\partial n/\partial c_2)_\mu{}^0 = (\partial n/\partial c_2)^0_{P,m_3} + \xi_3{}^0(\partial n/\partial c_3)_{P,m_2=0}(1 - c_3\bar{v}_3) \tag{33}$$

to within the same approximation as $(\partial\rho/\partial c_2)_\mu{}^0$.

For the study of interaction parameters in various systems, the experimentally determinable quantities Eqs. (31)–(33) can be used to derive $\xi_3{}^0$ or related parameters [97,99,102,103]. The method is convenient and capable of good precision, in view of recent advance in instrumental methods [104]. However, for the interpretation of sedimentation [105], small-angle x-ray scattering [40,106] or light scattering [107,108], it is advisable to use directly measured values of $(\partial\rho/\partial c_2)_\mu$ or $(\partial n/\partial c_2)_\mu$, respectively.

4. *Binding and Exclusion of Low Molecular Weight Components*

In solutions comprising a macromolecular and a number of low molecular weight components, binding of one of the latter to or exclusion from the macromolecular component may occur. Some of the principal solvent, water (component 1), may be bound to the nucleic acid. This is known as hydration and a quantitative assignment may depend on the experimental method used to describe it. Ionic components, such as NaCl or CsCl, may also bind to the macromolecular component, but this is less likely in nucleic acid than, for instance, in protein solutions. As a result of the strong polyelectrolyte nature (see next section) of nucleic acids Donnan ion exclusion is more likely to occur under most experimental conditions. We shall, in the following see how these physical phenomena are related to the formal thermodynamic parameters. We may write, on the basis of straightforward considerations [102]

$$(\partial\rho/\partial c_2)_\mu = (1 - \bar{v}_2{}^0\rho^0) + ([B3] - [E3])(1 - \bar{v}_3\rho^0) + [B1](1 - \bar{v}_1\rho^0) \tag{34}$$

where the partial volume $\bar{v}_2{}^0$ (derived from the density increment $(\partial\rho/\partial c_2)_{P,m}$ at constant composition) takes into account all volume changes due to solvation and electrostriction, [B1] and [B3] represent binding of components 1 and 3, and [E3] Donnan exclusion of component 3, all in grams per gram of nucleic acid. With the use of the easily derived relation

$$c_3(1 - \bar{v}_3\rho^0) + c_1(1 - \bar{v}_1\rho^0) = 0$$

which is true at vanishing nucleic acid concentration, comparison of Eqs. (34), (31a), and (32a) yields

$$\xi_3 = [B3] - [E3] - [B1]w_3 \tag{35a}$$

and

$$\xi_1' = [B1] - ([B3]/w_3) + ([E3]/w_3) \tag{35b}$$

Interesting results may be obtained [102] if a fourth nonionic component is introduced which changes the density but does not otherwise interact with component 2. In that case we may write

$$(\partial\rho/\partial c_2)_\mu^0 = (1 - \bar{v}_2^0\rho^0) + \xi_3(1 - \bar{v}_3\rho^0) + \xi_4(1 - \bar{v}_4\rho^0) \tag{36}$$

ξ_3 is given as before by Eq. (35a) and

$$\xi_4 = [B4] - [B1]w_4 \tag{37}$$

If the binding [B4] of component 4 to the macromolecular component 2 is negligible, then [B1] can be determined, as well as [B3] − [E3] in Eq. (35a).

C. POLYELECTROLYTE ASPECTS

1. *Repulsive Forces in Charged Chains*

In the present section we shall discuss properties of polynucleotide chains and helices arising from the interaction of the electrical charges residing on the phosphate groups in the repeating units of the sugar–phosphate backbone. By virtue of these charges polynucleotide chains are typical polyelectrolytes over a wide range of pH values. Around neutrality the charge is due to the phosphate groups only, and at extreme pH values ionization of the purine and pyrimidine bases may provide additional electrical charge. In 1924 Linderstrøm-Lang, barely one year after the publication of the Debye–Hückel interionic attraction theory, presented the first quantitative interpretation of the influence of electrostatic charge on the properties of globular proteins. For the last 25 years continuous effort has been invested in studying the properties of synthetic coiling polyelectrolytes and the present status of this field is well summarized in a recent encyclopedia article [6] and a review [109]. The polyelectrolyte properties of polynucleotide chains are closely related to the properties of ionophoric [110] synthetic coiling polyelectrolytes. The well-defined geometry of the stiff DNA double helix has led to the often repeated use of DNA as a model substance for the testing of polyelectrolyte theories applicable to stiff coils, almost rodlike in nature.

Of major importance in understanding nucleic acid structure is knowledge of the influence of intramolecular electrostatic repulsive forces on the stability of the DNA Watson–Crick double helical structure and its sensitivity to changes in ionic strength. It is expected, and also observed [93,94], that the structure is rather unstable in pure aqueous solutions (where it can only be maintained at low temperatures close to the freezing point of water), and its stability increases (as ascertained by the increase in the DNA melting temperature, T_m) with increase in ionic strength [111]; T_m, for a given DNA species, increasing with the logarithm of the molar salt concentration. The melting temperatures also increase with increasing GC content. We are therefore dealing with a complex phenomenon which involves ionic repulsion potentials, energy contributions of nonionic origin, as well as entropy terms pertaining to both macromolecular conformation and solvent structure. Schildkraut and Lifson [111] have shown how the contribution to the free energy due to charge–charge repulsion can be understood in terms of the Debye–Hückel approximation with a reduced charge parameter. This and other studies emphasize the necessity for a precise understanding of the polyelectrolyte properties of polynucleotide chains and helices. Two major aspects can be singled out in this respect: (1) the interaction of the charged polynucleotide backbone with small ions in solution, and (2) the influence of electrostatic interactions on such macromolecular properties as polynucleotide conformation or the stability of stacked and multistrand structures. Polyriboadenylic acid (poly A) for instance, which at neutral pH forms single-strand structures (in which the purine bases show a decreasing tendency to stack with increasing temperature) will form double-stranded chains [112,113] when electrostatic interactions are sufficiently altered by a decrease in pH.

2. The Cylindrical-Rod Cell Model

Theories for the polyelectrolyte behavior of coiling macromolecules have originally been based, with rather limited success, on the concept of either a sphere occupying the domain excluded by the macromolecule and penetrable to small ions, or on coiled chain models [6]. As early as 1951 Fuoss *et al.* [114] as well as Alfrey *et al.* [115] laid the foundation for what was to become the most successful model to date for the study of properties of polyelectrolyte systems. The cylindrical-rod cell model is based on the assumption that over limited distances along the chain electrical charges are arranged in a linear array with well-defined charge density, a characteristic parameter of the system. Cylindrical symmetry is observed and the distribution of the small ions in the electrostatic field of the macromolecular rod (assumed to be of infinite length) is calculated by suitable application of the Poisson–Boltzmann equation. Exact solutions have been given in the salt-free case [114,115] and the free energy and derived quantities have been evaluated [116]. In the presence of added salt Alexandrowicz and Katchalsky [117] solved the Poisson–Boltzmann equation by an approximation technique; their work has been extensively reviewed [118]. Gross and Strauss [119] presented a numerical solution of the Poisson–Boltzmann equation and also introduced the concept of site binding (of ions to the macromolecules) to account for the deviation of certain properties from the expected behavior. A fair amount of controversy has been generated in the efforts devoted to the understanding of polyelectrolyte behavior and interest, both in general and also direct toward an understanding of polynucleotide behavior, remains unabated.

It is evident that, whereas the application of the rodlike model may not be entirely satisfactory for highly coiling macromolecules, DNA represents an outstanding example of a macromolecule by which the reliability of the model can be extensively tested. The double helical arrangement provides a very stiff backbone, almost rodlike over many hundreds of angstroms, and the geometry of the phosphate charges is well defined by the classical B structure of Watson and Crick [120] or similar structures, closely related to it [121]. This fact, for instance, prompted Strauss *et al.* [122] to proceed toward an evaluation of the interactions of polyelectrolytes with simple electrolytes by studying the Donnan equilibria obtained with DNA in solutions of 1:1 electrolytes.

Whereas the properties depending on the interactions of charged macromolecules with the small ions surrounding them are now reasonably well understood (mainly in terms of the cylindrical rod model discussed above) the influence of charged groups on the overall conformation of macromolecules (with and without excluded volume) are far less amenable to exact theoretical analysis [66,123–125]. Again DNA, because of its very high inherent chain stiffness, appears to be in a different class from the more flexible polynucleotides and synthetic polyelectrolytes usually encountered. Electrostatic interactions are much more important in the latter than in the former case in determining overall polymer dimensions, both unperturbed and perturbed by the excluded volume. Electrostatic interactions decrease roughly with $\exp(-\kappa r)/r$, where $1/\kappa$ is the characteristic Debye radius and r is the distance between charges; $1/\kappa = 10$ Å for a 10^{-1} M NaCl solution and 100 Å for a 10^{-3} NaCl solution and for values $\kappa r \ll 1$ electrostatic interactions become negligible. Although in DNA the electrical charges along the chain are spaced quite closely (at a few angstroms distance) near neighbors do not contribute greatly to a stiffening of what are already rigid units; neither will electrostatic interactions acting across statistical elements, over an average of more than 1000 Å, markedly contribute to the stretching of the chain. In more highly coiling polyelectrolytes, on the other hand, changes in electrostatic potential appear to be the main shape-determining factor. Due to the higher inherent flexibility of these chains the statistical element may be only 50–100 Å in length. Shape changes in native DNA (estimated from changes in the intrinsic viscosity $[\eta]$ at low values of the rate of shear) with change in salt concentration are relatively small in moderate-size calf thymus DNA [126]. The now well-documented [127,128] increase in $[\eta]$ of high molecular weight coliphage DNA with decreasing ionic strength is appreciably smaller in the native DNA than in the corresponding denatured single-strand polynucleotide chains [128]. Shape changes in coiling polynucleotides (such as, for instance, RNA) are likely to occur and have indeed been observed [129] at all values of $1/\kappa$ larger than (or of the order of the statistical element). In the latter case it has been shown [130] that suppression of electrostatic repulsion by an increase in ionic strength makes it possible for RNA molecules to fold into characteristic base-paired secondary structures.

3. *Colligative Properties of Charged Rods*

We conclude this section by presenting the main results pertaining to interactions between small ions and charged macromolecules in a novel interpretation of the cylindrical rod model due to Manning [131]. With few exceptions his results are identical to the results previously discussed [114–119]. The interest arises in that a simple and intuitively attractive picture emerges from Manning's basic assumptions and that other important results may become apparent in further work [132]

For the specification of the colligative properties of dilute polyelectrolyte solutions obeying a "limiting law," Manning defines a critical parameter $\xi = e^2/DkTb$, where e is the protonic charge, D is the dielectric constant of the solvent, and b is the distance between charged groups on the infinitely thin cylinder (in the present exposition and in view of the fact that dilute solutions only are discussed, the finite radius of the polyelectrolyte backbone is not considered). For univalent fixed and mobile ions ξ equal to unity represents a critical value below which mobile ions are free in solution (subject to the limitations of the linearized Debye–Hückel law) but above which some of the mobile counterions will condense on the macromolecular chain and neutralize a fraction $(1 - \xi^{-1})$ of the polyion charge. The remaining free mobile ions in solution are assumed to be subject to the limitations of the linearized Debye–Hückel law. The "condensed" and "free" mobile ions are not identical with "bound" and "free" ions defined in previous theories. For water at 25°C ($D = 78.5$) the critical value $\xi = 1$ corresponds to a charge spacing, b, of 7.14 Å. For DNA, $b = 1.7$ Å, therefore the condition $\xi > 1$ ($\xi = 4.2$) applies and condensation of ions, in the range defined by Manning, always occurs.

For the case $\xi < 1$ of a mixed aqueous polyelectrolyte uniunivalent salt solution, a simple calculation yields

$$ln\gamma_{\pm} = ln\gamma_i = -\tfrac{1}{2}\xi X(X + 2)^{-1} \qquad i = 1, 2; \xi < 1 \tag{38}$$

where γ_i and $\gamma_{\pm}$ are the activity and mean activity coefficients of mobile ions, 1 and 2 refer to the counter- and coion, respectively, and $X = C_u/C_3$, the ratio of the polyelectrolyte equivalent to added salt concentration. In salt-free solutions ($X \to \infty$) and

$$ln\gamma_{1p} = -\tfrac{1}{2}\xi \qquad \xi < 1 \tag{39}$$

The osmotic coefficient ϕ, which is the ratio of the osmotic pressure to its ideal value is given by

$$\phi = 1 - \tfrac{1}{2}\xi X(X + 2)^{-1} \qquad \xi < 1 \qquad (40a)$$

and in salt-free solutions

$$\phi_p = 1 - \tfrac{1}{2}\xi \qquad \xi < 1 \qquad (40b)$$

in agreement with the result of Lifson and Katchalsky [116].

The Donnan distribution coefficient Γ', in the limit as both C_u and C_3 tend to zero is given by

$$\Gamma' = \tfrac{1}{2}(1 - \tfrac{1}{2}\xi) \qquad \xi < 1 \qquad (41)$$

in agreement with the result of Gross and Strauss [119]; Γ' is that part of Γ_3 which is due to Donnan exclusion only; numerically Γ' equals (M_2/M_3). [E3] [cf. Eq. (35a)], experimentally Γ' may be identified with $-(M_2/M_3)\xi_3$ at very low concentration and binding of component 3.

The case $\xi > 1$ is of much greater interest since in polynucleotide and DNA solutions this condition almost certainly applies. In this situation

$$\gamma_1 = (\xi^{-1}X + 1)(X + 1)^{-1} \exp\left[-\tfrac{1}{2}\xi^{-1}X/\xi^{-1}X + 2\right] \quad \xi > 1 \qquad (42a)$$

which reduces in the salt-free case to

$$\gamma_{1p} = \xi^{-1}e^{-1/2} \qquad \xi > 1 \qquad (42b)$$

The square of the mean activity of the mobile ions

$$\gamma_{\pm}^2 = (\xi^{-1}X + 1)(X + 1)^{-1} \exp\left[-\xi^{-1}X/(\xi^{-1}X + 2)\right] \qquad \xi > 1 \qquad (43)$$

and

$$\Gamma' = (4\xi)^{-1} \qquad \xi > 1 \qquad (44)$$

in agreement with the result of Gross and Strauss [119]

$$\phi = (\tfrac{1}{2}\xi^{-1}X + 2)/(X + 2) \qquad \xi > 1 \qquad (45a)$$

and in salt-free solutions

$$\phi_p = (2\xi)^{-1} \qquad \xi > 1 \qquad (45b)$$

in agreement with Lifson and Katchalsky [116].

A number of interesting correlations, generally known as "additivity rules" [118], were obtained by Manning [131], some of which do not agree with the previous deductions. These are beyond the scope of this review and the correlations have so far not been investigated in nucleic acid solutions.

The best-known correlations (borne out by experimental results), for all values of ξ, are

$$\Gamma' = \tfrac{1}{2}\phi_p \qquad (46)$$

and

$$\phi(C_u + 2C_3) = \phi_p C_u + 2C_3 \qquad (47)$$

We now conclude our brief discussion of some essential points in our present understanding on the influence of charged groups on polyelectrolyte properties. No satisfactory answer has been provided for the basic question why nature chose to endow the nucleic acid backbone with a high density of permanently charged groups, with yet unexplained role in either structure or function.

D. EQUILIBRIUM PROPERTIES

1. *Osmotic Pressure*

Equilibrium colligative properties are primarily important in that they yield the molecular weight M (or some average or averages, in the case of polydisperse systems) and the degree of interaction between macromolecular solute particles usually expressed as the second virial coefficient A_2. In addition, from the angular dependence of scattering of either x-rays or light, it is possible to derive information about the distribution of mass. Scattering methods, viewed as fluctuation phenomena, and the distribution of mass at equilibrium in the ultracentrifuge, can both be derived and expressed in terms of the derivative of the osmotic pressure. We have had occasion to discuss the subject in great detail [96] and will therefore be brief in this exposition; some newer results though will be presented.

Osmotic pressure as an experimental method is not very useful in the study of nucleic acid solutions. Whereas polypeptides and proteins range in molecular weight from a few thousand to a few hundred thousands daltons, which is a useful range for molecular weight determination by osmotic pressure, nucleic acids range in molecular weight (with few exceptions) from a few hundred thousand (including "sonicated" DNA, broken by sonic irradiation into fragments in which the double-helix structure is conserved) to over a hundred million daltons for phage DNA and considerably higher for bacterial and mammalian DNA. The osmotic pressure is proportional to the inverse of the molecular weight and therefore becomes unmeasurable for such solutes; on the other hand, both the intensity of scattering and the steepness of the concentration distribution in the ultracentrifuge increase with increasing molecular weight. These methods have their practical range of usefulness; unfortunately, the molecular weight of nucleic acids is so large (to the extent that intact macromolecules can sometimes not be obtained) that at the high end of their molecular weight scale, both methods collapse and special modifications have to be devised. One such method, for instance, is equilibrium sedimentation in a density gradient, which is a specialized application of the conventional equilibrium sedimentation method.

In the three-component system, which we have defined in Section II,B, the osmotic pressures (π) may be expanded in powers of the concentration c_2 to give

$$\pi/RTc_2 = M_2^{-1} + A_2c_2 + \cdots \tag{48}$$

Even if the nucleic acid is polydisperse with respect to molecular weight, the macromolecular component can still be treated as a unique component (small differences in GC content and density heterogeneity can only be picked up in such specialized applications as equilibrium sedimentation in a density gradient). In this case properly weighted averages are obtained. For instance, the molecular weight determined by osmotic pressure is the number average molecular weight [22] over all species, $M_n = c/\sum_i c_i/M_i$.

2. Light and Small-Angle X-Ray Scattering

The light scattering properties of nucleic acid solutions have been reviewed in a classical article by Geiduschek and Holtzer [39] and more recently by Eisenberg [29]; the latter article also contains a short discussion of small-angle x-ray scattering. The extension to multicomponent systems was reviewed by Casassa and Eisenberg [96]. For a discussion of small-angle x-ray scattering of multicomponent systems, cf. Timasheff [106] and Eisenberg and Cohen [40]. The main results only will be given below.

In the case of small, isotropic particles (or, in the case of large particles, in the extrapolation to zero angle) Eq. (11) applies. The right-hand side of Eq. (11) is recognized to be equal to $d\pi/dc_2$ [cf. Eq. (48)] divided by RT. For large particles Eq. (19) is to be used. The constant H, for multicomponent systems, is given by $H = (2\pi^2/N_{Av})(n^2/\lambda^4)(\partial n/\partial c_2)^2_{P,\mu_3}$; the use of the refractive index increment at constant pressure P and constant chemical potential μ_3, which is, in practice, indistinguishable from the corresponding quantity at constant chemical potential of all diffusible solutes [Eq. (33)] properly accounts for the multicomponent nature of the system. For the determination of the molecular weight, interaction between particles (as reflected in the value of the second virial coefficient), and parameters related to the distribution of mass (from the angular dependence of scattering), decomposition of the refractive index increments into the quantities composing it [Eq. (33)] is not required. The molecular weight is properly defined in terms of the concentration units employed [96]. For additional features deriving from polydispersity, optical anisotropy, coherent incident light, external orienting fields, and experimental details, we refer to Eisenberg [29], and for earlier work, Geiduschek and Holtzer [39].

Small-angle x-ray scattering in multicomponent systems can be viewed [40,106] in similar fashion to light scattering, on the basis of fluctuation theory. Still, the exact conditions under which fluctuation theory and requirements of local conservation of electroneutrality break down in charged systems (in particular when the fluctuation volumes are rather small as in the case of small-angle x-ray scattering), have not been subjected to a thorough critical analysis nor to a detailed experimental test. The basic equivalence between light and small-angle x-ray scattering in multicomponent systems can be easily established, provided the constant H in Eqs. (11) and (19) is, for the latter case, redefined as $H = (\partial\rho/\partial c_2)^2_{p,\mu_3} I_{el}/N_A$; where $(\partial\rho/\partial c_2)_{P,\mu_3} \sim (\partial\rho/\partial c_2)_\mu$, ρ is here the density in electrons per milliliter and I_{el} is the scattering of an electron.

3. *Equilibrium Sedimentation*

To derive the differential equation for the concentration distribution of the charged nucleic acid (component 2) in a compressible multicomponent system, at constant temperature, at equilibrium in a centrifugal field potential $\omega^2 r^2/2$, we may write [133] for the polymeric ion P, of "effective" charge $-\theta Z$, at position r in the ultracentrifuge,

$$d\mu_p - M_p\omega^2 r\, dr - \theta Z F\, d\psi = 0 \tag{49}$$

and for the "free" univalent counterion X

$$d\mu_X - M_X\omega^2 r\, dr + F\, d\psi = 0 \tag{50}$$

where F is the Faraday and ψ is the electrostatic potential. We multiply Eq. (50) by θZ and add it to Eq. (49) to yield

$$d(\mu_p + \theta Z\mu_X) - (M_p + \theta Z M_X)\omega^2 r\, dr = 0 \tag{51}$$

which is identical with Eq. (6.1) of Casassa and Eisenberg [96] because $\mu_p + \theta Z\mu_X = \mu_2$ and $M_p + \theta Z M_X = M_2$, the chemical potential and the molecular weight of the electroneutral polyelectrolyte component 2; ψ is eliminated when electroneutral components (defined as above or in any other way) are chosen in an equilibrium experiment (cf. footnote 12 on page 360 of Casassa and Eisenberg [96]. It is therefore not necessary to define θ precisely in the case of an equilibrium experiment. From Eq. (51) and a similar expression for the distribution of component 3, one may derive [96,105] the exact symmetrical expression

$$(d\ln c_J/dr)^{(r)} = \omega^2 r(\partial\rho/\partial c_J)_\mu^{(r)}/(\partial\pi/c_J)_\mu^{(r)} \tag{52}$$

where J refers to any component and the superscript (r) indicates values applicable at a distance r from the center of rotation of the centrifuge rotor. Equation (12) is a special case of Eq. (52), obtained by introducing the virial expansion [Eq. (48)] for the osmotic pressure derivative $(\partial\pi/\partial c_2)_\mu$ and the value for $(\partial\rho/\partial c_2)_\mu$ from Eq. (31a), as applicable [in Eq. (12)] to a simple two-component system. It is customary, even in the case of multicomponent systems, to maintain the conventional formulation of the buoyancy term $(1 - \bar{v}_2{}^0\rho^0)$; the latter term may formally be substituted by $(1 - \phi'\rho^0)$, which replaces $(\partial\rho/\partial c_2)_\mu$ in Eq. (52). The apparent specific volume ϕ' is not constant and it should be used with great circumspection only, as it is not simply related to the partial volume; its variation with solvent composition and density is not easily predictable.

The molecular weight M_2 in the limit of vanishing nucleic acid concentration is unambiguously defined in terms of the weight of material included in the definition of c_2. This is quite clear from the formulation we are using for the concentration distribution of component 2 in the ultracentrifuge

$$d\ln c_2/dr^2 = (\omega^2/2RT)M_2(\partial\rho/\partial c_2)_\mu{}^0$$

but is not apparent in the conventional formulation, with use of partial specific volumes. Note that $M_2(\partial\rho/\partial c_2)_\mu$ can be written as $1000\,(\partial\rho/\partial C_2)_\mu$ (Section II,B,2) where the density increment represents the change in solution density per mole of nucleic acid added, at constant μ. Binding or exclusion by the macromolecular component of any other component or species in solution has no effect on this expression as long as the number of moles of nucleic acid remains unchanged. The statement (both with respect to scattering and equilibrium sedimentation) sometimes encountered, that counterions which are "bound" to the DNA contribute to the molecular weight, but others, which are "free" do not contribute, is completely meaningless. Whether, for instance, counterions are, in the sense of Section II,C,3, condensed on the DNA rods or not, is completely irrelevant with respect to the molecular weight determination of the DNA. The basic interpretation of results from colligative experiments is always in terms of mole concentration; the molecular weight only appears in the process of transforming into practical concentration units.

For practical evaluation of equilibrium sedimentation results cf. the recent review article of Creeth and Pain [134].

4. *Equilibrium Sedimentation in a Density Gradient*

Equilibrium sedimentation experiments in multicomponent systems may be performed under two extreme limiting conditions. In the "conventional" case described above the experiment is performed at relatively low speeds and low concentration (0.2–1.0 M) of a "light" component 3 (NaCl for instance). Redistribution of component 3 in the ultracentrifuge is negligible and ρ^0 is practically constant across the cell; $(\partial\rho/\partial c_2)_\mu$ [Eqs. (31) and (32)] is assumed to be independent of distance r from the center of rotation and Eq. (52) may be integrated in various ways [134]. A mixed solvent, on the other hand, comprising a "heavy" component 3 (CsCl at concentrations approaching saturation) generates a density gradient $d\rho/dr$ at high velocities of the ultracentrifuge; $(\partial\rho/\partial c_2)_\mu$ continuously decreases with increase in r. Under properly chosen initial conditions the nucleic acid component 2, present in small concentrations, concentrates in a narrow band at equilibrium in the ultracentrifuge; at the center r_b of the band $(\partial\rho/\partial c_2)_\mu = 0$. The half-width of the band is inversely related to the square root of the molecular weight of the nucleic acid component. Meselson *et al.* [135] were the first to use this principle in their classical study of the Cs salt of DNA in aqueous CsCl solutions. For intact DNA of very high molecular weight, homogeneous with respect to density, this method is, in principle, capable of yielding molecular weights in a range which is inaccessible to the more conventional methods; many experimental difficulties connected with the method have recently been overcome; in particular Schmid and Hearst [136,137] could show that neglect of virial coefficients was a major cause of underestimation of molecular weights by this method in the past. The calculation of molecular weights from equilibrium sedimentation in a density gradient by conventional procedures requires [138–140] precise knowledge of partial specific volumes, interaction parameters, and their detailed dependence on solvent activity and pressure. It was therefore of some interest to search for a simpler procedure to verify the above rather elaborate treatments. In particular it should be possible to determine colligative properties, as in the case of conventional equilibrium sedimentation, without recourse to effective density gradients of various kinds, "solvated" molecular weights and "solvated" partial specific volumes which are rather ill-defined in the strict thermodynamic sense. We [96] expand $(\partial\rho/\partial c_2)_\mu$ about r_b to write

$$\left(\frac{\partial\rho}{\partial c_2}\right)_\mu = \chi\delta + 0(\delta^2) \tag{53}$$

where $\delta = r - r_b$ and

$$\chi \equiv \left[\frac{d}{dr}\left(\frac{\partial\rho}{\partial c_2}\right)_\mu\right]^{(r_b)} \tag{54}$$

is the gradient of $(\partial\rho/\partial c_2)_\mu$ at r_b. Equation (53) can now be substituted into Eq. (52). At vanishing DNA concentration a Gaussian distribution (with the usual neglect of higher terms) of the concentration $c_2^{(r)}$ about the center of the band r_b is obtained

$$c_2^{(r)} = c_2^{(r_b)} \exp\left(-\delta^2/2\sigma^2\right) \tag{55}$$

where the standard deviation

$$\sigma^2 = -RT/M_2\omega^2 r_b\chi \tag{56}$$

Virial corrections can be allowed for as recently shown by Schmid and Hearst [136]. The major problem is the precise determination of χ in an accessory experiment. To this end we shall consider isotopic substitution.

The use of labeled nucleic acid goes back to the important observation [141] that, upon isotopic substitution (^{15}N for ^{14}N in *E. coli* DNA), the center of the nucleic acid band in the CsCl gradient shifts a distance Δr corresponding to $\Delta\rho$ density units. Vinograd and Hearst [138] showed that with the labeled DNA the "effective" density gradient can be estimated if the solvation parameter ξ'_1 is known. Baldwin *et al.* [142] have described a method concerning the determination of relative molecular weights from the buoyant densities of hybrid labeled and nonlabeled λ DNA dimers. We could show [97,143] that the well-founded assumption that isotope substitution changes the weight but not the volume of nucleic acid molecules permits straightforward determination of M of high molecular weight DNA from the bandwidth in the CsCl gradient. The gradient χ is given by

$$\chi = -(\alpha - 1)/\Delta r \tag{57a}$$

$$= -[(\alpha - 1)/\Delta\rho](d\rho/dr)^{(r_b)} \tag{57b}$$

where α (equal to about $1 + (15/4 \times 441) = 1.0085$ for a DNA with 50% GC content, i.e., 15 nitrogens per 4 nucleotides) is the ratio of the weights (per nucleotide) of the two DNA species, and $(d\rho/dr)^{(r_b)}$ is the density gradient of the compressed salt (the physical density gradient [138]). Schmid and Hearst [137] carefully reinvestigated density shifts subsequent to isotope substitution in DNA; their experimentally determined parameter $(1 + \Gamma')/\beta_{eff}$ [Eq. (29), ref. 137] equals $(\chi\rho_b/\omega^2 r_b)$ in our derivation. The simple result from the isotope substitution experiment provides strong support for the earlier procedure of Vinograd and Hearst [138] and allows [137] reliable values of M_2 to be determined from density gradient experiments.

That $(\partial\rho/\partial c_2)_\mu$ vanishes at r_b permits the estimation of ξ_3 or ξ'_1 (if the partial specific volumes $\bar{v}_2{}^0$ and $\bar{v}_3$, respectively, $\bar{v}_1$, are known) under specified conditions in the ultracentrifuge. Vinograd and Hearst [138] in a series of papers, observed a unique relationship between ξ'_1 and water activity; their results were obtained by determining buoyant densities of DNA in a large number of mixed solvent systems, in which both the nature and the concentration of the salt component 3 was suitably varied.

E. Transport Phenomena

1. *Interaction with External Fields*

When we leave the well-chartered realm of equilibrium properties for the incompletely explored waters of transport phenomena we must reluctantly sacrifice considerable rigor for the benefit of being able to perform various experiments; these often feature the convenience of ease of execution, high experimental precision and reproducibility, and also sometimes the requirement for only minute quantities of scarce nucleic acids. In transport phenomena we study the interaction of external fields on the flow properties of nucleic acid solutions. In the ultracentrifuge the centrifugal force leads to translatory motion, which is characterized in terms of the sedimentation (velocity) coefficient. Solvent velocity gradients in concentric cylinder or capillary viscometers produce rotary motion of the macromolecules and the related energy dissipation per particle is characterized as the intrinsic viscosity $[\eta]$ at vanishing nucleic acid concentration. Orientation or deformation in the field of flow usually results in a decrease of $[\eta]$ with increasing rate of shear; this is known as non-Newtonian viscosity, and is either studied [126,144,145] or avoided (by working at extremely low values of the rate of shear). Orientation or deformation in a field of flow may be followed by optical methods such as streaming birefringence [146] (Maxwell effect) or linear flow dichroism [147], which rely on anisotropy of refractive index or absorption coefficients, respectively. The optical effects associated with electrical or magnetic fields are known as the Kerr effect or Cotton-Mouton effect, respectively [148]; both steady state or relaxation measurements may be performed. Scheludko and Stoylov [149] have presented preliminary experiments on the variation in the intensity of light scattered by solutions of DNA subjected to an electric field, and Stoylov and Sokerov [150] have described a method for the determination of rotational diffusion coefficients from light scattering in a transient electrical field.

The variation of the polarized components of fluorescence of rodlike particles bearing a fluorescent label upon partial orientation has been calculated for some special geometry of the dye–macromolecule complexes by Weill and Hornick [151]. The method was applied to rodlike fragments of DNA labeled by intercalated molecules of acridine orange. Hornick and Weill [152] also determined the anisotropy of electrical polarizability of rodlike fragments of DNA by the Kerr effect (combined with flow-birefringence), light scattering, dichroism, and fluorescence in an electric field. The values obtained were compared to the predictions of theories of longitudinal polarization of rigid polyelectrolytes.

The driving force for simple diffusion is the gradient of the concentration, or more precisely of the chemical potential, of the nucleic acid component, whereas self-diffusion is closely coupled to the random motion of the particles in the absence of a concentration gradient. The experimental determination of the former quantity has always been difficult, time-consuming, and expensive in terms of solutions and solutes required. The self-diffusion coefficients of high molecular weight nucleic acids can now be determined from the spectrum of the scattered light by the use of optical mixing spectroscopy and laser radiation with a number of techniques [153–156].

2. Sedimentation Velocity

In an external centrifugal field individual ions tend to move with unequal velocities depending on their buoyant weights and frictional resistance. The unequal displacement of the ions creates an internal electrostatic field which may affect the sedimentation properties of the system. The retardation by the counterion cloud of the motion of the macroion is known as the primary charge effect, and the effect due to unequal mobilities of small ions in solution is known as the secondary charge effect; the latter effect can be minimized, or eliminated, by the choice of a suitable added single electrolyte, such as NaCl, for instance. For recent theoretical descriptions of the charge effect in sedimentation (and diffusion) we refer to Alexandrowicz and Daniel [157,158] and Nagasawa and Eguchi [159]. We expect the primary charge effect to be of greatest importance at low simple salt and high macroion concentration. It can be largely minimized by extrapolation to vanishing nucleic acid concentrations, in sufficiently high (0.2–1 M) concentrations of simple salt (the high absorption coefficient of nucleic acids at 260 nm enables performance of sedimentation experiments at extremely low polymer concentrations, about 10^{-5} gm/ml, in ultracentrifuges equipped with UV absorption optics). Whether the charge effect is indeed completely eliminated under these conditions has been a subject of discussion [158,159]. It has been claimed [160] that charge effects persist under equilibrium conditions and affect the DNA distribution in equilibrium sedimentation in a density gradient; this is rather unlikely as was shown in Section II,D,3.

We discuss next the form taken by the sedimentation and diffusion equation for the case of a multicomponent system, in the absence of charge effects, with the formalism of irreversible thermodynamics [22,161]. The flow J_2 at time t of component 2 relative to the cell, i.e., the mass crossing unit area in unit time, is given by

$$J_2 = -L_{22}\frac{d(\bar{\mu}_2/M_2)}{dr} - L_{23}\frac{d(\bar{\mu}_3/M_3)}{dr} \tag{58}$$

where L_{22} and L_{23} are the Onsager phenomenological coefficients as defined by Hooyman [162], and $\bar{\mu}_2/M_2$ and $\bar{\mu}_3/M_3$ are the total chemical potentials

$(\bar{\mu}_i/M_i = \mu_i/M_i - \omega^2 r^2/2)$ per gram. The sedimentation constant s is usually defined as the velocity per unit field in a homogeneous solution (i.e., both dm_2/dr and dm_3/dr equal to zero), and the diffusion constants are derived from J_2 in the absence of the external field. In a polyelectrolyte–simple electrolyte system, when m_3 is fairly large and $m_2 \ll m_3$, the diffusion of simple electrolyte is much larger than that of the macromolecules. Under these conditions a plausible assumption is that, in the sedimentation experiment, while the solution just ahead of the boundary is homogeneous with respect to component 2, component 3 is essentially in equilibrium throughout the cell, i.e., $d\bar{\mu}_3 = 0$. With this assumption Eq. (58) simplifies considerably (the term with L_{23} disappears) and it is possible to derive [105]

$$(D_2/s_2)(\partial\rho/\partial m_2)_{P,\mu_3} = (\partial\pi/\partial m_2)_\mu \tag{59}$$

In the limit of vanishing concentration of component 2 and with the use of the virial expansion [Eq. (48)] Eq. (59) reduces to

$$\frac{D_2^0}{RTs_2^0}\left(\frac{\partial\rho}{\partial c_2}\right)^0_{P,\mu_3} = \frac{1}{M_2} \tag{60}$$

which is the analog for the three-component system of the familiar Svedberg equation. In this limiting form, charge effects are minimal or absent. We may introduce a frictional coefficient f_2 per particle, to write

$$s_2^0 = \frac{M_2}{f_2 N_{\mathrm{Av}}}\left(\frac{\partial\rho}{\partial c_2}\right)^0_{P,\mu_3} \tag{61}$$

This form of the sedimentation equation will be useful below in the interpretation of changes in frictional properties of DNA with binding of small ligands (proflavine, for instance [163]); $(\partial\rho/\partial c_2)_{P,\mu_3}$, or rather $(\partial\rho/\partial c_2)_\mu$, is the density increment at dialysis equilibrium with a buffer containing a given concentration of ligand. If c_2 is expressed in grams of DNA (without reference to the ligand bound) per milliliter, then M_2 is the molecular weight in grams of DNA (without bound ligand!), per mole. This is the same argument we have encountered with respect to equilibrium sedimentation (we recall that c_2/M_2 is the concentration in moles/ml). Any changes in frictional parameters are clearly reflected in the variation of f_2 with change in ligand binding.

3. Intrinsic Viscosity

The intrinsic viscosity $[\eta]$ of nucleic acid solutions has been studied for many years; the extensive field of the hydrodynamic properties of DNA has been reviewed by Bloomfield [1] and by Robins [3]. Intrinsic viscosity is one of the most sensitive size parameters of DNA available; with increasing molecular weight $[\eta]$ becomes increasingly dependent on the rate of shear

[126]. Zimm and Crothers [144] made a major contribution to the study of the rheology of DNA by introducing a simple rotation viscometer capable of operation at extremely low rates of shear; this allowed measurement of $[\eta]$ of DNA molecules with M as high as 10^8 daltons. With the cartesian diver viscometer [164] it was possible to carry these determinations to even higher molecular weight DNA, in a range in which almost all other methods fail. It could be shown [165,166] that, if an optical polarizing system is attached to the cartesian diver viscometer, sudden stopping of the flow and measurement of the elastic recoil enabled the performance of a retardation time experiment from which M can then be estimated by straightforward considerations. This method is believed to be useful for molecular weights at least as high as 10^{10} daltons. Various aspects of the viscometry of nucleic acid solutions have been very recently discussed by Zimm [167].

III. Specific Topics

A. Molecular Weight of DNA and Relations with Hydrodynamic Parameters

1. *Molecular Weight*

The determination of the molecular weight of DNA is not a trivial problem in view of the extraordinary range of molecular weights and sizes which are not encountered in any other macromolecular system, either natural or synthetic. For many years a great deal of effort was expended on the study of the mammalian calf thymus DNA (cf. the compilation of Eigner and Doty [71]), and it became clear only in recent years that one was dealing with rather ill-defined fragments of much larger structures; moreover molecular weight values for the best preparations of calf thymus DNA were grossly underestimated [34,168]. It is believed [169] that the size of the DNA in the mammalian chromosome may be in excess of 10^{10} daltons, possibly composed of units of size 5×10^8, which in turn may be comprised of smaller chains. No intact material of that size has been isolated, but much thought is being devoted toward devising procedures for the purification and analysis of such gigantic structures (e.g., methods for *in situ* removal of protein and other components by the action of suitable enzymes and reagents).

Somewhat smaller in size, but still awe-inspiring is the DNA from the nucleus of bacterial cells. Molecular weights of moderately homogeneous DNA [170] of about 2.7×10^9 and 2×10^9 daltons have been determined [165,171] for DNA isolated from *E. coli* and *B.subtilis*, respectively, and it is believed that the total DNA per bacterial cell (3.8 to 4.2×10^9 daltons) may indeed belong to one or two nuclear bodies only.

Quantitative results have come in recent years from the study of well-defined monodisperse DNA from *E. coli* phages and from the study of the virus particles themselves. We may say that an uncertainty of about 30% in the molecular weight, in the range from about 2×10^7 to 10^8 daltons, has been considerably narrowed down by recent work. In spite of considerable success in the determination of the molecular weights themselves, uncertainty still persists in the relationships between experimental quantities such as s and $[\eta]$ and M. This uncertainty is of particular importance when considering, as is often done, the use of such relationships in the extrapolation above and below the reliable calibration range. At present the few points available for coliphage DNA are reliable enough for the interpolation of unknown monodisperse DNA but they do not allow extrapolation to much higher and much lower molecular weight values. The low molecular weight range where it may be anticipated that DNA properties gradually change from those of stiff coils to rodlike behavior has so far been rather poorly explored. Representative DNA samples (often circular) in this range have values of M as low as $2–3 \times 10^6$ daltons (cf., for instance, the DNA of ϕX174 [172] or polyoma [173] virus); double-stranded linear fragments (obtained by sonic irradiation or shearing) have been studied [174,175] in a molecular weight range down to $1–4 \times 10^5$ daltons; not enough attention has usually been paid in the past to problems related to molecular weight distribution in these fragments. Continued investigation is warranted both for calibration purposes and for theoretical interest; we would like to ascertain whether the functional relationship derived from data on high molecular weight DNA indeed describes the properties of DNA in the limit of low values of M. This problem will be discussed in more detail in Section III,B in the context of DNA flexibility. Our present review on the status of molecular weight determinations of coliphage DNA will cover work which has appeared since 1968 and will be concerned with M of the DNA from T4, T5, and T7 phages; these have been prepared in highest purity and have received most of the attention. Reliable methods to determine molecular weights up to 10^8 daltons are therefore required.

True equilibrium methods firmly based in thermodynamic theory are light scattering and equilibrium sedimentation. Unfortunately both of these methods in their conventional applications are not very useful for DNA above a few million daltons in molecular weight. It could be shown [34,168] that, when light scattering measurements were extended to lower angles (10°), higher molecular weights were obtained for both calf thymus and T7 DNA than previously determined with the lowest scattering angle limited to 30°. (Recently Krasna *et al.* [176,177] could demonstrate (Fig. 2), by extending light scattering measurements to low angles (10°), that DNA decreased in molecular weight by more than a factor of 2 upon both acid and alkali denaturation, as expected for complete strand separation and some single-strand breaks. This

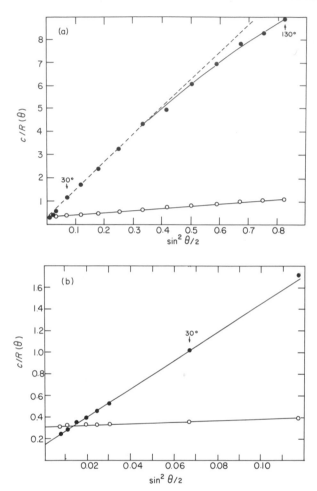

Fig. 2. Plot of $c/R(\theta)$ vs $\sin^2(\theta/2)$ from $10°$ to $130°$ for a single concentration of ($\bullet$) native T7 DNA at pH 6.8 and ($\circ$) denatured DNA at pH 2.5. Concentration of DNA was 19.5 μg/ml. (a) Intensities from $10°$ to $130°$ are plotted. A linear extrapolation for native DNA (dotted line) is drawn through data from $30°$ to $70°$. (b) Intensities from $10°$ to $40°$ are plotted. Linear extrapolations are drawn through all points. From Krasna *et al.* [176].

phenomenon was not observed when scattering measurements were restricted to values $\theta > 30°$, because of the curvature at low angles in the plot of $Hc/\Delta R(h)$ against h^2 for native DNA. The denatured samples are much more compact and no curvature is observed in the scattering range $10° < \theta < 30°$; correct molecular weights are therefore obtained for denatured DNA with conventional equipment with measurement limited to a lower value of $30°$.

Even though Harpst *et al.* [34] devoted considerable effort to measure scattered light below the conventional limit of 30°, their extrapolation in the case of T7 DNA is rather long, $P^{-1}(h)$ at the lowest angle equals about 2, whereas only for values $P^{-1}(h) < 1.2$ points in the plot against h^2 are expected to lie on the limiting curve [cf. Eq. (20)]. Both M_w and $(\langle R^2 \rangle)^{1/2}$ are therefore believed [178] to be overestimated in the evaluation of these light scattering data. The error in M_w is even larger when a lower value of dn/dc is used (the square of dn/dc determines H and therefore M, but the value of $\langle R^2 \rangle$ is not affected), such as $(\partial n/\partial c_2)_\mu = 0.166$ (at 5460 Å, 0.2 M NaCl)of Cohen and Eisenberg [97]. The latter value has recently been confirmed [177] in another laboratory. A light scattering study of λ DNA [180] reveals that, whereas from the angular dependence of scattering it is possible to distinguish clearly linear from circular chains, measurements to scattering angles as low as 10° are not adequate for unambiguous extrapolation of zero angle for molecular weight determinations in this range (3–4 × 10^7 daltons).

Conventional equilibrium sedimentation of huge DNA molecules is not practical. Bancroft and Freifelder [179] correctly remarked that it is not the huge *weight* but rather the large *size* of the DNA molecules which hampered convenient achievement of equilibrium conditions in the ultracentrifuge. They, and previously Davison and Freifelder [181], evaluated equilibrium sedimentation experiments of whole bacteriophage particles. Molecular weights of the phages and of the DNA were then determined [179] in a procedure which involves density determinations on the whole phage and concentration assignments (for partial volume determinations) on the basis of N and P analysis and protein and nucleic acid content. The methods appear trustworthy, if elaborate, but proper evaluation of error does not seem feasible. It is interesting, in this context, to mention the work of Weber *et al.* [182] who estimated a molecular weight of 1.05 ± 0.02 × 10^9 daltons for tipula irridescent virus in a gravity cell, in which the particles were allowed to assume equilibrium concentration distribution under the influence of gravity, under closely controlled temperature conditions (to avoid convection as a result of temperature gradients).

Equilibrium sedimentation in a density gradient appears to be a method of choice in the determination of high values of M, although molecular weights considerably too low had been evaluated in the past [183] by this method. It is mostly to the credit of Schmid and Hearst [136,137] that this excellent method, for homogeneous particles in which no band spreading (cf. Section II,D,4) as a result of density heterogeneity is to be expected, was fully rehabilitated. We have already discussed above that both proper use of a virial correction and a careful study of band shifts due to isotope substitution led to the desired result. Their use of the photoelectric scanner (Fig. 3) eliminated problems encountered in earlier studies with photographic methods; only minute amounts of DNA are required for a detailed study. It was finally established [137,184] that molecular weights of DNA from most coliphage viruses are somewhat lower than earlier estimates.

The evaluation of M by the measurement of s and D and their combined use in the Svedberg equation [Eq. (60)] is not an equilibrium method since it is

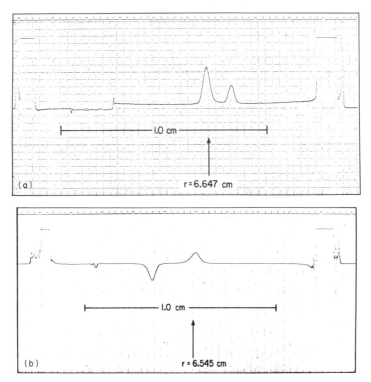

Fig. 3. Typical traces of photoelectric scanner from equilibrium sedimentation experiments in a density gradient with ^{14}N and ^{15}N isotopically substituted *E. coli* DNA; (a) 25,000 rpm at 8.5°C, Cs_2SO_4 gradient (b) trace obtained by putting DNA solutions of different densities in opposing sectors of a double sector cell; this run is at 35,000 rpm and 25°C in CsCl and the density difference is 0.0192 g/ml. From Schmid and Hearst [137].

based on two separate transport experiments. Sedimentation coefficients of phage particles and DNA molecules are fairly readily obtainable; diffusion coefficients, which until now could not readily be accurately determined, can now be obtained from a novel application of light scattering with coherent light, from the broadening of the central Rayleigh band. Dubin *et al.* [153] have determined by the new technique of optical mixing spectroscopy the diffusion coefficients of T4, T5, and T7 coliphages, their sedimentation coefficients as well as the molecular weights of the phage particles and their DNA (from the known percentage of DNA in the virus particles).

The combination of two other transport methods, namely sedimentation velocity and intrinsic viscosity, is less reliable because both translational and rotational frictional coefficients are involved (in the Svedberg equation only translational motion appears). This approach introduces a shape factor β, the Flory–Mandelkern parameter, [Eq. (7)] which depends, in a rather insensitive way, on the shape of the particles. Later we shall calculate β from the most reliable M, s, and $[\eta]$ data available.

For critical discussions about quantitative aspects of autoradiography, terminal labeling, zonal centrifugation (in sucrose gradients), and electron microscopy [185] we refer to two recent papers [136,184]. The authors conclude that none of these methods can provide absolute calibration of the DNA molecular weight scale, although considerable progress has been achieved in their critical application and certain procedures, properly calibrated relative to a known standard, may be useful. As late as 1969, considerably higher values for the molecular weights of T2 and T5 phage DNA than now discussed were obtained [186], although in the case of the lower molecular weight T7 DNA agreement was satisfactory.

Hearst *et al.* [187] in 1968 calculated molecular weight, excluded volume, and stiffness parameters for phage DNA's on the basis of values of s and $[\eta]$ and theoretical hydrodynamic calculations of persistent chains, before the later absolute molecular weight determinations became available. In particular a calculated value for β of 2.8×10^6 was used. That semiquantitative agreement was (within 10–20%) obtained with the now accepted molecular weights is gratifying.

2. *Relations with Hydrodynamic Parameters*

Whereas the narrowing down of the values of M for coliphage DNA to an uncertainty of not more than a few percent is indeed a worthy achievement, it has not been sufficient to establish definitive correlations between $[\eta]$ or s with M over a wider range (below 2.5×10^7 and above 10×10^7 daltons) of DNA molecular weights. In addition to the three closely bunched points for T4, T5, and T7 DNA the circular DNA of the ϕX174 virus (M about 3.6×10^6) has been used [178] for extending the low end of the scale, combined with a value of s estimated for the linear form from the value determined for the nicked double-stranded replicative form. Examination of plots of $\ln (s_{20,w}^0 - 2.8)$ vs $\ln M$ (Freifelder [184], Fig. 1) or of $s_{20,w}^0$ vs $M^{1/2}$ [178] (Fig. 1) shows that these are not strictly linear and that the correct relation between hydrodynamic properties and M has not been definitively established; clearly more experimental results are necessary, preferably over a wider range and for low values of M.

For Gaussian coils [Eqs. (4) and (6)] both $[\eta]$ and s are proportional to $M^{1/2}$. For very high molecular weight DNA both $[\eta]$ and s are thus expected to be proportional to $M^{1/2}$ (or to a power somewhat higher to 0.5 in the case of $[\eta]$ and slightly lower than 0.5 in the case of s, if the excluded volume is taken into account). For long cylindrical rods of constant (and negligible) cross section, $[\eta]$ is proportional to $M^2/\ln M$ (i.e., a power of M only slightly smaller than 2) and s is proportional to $\ln M$ (i.e., a very weak power of M only). For low molecular weight DNA Nicolaieff and Litzler have recently shown [188] (by an analysis of distribution of sedimentation coefficients as compared with distribution of lengths from electron microscopy for sonicated calf thymus DNA samples) that s is indeed a very weak function of M. A reasonable conclusion thus is that the exponents in the relationships between s and $[\eta]$ and M are functions of M. On the other hand, it has been assumed, both empirically [28], and on the basis of calculations of the hydrodynamics of stiff coils (Hearst *et al.* [68,76], for instance) that the exponent is invariant with M, but a nonvanishing intercept at $M = 0$ appears in the case of stiff coils. Thus the postulated relationship becomes

$$s - A = K'_s M^{a_s} \tag{62}$$

[compare with Eqs. (26) and (27)] and

$$[\eta] + B = K'_\eta M^{a_\eta} \tag{63}$$

where A and B are positive constants related to chain stiffness, a_s and a_η equal 0.5 for zero excluded volume and equal $0.5(1 - \varepsilon)$ and $0.5(1 + 3\varepsilon)$, respectively (cf. Section II,A,2) when excluded volume is considered. So far it has not been established that for DNA a_s and a_η are uniquely defined in terms of a common value of ε, although in the case of the much more flexible neutral poly A chains this appears [35] to be so. Ross and Scruggs [127] deduce that ε (as determined from measurements of $[\eta]$ at vanishing rates of shear) decreases with increase in ionic strength (ionic strength is known [6] to have strong influence on excluded volume properties in polyelectrolyte solutions), although true Θ conditions ($\varepsilon \approx 0$), such as have been reported in the study of poly A [35], are not observed. (In this context it is of interest that it has been shown only recently [34,136] that virial (intermolecular excluded volume) effects exist in DNA solutions, but a detailed study of the dependence of A_2 on ionic strength and temperature has not been undertaken). Different values of the statistical length are consistent with the data, depending on what corrections are made with respect to excluded volume. Hearst *et al.* [187] believe that ε is not affected by either ionic strength or temperature [189] and relate changes in s and $[\eta]$ with these variables to changes in the stiffness parameter only. We believe that that this may indeed be so for the temperature dependence of $[\eta]$ of short chains [175] in which the intramolecular excluded volume is inconsequential (a high exponent a_η and a low value a_s for short chains must be related to the rodlike nature of these chains, rather than to intramolecular volume), but in the case of long chains this is not necessarily so and the proper contribution of excluded volume (specifically the values of ε or of A_2) must be carefully considered in order to achieve a valid separation between chain stiffness and excluded volume properties.

Whereas the detailed functional relationship between hydrodynamic quantities and M cannot as yet be specified with sufficient authority (mainly because of paucity of data over a sufficiently wide range of M), we may examine a quantity repeatedly tested, namely, the Flory–Mandelkern parameter β, Eq. (7). We would like to know on the basis of reliable data available, the dependence of β on M, ionic strength and temperature; is Eq. (7) adequate for high molecular weight chains or should one rather use [70] an equivalent equation

$$\beta' = \frac{(s^0 - A)([\eta] + B)^{1/3}N_{\mathrm{Av}}}{M^{2/3}(1 - \bar{v}\rho^0)} = \beta\left(1 - \frac{A}{s^0}\right)\left(1 + \frac{B}{[\eta]}\right)^{1/3} \qquad (64)$$

based on relations similar to Eqs. (62) and (63)? We have assembled in Table II recent results for s, $[\eta]$ and M for coliphage virial DNA. The apparent specific volumes ϕ' (Section II,D,3), which replace the partial specific volumes $\bar{v}_2$ in the analysis of multicomponent systems have been calculated from the values of $(\partial\rho/\partial c)_\mu$ reported by Cohen and Eisenberg [97] for NaCl concentrations above 0.2 M by the equation

$$(\partial\rho/\partial c_2)_\mu = 1 - \phi'\rho$$

The viscosity and sedimentation measurements of Rosenberg and Studier [128] extend down to 0.0013 M NaCl. For salt concentrations below 0.2 M NaCl for which values of $(\partial\rho/\partial c_2)_\mu$ are not available, both $(\partial\rho/\partial c_2)_\mu$ and ϕ' were calculated from known values of $\bar{v}_2{}^0$, $\bar{v}_3$, and $\xi_3{}^0$ (cf. Fig. 4) by Eq. (31a). Note that $\bar{v}_2{}^0$ is definitely *not* the limit of ϕ' in very dilute NaCl solutions. This is clear from an examination of Eq. (31). It may also be concluded from the alternate formulation, Eq. (32a); although $(1 - \bar{v}_1\rho^0)$ tends to zero as $c_3 \to 0$, ξ_1^0 tends to infinity at the same time and the product $\xi_1^0(1 - \bar{v},)$ therefore stays finite.

In Table II we have tabulated s and $[\eta]$ from the work of Rosenberg and Studier [128] on T7 DNA, over a very wide range of ionic strengths. We have recalculated β with the value $M = 25.2 \times 10^6$ (as given by Schmid *et al.*

TABLE II

Calculation of Flory–Mandelkern Parameter β for *E. coli* Phage DNA

DNA	M^a $(\times 10^{-7})$	Na$^+$ (equiv/liter)	$s_{20,w}$ ^{b} (Svedberg)	$[\eta]^b$ (dl/gm)	ϕ' ^{d}	β^e $(\times 10^{-6})$	β' ^{f} $(\times 10^{-6})$
T7	2.52	0.0013	24.2	312	0.514	2.37	2.10
		0.0023	25.8	258	0.514	2.37	2.13
		0.0053	27.7	200	0.515	2.34	2.12
		0.01	28.8	172	0.516	2.32	2.11
		0.02	30.9	150	0.518	2.39	2.19
		0.05	31.2	132	0.522	2.33	2.15
		0.1	31.3	125	0.528	2.32	2.14
		0.2	31.4	119	0.540	2.35	2.17
		0.5	31.4	110	0.552	2.35	2.17
		1	31.5	100	0.565	2.35	2.18
T4	11.24	0.2	62.1^a	313^c	0.540	2.37	2.28
T5	7.03	0.2	51.8^a	240^c	0.540	2.48	2.37
T7	2.52	0.2	31.8^a	111^c	0.540	2.34	2.16
T4	11.24	0.2	63.4^g	293^h	0.540	2.36	2.33

Average 2.36 $\pm$ 0.04

[a] Values as given by Schmid *et al.* [178].
[b] Rosenberg and Studier [128].
[c] Interpolation by Hearst *et al.* [187] of data of Ross and Scruggs [127] for 0.2 ionic strength.
[d] Cohen and Eisenberg [97]; cf. also text.
[e] Eq. (7) with ϕ' instead of $\bar{v}_2^0$.
[f] Eq. (64), with ϕ' instead of $\bar{v}_2^0$.
[g] $s_{20,w}$ from value of s at 61°C, estimated from data of Gray and Hearst [189] between 5° and 49°C.
[h] Ross and Scruggs [127] at 61°C (0.937 of value at 25°C).

[178]) and ϕ' as described above (the values of $[\eta]$ of Rosenberg and Studier [128] are slightly different from the values of Ross and Scruggs [127], but this is not of major importance as the dependence of β is with $[\eta]^{1/3}$). We have also calculated β (Table II, column 7) for T4, T5, and T7 DNA with the values for s, $[\eta]$ and M as compiled by Schmid *et al.* [178] from the best available data; temperature dependence [127,189] has also been included. We find that β is remarkably constant (2.35 to 2.4 $\times$ 10^6) and no specific trend is observed with change in any of the parameters; it is significantly lower than the value 2.8 $\times$ 10^6 of Hearst *et al.* [187] calculated from the theory of the wormlike coil (they do allow variation of $\beta \times 10^{-6}$ between 2.65 and 2.85). We may accept $\beta \times 10^{-6}$ equal to 2.35–2.40 as a useful experimental parameter in the high molecular weight range; β' (Table II, column 8) with $A = 2.7$

and $B - 5$, did not exhibit the same constancy as β with the use of the same experimental results.

In a recent series of papers [190–193] Reinert and collaborators report new values for β [Eq. (7)] as well as for the parameters of Eqs. (62)–(64) for monodisperse DNA, on the basis of an elaborate evaluation of diffusion, sedimentation and viscosity data of highly polydisperse DNA, over a wide

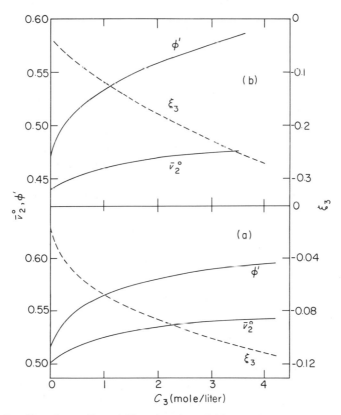

Fig. 4. Specific volumes ($\bar{v}_2$ and ϕ') and preferential interaction parameters ξ_3 in DNA solutions (a) NaDNA in NaCl; (b) CsDNA in CsCl. From Cohen and Eisenberg [97].

range of ionic strengths. (Their values are $A = 2.5$, $K'_s = 1.90 \times 10^{-2}$, $a_s = 0.435$, $B = 3.5$, $K'_\eta = 8.00 \times 10^{-4}$ and $a_s = 0.695$, valid for the homologous series of homogeneous native DNA at standard salt conditions over the molecular weight range from 0.3×10^6 to more than 115×10^6 daltons.) They use a high value (0.188 ml/gm at 436 nm) for dn/dc and a constant value for $\bar{v}$, independent of concentration of salt. They find [192] "a decrease of consistency" when the new values of Cohen and Eisenberg [97] for these accessory quantities are used, and therefore desisted from use of the latter in the evaluation of β. We believe that a detailed analysis cannot circumvent the result that ϕ', for instance (the quantity to be used in multicomponent systems instead of $\bar{v}^0$ in Eq. (7) for the evaluation of β), significantly increases with ionic strength (Table II). So-called consistency criteria are based on working concepts derived from experimental observations and are subject to constant reevaluation if more reliable data become available.

New calibration correlations for molecular weights of the covalently closed ring form and open relaxed rings of circular DNA by sedimentation analysis have been given by Böttger *et al.* [194].

The possibility of using the electrophoretic mobility of nucleic acids (RNA species, for instance) in polyacrylamide gels as a measure of the Stokes radius permits estimation of molecular weight by calibration with M of known species, if conformational effects can be eliminated. Staynov *et al.* [195] showed that formaldehyde in aqueous gels does not lead to disruption of all base-pairing in the experimental conditions. They developed a completely nonaqueous system, using formamide as a solvent, which causes complete loss of all polynucleotide structure, both base-pairing and -stacking. In this medium all polyribonucleotide chains are true homologs and the Stokes radius is a monotonic function of chain length; molecular weights in paucidisperse systems can be determined on no more than about 10 μg with good relative accuracy.

B. The Flexibility of DNA

1. *Persistence Length and Inherent Flexibility*

A problem somewhat related to the molecular weight and size determinations by either scattering or hydrodynamic methods is the flexibility of DNA. Most calculations have been based on the model of the persistent chain (Section II,A,3) and disparities in numerical values of the persistent length are due to uncertainties in the early experimental results, poor samples, and the use of approximate hydrodynamic and scattering theories. The inherent flexibility of DNA may be a significant parameter in the interpretation of the folding processes of DNA in higher structures, such as phage heads and chromosomes [195a]. The complexity of these processes exceeds manifestations found in carefully purified dilute solutions: interactions with basic proteins [195b] or subtle changes in structure related to partial separation of strands compound the difficulties encountered in quantitative comprehension. Deuterium and tritium exchange experiments [196] clearly indicate that DNA "breathes," in distinction to other much more tightly bound multichain linear polymers, such as collagen, for instance. Polycyclic dye intercalation [197] is another manifestation of the way DNA responds to interaction with outside agents. Enzymes often respond in similar ways to outside stimuli. Clearly the distinction is between structural and functional macromolecules, and DNA is a prominent member of the latter class.

It is thus important to ascertain the true flexibility of nucleic acids in terms of the energetics of the process. In single-chain structures we may eventually reach this goal by a complete analysis of hindered rotation around bonds (Section II,A,4). In helical structures, with more than one chain per helix, we require more information about the force constants in the stacked and hydrogen-bonded arrays, as well as extensive dynamical measurements [198] and a detailed theoretical frame. The persistence length is at best a parameter characteristic of the equilibrium configuration (typical transport methods such as viscosity and sedimentation are used in this sense as probes to characterize such equilibrium conformations) which can be related [175,189] to static elastic parameters of the chain. It may be that, whereas for fairly long chains the persistent chain represents an adequate description [33] of the equilibrium configuration of actual chains, it does not describe the local curvature of the chains, or the configuration of short chains [199]. It has been suggested, for instance, from the study [200–202] of the solution behavior of polybutylisocyanate (a rigid synthetic polymer) that the contour of "real" stiff molecules responds more radically to increasing length than is envisaged by the Kratky–Porod model: at low molecular weights the behavior is truly rod-like, in an intermediate molecular weight range "bent rod" or semicircle arc conformations predominate, and above this more pronounced curvature and change in direction may lead to persistent coil conformations (Kirste [47] distinguishes between persistence of direction, the Kratky–Porod persistence, and persistence of curvature). Recent experiments though (Section III,B,3) show that the Kratky–Porod chain describes correctly the behavior of short DNA chains. High molecular weight DNA and short chains will be discussed in separate sections below.

2. High Molecular Weight DNA

In 1953 Peterlin [53] estimated values of the persistence length a of DNA, on the basis of earlier data from Doty's laboratory, as between 260 and 606 Å; this result is historically interesting because it antedated the Watson–Crick structure. It is not excluded that more consistent scattering results with better characterized DNA samples would have enabled the early establishment of a double-strand structure for DNA, in particular if the mass per unit length (cf. Table 1, Peterlin [53]) would have been better defined.

The experimental difficulty in extrapolating to low enough values of h to obtain the correct value of M and $\langle R^2 \rangle^{1/2}$ for high molecular weight DNA prompted the investigation of the asymptotic behavior at high angles. Sadron *et al.* [203] showed that for sonicated low molecular weight DNA ($M = 6 \times 10^5$ daltons) the intercept of the asymptotic plot of the reciprocal scattering intensity vs sin $(\theta/2)$ is positive; this corresponds to a stiff, rodlike particle. From the slope of this plot [29] $M/L = 201$ daltons/Å is obtained, in agreement with the Watson and Crick B structure. In the case of higher molecular weight ($M = 8 \times 10^6$ daltons) calf thymus DNA the intercept of a similar plot is negative. This result was interpreted [203], following the theoretical work of Luzzati and Benoit [51], to mean that DNA is a zigzag chain, composed of more or less rectilinear segments the length of which is of the order of 4000 Å (folding of these segments was enhanced by the preparation of artificial DNA–histone complexes). Ptitsyn and Federov [204], on the other hand, in a view disputed by Sadron *et al.* [203], believe that the negative intercepts in these plots may indicate chain flexibility, which they evaluate on a calculation in which the Kratky–Porod limiting process is applied to the zigzag chain. They calculate values of 220–320 Å for the persistence length, *a*, of DNA chains, based on light scattering experiments from the laboratories of Doty and Sadron.

The difficulties associated with the experimental and theoretical aspects of light scattering led to reinvestigation of hydrodynamic results. Hearst and Stockmayer [68] estimated a value of 360 Å for *a* of high molecular weight DNA, in 0.2 *M* NaCl, from an analysis of *s* of the wormlike chain without excluded volume [Eqs. (26) and (27)]. Gray *et al.* [75] later took excluded volume into account and calculated 450 Å for *a* (for the excluded volume parameter ε they use $\varepsilon = 0.11$). More recently Hearst *et al.* [187] ascribe (from hydrodynamic measurements) a value of 0.072 ± 0.036 to ε, and 403 Å to *a* at 0.2 ionic strength. From viscosity data [127] over a wide NaCl concentration range they conclude that *a* increases from 332 Å at 1.0 *M* NaCl to 658 Å at 0.005 *M*, on the assumption that ε remains unchanged. Triebel *et al.* [205] believe that a value of 454 Å is satisfactory for the Kratky–Porod length of high molecular weight DNA. Ross and Scruggs [127] interpreted their own data by allowing ε to change (from 0.076 at 1.0 *M* NaCl to 0.195 at 0.005 *M*). The determination of the excluded volume parameter from a change in the exponent in the dependence of either *s* or [η] on *M* (over a limited range of *M*) is uncertain. Stiffness and excluded volume are mutually dependent, and it is preferable to obtain the volume exclusion parameter from independent thermodynamic studies [33]. Ross and Scruggs [127] also observed that above 1 *M* of either NaCl, KCl, or LiCl, [η] of T4 DNA does not depend on temperature (between 25° and 61°C) and that at lower ionic strength a negative temperature coefficient is observed. At 0.2 *M* salt concentration the decrease they observe in [η] (6.3%) between 25° and 61°C is identical with that reported by Cohen and Eisenberg [175] for a low molecular weight sample (5×10^5 daltons).

Gray and Hearst [189] also determined the flexibility of phage DNA's (T7, T5, T4) from their sedimentation behavior as a function of M and temperature. They assume constant volume exclusion and calculate a to decrease from 459 to 412 Å from 5 to 49°C. From the temperature dependence of a, viewed in the sense of an elastic bending parameter [45] they calculate the enthalpy and entropy of bending of the DNA double helix. The enthalpy for packing DNA into a bacteriophage head (according to the model of Kilkson and Maestre [206]) is calculated to be 10 kcal/100 Å length of DNA and the entropy of bending is negative. Mechanisms suggested include a DNA model in which wormlike coils are connected by universal joints, or uniformly stiff chains able to bend sharply at any point by breathing [196] in the sense that local regions of DNA become unstacked and their hydrogen bonds are temporarily broken. Gray and Hearst [189] also discussed the effect of glucosylation of T-even coliphage DNA on its molecular weight and sedimentation coefficient.

Hays *et al.* [33] have argued that the influence of excluded volume was overemphasized in persistence length calculations, and it may be ignored for molecular weights as high as 2.5×10^7 (T7 DNA). They calculate 600 ± 100 Å for a from hydrodynamic data and 900 ± 200 Å from light scattering results [34] at low angles. Sharp and Bloomfield [57] criticized the extrapolation to zero angle of Harpst *et al.* [34] and calculate $a = 450$ Å with $\varepsilon = 0.11$ from these data. Schmid *et al.* [178] also believe that the extrapolation of Harpst *et al.* [34] to zero angle (in the region from 10° to 25°) yields molecular weights which are about 20% too high; they also correct the experimental data using the value of Cohen and Eisenberg [97] for the refractive index increment. With $\varepsilon = 0$ they find that both hydrodynamic and light scattering results are well described by $a = 650 \pm 50$ Å. This value is confirmed by Ullmann [79,80] who calculates $a = 600$ Å from viscosity and $a = 700$ Å from sedimentation data of high molecular weight DNA on the basis of improved hydrodynamic calculations of the persistent chain without excluded volume. At present a value of about 600 to 700 Å appears to be reasonable for the persistence length of high molecular weight DNA, from both scattering and hydrodynamic studies.

Hays and Zimm [207] have discussed the effect of single-strand breaks ("nicks") on the flexibility of DNA in saline solutions, over a range of temperatures. Changes in DNA flexibility were followed by observing changes in [η] and s subsequent to treatment of mechanically degraded (sheared) DNA with pancreatic DNase. At room temperature even samples with more than one nick per persistence length showed no increased flexibility, but near the DNA melting temperature, [η] of nicked DNA was lower than that of intact DNA at the same temperature.

Sugi *et al.* [208] calculated the rigidity toward bending of double-stranded

DNA (from an analysis of the contour lengths and end-to-end distances in DNA electron micrographs prepared by the "diffusional method" of Lang and Coates [209]) to be about one-half an earlier estimate of Cohen and Eisenberg [175].

Finally, in recent studies Bram [210,211] concludes from wide-angle x-ray scattering of solutions and gels of calf thymus DNA that the secondary structure of DNA in 0.05–0.15 M monovalent salt is of the B kind. A structure which is derived from the B form by decreasing the turn angle to 0.575 radians (33°) and the pitch to 37 Å as compared to 0.628 radians (36°) and 32 Å in the B form is consistent with the results. (In similar diffraction work with nucleohistones the structure of the DNA corresponds to the B form.) Bram believes that ionic strength-dependent charge repulsion may determine the turn angle in DNA and relates his observations to the origin of tertiary turns in cyclic DNA molecules [212], optical rotatory dispersion changes with increasing salt concentration and the related decrease in hydration with decreased water activity [213], and other phenomena. We have earlier mentioned (Section II,C,2) that electrostatic repulsion at salt concentrations above 10^{-3} M monovalent salt is not likely to affect chain dimensions by acting across distances of the order of the persistence length. The present mechanism, though, of a salt concentration-dependent dynamic local structure, in which charge repulsion between closely spaced charges may play a significant role in the determination of torsional properties of the DNA double helix, without affecting the mass per unit length, M/L, merits further study.

Bram also concludes [214] from the shape of wide-angle x-ray scattering curves that the secondary structure of DNA depends on base composition and that unlike DNA of low and moderate AT content, DNA very rich in AT does not adopt the B conformation; also poly dI·poly dC, poly d(I–C)·d(I–C) and poly dG·poly dC have structures dissimilar to the A, B, and C forms of DNA, whereas poly rI·poly rC is like the A'-RNA form [215].

3. *Low Molecular Weight DNA*

DNA can be broken into shorter fragments, without loss of double-helix structure by mechanical shear [216] or sonic irradiation [174]. Cohen and Eisenberg [175] deduced from light scattering and hydrodynamic studies that the structures of Na and CsDNA are identical, and that, by proper interpretation of changes in s and $[\eta]$ of these nearly rodlike structures, binding of proflavin and lengthening of DNA may be closely related [163]. The sonicated samples were shown by electron microscopy to have a rather narrow distribution of lengths. Calculation [199] of a formal persistence length from both light scattering and sedimentation yielded values of between 1500 and

1900 Å which are largely in excess of the values (600–700 Å) now accepted for high molecular DNA. The problem raised warranted detailed reinvestigation. Measurements on short pieces of DNA ($2 \times 10^5 < M < 2 \times 10^6$) are convenient because they fall in a range in which experimental problems in light scattering and hydrodynamical [217,218] studies can be satisfactorily tackled; polydispersity can be reduced by fractionation and can also be accounted for properly. The major difficulty is that, in order to determine a flexibility parameter, small deviations from strictly rodlike behavior must be very precisely defined.

On the basis of careful light scattering, viscosity, sedimentation, and electron microscopy measurements on sonicated calf thymus DNA, fractionated by zonal centrifugation in sucrose gradients [219], Jamie Godfrey (work recently completed in our Laboratory) finds that a is close to 600 Å for molecular weights from 0.25×10^6 to about 1.2×10^6. In the calculation for a for these narrow fractions ($M_w/M_n < 1.04$) proper polydispersity corrections and weighting factors have been considered. Godfrey also finds that optical anisotropy corrections are considerably lower than reported in the literature and therefore corrections due to anisotropy are nearly always negligible in the calculation of M_w and the mean square radius from light scattering. These preliminary calculations, if confirmed, provide evidence for the applicability of the Kratky–Porod model down to DNA chains of about 1300 Å contour length. Significant measurements on shorter chains for the determination of persistence lengths do not appear feasible.

C. INTERACTION OF NUCLEIC ACIDS WITH SMALL MOLECULES

1. *Donnan Effect and Hydration*

We have in Section II,B defined various thermodynamic interaction parameters and their relation to physical binding of small molecules to the nucleic acid moiety. So-called net hydration, as expressed by the formal interaction parameter ξ_1' [Eq. (35b)], increases strongly with decreasing salt molality w_3, with no required relationship to the physical binding [B1] of water to the nucleic acid. This is due to the fact that because of Donnan exclusion [E3] the molality $w_3 < w_3^*$; net hydration, then, incorporates a fictitious shell of water (increasing with decreasing w_3) created by "compression" of the simple salt in the nucleic acid solution from the molality w_3 to the molality w_3^* of the equilibrium dialysis polymer-free simple salt solution. Clearly the physical concept of hydration should not be associated with this phenomenon. The measured degree of physical hydration may vary with various experimental methods because of interpretative difficulties and a certain degree of arbitrariness in the precise definition of the tightness of water binding; yet a realistic nucleic acid–water interaction must necessarily be included in any definition of hydration. No such element is contained in the definition of the purely phenomenological concept of net hydration and its identification with physical binding is not immediate. Some dominant factors which influence net hydration have been discussed by Hearst [100], who also estimated the relative importance of the electrostatic contribution (which leads to Donnan exclusion) to it.

Costello and Baldwin [222] showed that the banding density of phage λ varies with the activity of water when the phage particles are banded in a series of different cesium salts; the total net hydration of the phage is approximately equal to the net hydrations of free λ DNA and λ salts, all measured at the same water activity. The simple additivity approximation is not adequate, however, to explain the banding density differences between a deletion mutant and phage λ in the different cesium salts. The density differences apparently are affected by a restriction of DNA hydration inside the phage head, which depends both on water activity and on DNA length. It appears possible to estimate the size of a DNA deletion from the phage sedimentation coefficient.

Tunis and Hearst [220] have recently presented an excellent discussion on hydration studies and mechanisms of hydration in DNA solutions. They also measured the net hydration, its dependence upon base composition, and the stability of DNA in concentrated trifluoroacetate solution. They found that the buoyant density for both native and denatured DNA increases with increasing temperature, although potassium trifluoroacetate to some degree inhibits the recovery of secondary structure of denatured DNA upon thermal quenching; the destabilizing effect of trifluoroacetate has been related to the loss of (net) hydration on denaturation (we shall see below that physical hydration, as derived from calorimetric measurements, is believed to increase slightly upon denaturation). Extrapolation to zero water activity showed the dependence of the partial specific volume on base composition to be very small for CsDNA and zero for LiDNA. In addition, Chapman and Sturtevant [221] believe that the partial specific volumes of anhydrous native and denatured DNA are equal. The differences seen at finite water activity (corresponding to lower ionic strength of the simple salt component 3) are due to a combination of physical hydration [B1], ion exclusion [E3], and binding [B3] as expressed in the preferential interaction parameters of Eqs. (35).

An analysis of the effect of pressure on the DNA melting transition is instructive. Measurements of the buoyant density of native and denatured DNA by Schildkraut *et al.* [223] in 7.8 M CsCl at room temperature have led to the conclusion that the density increases with strand separation by 0.015–0.017 gm/ml at a buoyant density of about 1.7 gm/ml; this was interpreted to correspond to a decrease in specific volume of about 0.005 ml/gm upon DNA denaturation. It was therefore conjectured [224] that increasing pressure should lead to DNA denaturation at lower temperature and that the phenomenon might have biological significance. Quite to the contrary it was found [224] (Fig. 5) that an increase in pressure up to 2700 atm [225] promoted moderate stabilization of the double-helix structure. The apparent contradiction is due to the fact that buoyant densities relate to $(\partial \rho / \partial c)_\mu$ (cf. Section II,D,4), a density increment which includes (cf. Section II,B,3) changes in density deriving from redistribution of components. In the application of high pressures in a closed system true volume changes, which may be related to

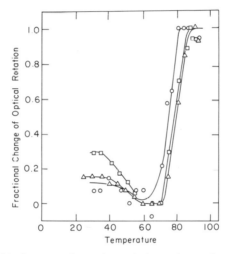

Fig. 5. Relationship between change in optical rotation and temperature at different pressures, for calf thymus DNA in 0.03 *M* NaCl, 0.001 *M* EDTA; (○) atmospheric pressure; (□) 10,000 psi; (△) 20,000 psi. From Weida and Gill [224].

changes in hydration with temperature and upon melting, determine the stability of the system. It has been shown [221] (Fig. 6) by straightforward dilatometry at neutral pH, that the specific volume of DNA indeed increases with temperature; the increase, though, proceeds smoothly throughout the denaturation region and surprisingly enough no discontinuities (related to changes in hydration) upon melting may be discerned. It therefore appears from these results that hydration of DNA decreases with increase in temperature but is not significantly related to double-helix structure. By implication the site of water binding are the phosphate groups, a conclusion acceptable to some [226,227], but not to other [228] workers who believe in a more specific role of water in the stabilization of the double-helix structure by the presence of hydrogen-bonded and -stacked water structures. The rather small influence of pressure (Fig. 5) on the melting transition supports the view that hydration does not play a major role in the stabilization of the double-helix structure. The effect on buoyant densities of strand separation is more likely due to changes in preferential interactions rather than in the specific volume deriving from a partial loss of the hydration shell.

Contrary to the results observed in the denaturation of DNA it was found [229,230] that helix formation of poly A and poly U is accompanied by an increase in specific volume and this opposite effect also carries over [230] to density gradient experiments. From the pressure dependence of the helix–coil

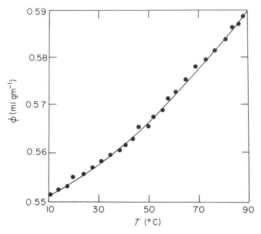

Fig. 6. Typical dilatometric data of solution of 0.732% calf thymus DNA in pH 7 phosphate buffer ($Na^+ = 0.013$ M) containing 10^{-3} M EDTA, ionic strength 0.02. From Chapman and Sturtevant [221].

transition temperature Gunter and Gunter [231] find the volume changes to be -0.96 ml/mole of nucleotide base pairs for the poly (A + U) transition, $+0.35$ ml/mole for the poly (A + 2U) transition and $+2.7$ ml/mole for the DNA transition.

Chapman and Sturtevant [221] also studied the change in the apparent molal volume of DNA on thermal denaturation in carbonate buffer at the alkaline pH 11. The major part of the volume changes observed (in the presence of various counterions) could be related to a volume change expected to accompany the transfer of protons from the bases guanine and thymine to carbonate ions; as at neutral pH, the volume change directly due to the change in shape of the nucleic acid molecules upon denaturation is so small as to be experimentally undetectable. The general increase in the volume–temperature coefficient with increasing temperature was attributed to decreased electrostriction of water at higher temperatures.

Falk and co-workers [226,227] have studied the hydration of sodium and lithium salts of DNA and of polynucleotides as a function of relative humidity by gravimetric measurements. The hydration curves of the polynucleotides so far investigated [DNA, soluble RNA, two forms of polyriboadenylic acid (poly A)] is in agreement with the suggestion that the ionic phosphate groups are primary hydration sites in polynucleotides and that the hydration is not strongly affected by the nature of the cation and by the detailed secondary structure. The data are interpreted to yield a value of about 2.0 molecules of water per nucleotide for the first layer of adsorption for all nucleotides investigated; the absence of distinct plateaus, such as occur for the hydration and dehydration curves of some compounds of low molecular weight which contain the phosphate group, indicates that no well-defined hydrates are formed.

Privalov and Mrevlishvili [232] have reviewed NMR and other work on the hydration of macromolecules in the native and denatured states. In their own experimental work they determine the heat capacity of DNA in the presence of different amounts of water, in the temperature range 5 to −35°C. Addition of water below 0.5 gm/gm dry wt DNA does not lead to any absorption of heat in the region around 0°C (Fig. 7). It is presumed that this water

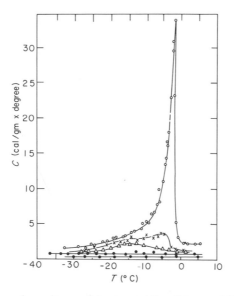

Fig. 7. Temperature dependence of heat capacity of 1 gm of DNA in presence of different amounts of water: (●) 0.05 gm; (△) 0.75 gm; (×) 1.00 gm; (○) 2.00 gm. From Privalov and Mrevlishvili [232].

does not freeze on cooling or fuse upon heating, and is therefore believed to
be present in an ordered state. Upon addition of larger amounts of water a
small heat absorption hump appears below 0°C indicating water capable of
freezing and melting, but still under the influence of the charged macromolecule. With further increase in the amount of water a well-developed peak
appears and shifts toward 0°C, indicating the formation of layers of water
less influenced by the macromolecules. The authors conclude that the hydration corresponds to 0.61 gm water/gm DNA (11.6 molecules of water per
nucleotide) a figure largely in excess of the value of Falk [226,227] (about 2
molecules of water per nucleotide) or the value of about 6.7 molecules of
water per nucleotide derived by Wang [233] from self-diffusion studies. Similar high values were obtained [234] by the same method for the hydration of
transfer RNA. Privalov and Mrevlishvili [232] also report that hydration
increases upon denaturation and discuss conflicting evidence from other work
with respect to this rather small effect (they do believe this confirms Kauzman's [235] view that denaturation in general leads to increase in hydration
since the number of contacts of nonpolar group with water increases in this
case). Chattoraj and Bull recently reported on hydration properties of DNA
in solutions comprising different salts or neutral substances, such as urea and
sucrose, in isopiestic [236] and calorimetric [237] studies.

The hydrogen-exchange kinetics of native DNA have been studied as a
function of pH and salt concentration [196,238,239]. Of the labile hydrogens
only hydrogens involved in interchain hydrogen bonds in DNA exchange
sufficiently slowly ($t_{1/2} > 50$ sec) to be visualized by tritium–Sephadex gel
filtration techniques. Recently though Hanson [240], by use of a high-speed
chromatographic technique in which data points were obtained as early as
6 secs, reported a new class of slowly exchanging hydrogens in a variety of
DNA's, which exchange at a rate faster than that of those slowly exchanging
hydrogens previously observed in nucleic acids. He concluded that exchangeable hydrogens include not only those involved in base-pairing hydrogen
bonds, but the other amino hydrogens of the DNA bases as well; implications
of this finding were discussed in terms of DNA hydration in general, and
dynamic properties of the amino group in particular. Englander and von
Hippel [241] report similar results and comment on the fact that these hydrogens exchange many times slower than expected from the exchange times in
the free bases or in single-stranded nucleic acids. In another study Englander
et al. [242] report on hydrogen exchange in some polynucleotides and nucleic
acids. A different approach was taken by Killion and Reyerson [243] who
made a series of adsorption–desorption cycles of D_2O on salmon sperm DNA
in order to measure its total equilibrium deuterium exchange. Gravimetric
measurements showed the total exchange to be nearly twice the amount required for complete exchange of all labile hydrogens on the macromolecule.

This amount would accommodate at least 1.7 molecules of tightly bound water per nucleotide and is believed [243] to support the arguments of Lewin [228] that water is associated with DNA and its structure, although quantitative proof for this association and its exact nature is lacking.

The evidence with respect to hydration from nuclear magnetic resonance studies is not very satisfactory. Gordon and co-workers [244] investigated the absorption area, linewidth, and chemical shift of the water hydroxyl proton in DNA solutions, over the range $10°-90°C$—above the denaturation temperature of DNA; the results indicated that the amount of water bound to DNA is very small, and is not measurably changed during denaturation. The correlation of the denaturation temperature of DNA with the structure of water and D_2O were measured by the chemical shift of the hydroxyl proton resonance. The presence of up to 10% DNA did not cause any measurable change in water structure. Water structure was found to be a major factor in determining DNA stability, although this is not easily correlated with NMR studies. In sodium perchlorate solutions, for instance, both ions shift the proton resonance in water to higher fields; however Na^+ has an ordering effect on water, through hydration, whereas ClO_4^- disorders water in correspondance to an increase in temperature. Nuclear magnetic resonance studies of hydration are therefore not easily interpretable, are not very sensitive (in view of the small amount of water bound), and may be subject to experimental artifacts.

Lubas and co-workers [245] discuss the importance of dynamic equilibrium between water molecules in the hydration lattice as detected by a NMR spin–echo technique. Migchelsen and co-workers [246] interpreted the splitting of the water lines in NMR spectra of oriented DNA to indicate the existence of chainlike structures of water molecules in which rotations are anisotropic as a result of the anisotropic distribution of hydrogen bonds. Kuntz and co-workers [247] detect "unfrozen" (non-icelike) water by a NMR signal identified with water of hydration of proteins (0.3–0.5 gm/gm protein) and transfer RNA (1.7 gm/gm nucleic acid); the latter number corresponds to 31 molecules of water and is far in excess of other physical hydration numbers reported. A very recent slow-neutron scattering study [248] of oriented NaDNA supports the idea that water molecules are bound to hydrogen-bonding sites on the Na DNA double helix but does not support the suggestion [246] that the fit of the axial repeating distance of the macromolecule to a multiple of 4.7 Å (the assumed second-neighbor distance in water) acts as a stabilizing factor for longitudinal water chains.

Glasel [249] has described deuteron magnetic spin lattice relaxation experiments on D_2O solutions of polyuridylic and polyadenylic acid; general classes of interactions are discussed on the basis of these and other results as are the effects of geometrical fluctuations of polymer conformation upon polymer–water interactions. The competing effects of polymer counterion and polymer water interaction are believed to be of vital importance in mediating polymer conformation.

In a theoretical study [250] Khanagov analyzes spectra of proton and deuteron magnetic resonance of water molecules in DNA on the basis of a model of the orientation symmetry of water molecules in biopolymers. Edzes *et al.* showed [251] that the NMR spectra of ^7Li and ^{23}Na in oriented DNA fibers are split into three lines (depending on the water and salt content, and on the crystal form of the DNA) due to quadrupolar interactions. These observations are relevant in terms of an alternative interpretation of NMR signals in biological tissues, which was recently proposed by Shporer and Civan [252], and states that all the sodium present yields a signal of which only 40% is easily detected due to quadrupolar effects. This is contrary to the interpretation of these experiments in terms of a "free" or "bound" fraction of sodium [253], where the bound fraction is believed to produce a signal which is too broad to be detected in the conventional way.

If the careful reader has discovered inconsistencies in the above discussion, this is due in large part to conflicting results of the various studies reported. We must conclude that our knowledge about hydration of nucleic acids is not in a very satisfactory state. It can be said with some assurance that two molecules of water are firmly associated with each nucleotide, probably with the phosphate group. Up to six molecules are less firmly bound and may fill the grooves of the double helix. Any further water molecules are only very loosely connected to the nucleic acid structure and are not easily distinguished from bulk water. Net hydration may assume any value, depending on simple salt concentration (or water activity) and may be only in small part related to physical hydration [cf. Eq. (35b)]. Changes in hydration with denaturation are small and not measurable by most methods.

2. *Interactions with Small Ions*

Whereas in the case of hydration it has not been possible to date to identify clearly the sites of water binding, somewhat less hesitant statements may be proffered concerning interactions with small ions, although definite conclusions as to the role of both cations and anions with respect to nucleic acid structure and function are still lacking. It would not be possible to discuss here the wealth of contributions, both experimental and theoretical, which deal (often in conflicting fashion) with this important topic. Felsenfeld and Miles [87] as well as von Hippel and Schleich [254] have recently discussed many aspects of interactions of nucleic acids with ions and neutral salts, and we shall try here to provide the reader with guidelines to the newer (since 1966), and some of the older literature.

Nucleic acids, at neutral pH, are characterized by the high electrostatic potential residing on the phosphate residues. The random or stacked, free or hydrogen-bonded bases (as the case may be) do not carry electrostatic charges at neutral pH. The counterions to the phosphate group are under the strong influence of the electrostatic potential (Section II,C). This interaction, which leads to effective reduction of the electrostatic potential, is nonspecific in nature, depends on ion-charge and ion-radius only, and may be estimated by polyelectrolyte theory. A different kind of binding is site binding [255] to the phosphate groups which involves close interaction between the backbone and mobile counterions, with or without concomitant release of the neutral hydration shell (depending on the strength of the interaction). In the nucleic acid literature, site binding and specific binding are not clearly distinguished. It seems appropriate to define site binding, as opposed to ion-cloud interaction, whenever an ion pair is formed from which the exchange of the counterion with the counterions in the bulk of the solution is considerably slowed down. Specific binding should be restricted to the very strong site binding in which the specific nature of the counterion (and not its electronic charge only) plays a significant role. Another type of binding is by chelation to the bases, and finally it is possible to have chelation by metal ions between the bases and the phosphate groups. In the latter two cases tautomerization of the bases, or the effect of ion binding on the hydrogen bonding scheme, may have profound influence on nucleic acid structure and optical or magnetic properties. In the case of binding to phosphate ions these latter properties may be affected, but to a small extent only. This is a consequence of the fact that, because of charge decrease (or even possibly charge reversal) upon binding of divalent ions to the phosphate groups stacking or hydrogen bonding of bases (with subsequent change in optical or magnetic properties) may occur, while it is prevented in the absence of these ligands because of electrostatic repulsion on the same or on opposite polynucleotide strand. Binding and specificity of binding, as well as such properties as dependence of melting transition and its breadth with base composition, often change with change in ionic strength.

Daune [256] has recently extended the theory of Alexandrowicz and Katchalsky to a calculation of the stability of DNA double helices in solutions containing monovalent and divalent neutral salts; he has also discussed [257] the biological significance of ion binding of some divalent metal cations, notably Mg^{2+} and Mn^{2+}, and has represented binding of metallic ions according to the schemes shown in (Fig. 8), which are presented as illustrations, without necessarily exhausting all the possibilities. The electric field produced by the polyion may, in conjunction with that of the compensating counterion, create a high polarization of the hydration water. A partial ionization (local hydrolysis) may then transform the phosphate into a true binding site for the counterion as would be the case for the Li^+ or Mg^{2+} ions (Fig. 8a and b).

At low ionic strength the metal cation (Mn^{2+}, Zn^{2+}, Co^{2+}, Fe^{2+}, and Cu^{2+}) may form (Fig. 8c) a bridge between the phosphate ion and a purine site (presumably N-7) on the base. The metal cation forms an internal chelate between the nitrogen N-7 and a neighboring NH_2 or C=O group (Fig. 8d). Metals like Ag^+ or Cu^+ complex between succeeding bases (Fig. 8e) under steric conditions more favorable than Fig. 8f. Finally, a complex may form between two bases on the same base pair with the elimination of a proton. This has been suggested by Jensen and Davidson [258] for Ag^+ and by Yamane and Davidson [259] for Hg^{2+} complexes, the metal ion taking the place of a hydrogen bond. Mercury ion Hg^{2+} is preferentially bound between adenine and thymine or better between thymine and thymine, but Cu^{2+} seems to be

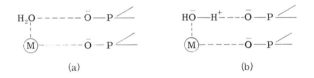

(a)   (b)

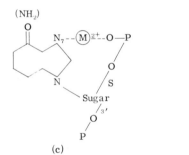

(c)   (d)

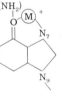

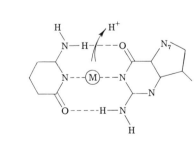

(e)   (f)

Fig. 8. Models of possible fixation of cations on DNA (details in text). From Daune [256].

intercalated between two guanine residues or between guanine and cytosine in the same strand.

Whereas the cationic counterions to the negatively charged nucleic acids are expected to be closely attracted to the macromolecular chains, and therefore subject to specific interactions, the concentration of negatively charged anionic coions is likely to be lower in the immediate vicinity of the nucleic acids than in the bulk of the solution. The influence of anions on DNA stability is believed to be due to their influence on water structure, and may therefore be called an environmental effect. In an extensive study of the thermal denaturation temperature of DNA at neutral pH Hamaguchi and Geiduschek [260] found that at high salt concentration (4 M) denaturing power increases in the order CHO_2^-, Cl^-, $Br^- < CH_3COO^- < I^- < ClO_4 < CF_3COO^- < CNS^- < CCl_3COO^-$; on the other hand $(CH_3)_4NCl^-$ does not denature DNA even at 15.7 m. The broadness of the melting transition is also affected. Cation effects on the denaturation of DNA were studied by Dove and Davidson [261] as a function of pH and salt concentration. Supporting electrolytes comprising an alkali metal ion may increase the denaturation temperature up to an ionic strength of at least 1 M (presumably because of a decrease in electrostatic repulsions), whereas at higher ionic strength the melting temperature may again decrease because of anion effects. The cations Mg^{2+}, Co^{2+}, and Ag^+ raise and broaden the melting transition considerably. According to Eichhorn [262], Cu^{2+}, Cd^{2+}, and Pb^{2+} reduce the thermal stability of DNA in solution, while in the presence of Mg^{2+}, Ca^{2+}, Ba^{2+}, Mn^{2+}, Co^{2+}, Ni^{2+}, and Zn^{2+} the heat stability of DNA rises. Kuznetsov *et al.* [263] found that the stabilization of DNA structure is related to the affinity of the DNA for various alkali metal ions and decreases in the order $Li^+ > Na^+ > Rb^+ > K^+ > NH_4^+ > Cs^+$ at 0.03 ionic strength. The interpretation, which is related to the hydration energy of the counterions is complicated by the fact that at different ionic strengths, reversals of order may occur. In more recent studies Shapiro *et al.* [264] measured the relative affinities of small weakly bound cations for DNA by dialysis equilibrium experiments carried out by allowing two DNA's of different base composition to compete with one another for a strongly bound "test" cation such as Ca^{2+} or spermine. It was found that Na^+, Li^+, Cs^+, and K^+, as well as arginine, lysine, and tetralysine, are bound equally tightly to DNA of any base composition. Tetramethyl- and some other tetraalkylammonium ions are bound more tightly to adenine–thymine-rich than to guanine–cytosine-rich DNA's. In a study of the optical rotatory dispersion of DNA in strong salt solutions, Tunis and Hearst [213] found that the order of effectiveness of ions in decreasing rotation is $Li^+ > Cs^+ > Na^+$ for cations and $Cl^- > Br^-$ for anions; formate and trifluoroacetate have a larger effect than chloride and there is no correlation with water activity. This cannot be correlated with either changes

in water structure or with the ability of the salts to lower the melting temperature of the DNA. There may be a correlation with the fine structure of DNA
in solution, as evidence by changes in the number of base pairs per turn in
the Watson–Crick helix (cf. Wang *et al.* [212]) with change in ionic environment.

The effects of aqueous neutral salt solutions on the melting temperatures
of DNA's ranging in base composition from 0 to 72 mole% GC, has been
studied by Gruenwedel *et al.* [265]; Auer and Alexandrowicz [93] have investigated properties of sodium DNA in salt-free solution. Okubo and Ise [266]
report mean and single-ion activity coefficients and transference numbers of
sodium DNA in aqueous solution. In a small-angle x-ray scattering study
Pilz *et al* [267] found that the radius of gyration of phenylalanine specific
transfer RNA is 23.6 Å with Li^+, 26.2 Å with Cs^+ and 26.3 Å with Ba^{2+}
counterions. Electrostatic effects on polynucleotide transitions in acidic and
alkaline solutions, have also been investigated [268–270]. Using electrophoretic mobility measurements of calf thymus DNA in the presence of weakly
interacting $(CH_3)_4N^+$ ions over a range of values of ionic strength, Ross and
Scruggs [271] have demonstrated [268–270] that the binding order to DNA is
$Li^+ > Na^+ > K^+$ for the alkali metal ions and $Mn^{2+} > Mg^{2+} > Ca^{2+}$ for
the divalent ions studied. In the temperature range 5°–69°C Vinograd *et al.*
[272] find the buoyant density of bacterial and viral DNA increases in CsCl
solutions in the ultracentrifuge. Aubel-Sadron *et al.* [273] and Ebel *et al.*
[274] have shown that it is possible to obtain quaternary ammonium salts of
various nucleic acids, which are soluble in organic solvents. The biological
activity of nucleic acids recovered as sodium salts from the ethanol or dimethylformamide solutions of these salts suggests the complete or partial renaturation of the secondary structure. Conformational changes of DNA and
polydeoxynucleotides in ethyleneglycol–water mixtures have been studied by
Green and Mahler [275]. Lewin [276] has discussed the behavior at high
concentrations of urea. Harrington [277] studied the influence of aqueous
glycerol solvents on the opticohydrodynamic properties of T2 bacteriophage
DNA and Londos-Gagliardi *et al.* [278] report that molecules of native DNA
become more flexible at high concentrations of sucrose, whereas there is no
change in the conformation of sonicated DNA.

When discussing the stability of nucleic acids and their interactions with
neutral salts in various media one should mention the reaction with formaldehyde [279] which has been used as a reagent for maintaining nucleic
acids in a random state [280], and for studying dynamic aspects of DNA
structure [281,282] (cf. the use of dimethylsulfoxide [283] as a solvent for this
purpose) or as a tool for the quantitative estimation of the fraction of amino
groups involved in hydrogen bonding [284]. The reaction has been studied
both under reversible and more stringent, irreversible, conditions [285]; it

has wide significance with respect of local unwinding of DNA during RNA synthesis [286], for instance. We also mention chemical modifications such as the study by Leng *et al.* [287] on the effects of methylation on the thermal stability of DNA and the interactions with spermine and spermidine.

We now return to the main topic of this section and proceed to a discussion of the interactions with divalent ions.

Lyons and Kotin [288] have studied the interaction of magnesium ion with DNA; they concluded, on the basis of the absence of shifts in the ultraviolet absorption and the proton magnetic resonance spectrum, that magnesium ion is essentially bound to the phosphate site, and not to the bases. They also find that an excess of magnesium ion lowers the thermal stability of the DNA; the formation of ion links by site bound magnesium is envisaged, which eventually may lead to denaturation, aggregation, and precipitation of the disordered DNA. A charge reversal mechanism (with chloride ions presumably acting as counterions) was postulated but was not established. Previously Eisinger and co-workers [289] had shown that purine polynucleotides [polyadenylic (poly A) and polyinosinic (poly I) acids] which may form ordered structures are precipitated by magnesium, whereas pyrimidine polynucleotides [polycytidilic (poly C) and polyurydilic (poly U) acids], which do not form ordered structures, are not precipitated. Thus it appears that, although magnesium is not bound to the bases, the nature of the bases and the secondary structure of the polynucleotide are not without influence on the stability of the complexes with this ion. Skerjanc and Strauss [290] studied the interaction between magnesium ion and DNA in aqueous solution by dialysis equilibrium, viscosity and dilatometric techniques, over a wide range of concentrations, in the MgDNA–MgCl$_2$ and DNA–MgCl$_2$–tetramethylammonium systems; it was found that the association of magnesium ion is three orders of magnitude weaker with DNA than with polyphosphates. The results follow the mass action law if account is taken of the effects of electrostatic potential at the surface of the DNA molecule. The special structure of DNA, which prevents close contact by magnesium with more than one phosphate group, is held responsible for the slight volume increase (6.4–8.7 ml/mole) upon loss of solvation, as the result of the reaction $Mg^{2+} + -PO_4^- \rightarrow PO_4Mg^+$. The slight viscosity decrease upon magnesium binding indicated that the rather rigid structure of DNA was maintained. Sander and Ts'o [291] determined the activity of Mg^{2+} ions in the presence of nucleic acids with a cation specific electrode, and introduced an exponential interference factor in their analysis of site binding to the phosphate. They find that the intrinsic constants for Mg^{2+} binding by nucleic acids in neutral sodium phosphate buffer follow the order tRNA > poly A, poly A·poly U, poly I·poly C, native DNA, stabilized stored denatured DNA > freshly denatured DNA, poly U, poly C ≫ monomers. Clearly conformation and the nature of the nucleic acid determines binding, although the binding itself is to the phosphate groups.

Willemsen and van Os [292] find, by the use of the metallochromic indicator calgamite, that there is no difference in the binding of Mg^{2+} to poly A or to poly U. Intrinsic binding constants, evaluated by taking into account the electrostatic interaction between the binding sites, agree with the results of some, but not of other workers. Small changes in optical properties of both polynucleotides are related to changes in secondary structure following magnesium binding. Walters and van Os [293] have studied the binding of magnesium to the phosphate groups of yeast ribosomes. Willick and Kay [294] report on magnesium-induced conformational changes in transfer RNA as measured by circular dichroism.

Krakauer has investigated the binding [295] of Mg^{2+} ions to poly A, to poly U and to their complexes and determined calorimetrically [296] the heats of binding of Mg^{2+} in the same systems. Vijayendran and Vold report [297] buoyant density as well as sedimentation studies of DNA in concentrated solutions of magnesium salts. Rialdi *et al.* determined [298] by calorimetry the thermodynamics of Mg^{2+} binding to yeast phenylalanine transfer RNA. Bauer reports that native PM-2 DNA, a closed circular double-stranded molecule isolated from mature bacteriophage PM-2, is unwound by up to 6% by the addition of the neutral denaturing salt magnesium perchlorate [299] as well as by up to 3.9% as a result of the addition of $NaClO_4$ [300] to 7.2 M at 20°C. The interaction of Na^+ with poly A, poly U, poly A · poly U and poly A · 2 poly U has been investigated [301] by means of potentiometry and by means of a linked function analysis of its effect on the binding of Mg^{2+} ions.

Native DNA binds mercuric ion reversibly at pH 9; Nandi *et al.* [302] find that one type of complex forms to a maximum mercury to nucleotide ratio of 0.5 (irrespective of base composition of DNA) and a second type of complex to a ratio of 1:1. [For synthetic or natural (crab deoxyadenylyl-thymidine) alternating copolymer (dAT) the first complex saturates at one mercury per four nucleotides.] Transforming activity of native pneumococcal DNA is not affected by a cycle of mercuration and demercuration, AT-rich DNA's bind more strongly than do GC-rich ones, and denatured DNA more strongly than native: the evidence indicates that the mercuric ion is bound to the base moieties rather than to the phosphates and possible structures have been suggested. Upon Hg^{2+} binding to DNA protons are released and the equilibrium is consequently pH-sensitive. Fractionation procedures have been devised for DNA mercury complexes in Cs_2SO_4 density gradients [303]; thus the crab dAT has been separated from the main component of crab DNA and λdg DNA halves have been separated [304] by the same procedure. Luck and Zimmer [305] concur that the binding is to TT and

AT base pairs, on the basis of absorbence and optical rotatory dispersion measurements of DNA–Hg^{2+} complexes with varying base composition. Gruenwedel and Lu [306] have studied the reaction between DNA and methylmercuric ion both spectrophotometrically and by buoyant density measurements. Further information on the nature of the binding of mercuric ion to DNA results from the effect of the ion on the DNA–methylene blue complex [307] and, from crystal and molecular structure of 2:1 complexes of uracil and dihydrouracil–mercuric chloride [308].

The binding of cupric ion (Cu^{2+}) to DNA has been studied by many authors and only a few examples of recent work will be given. Schreiber and Daune [309] conclude from spectrophotometry, melting profiles, and hydrodynamic techniques (in 0.1 M Na ClO_4 and at pH 5.6) that at least two types of sites are available for Cu^{2+}. The first one, where Cu^{2+} is chelating N-7 of purines to phosphate, is observed only at low ionic strength and destabilizes the double helix. The second exists mainly at 0.1 M or higher ionic strength and could be attributed to two successive guanine residues in the same strand; similar behavior was found for other divalent cations, e.g., Fe^{2+}, Mn^{2+}, and Co^{2+}. Eichhorn and Shin [310] discuss the ability of Cu^{2+} and Zn^{2+}, and other metal ions to bring about the unwinding and rewinding of the DNA double helix. Metal ions may thus be placed in the sequence $Mg^{2+} < CO^{2+} < Ni^{2+} < Mn^{2+} < Zn^{2+} < Cd^{2+} < Cu^{2+}$ according to their ability to influence DNA structure. The series is an outcome of the competing affinities for either phosphate or base, which may also depend on experimental conditions. Reaction with phosphate means stabilization of ordered structures but cleavage of phosphodiester bonds with polyribonucleotides at high temperatures (and is more evident with the metals at the left of the series), whereas reaction with bases means destabilization of ordered structures (and increases as we follow the series to the right). Berger and Eichhorn [311] have used paramagnetic broadening of the PMR peaks of nucleotides and polynucleotides to determine Cu^{2+} binding sites. Zimmer *et al.* [312] have used spectrophotometric, sedimentation, infrared, optical rotatory dispersion, and circular dichroism methods to demonstrate the structural changes in DNA induced by the interaction of Cu^{2+} with bases and to elucidate the complex binding sites. Miller and Bach [313] have determined the equilibrium binding constants of Cd^{2+} and Cu^{2+} to native and denatured DNA polarographically, as well as the transport of an anionic depolarizer [314] [Cu ethylenediamine tetraethyl acetic acid (CuEDTA)] through DNA adsorbed on a polarized mercury surface. Liebe and Stuehr studied copper(II)–DNA denaturation [315]. Sundaralingam and Carrabine determined crystal and molecular structures of guanine and cytosine copper(II) chloride complexes [316] in view of specifying copper binding sites and the mechanism of GC selective denaturation of DNA.

From a NMR study of the ^{31}P resonance in yeast RNA and adenosine 5'-monophosphate (AMP) in aqueous solutions containing paramagnetic ions, Shulman *et al.* [317] conclude that Mn^{2+} and Co^{2+} ions bind to the phosphate group; this finding is not contradictory to the less direct conclusions of Eichhorn and Shin [310]. Peacocke *et al.* [318] have presented a study, over a range of temperatures and magnetic field strengths, of the spin relaxation of water protons in aqueous solutions of *E. coli* ribosomal RNA containing Mn^{2+} ions.

Complexes of silver ion with DNA, natural and synthetic polynucleotides have been studied by Jensen and Davidson [319], by Daune *et al.* [320], and by Wilhelm and Daune [321], by spectrophotometry, thermal transitions, potentiometry, and density gradient ultracentrifugation. Silver forms complexes by interactions with the purine and/or pyrimidine bases, rather than by interaction with the phosphates. The binding is complex and many types of binding have been considered. The strength of the binding increases with GC content; denatured DNA binds more strongly than does native DNA, and the binding is reversible. The large buoyant density changes and the selective nature of the complexing reaction make it possible to perform good separations between native and denatured DNA or between GC-rich and GC-poor native DNA's by density gradient centrifugation.

Various di- and trivalent metals ions are able to bring about the depolymerization of polyribonucleotides and RNA at neutral pH. Zinc(II) causes a much more rapid reaction than any of the other metals of the first transition series, and the rate depends greatly on the nature of the polymer [322,323]. Degradation by metal ions proceeds through chelation of the metal between the phosphate and the 2'-hydroxyl and therefore DNA is not readily degraded by metal ion.

3. *Binding of Hydrocarbons and Dyes*

The interaction of nucleic acids with antibacterial, mutagenic dyes, and antibiotics on the one hand, and carcinogenic hydrocarbons on the other, has long engaged the attention of the biologists and the molecular biophysicists. The use of aminoacridines for instance, as histological stains antedates their current use as antibacterial agents and the name of the chromosomes themselves is due to the ability of their DNA to be stained by various dyes. This field has been reviewed in recent years and we would again like to mention here only some very recent results. For a review of the older literature the reader is referred to the reviews by Lerman [324] on acridine mutagens and DNA structure; by Blake and Peacocke [325] and Peacocke [326] on the interaction of aminoacridines with nucleic acids and polynucleotides; by Waring [327] on drugs which affect the structure and function of DNA; by

Jordan [328] on the interaction of heterocyclic compounds with DNA; by Van Duuren *et al.* [329] on the interaction of mutagenic and carcinogenic agents with nucleic acids; finally we refer to a discussion on the interaction between dyes and nucleic acids [330] at the 5th Jena symposium of molecular biology.

It can be said with some assurance that, if some requirements of size and planarity are satisfied, heterocyclic dyes, antibiotics, and hydrocarbons may intercalate into the DNA active structure or bind in a somewhat less specified way to single-chain polynucleotides. This is known as strong binding, or intercalation, and is a slower process than the weaker but faster binding to an outside region of the nucleic acid structure. Further to structural, steric, and electron distribution requirements the relative distribution of binding on the various "sites" available is also dependent on ionic strength. In addition *in vivo* systems contain protein as well which is expected to play an important role in the overall interaction. This may be a clue to the modulation of DNA polymerase performance, for instance, as influenced by proflavine in replicating systems [331]. It is, therefore, not very surprising that, although we know a fair amount about dye, drug, and hydrocarbon interaction with DNA, the final word on molecular mutagenesis and chemical carcinogenesis is yet to come.

Nagata *et al.* [332] have studied by flow dichroism the *in vitro* interaction of polynuclear hydrocarbons with calf thymus DNA; 3,4-benzpyrene, pyrene, phenantrene, and 4-nitroquinoline 1-oxide (4-NQO) oriented parallel to the direction of flow, whereas 20-methylcholantrene, tetracene, pentacene, and coronene produced a perpendicular orientation. Purines were suggested for 4-NQO binding and the amount of binding and carcinogenicity were found to be related. Actinomycin D, several acridine dyes, and methylene blue were oriented parallel to the DNA bases, as expected from the intercalation mechanism proposed by Lerman [324]. Green and McCarter [333] measured the characteristic changes in intensity of the polarized fluorescence emitted by various aromatic hydrocarbons in aqueous DNA solutions when subjected to flow orientation and found that the hydrocarbons have a definite orientation with respect to the DNA helix axis. Isenberg *et al.* [334] and Craig and Isenberg [335] have tested the intercalation model by proposing a size criterion which states that hydrocarbons small enough to intercalate into DNA and well protected from contact with the medium, are found to bind to DNA; those that are too large will not. Predictions based on the size criterion were found valid in all cases tested (cf. Table III). Poly A, neutral pH form, did not complex any of the hydrocarbons tested, and, somewhat surprisingly, denatured DNA solubilized pyrene, phenanthrene, and benzpyrene to a somewhat greater extent (one hydrocarbon bound per 250,170, and 600 base pairs, respectively) than native DNA (one hydrocarbon bound per 330,330, and

TABLE III

SOLUBILITY OF HYDROCARBONS IN SOLUTIONS OF DNA[a]

	Structure	$10^4 \times$ moles hydrocarbon per mole DNA phosphate	
		Absorption	Extraction
Group A			
Class 1			
Anthracene		8.3	8.8
1,2-Benzanthracene		4.0	5.1
1,2,3,4-Dibenzanthracene		1.4	4.0
Class 2			
1,2,5,6-Dibenzanthracene		< 0.72	< 0.72
Tetracene		< 0.10	< 0.10
Pentacene		< 0.10	< 0.10
Group B			
Class 1			
Anthracene		8.3	8.8
9-Methylanthracene	CH₃	7.0	8.0

[a] From Craig and Isenberg [335].

TABLE III (*continued*)

	Structure	$10^4 \times$ moles hydrocarbon per mole DNA phosphate	
		Absorption	Extraction
Class 2			
9-Phenylanthracene		< 0.1	< 0.1
Group C			
Class 1			
Pyrene		15	15
3,4-Benzpyrene		1.5	1.5
Class 2			
1,2,3,4-Dibenzpyrene		< 0.6	< 0.6
1,2,6,7-Dibenzpyrene		< 0.36	< 0.36
Group D			
20-Methylcholanthrene		< 0.1	< 0.1
Coronene		0.1	1.0

243

4000 base pairs, respectively). Lesko *et al.* [336] and Hoffmann *et al.* [337] demonstrated the formation of a chemical linkage of polycyclic hydrocarbons with DNA in an iodine-induced reaction; 3,4-benzpyrene was introduced at the rate of one molecule per 300 bases of calf thymus DNA and 100 bases of *Bacillus subtilis* SB-19 transforming DNA. Under identical conditions, carcinogenic hydrocarbons (3,4-benzpyrene, for instance) are manifold more reactive than their noncarcinogenic isomers or analogs (1,2-benzpyrene, for instance). In synthetic polynucleotides 3,4-benzpyrene is linked preferentially to purine polynucleotides with poly G being the most reactive.

Exposure of solutions of DNA containing intercalated molecules of benz[*a*]pyrene or pyrene to γ-irradiation resulted in covalent binding to the extent of one hydrocarbon molecule for every 100 and 270 DNA nucleotides, respectively; the template function of the DNA was inhibited as well [338]. Cavalieri and Calvin [339] believe that the chemical reactivity and presumably the carcinogenic activity induced in aromatic hydrocarbons by hydroxylating enzymes may be due to the generation of electrophilic centers in some (benz[*a*]pyrene) or nucleophilic centers in others (7,12-dimethyl benz[*a*]anthracene or 3-methylcholantrene). Fuchs and Daune [340] reacted DNA from various sources with *N*-acetoxy-N_2-acetylaminofluorene and studied the physical properties of the modified DNA. Circular dichroism and melting curve analysis showed that modified bases are shifted outside the double helix, while the fixed carcinogen is inserted. Viscosity and light-scattering studies indicated that the fixation of the carcinogen induces hinge points in the DNA molecule. The existence of crosslinks in DNA reacted with carcinogen, and the importance of varying amounts of guanine in different DNA samples, was demonstrated by helix–coil transition and light scattering. In other studies [341,342] alterations in the structure of DNA by chemical methylation with the mutagenic and carcinogenic methylating reagent dimethyl sulfate have been investigated.

In distinction to the neutral hydrocarbons just discussed, which are bound in rather low amounts, the cationic heterocyclic dyes, antibiotics, and their derivatives are bound to a much higher extent. At low values of the ionic strength binding on the outside of the nucleic acid is favored and one-to-one complexes (with respect to the phosphate groups) may be achieved [343]. At higher values of the ionic strength, intercalation appears to be favored and proceeds up to an apparent limit of one ligand molecule per two base pairs; this limit results from the fact that, whereas intercalation proceeds between any set of base pairs, the sites close to an already occupied site—the nearest neighbors—are strongly deactivated.

Actinomycin (Fig. 9) binds to DNA and blocks the synthesis of messenger RNA; Müller and Crothers [344] report equilibrium, kinetic, and hydrodynamic studies of the binding to DNA of actinomycin, its derivatives, and some simpler analogs. They believe that the actinomycin chromophore is intercalated between the base pairs in the DNA complex and that binding can occur adjacent to any GC pair, but binding at a given site produces a distortion of the helix that greatly disfavors binding of another actinomycin closer than six base pairs away. An interesting observation (Fig. 10) concerns the intrinsic viscosity ratio $[\eta]_r/[\eta]_{r=0}$ (r is the degree of binding ligand per nucleotide) vs molecular weight M of samples obtained by sonic irradiation. At low values of M the ratio increases as expected from chain lengthening by dye intercalation. At high values of M the viscosity ratio decreases, which may be related to secondary interactions between the nonintercalated peptide

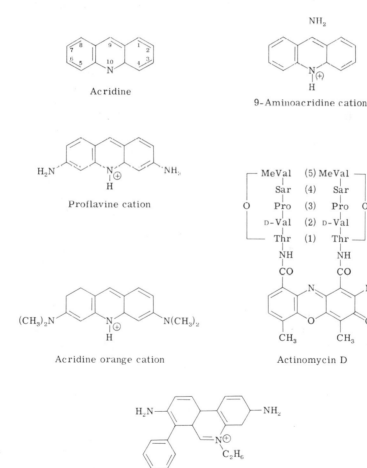

Fig. 9. Formulas of heterocyclic dyes and related compounds.

chains of actinomycin, leading to contraction of the DNA coils. This finding is a good example of how opposite conclusions could be reached from similar molecular processes, if the detailed properties of the system are not carefully defined.

Wells and Larson [345] have studied the ability of 17 different DNA's to bind actinomycin D by equilibrium dialysis, *in vitro* transcription, and analytical buoyant centrifugation. They conclude, among other results, that the presence of deoxyguanylic acid in a DNA is neither necessary nor sufficient for complex formation. Many specific sequences of isomeric DNA's have been tested. The results were discussed in relation to the hydrogen-bonded "outside bonding" model of Hamilton *et al.* [346] and the intercalation model

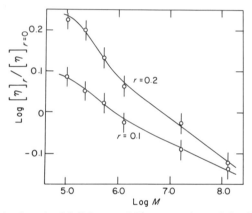

Fig. 10. Intrinsic viscosity $[\eta]_r/[\eta]_{r=0}$ vs M for two values of the degree of binding, r, of actinomycin to DNA. From Müller and Crothers [344].

of Müller and Crothers [344]. The data are not consistent with the hydrogen-bonded model. Gurskii [347], on the other hand, does not accept the intercalation model and cites stereochemical and other evidence for an actinomycin DNA complex in which the actinomycin chromophore and the two peptide chains are wrapped in the small groove of DNA. Absorption, optical rotatory dispersion, and circular dichroism studies of actinomycin DNA complexes were reported by Courtois *et al.* [348].

By converting λ DNA to the covalently closed form in the presence of actinomycin D, and subsequently determining the degree of superhelicity of the DNA after removal of the ligand, Wang [349] showed that actinomycin D unwinds the DNA helix by an angle identical to the unwinding angle of ethidium; this result supports the intercalating model of Müller and Crothers [344]. From the study of the three-dimensional structure of a crystalline complex containing actinomycin D and deoxyguanosine Sobell *et al.* [350]

conclude that the phenoxazone ring system of actinomycin intercalates into the DNA helix, while deoxyguanosine residues interact with both cyclic peptides through specific hydrogen bonds—a detailed pattern of recognition between naturally occurring operators and repressors is suggested. Binding studies [351] of actinomycin D and 11 different deoxyribonucleotides support the intercalation model and show that actinomycin D has a preference for GC sequences of DNA as potential binding sites. Daunomycin, a glycosidic anthracycline antibiotic from *Streptomyces peucetius*, used in the treatment of acute leukemia and solid tumors in man, also interacts with DNA by an intercalation process [352].

The binding of 16 drugs to closed circular DNA has been studied by Waring [353] to correlate intercalation with uncoiling of the double helix. Eight drugs, which are believed to bind by intercalation, elicit the same characteristic fall-and-rise response in the sedimentation coefficient of closed circular DNA molecules. Nonintercalating substances were found to have little effect on the sedimentation coefficient of closed circular DNA. Wawra *et al.* [354] studied sonicated NaDNA complexed with actinomine (in dilute NaCl solutions) by small-angle x-ray scattering and found that (for very short DNA fragments of molecular weight about 10^5) for an average binding of one actinomine molecule per 7.1 base pairs, the DNA is extended by 18%. This corroborates the intercalation hypothesis.

The interaction of ethidium bromide (Fig. 9) and its analogs with a variety of circular DNA's have been extensively studied by Vinograd and collaborators [355] by equilibrium sedimentation in CsCl density gradients. Sutherland and Sutherland [356] have shown that ethidium bromide inhibits the formation of pyrimidine dimers in DNA induced by ultraviolet light; they have used sensitized fluorescence as a probe of excited states in studies of pyrimidine dimer inhibition. An interesting contribution by Wahl *et al.* [357] presents evidence (by decay of fluorescence emission anisotropy of the ethidium bromide–DNA complex for high ratios of nucleic acid to dye) for an internal oscillatory Brownian motion in DNA. The amplitude of the oscillation is found to be equal to 35° and the relaxation time equal to 28 nsec. It is interesting to compare this evidence with the ionic strength-dependent family of equilibrium B structures (with displaced turn angles between bases, cf. Section III,B,2.) postulated by Bram [211], from wide-angle x-ray scattering.

Genest and Wahl [358] find that the slope of the fluorescence anisotropy increases with increasing ratio of dye to nucleic acid. Paoletti and LePecq [359] believe, from an analysis of the resonance energy transfer between ethidium bromide molecules bound to nucleic acids, that the change of torsion of the DNA helix caused by intercalation, winds, rather than unwinds the DNA helix, contrary to what is generally supposed. They propose a modified intercalation model which is tested [360] by determining the amount of various drugs necessary to relax the supercoiled DNA from PM-2 bacteriophage (the proportion of drug truly intercalated is controlled by measurement of the length increase of short helices of sonicated DNA [163]). The modified intercalation model leads to winding equal to 12° per ethidium bromide inter-calated, 8°–9° for proflavine and quinacrine, 7° for methoxyellipticine, and only 4° for daunomycin. Direct examination by electron microscopy [361,362] of ethidium bromide–DNA complexes reveals contour length increments of up to about 27% at saturation. Deoxynucleoproteins from calf thymus bind less ethidium bromide than the DNA and, as the protein is selectively re-moved, an increase in association constant as well as an increase in the number of available binding sites is observed [363]; the results reflect important struc-tural differences between native and artificial nucleoproteins and the con-stituent DNA's.

Proflavine (Fig. 9) has been a favorite dye for binding studies to DNA for many years. Weill and co-workers [364–367] have studied the quantum yield, lifetime of fluorescence, and heterogeneity of fixation sites by proflavine to DNA [the binding of acridine orange (Fig. 9) to DNA results in an increase of the quantum yield of fluorescence of the bound dye with respect to the free dyes, whereas the binding of proflavine under the same conditions results in a large decrease of quantum yield]. Sharp and Bloomfield [368] have studied the binding of proflavine and ethidium bromide to two forms of T2 bacterio-phage with different sedimentation coefficients. Jordan and Sansom [369] have studied the influence of temperature on the binding of 9-aminoacridine and of proflavine to *E. coli* DNA in 10^{-3} *M* NaCl solution by a spectrophoto-metric technique, and point out the inadequacy of the expression normally used for the determination of the extent of binding at temperatures above which dissociation of the double helix occurs. Armstrong *et al.* [370] believe, from studies by equilibrium dialysis, absorption spectrophotometry, and low shear viscosity at several ionic strengths, and also from the interaction of acridine orange and proflavine with DNA at pH 6.5, that the total number of possible interaction sites is fixed *a priori* to include every site between succes-sive DNA pairs. An intercalated dye cation inhibits intercalation at the two sites immediately adjoining the occupied one, and an intercalated dye cation may associate with a nonintercalated dye cation to produce a spectroscopically distinct bound dimer.

Cohen and Eisenberg [163] have made a viscosity and sedimentation study of sonicated DNA–proflavine complexes at 0.2 ionic strength, pH 6.8 at 25°C. It was found that $[\eta]_r/[\eta]_{r=0}$ increased, whereas $s_r/s_{r=0}$ decreased with increasing value of r in the range $0 < r < 0.13$ (Fig. 11a). It was possible to show that for short, almost rodlike fragments, the ratio $L_r/L_{r=0}$ of the length of the DNA at binding ratios r and in the absence of dye, may be calculated from the sedimentation [cf. Eq. (61)] and viscosity data by

$$\frac{L_r}{L_{r=0}} = \frac{s_{r=0}}{s_r} \frac{(\partial\rho/\partial c_2)_{\mu,r}}{(\partial\rho/\partial c_2)_{\mu,r=0}} \frac{\ln 2p_r}{\ln 2p_{r=0}} \tag{65}$$

$$= \left[\frac{[\eta]_r}{[\eta]_{r=0}} \frac{f(p)_{r=0}}{f(p)_r}\right]^{1/3} \tag{66}$$

where p is the axial ratio of the DNA rods and $f(p)$ is a weak function of p. The concentration c_2 in $(\partial\rho/\partial c_2)_{\mu,r}$ refers to grams of DNA per milliliter, without reference to dye binding (cf. Section II,E,2). The same definition of c_2 is used in the calculation of $[\eta]_r \equiv \lim (\eta_{sp}/c)$, at $c_2 \to 0$, and the molecular

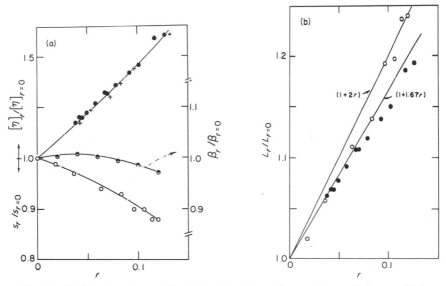

Fig. 11. (a) Relative change of intrinsic viscosity, sedimentation coefficient, and Mandelkern–Flory parameter β, with change in degree of binding, r, of proflavine to sonicated calf thymus DNA; pH 6.8, ionic strength 0.2. (b) Relative length increase $L_r/L_{r=0}$ as a function of r. (●) Calculated from $[\eta]_r/[\eta]_{r=0}$; (○) calculated from $s_r/s_{r=0}$. From Cohen and Eisenberg [163].

weight of the DNA–dye complex does not appear in the above expressions. Calculation of $L_r/L_{r=0}$ from the data of Fig. 11a (cf. Fig. 11b) leads to the conclusion that the lengthening of the DNA chains corresponds to 84% of the lengthening expected from intercalation of all the proflavine bound. The result is in good accord with Lerman's intercalation hypothesis if we take in account the result of Li and Crothers [371] who find, from relaxation studies (by temperature jump) of the proflavine–DNA complex that most of the proflavine is intercalated, but some of the ligand is bound to the outside of the DNA chains. Schmechel and Crothers [372] found a similar result from the kinetics and hydrodynamic studies of the complex proflavine with poly A·poly U; Sakoda *et al.* [373] also studied the kinetics (by stopped-flow methods) of interaction between acridine orange and DNA. We also refer to a statistical mechanical analysis [374] of binding of acridines to DNA to a theoretical study [375] of the intercalation model, the evaluation of the Kramers–Kronig relationships in linear birefringence and dichroism, as applied to the DNA–proflavine complex [376], the stronger intercalation in AT-rich rather than GC-rich DNA's [377], the quantitative analysis of proflavine binding to poly A, poly U, and transfer RNA [378], proflavine binding to yeast rRNA and ribosomes [379], and the role of histones in the interaction of proflavine with DNA chromatin [380]. Reinert [380a] reports the AT cluster-specific elongation and stiffening of DNA induced by the oligopeptide antibiotic netropsin.

D. POLYNUCLEOTIDE CONFORMATION

The ability of the bases to undergo stacking interactions and to form hydrogen bonds is, within the limitations of the steric restrictions of the sugar–phosphate backbone chain (cf. Section II,A,4), responsible for the various stable conformations which can be assumed by nucleic acid chains in aqueous and nonaqueous solutions. Whereas it was earlier believed that hydrogen bonds (in the Watson–Crick bonding scheme, for instance) constitute the main factor in the stabilization of nucleic acid structures, it is now commonly accepted that hydrophobic stacking properties of the bases constitute a major stabilizing force in certain structures (such as, for instance, single-stranded helical poly A [85,381–386] in aqueous solutions, at low temperatures and neutral pH), even in the complete absence of inter- or intrachain hydrogen bonds. Vapor pressures of purine and pyrimidine nucleosides were determined by Ts'o [387] and co-workers and osmotic coefficients calculated from the data. It was concluded that the tendency of purine nucleosides to associate is much greater than that of pyrimidine nucleosides which, in turn, is greater than that of urea, one of the best hydrogen-bonding agents in water. The stacking properties of nucleotides and polynucleotides in aqueous solutions (as well as in ethyleneglycol [386]), as followed by studies of solvent partition [388], nuclear magnetic resonance [387], hypochromicity [85,381,383,384], optical rotatory dispersion [381,382], and circular dichroism [384,385] have been extensively studied. Following the recent availability of a number of polydeoxyribonucleotides containing well defined repeating nucleotide sequences, Wells and co-workers [345,389] could show that DNA's which have the same base composition but different nucleotide sequences (sequence isomers) differ with respect to their thermal helix coil transition as well as their buoyant densities as determined in CsCl density gradients (Fig. 12).

Fascinating possibilities arise to identify the behavior of specific nonrandom sequences in natural DNA chains in relation to the properties of the synthetic model systems. Gray and Tinoco [390] have presented a theoretical scheme which relates sequence-dependent properties of long polynucleotides to the properties of a limited number of constituent polynucleotide subunits.

In a very different context, Neumann and Katchalsky [391,392] have shown that salt concentration-dependent conformational changes leading to hysteresis and metastability loops in the titration behavior of RNA [393] and certain model systems (complexes [394] of poly A and poly U) and have proposed these as models for a possible biological memory mechanism. It was found [395] that electric impulses can release metastable conformational states in

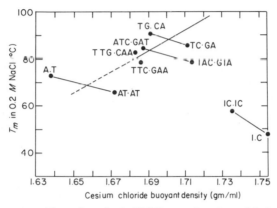

Fig. 12. Base composition of synthetic DNA's as a function of their Cs_2SO_4 buoyant density values. The solid diagonal line represents the values found for a variety of naturally occurring DNA's of different base composition. From Wells *et al.* [389].

biopolymers, the field intensities applied experimentally in inducing long-lived structural changes correspond to those measured across nerve membranes.

A salt-induced cooperative conformational change of a synthetic DNA (poly dG·dC) at 25°C and neutral pH, when the salt concentration is increased to 2.5 M NaCl, 1.8 M NaClO$_4$, or 0.7 M MgCl$_2$, has been reported by Pohl and Jovin [396]. We have already mentioned earlier the topical interest in the now well investigated amino acid specific transfer RNA conformations [7–9,397]

Out of the wealth of nucleic acid conformational problems, we only dwell on one rather limited aspect (and yet incompletely resolved). We refer to the problem of the conformation of the single homopolynucleotide chain and to transitions which it may undergo as a result of changes in temperature or other environmental factors.

Polyadenylic acid (poly A), for instance, can exist in a number of structures. At acidic pH double helical forms have been described [398,399], at neutral (and alkaline) pH it is believed that the structure is single-stranded, and at low temperatures organized into a helical form with stacked bases. Mostly optical [381–384] flow dichroic [400] and small-angle x-ray scattering [401,402] evidence has been presented. With increasing temperatures the poly A single helix (which is believed [403] to correspond to an RNA-11 [404] structure) undergoes a noncooperative structural transition to a less ordered form. Considerably less information is available about the exact nature of the latter; customarily it was called a random coil, but it is more appropriate to acknowledge that in both polypeptide and polynucleotide chains the less ordered forms may have some residual short-range regularity. Short-range order can be ascertained by optical and NMR methods. The noncooperative optical transition of the poly A single-chain helix with increase in temperature has been interpreted in terms of a two state process. On the other hand, Davis and Tinoco [405] have argued, on the basis of ORD, hypochromicity, and PMR experiments, in addition to theoretical considerations, that the bases in poly A may never unstack but rather undergo parallel oscillations in a dynamic process, increasing in amplitude with increase in temperature, throughout the whole transition range (cf. Schneider and Harris [406] for a related model). It really comes down to the question, posed by Davis and Tinoco, whether neutral poly A, at the middle of the melting transition is composed of one-half stacked and one-half unstacked bases, or whether all of the bases are half-stacked at this point. Powell *et al.* [407] have reexamined the temperature-dependence of the optical properties of dimers in aqueous solutions, avoiding the very high concentrations of LiCl used by Davis and Tinoco [405], and have applied to the data a number of criteria to distinguish between a two-state and a multistate model; they find that the data are most readily compatible with an effective two-state system.

What can be learned from scattering properties of polynucleotide solutions? Small-angle x-ray scattering is sensitive to short distances along the polynucleotide chains. For structures to which a rodlike character can be assigned over reasonable values of the scattering parameter h (Section II,A,3), the mass per unit length M/L may be evaluated from the scattering curves. These values can then be interpreted in terms of a molecular conformational model. In view of the fact that the polynucleotide chain appears to be inherently stiff as a result of hindered backbone rotation, neither interphosphate distances nor the M/L are expected to be largely affected by the removal of stacking interactions. Small-angle x-ray scattering, therefore, does not promise to be a sensitive index of changes in polynucleotide structure. Light scattering, in distinction to small-angle x-ray scattering, probes over much larger distances of the macromolecular structure. From the angular dependence of scattering we may determine the mean square radius $\langle R^2 \rangle$, the value of which is determined by both short-range and long-range interactions. The latter interactions may be minimized (or eliminated) by a felicitous choice of solvent, salt concentration, or temperature (under these conditions solution properties of a partially ordered polynucleotide chain may approximate the behavior of a random coil, although the random element will be much larger than the basic polynucleotide repeating unit). In the absence of long-range interactions the effect of short-range interactions on polynucleotide chain dimensions may then be evaluated on the basis of various reasonable considerations. A definitive answer with respect to the uniqueness of the structure derived cannot be given. It is left to the ingenuity of the chemist to devise experiments and to investigate related systems in order to arrive at unambiguous conclusions. The scattering and hydrodynamic evidence in the case of poly A (see below) favors a rigid structure at low temperatures which rapidly collapses with increase in temperature. As a general remark it may be said that, whereas the spectroscopic evidence is sensitive to small changes at low degrees of stacking, changes in scattering and hydrodynamic properties are most noticeable at high degrees of stacking. We may illustrate this by pointing out that, in the extreme case of a rigid, fully stacked chain, only one extended configuration is possible, which may be drastically changed by the introduction of one hinge point subsequent to the loss of one stacking interaction; the optical properties, though, will hardly be affected.

Eisenberg and Felsenfeld [35] have studied the temperature-dependent conformation and phase separation behavior of poly A at neutral pH, by light scattering, viscosity, and sedimentation. Long-range interactions in solution were eliminated by working close to the "ideal" Θ temperature. The temperature dependence of $\langle R^2 \rangle^{1/2}$ of poly A vs the fraction of stacking is shown in Fig. 13 (the calculated curves correspond to a somewhat improved calculation by Inners and Felsenfeld [408] of the conformational transition). Polyribouridilic acid (poly U) at neutral pH is unstacked at temperatures above 15°C; at lower temperatures a sharp optical transition occurs [409], the midpoint of which is dependent to some extent on ionic conditions. The transition to the ordered structure at low temperatures proceeds without change in molecular weight and is believed to correspond to a folded back hairpinlike structure [410], which is known to exist also in the case of other polynucleotides [86] and for transfer RNA [7–9,397]. No further optical transition is observed at higher temperatures in solutions of presumably unstacked poly U. Inners and Felsenfeld [408] have shown (cf. Fig. 13) that the dimensions of the unstacked poly U are very close to those of poly A at high temperatures, when all stacking interactions have been effectively eliminated. They find for the characteristic ratio C_x [Eq. (28)] of poly U a

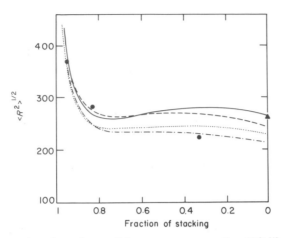

Fig. 13. Temperature dependence of the mean square radius $\langle R^2 \rangle^{1/2}$ of poly A. The dependence of $\langle R^2 \rangle^{1/2}$ upon fraction of bases stacked, calculated for a hypothetical chain containing 1740 nucleotides, for 4 possible sets of restrictions or torsional angles in the unstacked form; (●) poly A experimental values (Eisenberg and Felsenfeld [35]) of unperturbed dimensions of fraction with $Z = 1740$ corrected for heterogeneity; (▲) value of $\langle R_0^2 \rangle^{1/2}$ to be expected for a homogeneous fraction of poly U with $Z = 1740$. From Inners and Felsenfeld [408].

value of 17.6, corresponding to a quite highly extended coil. Only slightly smaller values are found by extrapolation of the data for poly A to the completely unstacked form, suggesting that the degree of extension of an unstacked polynucleotide does not significantly depend upon the nature of the base. To further elucidate the role of the bases in single-stranded polynucleotide conformation Achter and Felsenfeld [411] have studied (by sedimentation velocity and sedimentation equilibrium) apurinic acid, a single-stranded polydeoxyribonucleotide, from which almost half of the bases have been removed. The unperturbed coil dimensions of apurinic acid were found to be essentially identical to those of poly U of the same degree of polymerization. They therefore concluded that the considerable rigidity of polynucleotides is conferred not by residual, undetected base stacking, but by restrictions in rotation about the bonds of the backbone. They also believe that the rigidity of the ribose–phosphate backbone cannot be attributed to interactions involving the 2′-OH group. These conclusions on polynucleotide conformation are in good accord with the conformational considerations presented in Section II,A,4, and in particular with the recent more detailed conformational study of Olson and Flory [88]. A better understanding of the basic character of the polynucleotide chain will no doubt lead to further advances with respect to the specific behavior of more elaborate polynucleotide structures of biological interest.

Acknowledgements

I am grateful to Dr. Jamie Godfrey and Dr. Robert Josephs for reading this manuscript and for their critical comments, and to Mrs. S. Gibraltar for patiently typing it.

Work of the author quoted here was generously supported by Project No. 06–059–1 under the Special International Research Program of the National Institutes of Health, U.S. Public Health Service.

References

1. V. A. Bloomfield, *Macromol. Rev.* **3**, 255 (1968).
2. C. Sadron, *Bull. Soc. Chim. Biol.* **50**, 233 (1968).
3. A. B. Robins, *Biorheology* **3**, 153 (1966).
4. J. Josse and J. Eigner, *Annu. Rev. Biochem.* **35**, 789 (1966).
5. P. A. Edwards and K. V. Shooter, *Quart. Rev., Chem. Soc.* **19**, 369 (1965).
6. R. W. Armstrong and U. P. Strauss, *Encyl. Polym. Sci. Technol.* **10**, 781 (1969).
7. I. Pilz, O. Kratky, F. Cramer, F. von der Haar and E. Schlimme, *Eur. J. Biochem.* **15**, 401 (1970).
8. F. Cramer, *Progr. Nucl. Acid Res. Mol. Biol.* **11**, 391 (1971).
9. D. J. Abraham, *J. Theor. Biol.* **30**, 83 (1971).
10. W. Bauer and J. Vinograd, *J. Mol. Biol.* **47**, 419 (1970); **54**, 281 (1970).
11. D. J. Jolly and A. M. Campbell, *Biochem. J.* **128**, 569 (1972).
12. P. H. von Hippel and J. D. McGhee, *Annu. Rev. Biochem.* **41**, 231 (1972).
13. L. S. Lerman, *Proc. Nat. Acad. Sci. U.S.* **68**, 1886 (1971).
14. E. Dore, C. Frontali, and E. Gratton, *Biopolymers* **11**, 443 (1972).
15. Yu. M. Evdokimov, A. L. Platonov, A. S. Tikhonenko, and Ya. M. Varshavsky, FEBS *Lett.* **23**, 180 (1972).
16. J. Blamire, D. R. Cryer, D. B. Finkelstein, and J. Marmur, *J. Mol. Biol.* **67**, 11 (1972).
17. R. Kolodner and K. K. Tewari, *Proc. Nat. Acad. Sci. U.S.* **69**, 1830 (1972).
18. M. Edelman, I. M. Verma, R. Herzog, E. Galun, and U. Z. Littauer, *Eur. J. Biochem.* **19**, 372 (1971).
19. W. Kuhn, *Kolloid. Z.* **68**, 2 (1934).
20. E. Guth and H. Mark, *Monatsh. Chem.* **65**, 93 (1934).
21. P. Debye, *J. Phys. Colloid Chem.* **51**, 18 (1947).
22. C. Tanford, "Physical Chemistry of Macromolecules." Wiley, New York, 1961.
23. P. J. Flory and T. G. Fox, Jr., *J. Amer. Chem. Soc.* **73**, 1904 (1951).
24. P. J. Flory, "Principles of Polymer Chemistry." Cornell Univ. Press, Ithaca, New York, 1953.
25. L. Mandelkern and P. J. Flory, *J. Chem. Phys.* **20**, 212 (1952).
26. C. W. Pyun and M. Fixman, *J. Chem. Phys.* **44**, 2107 (1966).
27. B. H. Zimm, W. H. Stockmayer, and M. Fixman, *J. Chem. Phys.* **21**, 1716 (1953).
28. V. A. Bloomfield and B. H. Zimm, *J. Chem. Phys.* **44**, 315 (1966).
29. H. Eisenberg, *in* "Procedures in Nucleic Acids" (G. L. Cantoni and D. R. Davies, ed.), Vol. 2, p. 137. Harper, New York, 1971.
30. G. C. Berry and E. F. Casassa, *Ann. N.Y. Acad. Sci.* **155**, 507 (1969); *Macromol. Rev.* **4**, 1 (1970).
31. W. H. Stockmayer, *J. Polym. Sci.* **15**, 595 (1955).

32. M. Fixman, *J. Chem. Phys.* **23**, 1656 (1955).
33. J. B. Hays, M. E. Magar, and B. H. Zimm, *Biopolymers* **8**, 531 (1969).
34. J. Harpst, A. Krasna, and B. H. Zimm, *Biopolymers* **6**, 585 and 595 (1968).
35. H. Eisenberg and G. Felsenfeld, *J. Mol. Biol.* **30**, 17 (1967).
36. W. H. Stockmayer, *Makromol. Chem.* **35**, 541 (1970).
37. D. Patterson, *Macromolecules* **2**, 672 (1969).
38. B. H. Zimm, *J. Chem. Phys.* **16**, 1093 and 1099 (1948).
39. E. P. Geiduschek and A. Holtzer, *Advan. Biol. Medi. Phys.* **6**, 431 (1958).
40. H. Eisenberg and G. Cohen, *J. Mol. Biol.* **37**, 355 (1968); erratum, **42**, 607 (1969).
41. G. Porod, *Monatsh. Chem.* **80**, 251 (1949).
42. O. Kratky and G. Porod, *Rec. Trav. Chim. Pays-Bas* **68**, 1106 (1949).
43. J. J. Hermans and R. Ullman, *Physica (Utrecht)* **18**, 951 (1952).
44. P. J. Flory, "Statistical Mechanics of Chain Molecules." Wiley (Interscience), New York, 1969.
45. L. Landau and E. Lifshitz, "Statistical Physics," pp. 478–482. Pergamon, Oxford, 1958.
46. P. Bugl and S. Fujita, *J. Chem. Phys.* **50**, 3137 (1969).
47. R. G. Kirste, *Kolloid-Z. Z. Polym.* **244**, 290 (1971).
48. G. Porod, *J. Polym. Sci.* **10**, 157 (1953).
49. A. Guinier, *Ann. Phys. (Paris)* [11] **12**, 161 (1939).
50. J. Hermans, Jr. and J. J. Hermans, *J. Phys. Chem.* **62**, 1543 (1958).
51. V. Luzzati and H. Benoit, *Acta Crystallogr.* **14**, 297 (1961).
52. H. E. Daniels, *Proc. Roy. Soc. Edinburgh* **63**, 290 (1952).
53. A. Peterlin, *Makromol. Chem.* **9**, 244 (1953).
54. A. Peterlin, *in* "Electromagnetic Scattering" (M. Kerker, ed.), p. 357. Macmillan, New York, 1963.
55. H. Benoit and P. Doty, *J. Phys. Chem.* **57**, 958 (1953).
56. R. G. Kirste, *Discuss. Faraday Soc.* **49**, 51 (1970).
57. P. Sharp and V. A. Bloomfield, *Biopolymers* **6**, 1201 (1968).
58. P. Sharp and V. A. Bloomfield, *J. Chem. Phys.* **49**, 4564 (1968).
59. C. Domb, J. Gillis, and G. Wilmers, *Proc. Phys. Soc., London* **85**, 625 (1965).
60. M. E. Fisher, *J. Chem. Phys.* **44**, 616 (1966).
61. J. Mazur, *J. Chem. Phys.* **43**, 4354 (1965); D. McIntyre, J. Mazur, and A. M. Wims, *ibid.* **49**, 2887 and 2896 (1968).
62. M. M. Kumbar, *J. Chem. Phys.* **55**, 5046 (1971).
63. I. C. Sanchez and C. von Frankenberg, *Macromolecules* **2**, 666 (1969).
64. Y. Tagami, *Macromolecules* **2**, 8 (1969).
65. A. Horta, *Macromolecules* **3**, 371 (1970; A. Horta, *Makromol. Chem.* **154**, 63 (1972).
66. R. Yeh and A. Isihara, *J. Polym. Sci., Part A-1* **8**, 861 (1970); *Part A-2* **9**, 373 (1971).
67. A. Peterlin, *J. Polym. Sci.* **8**, 173 (1952).
68. J. E. Hearst and W. H. Stockmayer, *J. Chem. Phys.* **37**, 1425 (1962).
69. H. Kuhn, W. Kuhn, and A. Silberberg, *J. Polym. Sci.* **14**, 193 (1954).
70. D. M. Crothers and B. H. Zimm, *J. Mol. Biol.* **12**, 525 (1965).
71. J. Eigner and P. Doty, *J. Mol. Biol.* **12**, 549 (1965).
72. O. B. Ptitsyn and Yu. E. Eizner, *Vysokomol. Soedin.* **3**, 1863 (1961).
73. Yu. E. Eizner and O. B. Ptitsyn, *Vysokomol. Soedin.* **4**, 1725 (1962).
74. B. H. Zimm, *J. Chem. Phys.* **24**, 269 (1956).
75. H. B. Gray, Jr., V. A. Bloomfield, and J. E. Hearst, *J. Chem. Phys.* **46**, 1493 (1967).
76. J. E. Hearst, E. Beals, and R. A. Harris, *J. Chem. Phys.* **45**, 3106 (1966); **48**, 5371 (1968); I. Noda and J. E. Hearst, *ibid.* **54**, 2342 (1971).

77. R. A. Harris and J. E. Hearst, *J. Chem. Phys.* **44**, 2595 (1966).
78. P. E. Rouse, Jr., *J. Chem. Phys.* **21**, 1272 (1953).
79. R. Ullman, *J. Chem. Phys.* **49**, 5486 (1968); **53**, 1734 (1970).
80. R. Ullman, *Macromolecules* **2**, 27 (1969); *J. Phys. Chem.* **76**, 1755 (1972).
81. R. G. Kirste, *Macromol. Chem.* **101**, 91 (1967); *in* "Small Angle X-Ray Scattering" (H. Brumberger, ed.) p. 33. Gordon & Breach, New York, 1967.
82. P. J. Flory and R. L. Jernigan, *J. Amer. Chem. Soc.* **90**, 3128 (1968).
83. T. M. Birstein and O. B. Ptitsyn, "Conformation of Macromolecules." Wiley (Interscience), New York, 1966.
84. R. A. Scott, *Biopolymers* **6**, 625 (1968).
85. M. Leng and G. Felsenfeld, *J. Mol. Biol.* **15**, 455 (1966).
86. E. L. Elson, I. E. Scheffler, and R. L. Baldwin, *J. Mol. Biol.* **54**, 401 (1970).
87. G. Felsenfeld and H. T. Miles, *Annu. Rev. Biochem.* **36**, 407 (1967).
88. W. K. Olson and P. J. Flory, *Biopolymers* **11**, 25 and 57 (1972).
89. M. Sundaralingam, *Biopolymers* **7**, 821 (1969); C. E. Bugg, J. M. Thomas, M. Sundaralingam, and S. T. Rao, *Biopolymers* **10**, 175 (1971).
90. V. Sasisekharan, A. V. Lakshminarayanan, and G. N. Ramachandran, *in* "Conformation of Biopolymers" (G. N. Ramachandran, ed.,), Vol. 2, p. 641. Academic Press, New York, 1967.
91. A. V. Lakshminarayanan and V. Sasisekharan, *Biochim. Biophys. Acta* **204**, 49 (1970).
92. M. V. Volkenstein, "Configurational Statistics of Polymeric Chains." Wiley (Interscience), New York, 1963.
93. H. E. Auer and Z. Alexandrowicz, *Biopolymers* **8**, 1 (1969).
94. D. O. Jordan, "The Chemistry of Nucleic Acids," p. 242. Butterworth, London, 1960.
95. H. J. Lin and E. Chargaff, *Biochim. Biophys. Acta* **145**, 398 (1967).
96. E. F. Casassa and H. Eisenberg, *Advan. Protein Chem.* **19**, 287 (1964).
97. G. Cohen and H. Eisenberg, *Biopolymers* **6**, 1077 (1968).
98. E. F. Kirby Hade and C. Tanford, *J. Amer. Chem. Soc.* **89**, 5034 (1967).
99. M. E. Noelken and S. N. Timasheff, *J. Biol. Chem.* **242**, 5080 (1967).
100. J. E. Hearst, *Biopolymers* **3**, 57 (1965).
101. N. Imai and H. Eisenberg, *J. Chem. Phys.* **44**, 130 (1966).
102. E. Reisler, Y. Haik, and H. Eisenberg, in preparation.
103. E. Reisler and H. Eisenberg, *Biochemistry* **8**, 4572 (1969).
104. O. Kratky, H. Leopold, and H. Stabinger, *Z. Angew. Phys.* **27**, 273 (1969).
105. H. Eisenberg, *J. Chem. Phys.* **36**, 1837 (1962).
106. S. N. Timasheff, *in* "Electromagnetic Scattering" (M. Kerker, ed.), p. 337. Pergamon, Oxford, 1963.
107. E. F. Casassa and H. Eisenberg, *J. Phys. Chem.* **65**, 427 (1961).
108. A. Vrij and J. T. G. Overbeek, *J. Colloid Sci.* **17**, 570 (1962).
109. V. Crescenzi, *Advan. Polym. Sci.* **5**, 358 (1968).
110. R. M. Fuoss, *J. Chem. Educ.* **32**, 527 (1955).
111. C. Schildkraut and S. Lifson, *Biopolymers* **3**, 195 (1965).
112. J. T. Finch and A. Klug, *J. Mol. Biol.* **46**, 597 (1969).
113. R. Steiner and D. B. S. Millar, *in* "Biological Polyelectrolytes" (A. Veis, ed.), Chapter 2, p. 65. Dekker, New York, 1970.
114. R. M. Fuoss, A. Katchalsky, and S. Lifson, *Proc. Nat. Acad. Sci. U.S.* **37**, 579 (1951).
115. T. Alfrey, Jr., P. W. Berg, and H. Morawetz, *J. Polym. Sci.* **7**, 543 (1951).
116. S. Lifson and A. Katchalsky, *J. Polym. Sci.* **13**, 43 (1954).

117. Z. Alexandrowicz and A. Katchalsky, *J. Polym. Sci., Part A* **1**, 3231 (1963).
118. A. Katchalsky, Z. Alexandrowicz, and O. Kedem, *in* "Chemical Physics of Ionic Solutions" (B.E. Conway and R. G. Barradas, eds.), p. 295. Wiley, New York, 1966.
119. L. M. Gross and U. P. Strauss, *in* "Chemical Physics of Ionic Solutions" (B. E. Conway and R. G. Barradas, eds.), p. 361. Wiley, New York, 1966.
120. R. Langridge, D. A. Marvin, W. E. Seeds, H. R. Wilson, C. W. Hooper, M. H. F. Wilkins, and L. D. Hamilton, *J. Mol. Biol.* **2**, 38 (1960).
121. S. Arnott, M. H. F. Wilkins, H. R. Wilson, and L. D. Hamilton, *J. Mol. Biol.* **27**, 535 (1967).
122. U. P. Strauss, C. Helfgott, and H. Pink, *J. Phys. Chem.* **71**, 2550 (1967).
123. Z. Alexandrowicz, *J. Chem. Phys.* **47**, 4377 (1967); *J. Phys. Chem.* **75**, 442 (1971).
124. F. Oosawa, *Biopolymers* **6**, 145 (1968).
125. I. Noda, T. Tsuge, and M. Nagasawa, *J. Phys. Chem.* **74**, 710 (1970).
126. H. Eisenberg, *J. Polym. Sci.* **25**, 257 (1957).
127. P. D. Ross and R. L. Scruggs, *Biopolymers* **6**, 1005 (1968).
128. A. H. Rosenberg and F. W. Studier, *Biopolymers* **7**, 765 (1969).
129. U. Z. Littauer and H. Eisenberg, *Biochim. Biophys. Acta* **32**, 320 (1959).
130. R. A. Cox and U. Z. Littauer, *Biochim. Biophys. Acta* **61**, 197 (1962).
131. G. Manning *J. Chem. Phys.* **51**, 924, 934, and 3249 (1969).
132. G. Manning, *Annu. Rev. Phys. Chem.* **23**, 117 (1972).
133. S. Yeandle, *Proc. Nat. Acad. Sci. U.S.* **45**, 184 (1959).
134. J. M. Creeth and R. H. Pain, *Progr. Biophys. Mol. Biol.* **17**, 217 (1967).
135. M. Meselson, F. W. Stahl, and J. Vinograd, *Proc. Nat. Acad. Sci. U.S.* **43**, 581 (1957).
136. C. W. Schmid and J. E. Hearst, *J. Mol. Biol.* **44**, 143 (1969).
137. C. W. Schmid and J. E. Hearst, *Biopolymers* **10**, 1901 (1971).
138. J. Vinograd and J. E. Hearst, *Progr. Chem. Org. Nat. Prod.* **20**, 372 (1962).
139. H. Fujita "Mathematical Theory of Sedimentation Analysis," Chapter 5. Academic Press, New York, 1962.
140. J. J. Hermans and H. A. Ende, *in* "Newer Methods of Polymer Characterization" (B. Ke, ed.), Chapter 13, p. 525. Wiley (Interscience), New York, 1964.
141. M. Meselson and F. W. Stahl, *Proc. Nat. Acad. Sci. U.S.* **44**, 671 (1958).
142. R. L. Baldwin, P. Barrand, A. Fritsch, D. A. Goldthwait, and F. Jacob, *J. Mol. Biol.* **17**, 343 (1966).
143. H. Eisenberg, *Biopolymers* **5**, 681 (1967).
144. B. H. Zimm and D. M. Crothers, *Proc. Nat. Acad. Sci. U.S.* **48**, 905 (1962).
145. O. C. C. Lin, *Macromolecules* **3**, 81 (1970).
146. R. E. Harrington, *Biopolymers* **9**, 141 and 159 (1970).
147. P. R. Callis and N. Davidson, *Biopolymers* **7**, 335 (1969). **8**, 379 (1969).
148. G. Weill, C. Hornick, and S. Stoylov, *J. Chim. Phys.* **65**, 182 (1968).
149. A. Scheludko and S. Stoylov, *Biopolymers* **5**, 723 (1967).
150. S. P. Stoylov and S. Sokerov, *J. Colloid Interface Sci.* **24**, 235 (1967).
151. G. Weill and C. Hornick, *Biopolymers* **10**, 2029 (1971).
152. C. Hornick and G. Weill, *Biopolymers* **10**, 2345 (1971).
153. S. B. Dubin, G. B. Benedek, F. C. Bancroft, and D. Freifelder, *J. Mol. Biol.* **54**, 547 (1970).
154. W. L. Peticolas, *Advan. Polym. Sci.* **9**, 286 (1972).
155. N. C. Ford, *Chem. Scr.* **2**, 193 (1972).
156. P. N. Pusey, D. W. Schaefer, D. E. Koppel, R. D. Camerini-Otero and R. M. Franklin, *J. Phys. (Paris)* **33**, *C1–163* (1972).
157. Z. Alexandrowicz and E. Daniel, *Biopolymers* **1**, 447 (1963).

158. Z. Alexandrowicz and E. Daniel, *Biopolymers* **6**, 1500 (1968).
159. M. Nagasawa and Y. Eguchi, *J. Phys. Chem.* **71**, 880 (1967).
160. E. Daniel, *Biopolymers* **7**, 359 (1969); **8**, 553 (1969).
161. A. Katchalsky and P. F. Curran, "Nonequilibrium Thermodynamics in Biophysics." Harvard Univ. Press, Cambridge, Massachusetts. 1965.
162. G. J. Hooyman, *Physica (Utrecht)* **22**, 751 and 761 (1956).
163. G. Cohen and H. Eisenberg, *Biopolymers* **8**, 45 (1969).
164. S. J. Gill and D. S. Thompson, *Proc. Nat. Acad. Sci. U.S.* **57**, 562 (1967).
165. R. E. Chapman, Jr., L. C. Klotz, D. S. Thompson and B. H. Zimm, *Macromolecules* **2**, 637 (1969).
166. L. C. Klotz and B. H. Zimm, *Macromolecules* (**5**, 471 (1972)).
167. B. H. Zimm, *in* "Procedures in Nucleic Acids" (G. L. Cantoni and D. R. Davies, eds.), Vol. 2, p. 245. Harper, New York, 1971.
168. D. Froelich, C. Strazielle, G. Bernardi, and H. Benoit, *Biophys. J.* **3**, 115 (1963).
169. J. T. Lett, E. S. Klucis, and C. Sun, *Biophys. J.* **10**, 277 (1970).
170. H. R. Massie and B. H. Zimm, *Proc. Nat. Acad. Sci. U.S.* **54**, 1636 (1965).
171. L. C. Klotz and B. H. Zimm, *J. Mol. Biol.* **72**, 779 (1972).
172. A. Burton and R. L. Sinsheimer, *J. Mol. Biol.* **14**, 327 (1962).
173. J. Vinograd, J. Lebowitz, R. Radloff, R. Watson, and P. Laipis, *Proc. Nat. Acad. Sci. U.S.* **53**, 1104 (1965).
174. A. R. Peacocke and N. J. Pritchard, *Biopolymers* **6**, 605 (1968).
175. G. Cohen and H. Eisenberg, *Biopolymers* **4**, 429 (1966).
176. A. I. Krasna, J. R. Dawson, and J. A. Harpst, *Biopolymers* **9**, 1017 (1970).
177. A. I. Krasna, *Biopolymers* **9**, 1029 (1970).
178. C. W. Schmid, F. P. Rinehart, and J. E. Hearst, *Biopolymers* **10**, 883 (1971).
179. F. C. Bancroft and D. Freifelder, *J. Mol. Biol.* **54**, 537 (1970).
180. J. R. Dawson and J. A. Harpst, *Biopolymers* **10**, 2499 (1971).
181. P. F. Davison and D. Freifelder, *J. Mol. Biol.* **5**, 635 (1962).
182. F. N. Weber, Jr., D. W. Kupke, and J. W. Beams, *Science* **139**, 837 (1963).
183. C. A. Thomas and T. Pinkerton, *J. Mol. Biol.* **5**, 356 (1962).
184. D. Freifelder, *J. Mol. Biol.* **54**, 567 (1970).
185. D. Lang, *J. Mol. Biol.* **54**, 557 (1970).
186. S. B. Leighton and I. Rubenstein, *J. Mol. Biol.* **46**, 313 (1969).
187. J. E. Hearst, C. W. Schmid, and F. P. Rinehart, *Macromolecules* **1**, 491 (1968).
188. A. Nicolaieff and R. Litzler, *Biopolymers* **8**, 181 (1969).
189. H. B. Gray and J. E. Hearst, *J. Mol. Biol.* **35**, 111 (1968).
190. J. Strassburger and K. E. Reinert, *Biopolymers* **10**, 263 (1971).
191. K. E. Reinert, *Biopolymers* **10**, 275 (1971).
192. K. E. Reinert, J. Strassburger, and H. Triebel, *Biopolymers* **10**, 285 (1971).
193. H. Triebel and K. E. Reinert, *Biopolymers* **10**, 827 (1971).
194. M. Böttger, D. Bierwolf, V. Wunderlich, and A. Graffi, *Biochim. Biophys. Acta* **232**, 21 (1971).
195. D. Z. Staynov, J. C. Pinder, and W. B. Gratzer, *Nature (London), New Biol.* **235**, 108 (1972).
195a. S. M. Klimenko, T. I. Tikchonenko, and V. M. Andreev, *J. Mol. Biol.* **23**, 523 (1967).
195b M. Haynes, R. A. Garrett, and W. Gratzer, *Biochemistry* **9**, 4410 (1970).
196. B. McConnell and P. H. von Hippel, *J. Mol. Biol.* **50**, 297 and 317 (1970).
197. R. W. Armstrong, T. Kurucsev, and U. P. Strauss, *J. Amer. Chem. Soc.* **92**, 3174 (1970).
198. H. H. Meyer, W. F. Pfeiffer, and J. D. Ferry, *Biopolymers* **5**, 123 (1967).

199. H. Eisenberg, *Biopolymers* **8**, 545 (1969).
200. S. B. Dev, R. Y. Lochhead, and A. M. North, *Discuss. Faraday Soc.* **49**, 244 (1970).
201. U. Shmueli, W. Traub, and K. Rosenheck, *J. Polym. Sci., Part A-2* **7**, 515 (1969).
202. J. B. Milstien and E. Charney, *Macromolecules* **2**, 678 (1969).
203. C. Sadron, J. Pouyet, A. M. Freund, and M. Champagne, *J. Chim. Phys.* **62**, 1187 (1965).
204. O. B. Ptitsyn and B. A. Fedorov, *Biofizika* **8**, 659 (1963).
205. H. Triebel, K. E. Reinert, and J. Strassburger, *Biopolymers* **10**, 2619 (1971).
206. R. Kilkson and M. F. Maestre, *Nature (London)* **195**, 494 (1962).
207. J. B. Hays and B. H. Zimm, *J. Mol. Biol.* **48**, 297 (1970).
208. M. Sugi, M. Fuke, and A. Wada, *Polym. J.* **1**, 457 (1970).
209. D. Lang and P. Coates, *J. Mol. Biol.* **36**, 137 (1968).
210. S. Bram and W. W. Beeman, *J. Mol. Biol.* **55**, 311 (1971).
211. S. Bram, *J. Mol. Biol.* **58**, 277 (1971).
212. J. C. Wang, D. Baumgarten, and B. M. Olivera, *Proc. Nat. Acad. Sci. U.S.* **58**, 1852 (1967).
213. M-J. B. Tunis and J. E. Hearst, *Biopolymers* **6**, 1218 (1968).
214. S. Bram, *Nature (London), New Biol.* **232**, 174 (1971).
215. S. Bram, *Nature (London), New Biol.* **233**, 161 (1971).
216. J. V. Champion and P. F. North, *J. Chim. Phys.* **68**, 1585 (1971).
217. R. E. Harrington, *J. Amer. Chem. Soc.* **92**, 6957 (1970).
218. H. Benoit, L. Freund, and G. Spach, *in* "Poly α-Amino Acids" (G. D. Fasman, ed.), Chapter 3, p. 105. Dekker, New York, 1967.
219. H. B. Halsall and V. N. Schumaker, *Nature (London)* **221**, 772 (1969).
220. M-J. B. Tunis and J. E. Hearst, *Biopolymers* **6**, 1325 and 1345 (1968).
221. R. E. Chapman, Jr. and J. M. Sturtevant, *Biopolymers* **7**, 527 (1969); **9**, 445 (1970).
222. R. C. Costello and R. L. Baldwin, *Biopolymers*, **11**, 2147 (1972).
223. C. Schildkraut, J. Marmur, and P. Doty, *J. Mol. Biol.* **3**, 595 (1961).
224. B. Weida and S. J. Gill, *Biochim. Biophys. Acta* **112**, 179 (1966).
225. C. G. Heden, T. Lindahl, and I. Toplin, *Acta Chem. Scand.* **18**, 1150 (1964).
226. M. Falk, K. A. Hartmann, and R. C. Lord, *J. Amer. Chem. Soc.* **84**, 3843 (1962); **85**, 387 and 391 (1963).
227. M. Falk, *Can. J. Chem.* **44**, 1107 (1966).
228. S. Lewin, *J. Theor. Biol.* **17**, 181 (1967).
229. F. Hughes and R. F. Steiner, *Biopolymers* **4**, 1081 (1966).
230. H. Noguchi, S. R. Aryo, and J. T. Yang, *Biopolymers* **10**, 2491 (1971).
231. T. E. Gunter and K. K. Gunter, *Biopolymers* **11**, 667 (1972).
232. P. L. Privalov and G. M. Mrevlishvili, *Biofizika* **12**, 22 (1967).
233. J. H. Wang, *J. Amer. Chem. Soc.* **77**, 258 (1955).
234. N. G. Bakradze, D. R. Monaselidze, G. M. Mrevlishvili, A. D. Bibikova, and L. L. Kisselev, *Biochim. Biophys. Acta* **238**, 161 (1971).
235. W. Kauzman, *Advan. Protein Chem.* **14**, 1 (1959).
236. D. K. Chattoraj and H. B. Bull, *Arch. Biochem. Biophys.* **142**, 363 (1971).
237. D. K. Chattoraj and H. B. Bull, *J. Colloid Interface Sci.* **35**, 220 (1971).
238. M. P. Printz and P. H. von Hippel, *Biochemistry* **7**, 3194 (1968).
239. R. E. Bird, K. G. Lark, B. Curnutte, and J. E. Maxfeld, *Nature (London)* **225**, 1043 (1970).
240. C. V. Hanson, *J. Mol. Biol.* **58**, 847 (1971).
241. J. J. Englander and P. H. von Hippel, *J. Mol. Biol.* **63**, 171 (1972).
242. J. J. Englander, N. R. Kallenbach, and S. W. Englander, *J. Mol. Biol.* **63**, 153 (1972).

243. P. J. Killion and L. H. Reyerson, *J. Colloid Interface Sci.* **22**, 582 (1966).
244. D. E. Gordon, B. Curnutte, Jr., and K. G. Lark, *J. Mol. Biol.* **13**, 571 (1965).
245. B. Lubas, T. Wilczok, and O. K. Daszkiewicz, *Biopolymers* **5**, 967 (1967).
246. C. Migchelsen, H. J. C. Berendsen, and A. Rupprecht, *J. Mol. Biol.* **37**, 235 (1968).
247. I. D. Kuntz, Jr., T. S. Brassfield, G. D. Law, and G. V. Purcell, *Science* **163**, 1329 (1969).
248. U. Dahlborg and A. Rupprecht, *Biopolymers* **10**, 849 (1971).
249. J. A. Glasel, *J. Amer. Chem. Soc.* **92**, 375 (1970).
250. A. A. Khanagov, *Biopolymers* **10**, 789 (1971).
251. H. F. Edzes, A. Rupprecht, and F. C. Berendsen, *Biochem. Biophys. Res. Commun.* **46**, 790 (1972).
252. M. Shporer and M. M. Civan, *Biophys. J.* **12**, 114 (1972).
253. F. W. Cope, *Biophys. J.* **10**, 843 (1970).
254. P. H. von Hippel and T. Schleich, *in* "Structure and Stability of Biological Macromolecules" (S. N. Timasheff and G. D. Fasman, eds.), p. 417. Dekker, New York, 1969.
255. U. P. Strauss and Y. P. Leung, *J. Amer. Chem. Soc.* **87**, 1476 (1965).
256. M. Daune, *Biopolymers* **7**, 659 (1969).
257. M. Daune, *Stud. Biophys.* **24/25**, 287 (1970).
258. R. H. Jensen and N. Davidson, *Biopolymers* **4**, 17 (1966).
259. T. Yamane and N. Davidson, *J. Amer. Chem. Soc.* **83**, 2599 (1961).
260. K. Hamaguchi and E. P. Geiduschek, *J. Amer. Chem. Soc.* **84**, 1329 (1962).
261. W. F. Dove and N. Davidson, *J. Mol. Biol.* **5**, 467 (1962).
262. G. L. Eichhorn, *Nature (London)* **194**, 474 (1962).
263. I. A. Kuznetsov, A. N. Mezentsev, Yu. Sh. Moshkovskii, and A. S. Lukanin, *Biofizika* **12**, 373 (1967).
264. J. T. Shapiro, B. S. Stannard, and G. Felsenfeld, *Biochemistry* **8**, 3233 (1969).
265. D. W. Gruenwedel, C. H. Hsu, and D. S. Lu, *Biopolymers* **10**, 47 (1971).
266. T. Okubo and N. Ise, *Macromolecules* **2**, 407 (1969).
267. I. Pilz, O. Kratky, F. von der Haar, and F. Cramer, *Eur. J. Biochem.* **18**, 436 (1971)
268. M. T. Record, Jr., *Biopolymers* **5**, 975 and 993 (1967).
269. C. Delisi and D. M. Crothers, *Biopolymers* **10**, 2323 (1971).
270. M. Nagasawa and Y. Muroga, *Biopolymers* **11**, 461 (1972).
271. P. D. Ross and R. L. Scruggs, *Biopolymers* **2**, 79 and 231 (1964).
272. J. Vinograd, R. Greenwald, and J. E. Hearst, *Biopolymers* **3**, 109 (1965).
273. G. Aubel-Sadron, G. Beck, J.-P. Ebel, and C. Sadron, *Biochim. Biophys. Acta* **42**, 542 (1960).
274. J. P. Ebel, G. Aubel-Sadron, J. H. Weil, G. Beck, and L. Hirth, *Biochim. Biophys. Acta* **108**, 30 (1965).
275. G. Green and H. R. Mahler, *Biochemistry* **10**, 2200 (1971).
276. S. Lewin, *Lab. Pract.* **13**, 400 (1964).
277. R. E. Harrington, *Biopolymers* **10**, 337 (1971).
278. D. Londos-Gagliardi, G. Serros, and G. Aubel-Sadron, *J. Chim. Phys.* **68**, 671 (1971).
279. Ye. M. Gubarev, A. M. Poverennyi, and T. L. Aleinikova, *Biofizika* **9**, 434 (1964).
280. H. Boedtker, *J. Mol. Biol.* **35**, 61 (1968).
281. P. H. von Hippel and K. Y. Wong, *J. Mol. Biol.* **61**, 587 (1971).
282. H. Utiyama and P. Doty, *Biochemistry* **10**, 1254 (1971).
283. J. H. Strauss, R. B. Kelly, and R. L. Sinsheimer, *Biopolymers* **6**, 793 (1968).
284. C. Stevens and A. Rosenfeld, *Biochemistry* **5**, 2714 (1966).

285. C. J. Collins and W. R. Guild, *Biochim. Biophys. Acta* **157**, 107 (1968)
286. Yu. N. Kosaganov, M. I. Zarudnaja, Yu. S. Lazurkin, and M. D. Frank-Kamenet-skii, *Nature (London), New Biol.* **231**, 212 (1971).
287. M. Leng, C. Rosilio, and J. Boudet, *Biochim. Biophys. Acta* **174**, 574 (1969).
288. J. W. Lyons and L. Kotin, *J. Amer. Chem. Soc.* **87**, 1781 (1965).
289. J. Eisinger, F. Fawaz-Estrup, and R. G. Shulman, *Biochim Biophys. Acta* **72**, 120 (1963).
290. J. Skerjanc and U. P. Strauss, *J. Amer. Chem. Soc.* **90**, 3081 (1968).
291. S. Sander and P. O. P. Ts'o, *J. Mol. Biol.* **55**, 1 (1971).
292. A. M. Willemsen and G. A. J. van Os, *Biopolymers* **10**, 945 (1971).
293. J. A. L. I. Walters and G. A. J. van Os, *Biopolymers* **10**, 11 (1971).
294. G. E. Willick and C. M. Kay, *Biochemistry* **10**, 2216 (1971).
295. H. Krakauer, *Biopolymers* **10**, 2459 (1971).
296. H. Krakauer, *Biopolymers* **11**, 811 (1972).
297. B. R. Vijayendran and R. D. Vold, *Biopolymers* **9**, 1391 (1970); **10**, 991 (1971).
298. G. Rialdi, J. Levy, and R. Biltonen, *Biochemistry* **11**, 2472 (1972).
299. W. R. Bauer, *Biochemistry* **11**, 2915 (1972).
300. W. R. Bauer, *J. Mol. Biol.* **67**, 183 (1972).
301. B. G. Archer, C. L. Craney, and H. Krakauer, *Biopolymers* **11**, 781 (1972).
302. U. S. Nandi, J. C. Wang, and N. Davidson, *Biochemistry* **4**, 1687 (1965).
303. N. Davidson and J. C. Wang, *in* "Structural Chemistry and Molecular Biology" (A. Rich and N. Davidson, eds.), p. 430. Freeman, San Francisco, California, 1968.
304. J. C. Wang, U. S. Nandi, D. S. Hogness, and N. Davidson, *Biochemistry* **4**, 1687 (1965).
305. G. Luck and C. Zimmer, *Stud. Biophys.* **24/25**, 307 (1970).
306. D. W. Gruenwedel and D. S. Lu, *Biochem. Biophys. Res. Commun.* **40**, 542 (1970).
307. B. Bhattacharga, N. D. Gupta, and U. S. Nandi, *Biopolymers* **10**, 2625 (1971).
308. J. A. Carrabine and M. Sundaralingam, *Biochemistry* **10**, 292 (1972).
309. J. P. Schreiber and M. Daune, *Biopolymers* **8**, 139 (1969).
310. G. L. Eichhorn and Y. A. Shin, *J. Amer. Chem. Soc.* **90**, 7323 (1968).
311. N. A. Berger and G. L. Eichhorn, *Biochemistry* **10**, 1847 and 1857 (1971).
312. C. Zimmer, G. Luck, H. Fritzsche, and H. Triebel, *Biopolymers* **10**, 441 (1971).
313. I. R. Miller and D. Bach, *Biopolymers* **6**, 169 (1968).
314. I. R. Miller and D. Bach, *Biopolymers* **4**, 705 (1966).
315. D. C. Liebe and J. E. Stuehr, *Biopolymers* **11**, 145 and 167 (1972).
316. M. Sundaralingam and J. A. Carrabine, *J. Mol. Biol.* **61**, 287 (1971).
317. R. G. Shulman, H. Sternlicht, and B. J. Wyluda, *J. Chem. Phys.* **43**, 3116 (1965).
318. A. R. Peacocke, R. E. Richards, B. Sheard, *Mol. Phys.* **16**, 177 (1969).
319. R. H. Jensen and N. Davidson, *Biopolymers* **4**, 17 (1966).
320. M. Daune, C. A. Dekker, and H. K. Schachman, *Biopolymers* **4**, 51 (1966).
321. F. X. Wilhelm and M. Daune, *Biopolymers* **8**, 12 (1969).
322. J. J. Butzow and G. L. Eichhorn, *Biopolymers* **3**, 95 (1965); *Biochemistry* **10**, 2019 (1971).
323. G. L. Eichhorn, E. Tarien, and J. J. Butzow, *Biochemistry* **10**, 2014 (1971).
324. L. S. Lerman, *J. Cell. Comp. Physiol.* **64**, Suppl. 1, 1 (1964).
325. A. Blake and A. R. Peacocke, *Biopolymers* **6**, 1225 (1968).
326. A. R. Peacocke, *Stud. Biophys.* **24/25**, 213 (1970).
327. M. J. Waring, *Nature (London)* **219**, 1320 (1968).
328. D. O. Jordan, *in* "Molecular Associations in Biology" (B. Pullman, ed.), p. 221 Academic Press, New York, 1968.

329. B. L. Van Duuren, B. M. Goldschmidt, and H. H. Seltzman, *Ann. N.Y. Acad. Sci.* **153**, 744 (1969).
330. cf. *Stud. Biophys.* **24/25**, 447 (1970).
331. J. A. McCarter, N. Kadohama, and C. Tsiapalis, *Can. J. Biochem.* **47**, 391 (1969).
332. C. Nagata, M. Kodama, Y. Tagashira, and A. Imamura, *Biopolymers* **4**, 409 (1966).
333. B. Green and J. A. McCarter, *J. Mol. Biol.* **29**, 447 (1967).
334. I. Isenberg, S. L. Baird, and R. Bersohn, *Ann. N.Y. Acad. Sci.* **153**, 780 (1969).
335. M. Craig and I. Isenberg, *Biopolymers* **9**, 689 (1970).
336. S. A. Lesko, Jr., P. O. P. Ts'o, and R. S. Umans, *Biochemistry* **8**, 2291 (1969).
337. H. D. Hoffmann, S. A. Lesko, Jr., and P. O. P. Ts'o, *Biochemistry* **9**, 2594 (1970).
338. E. W. Chan and J. K. Ball, *Biochim. Biophys. Acta* **238**, 461 (1971).
339. E. Cavalieri and M. Calvin, *Proc. Nat. Acad. Sci. U.S.* **68**, 1251 (1971).
340. R. Fuchs and M. Daune, *Biochemistry* **11**, 2659 (1972).
341. E. L. Uhlenhopp and A. I. Krasna, *Biochemistry* **10**, 3290 (1970).
342. J. Ramstein, C. Hélène, and M. Leng, *Eur. J. Biochem.* **21**, 125 (1971).
343. A. L. Stone, L. G. Childers, and D. F. Bradley, *Trans. Faraday Soc.* **66**, 3081 (1970).
344. W. Müller and D. M. Crothers, *J. Mol. Biol.* **35**, 251 (1968).
345. R. D. Wells and J. E. Larson, *J. Mol. Biol.* **49**, 319 (1970); *J. Biol. Chem.* **247**, 3405 (1972).
346. L. D. Hamilton, W. Fuller, and E. Reich, *Nature (London)* **198**, 538 (1963).
347. G. V. Gurskii, *Biofizika* **11**, 737 (1966); *Stud. Biophys.* **24/25**, 265 (1970).
348. Y. Courtois, W. Guschlbauer, and P. Fromageot, *Eur. J. Biochem.* **6**, 106 (1968).
349. J. C. Wang, *Biochim. Biophys. Acta* **232**, 246 (1971).
350. H. M. Sobell, S. C. Jain, T. D. Sakore, and C. E. Nordman, *Nature (London)* **231**, 200 (1971).
351. T. R. Krugh, *Proc. Nat. Acad. Sci. U.S.* **69**, 1911 (1972).
352. W. J. Pigram, W. Fuller, and L. D. Hamilton, *Nature (London), New Biol.* **235**, 17 (1972).
353. M. J. Waring, *Stud. Biophys.* **24/25**, 257 (1970).
354. H. Wawra, W. Müller, and O. Kratky, *Makromol. Chem.* **139**, 83 (1970).
355. B. Hudson, W. B. Upholt, J. Devinny, and J. Vinograd, *Proc. Nat. Acad. Sci. U.S.* **62**, 813 (1969).
356. J. C. Sutherland and B. M. Sutherland, *Biopolymers* **9**, 639 (1970).
357. P. Wahl, J. Paoletti, and J.-B. Le Pecq, *Proc. Nat. Acad. Sci. U.S.* **65**, 417 (1970).
358. D. Genest and P. Wahl, *Biochim. Biophys. Acta* **259**, 175 (1972).
359. J. Paoletti and J.-B. LePecq, *J. Mol. Biol.* **59**, 43 (1971); *Biochimie* **53**, 969 (1971).
360. J. M. Saucier, B. Festy, and J.-B. LePecq, *Biochimie* **53**, 973 (1971).
361. D. Lang, *Phil. Trans. Roy. Soc. London, B Ser.* **261**, 151 (1971).
362. D. Freifelder, *J. Mol. Biol.* **60**, 401 (1971).
363. L. M. Augerer and E. N. Mondrianakis, *J. Mol. Biol.* **63**, 505 (1972).
364. G. Weill, *Biopolymers* **3**, 567 (1965).
365. J. C. Thomes, G. Weill, and M. Daune, *Biopolymers* **8**, 647 (1969).
366. R. Bidet, J. Chambron, and G. Weill, *Biopolymers* **10**, 225 (1971).
367. M. Kaufmann and G. Weill, *Biopolymers* **10**, 1983 (1971).
368. P. A. Sharp and V. A. Bloomfield, *Biochem. Biophys. Res. Commun.* **39**, 407 (1970).
369. D. O. Jordan and L. N. Sansom, *Biopolymers* **10**, 399 (1971).
370. R. W. Armstrong, T. Kurucsev, and U. P. Strauss, *J. Amer. Chem. Soc.* **92**, 3174 (1970).
371. H. J. Li and D. M. Crothers, *J. Mol. Biol.* **39**, 461 (1969).
372. D. E. U. Schmechel and D. M. Crothers, *Biopolymers* **10**, 465 (1971).

373. M. Sakoda, K. Hiromi, and K. Akasaka, *Biopolymers* **10**, 1003 (1971).
374. D. F. Bradley and S. Lifson, *in* "Molecular Associations in Biology" (B. Pullman, ed.), p. 261. Academic Press, New York, 1968.
375. M. Gilbert and P. Claverie, *in* "Molecular Associations in Biology (B. Pullman. ed.), p. 745. Academic Press, New York, 1968; *J. Theor. Biol.* **18**, 330 (1968).
376. C. Houssier and H. G. Kuball, *Biopolymers* **10**, 2421 (1971).
377. J. Ramstein, M. Dourlent, and M. Leng, *Biochem. Biophys. Res. Commun.* **47**, 874 (1972).
378. M. Dourlent and C. Hélène, *Eur. J. Biochem.* **23**, 86 (1971).
379. M. Schoentjes and E. Fredericq, *Biopolymers* **11**, 361 (1972).
380. J. J. Laurence and M. Louis, *Biochim. Biophys. Acta* **272**, 231 (1972).
380a. K. E. Reinert, *J. Mol. Biol.* **72**, 593 (1972).
381. D. N. Holcomb and I. Tinoco, Jr., *Biopolymers* **3**, 121 (1965).
382. D. Poland, J. N. Vournakis, and H. A. Scheraga, *Biopolymers* **4**, 223 (1966).
383. M. Leng and A. M. Michelson, *Biochim. Biophys. Acta* **155**, 91 (1968).
384. A. M. Michelson, *in* "Molecular Associations in Biology" (B. Pullman, ed.), p. 93. Academic Press, New York, 1968.
385. J. Brahms, *J. Chim. Phys.* **65**, 105 (1969).
386. S. Hanlon and E. O. Major, *Biochemistry* **7**, 4350 (1968).
387. P. O. P. Ts'o, *Ann. N.Y. Acad. Sci.* **153**, 785 (1969).
388. W. B. Gratzer, *Eur. J. Biochem.* **10**, 184 (1969).
389. R. D. Wells, J. E. Larson, R. C. Grant, B. E. Shortle, and C. R. Cantor, *J. Mol. Biol.* **54**, 465 (1970).
390. D. M. Gray and I. Tinoco, Jr., *Biopolymers* **9**, 223 (1970).
391. E. Neumann and A. Katchalsky, *Ber. Bunsenges. Phys. Chem.* **74**, 868 (1970).
392. A. Katchalsky and E. Neumann, *Int. J. Neurosci.* **3**, 175 (1972).
393. R. A. Cox and A. Katchalsky, *Biochem. J.* **126**, 1039 (1972).
394. J. C. Thrierr and M. Leng, *Biochim. Biophys. Acta.* **272**, 238 (1972).
395. E. Neumann and A. Katchalsky, *Proc. Nat. Acad. Sci. U.S.* **69**, 993 (1972).
396. F. M. Pohl and T. M. Jovin, *J. Mol. Biol.* **67**, 375 (1972).
397. S. Arnott, *Progr. Biophys. Mol. Biol.* **22**, 181 (1971).
398. D. N. Holcomb and S. N. Timasheff, *Biopolymers* **6**, 513 (1968).
399. J. T. Finch and A. Klug, *J. Mol. Biol.* **46**, 597 (1969).
400. A. Wada, *in* "Conformation of Biopolymers" (G. N. Ramachandran, ed.), p. 655. Academic Press, New York, 1967; A. Wada, I. Kawata, and K. I. Miura, *Biopolymers* **10**, 1153 (1971).
401. V. Luzzati, J. Witz, and A. Mathis, *in* "Genetic Elements" (D. Shugar, ed.), p. 41. Academic Press, New York, 1967.
402. A. Gulik, H. Inoue, and V. Luzzati, *J. Mol. Biol.* **53**, 221 (1970).
403. B. Sommer and J. Jortner, *in* "Quantum Aspects of Heterocyclic Compounds in Chemistry and Biochemistry" (E. D. Bergmann and B. Pullman, eds.), p. 413. Isr. Acad. Sci. Humanities, Jerusalem, 1970; *J. Chem. Phys.* **49**, 3919 (1968).
404. S. Arnott, *Progr. Biophys. Mol. Biol.* **21**, 265 (1970).
405. R. C. Davis and I. Tinoco, Jr., *Biopolymers* **6**, 223 (1968).
406. A. S. Schneider and R. A. Harris, *J. Chem. Phys.* **50**, 5204 (1969).
407. J. T. Powell, E. G. Richards, and W. B. Gratzer, *Biopolymers* **11**, 235 (1972).
408. L. D. Inners and G. Felsenfeld, *J. Mol. Biol.* **50**, 373 (1970).
409. J. C. Thrierr and M. Leng, *Biochim. Biophys. Acta* **182**, 575 (1969).
410. J. C. Thrierr, M. Dourlent, and M. Leng, *J. Mol. Biol.* **58**, 815 (1971).
411. E. K. Achter and G. Felsenfeld, *Biopolymers* **10**, 1625 (1971).

4

CIRCULAR DNA

William Bauer and Jerome Vinograd

I. General Introduction

Circular DNA is distinguished from other DNA's in that it contains a circularly continuous chain axis. The arrangement of the atoms and covalent bonds about this axis may be used to distinguish further at least three distinct types of circular molecules. The *single-stranded circles*, such as those found in the bacteriophages ϕX174 and M13, possess an axis of circularity which may be considered to pass along the phosphodiester–sugar backbone chain. The circular axis in this case is lost upon the scission of one covalent chain link with the formation of a linear molecule.

The second physically distinct class of circular molecules is termed *open circular DNA*. These molecules are composed of a duplex helix, and the axis

265

of circularity may be regarded as collinear with the duplex helix axis. These molecules are further characterized by the presence of at least one chain discontinuity. Such molecules can be formed by the cyclization of linear DNA's possessing complementary single-stranded end regions. Examples of such open circles are those prepared from bacteriophage λ DNA's and from terminally redundant DNA's after exonucleolytic digestion and from cohesive-ended fragments produced by the *E. coli* R1 restriction enzyme. A second type of open circle is formed by the introduction of one or more chain scissions into a closed double-stranded duplex, as described below. These nicked circular duplexes serve as reference materials in experiments which illustrate the special properties of closed DNA, and we will discuss them primarily in this context.

We will be principally concerned with an exposition of the special features of the third class of circular DNA's, *closed duplex DNA*, in which covalent chain scissions are absent. As will be detailed in the sections to follow, the covalent continuity of both strands about the axis of circularity results in the modification of many of the characteristic properties associated with DNA, as well as in the introduction of features not found in linear DNA or among representatives of the other two categories of circular DNA. A review of research in this field has recently appeared [1], containing a comprehensive review of the literature.

II. The Topological Constraint

When two circles made of impenetrable material are interlocked, physical separation is impossible without severing one of the circles. The combined circles are further characterized by a topological constant which indicates the extent of interwinding. This is illustrated in Fig. 1 for two such interwound circles which contain one, two, and ten interwindings. One of the circles remains in a plane in Fig. 1a–c, and the other circle is wound about it. The arrangement of turns in Fig. 1c is topologically equivalent to the symmetrical configuration shown in Fig. 1d, in which the two circles wind about a common helix axis. If the helix axis is confined to the plane of the paper, the number of plane-projected crossover points between the two strands is necessarily an even integer and is equal to twice the number of interwindings, which we will refer to as *turns*. The number of such turns, as counted with the axis so constrained, is an invariant quantity which characterizes any such system of interlocked rings. This *topological winding number* may be defined for paired covalently closed single-stranded DNA circles, as in closed duplex DNA, or for any other system of closed impenetrable curves.

While it is impossible to change the topological winding number, the

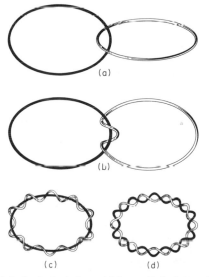

Fig. 1. Examples of interlocked circles which are linked together after interwinding (a) once, (b) twice, and (c) ten times. Structures (c) and (d) are topologically equivalent.

physical disposition of the circles as they turn about one another is determined by other considerations. Figure 2 represents several possible dispositions of the turns for a model system with a topological winding number of zero. The introduction of right-handed duplex turns into the right side, three in this case, requires the introduction of exactly the same number of left-handed turns (Fig. 2b). We adopt the convention that right-handed turns in structures with a planar axis have a positive sign. The compensating left-handed turns, however, can be absorbed in higher-order windings of the helical axis itself. Two such higher-order windings are illustrated in Fig. 2c and d. These are termed the *toroidal* and *interwound* modes, respectively. The former is a straightforward winding of the duplex axis about a circular tertiary or *super-helix* axis, and superhelical turns of a right-handed sense are correspondingly associated with a positive sign. The number of superhelical turns in a pure toroidal superhelix is also necessarily an integer. The planar winding shown in Fig. 2b may be transferred without topological restriction and with no change of sign to the superhelical windings in Fig. 2c. Here the three compensating superhelical turns are seen to be left-handed and negative.

The interwound superhelix, which is illustrated in Fig. 2d, is more complex in at least three respects. In the first place, the superhelix axis is linear and therefore contains end regions. The winding in these end regions is necessarily different from that about the remainder of the superhelix axis. Second, the

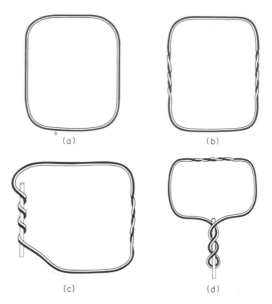

Fig. 2. An illustration of the formation of superhelical turns in a closed duplex consisting of two noninterwound strands. The topological winding number, α, in these structures has a value of zero. (a) Two noninterwound strands. The duplex winding number, β, and the superhelical winding number, τ, have values of zero. (b) The noninterwound structure in (a) has been modified by the introduction of three right-handed duplex turns and three compensatory left-handed duplex turns. The winding numbers $\alpha = \beta = \tau = 0$. (c) Three right-handed duplex turns and three left-handed superhelical turns have been introduced into the duplex in (a). The separate strands in the superhelical region rotate about each other. In this arrangement $\alpha = 0$, $\beta = +3$, and $\tau = -3$. The superhelical turns may be regarded as forming a section of a toroidal superhelix. (d) Three right-handed duplex turns and three right-handed interwound superhelical turns have been introduced into the duplex in (a). The paired strands in the interwound superhelix do not rotate around each other. In this arrangement, $\alpha = 0$, $\beta = +3$, and $\tau = -3$. (From Vinograd *et al.* [22]; reproduced by permission of *J. Mol. Biol.*)

winding about the interwound superhelix axis is not necessarily an integer, but may have an arbitrary fractional value. And third, the constancy of the topological winding number for the entire interlocked circular complex requires that a negative sign be associated with superhelical turns which have a right-handed orientation about the linear interwound superhelix axis. This change in sign is due to a 90° shift in the orientation of the interwound superhelix axis with respect to the viewing axis for the duplex and is discussed in greater detail by Bauer and Vinograd [2].

With the aid of the above considerations and sign conventions, a mathematical relationship was proposed by Vinograd and Lebowitz [3]

$$\alpha = \beta + \tau \tag{1}$$

where α (the topological winding number) is the integer obtained by taking one-half the number of crossovers as counted in the planar representation of Fig. 1; τ (the number of superhelical turns) is the number of times the duplex axis winds about the superhelix axis (either interwound or toroidal) when the superhelix axis is constrained to a plane; and β (the duplex winding number) is the residual number of windings accounted for by winding of one circle around the other in the duplex. This formulation takes no account of still higher-order windings, those which may result from net winding of the superhelix axis. Equation (1) has been proved analytically by Fuller [4], and it has been discussed in greater detail by several others [5–7].

The application of the above definitions and concepts to closed duplex DNA is most conveniently made in terms of comparison with the known amount of interstrand winding in the Watson–Crick B form, for which we take the constant β^0 to be equal to exactly 1 duplex turn per 10 base pairs. On an intensive basis, Eq. (1) becomes

$$A = B + \sigma \tag{2}$$

where A, the *topological winding density*, is equal to α/β^0; B, the *duplex winding density*, is equal to β/β^0; and σ, the *superhelix density*, is equal to τ/β^0. The winding in the superhelix mode may also be described in terms of the *degree of supercoiling*, which is equal to $100 \times |\sigma|$. The sign of the superhelix density must then be separately expressed if necessary.

Since it is only α that is a constant of the molecule, both β and τ are subject to topologically unrestricted variation provided only that their sum remains fixed. In view of the fact that a variety of factors is known to influence the values of these parameters for a given closed circular molecule, as discussed in the sections to follow, it is convenient to define a reference state for the designation of superhelix densities. It has been suggested [8] that a suitable reference state be 2.85 M CsCl at 20°C and neutral pH, and we will use the symbols σ_0, τ_0, B_0, and β_0 to refer specifically to this state.

III. The Special Thermodynamic Properties of Closed Double-Stranded DNA

The topological considerations discussed in the preceding section result in special restrictions on closed double-stranded DNA which are absent in linear or open circular molecules. These extra constraints profoundly influence the physical and chemical nature of the molecule and, in particular, result in an altered free energy of formation which depends upon the magnitude of the superhelix density. The potential interconvertibility of the duplex and super-helical interwindings subject to the constraint of Eq. (1) means that the most stable structural form of a closed double-stranded DNA is not determined by winding considerations alone. The structure which results under any given set of conditions is expected to be that which minimizes the total energy of the system composed of the nucleic acid and its surrounding solvent.

We expect the dominant forces tending to maintain the integrity of the duplex structure to be operative in closed duplex DNA in much the same way as in its open counterparts. Likewise, an energy barrier must be overcome in closed DNA prior to a reduction in the duplex winding.*

All known naturally occurring closed duplex DNA molecules behave as though the strands had been covalently closed when less than the optimum number of duplex turns had been incorporated. The formation of a superhelix is, therefore, preferred to the alternative reduced duplex winding density.† A superhelix is, however, a more highly organized DNA structure, the formation of which is expected to decrease the entropy of the system [9,10]. In addition to the entropic factor, the bending and torsional stresses on the duplex which potentially accompany the superhelical constraint lead to the anticipation that the superhelical form of closed duplex DNA is energetically unfavorable compared to a closed molecule of the same primary structure but which is relaxed. This latter species is closely approximated by the nicked, open circular derivative of closed circular DNA, and we correspondingly expect the release of energy to accompany the conversion of the closed superhelical molecule to the nicked form. By extension, any chemical interaction which results in a diminution in the number of turns is expected to be energetically favored compared to the same reaction when conducted with a DNA containing no or fewer superhelical turns and, conversely, interactions which increase the superhelical windings are thermodynamically inhibited.

The above effects have been most dramatically demonstrated in studies of the reaction between closed duplex DNA's and intercalating dyes and will be discussed in a later section. Bauer and Vinograd [6] showed that the binding isotherm of the dye ethidium bromide with closed circular SV40 DNA is perturbed relative to the binding to the corresponding open circular DNA. By analyzing the extent of this perturbation as a function of the number of

---

* Wang [5] has suggested that the stress associated with the superhelix free energy might produce a small reduction in the average duplex rotation angle. Although small differences in circular dichroism between superhelical and relaxed DNA's have been detected [8a] the interpretation of these effects is presently uncertain. Dean and Lebowitz [8b] have proposed that a small fraction of the winding deficiency in PM2 and ϕXRF DNA's may be accounted for in terms of a disordered duplex region containing bases with an elevated reactivity toward formaldehyde.

† This description has recently been questioned by Paoletti and LePecq [8c], who interpret the results of fluorescence depolarization experiments with ethidium bromide in closed DNA in terms of an overwinding of the native closed duplex. This interpretation does not appear to be supported by recent investigations (Denhart [69]; Schmir *et al.* [70]).

superhelical turns, they were able to estimate that the formation of the 15 superhelical turns in the above DNA molecule of contour length 1.5 μm is accompanied by the expenditure of approximately 100 kcal/mole DNA. The same amount of energy is, of course, liberated upon elimination of the superhelical winding. In the case of the binding of an intercalating dye, the energy is available to either facilitate or inhibit the reaction, depending upon whether superhelical turns are being formed or removed in the region of the isotherm in question.

IV. The Tertiary Structure

The presence of tertiary winding in closed duplex DNA results in a more compact shape for the molecule than for its relaxed circular counterpart. This increased compactness is translated into a decreased effective hydrodynamic volume and consequent altered sedimentation and viscometric behavior. The extent of this altered hydrodynamic behavior depends upon the superhelix density and, therefore, upon any factor to which the superhelix density is sensitive.

Native closed circular polyoma DNA, for example, sediments with a sedimentation coefficient of 21 S in contrast to the open form, which moves approximately 20% slower. The introduction of a single-strand scission in this closed DNA of 3.3% supercoiling results in the irreversible conversion to the slower open species. The relative sedimentation velocity as a function of superhelix density for enzymically closed SV40 DNA, with a molecular weight of 3.1×10^6 and $\beta_0 = 480$, and for enzymically closed $\lambda b_2 b_5 c$ DNA, molecular weight of 25×10^6 and $\beta_0 = 3800$, are shown in Fig. 3. In each case the sedimentation coefficient varies in a complex manner with σ_0. As the degree of supercoiling increases from zero to approximately 2%, the sedimentation coefficient rises steeply and, in the case of $\lambda b_2 b_5 c$, approximately linearly. It then passes through a maximum and declines to a local minimum at approximately 3.5–4% supercoiling, and finally, in the case of SV40 DNA rises to the highest degree of supercoiling studied, 8%. Although the experimental results for $\lambda b_2 b_5 c$ DNA do not extend to a degree of supercoiling greater than 2.6%, similar studies in the presence of ethidium bromide (Section VI) reveal a behavior qualitatively similar to that of SV40 DNA at high degrees of supercoiling.

The initial increase in sedimentation coefficient as the supercoiling increases from zero to approximately 2% has been interpreted [8] as a progressive decrease in equivalent hydrodynamic volume in an essentially randomly coiled molecule. In the fully relaxed molecule the chain segments fluctuate in position in response to thermal agitation, subject to the usual constraint

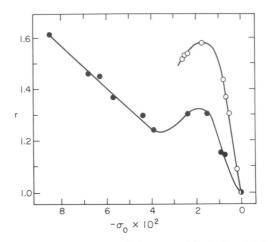

Fig. 3. The relative sedimentation coefficient, $r = (s_{20,w}^0)$ closed$/(s_{20,w}^0)$ nicked, of two closed circular DNA's as a function of superhelix density. (●) SV40 DNA, 2.83 M CsCl, 0.01 M tris, pH 8.0, 20°C [8]; (○) $\lambda b_2 b_5 c$ DNA, 3.0 M CsCl, 0.01 M Na$_3$ EDTA, 20°C [5].

of segmental connectedness and to the further requirement that no ends be present. The supercoils that are introduced in this region are not localized in position and extend over the whole molecule. The effect of these first few supercoils is to restrict the molecule in such a way that the radius of gyration is reduced without the introduction of a preferred axis for supercoiling. The representation of the closed duplex DNA molecule as a rigid superhelix is especially unwarranted and misleading in this region. Diagrammatic representations of supercoiled molecules presenting this arrangement, such as in Fig. 2, should be regarded as figurative.

   At low degrees of supercoiling the resultant change in the curvature of the DNA molecule is not great and the occurrence of highly curved local regions of the duplex is unnecessary. Further increase in the number of supercoils would eventually result in the requirement for high local curvatures in an otherwise freely fluctuating molecule. Such high local curvatures are expected to be energetically unfavorable and to become more so with the square of the curvature. This restriction on the allowed fluctuations results in a more regularly arranged disposition of the molecular segments and in the development of a geometric form in which a greater probability is assigned to structures of lower curvature. The interwound superhelix and the toroidal superhelix are such structures. No experimental evidence is presently available to indicate which of these two forms, or what combination of them, most nearly represents the structure of closed duplex DNA in solution. The requirement for minimization of the total energy, however, suggests that the interwound superhelix will be preferred. The duplex is bent in both models but the pure interwound form can, in principle, be constructed without the additional stress of torsion on the duplex, neglecting end effects. Some evidence is provided for this point of view by the appearance in electron micrographs of closed circular molecules which are sufficiently supercoiled. It is perhaps worth noting that models for closed circular DNA, constructed of elastic tubes or rods, invariably assume the interwound form.

The progressive elongation of the closed molecule, associated with the transition from a coil with symmetric shape to a rodlike structure leads to an increase in the frictional coefficient and to a consequent decrease in the sedimentation coefficient. This effect is seen in Fig. 3 in the region between 2 and 4% supercoiling. At still higher degrees of supercoiling the sedimentation coefficient again increases, an indication that the frictional coefficient is decreasing. This latter event has been attributed to the formation of branched superhelical structures, an interpretation which is substantiated by the appearance of the molecules in the electron microscope [8].

V. The Helix-Coil Transition

The helix–coil transition in linear DNA is a cooperative process in which the ordered duplex structure is lost. The process begins in the AT-rich regions, with the formation of internal loops and unpaired end regions. This melting process is accompanied by progressive unwinding of the turns originally present in the duplex. The melting process occurs in a characteristic temperature range with a mean temperature, T_m. This temperature may be changed by the addition of organic solvents [11,12], by a change in the ionic strength [13,14], by the binding of reagents which prefer either the helix or coil form [15–17], or by certain concentrated salt solutions [18]. Alternatively, chemical reagents such as formaldehyde [19,20], acids, and bases [20] may be used to react covalently with the DNA bases so as to destabilize the duplex. The end result of the melting of linear DNA is the production of two single strands which have separated from each other. A fundamentally similar process occurs in the melting of singly nicked circular DNA, except that the end products are a single-stranded linear and a single-stranded circular molecule.

The melting behavior of closed duplex DNA is altered in several important respects due to the topological restraint and to the presence of supercoiling in the native molecule. All naturally occurring closed duplex DNA's are topologically underwound and contain supercoils which are removed in the early stages of the melting process, as illustrated in Fig. 4. The positive free energy associated with the native supercoiling acts as a destabilizing influence and facilitates the melting transition. The commencement of melting in turn brings about a reduction in the supercoiling of the molecule. This process continues until an equivalent portion of the duplex is melted and all of the supercoils are removed. Further melting of the relaxed, partially denatured molecule leads to supercoils of the opposite handedness and winding up of the denatured regions. As melting proceeds to completion, the closed molecule eventually loses its ordered secondary structure, but still retains the original topological winding number (cf. Fig. 8).

This behavior may be monitored by following the sedimentation coefficient in the course of the alkali- or acid-induced melting transition [3,21]. The corresponding variations in sedimentation coefficient may be used as an index

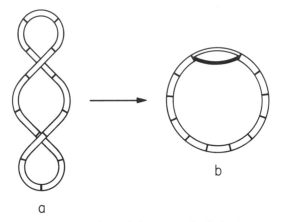

Fig. 4. Diagrammatic representation of the removal of the two superhelical turns initially present in a closed circular DNA of topological winding number $\alpha = 11$. An unwound denatured region containing 20 base pairs is formed in this process. (See also the legend to Fig. 8.)

of the conformational changes which accompany the melting process. The sedimentation velocity–pH titration of polyoma DNA is shown in Fig. 5. The sedimentation coefficient of the closed form is constant and 25% higher than that of the singly nicked molecule as the pH is increased to approximately 11.4. The sedimentation coefficient of the closed form then decreases as the early melting occurs, between pH 11.4 and 11.6. This initial decrease is followed by a rise as supercoils of the opposite handedness are introduced. The nicked molecule begins to titrate in the neighborhood of pH 11.6–11.8, and at pH 11.8 strand separation occurs with the appearance of single-stranded circular and linear molecules. The continuation of the titration in the closed molecule is accompanied by a steady rise in the sedimentation coefficient up to a high plateau value which is 3.3 times as large as that of the linear single strand.

The material sedimenting in the upper velocity plateau region has been termed a *double-stranded cyclic coil*. The titration of the thymine and guanine ring protons leads to the complete loss of the original duplex structure. The structure then consists of two intertwined single strands, no longer in register, which are, however, still constrained to wind around each other α times. The winding requirement reduces the radius of gyration of the single-stranded circles, each of which would otherwise be able to occupy a larger region of

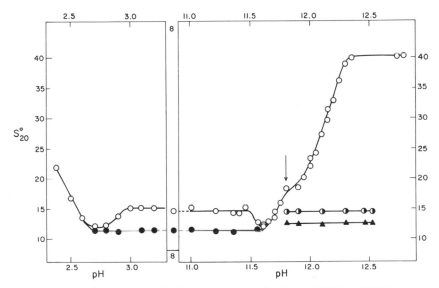

Fig. 5. The sedimentation velocity–pH titration of polyoma DNA at 20°C in aqueous CsCl, ρ = 1.35 gm/ml. (○) Closed duplex; (●) nicked duplex; (◖) circular single strands; (▲) linear single strands. Sedimentation coefficients are not corrected for solvent viscosity or buoyant effects. The arrow indicates the pH at which strand separation occurs in the nicked molecule. (From Vinograd and Lebowitz [3]; reproduced by permission of Rockefeller University Press.)

space. The frictional coefficient is, therefore, reduced and the sedimentation coefficient increased.

The alkaline titration of closed DNA may also be monitored by measuring the accompanying changes in buoyant density. The buoyant shifts, which are insensitive to the shape of the molecule, are a measure of the degree of titration and may be taken as an indication of the fraction of the molecule melted. The removal of a ring proton from the guanine and thymine bases is accompanied by cesium salt formation, with a corresponding increase in the buoyant density of the molecule. The alkaline buoyant titration of polyoma

DNA (Fig. 6) reveals the effect of the topological restraint upon the extent of the pH-induced melting [22]. As the pH increases from neutral to 11.4, the buoyant densities of the closed and nicked circular forms remain identical. A further increase in pH from 11.4 to 11.7 results in the early melting of approximately 3% of the closed molecule, whose buoyant density correspondingly increases, while the nicked molecule remains stable and untitrated.

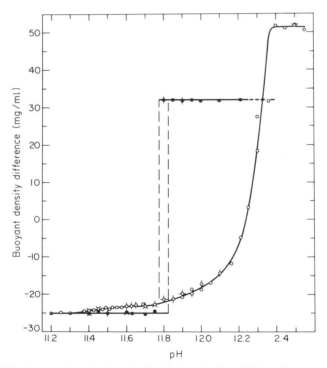

Fig. 6. The buoyant density titration in alkaline CsCl, at 20°C, of intact polyoma DNA I (○), and singly nicked polyoma DNA II (●). The buoyant density results are given as buoyant density differences between polyoma DNA and two markers. The dashed lines enclose the region in which the abrupt titration of II occurs. The tails on the data points indicate the number of overlapping results. (From Vinograd *et al.* [22]; reproduced by permission of *J. Mol. Biol.*)

This early melting is facilitated by the release of the free energy stored in the native superhelical turns. As the pH is increased further, the nicked molecule melts in a highly cooperative fashion at pH 11.8, while the closed molecule continues to titrate slowly. This melting of the closed form continues with increasing sharpness until at pH 12.4 the reaction is complete and a plateau in buoyant density is reached.

The fully titrated closed and nicked DNA's differ in buoyant density by 18 mg/ml. The buoyant densities of the untitrated forms are indistinguishable. The difference in alkaline buoyant densities has been attributed to a reduced extent of base-pairing and base-stacking in the closed molecule resulting from the restrictions imposed by the topological constraint [22].

The effects of the topological constraint are also evident in the thermally induced helix–coil transition as monitored by absorbence changes [3,22]. The midpoint of the transition in 7.2 M NaClO$_4$, a solvent which lowers the melting temperature of DNA, is 73°C as compared with 48°C for the nicked form. The corresponding melting temperatures in SSC (0.15 M NaCl, 0.015 M Na citrate, pH 7.0) are 107° and 89°C, respectively. The end of the transition for the closed molecule is at 86°C in NaClO$_4$ and calculated to be at 117°C in SSC. For the nicked species, the transition is essentially completed by 54°C in NaClO$_4$ and by 94°C in SSC. The early melting of the closed DNA is not observed in concentrated NaClO$_4$ solutions due to the nearly 4% relaxation of the duplex winding induced by this solvent system [23].

The melting profile of closed circular polyoma DNA is substantially broader than that of the nicked form. The increased breadth indicates that the cooperativity of melting is significantly reduced. A major contributor to this effect is the fact that the denatured regions in closed DNA contain strands that are still constrained to wind around each other with an average winding of approximately 1 turn per 10 base pairs. This restriction reduces the configurational entropy in the denatured regions, which in turn reduces the advantage of loop expansion relative to nucleation.

The detailed sequence of conformation changes which occur as melting proceeds beyond the relaxed state has not yet been investigated. A qualitative insight is gained, however, upon inspection of the dependence of the sedimentation coefficient upon the fraction of base pairs melted, shown in Fig. 7, obtained by combining the data in Figs. 5 and 6. The complete removal of the negative superhelical turns occurs at the minimum in the curve after about 2.5% of the titration. Beyond the minimum the sedimentation coefficient rises steeply and with steadily decreasing slope. The first 50% of the rise in s from the minimum occurs after a further 11.5% of the titration has occurred. The remaining change in s occurs in the course of the last 86% of the titration. The final 50% of the titration accounts for only the last 10% of the net change in s from its minimum value.

The changing molecular conformations which are believed to occur during the late helix–coil transition in closed DNA are shown diagrammatically in Fig. 8, a continuation of the process represented in Fig. 4. The late transition begins with the fully relaxed circular duplex containing one or more denatured regions. This species is diagrammatically represented in Fig. 8a for the case in which a single denatured region is formed [21]. In solution it would be expected that the single-stranded regions would be collapsed in high salt and that the duplex portion of the molecule would be irregularly disposed in space and time in accordance with the statistical theory of polymer chains. The continuation of the melting process results in an increase in the fraction of the molecule which is in the denatured state. This is illustrated in Fig. 8b for a further 46% denaturation in which the denaturation is represented in a single region and in which no higher-order winding is present. This model

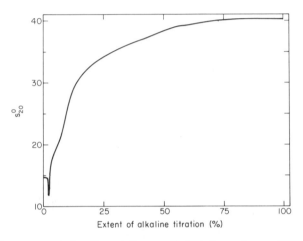

Fig. 7. The uncorrected sedimentation coefficient for polyoma DNA plotted as a function of the degree of titration as measured by changes in the buoyant density. The sedimentation coefficients were measured in 1.35 *M* CsCl at 20°C. The curve reveals the significant conformational changes which occur during the titration of closed DNA. No attempt has been made to correct the sedimentation coefficients for the change in mass that occurs during the titration.

requires that the disordered region contain about 10 base pairs per turn, and a correspondingly low configurational entropy. An entropy reduction of this nature is unlikely in the absence of the formation of stabilizing structures which could contribute compensating entropic or enthalpic terms. It is therefore to be expected that the interstrand winding in the denatured region represented in Fig. 8b will diminish to some lower value. Reduction of the interstrand winding in the coil will occur with the formation of positive higher-order turns in both the duplex and denatured regions. This is illustrated in Fig. 8c for a hypothetical molecule containing a linear higher-order axis and in which the native and denatured regions are kept separate for convenience. The completion of the melting process in the closed molecule removes the duplex region and gives rise eventually to a structure, illustrated in Fig. 8e with the α turns distributed between local interstrand winding and higher-order winding.

Linear DNA molecules, after partial melting, renature very rapidly and monomolecularly when brought into an environment in which the intact duplex is again the stable form. Closed double-stranded DNA behaves in a similar fashion. When linear DNA is completely melted and strand separation has occurred, renaturation is bimolecular and proceeds at slower rates which depend upon genetic complexity, molecular weight, and base composition

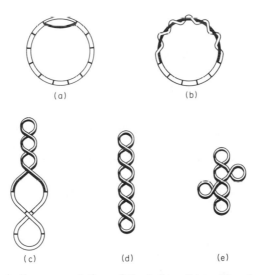

Fig. 8. Diagrammatic representation of the helix-coil transition in a closed circular DNA with topological winding number α = 11. A region in the duplex between adjoining bars, ⊏⊐⊏ contains 10 base pairs and 1 right-handed turn. Denatured or coiled regions are indicated by intertwined solid and hollow lines, and two crossovers of these single strands are required for each turn. A single crossover of two paired strands, in tertiary winding, is required for a turn. The figure has been drawn with the assumption that duplex and denatured DNA's have identical length per unit mass. (a) Formation of an unwound denatured region, as in Fig. 4, with reduction of β from an initial value of 13 to 11 and an increase in τ from -2 to 0. (b) Continuation of the denaturation process to the state in which $\beta_h = 5$ and $\beta_o = 6$, with no higher-order winding in either the duplex or the denatured region. The subscripts h and c refer to the helix and coil regions, respectively. (c) An alternative and more probable configuration of the structure shown in (b). Here, higher-order winding in both the helix and coil regions has occurred with a consequent reduction in β_o. The figure is drawn with the β_h remaining at 5 and $\beta_o = 2$, $\tau_h = 1$ and $\tau_o = 3$. These latter two winding numbers represent the tertiary winding numbers in the helix and coil regions, respectively. (d) Completion of the denaturation with elimination of the helical region. (e) The structure shown here contains tertiary windings which are branched, but does not show the still higher-order windings which might also occur.

[24]. The large difference in T_m between closed and open DNA's allows the ready attainment of conditions in which linear DNA's are denatured and strand-separated, while closed DNA's are still partially in the duplex state. This provides a basis for the chromatographic separation of closed from nicked and linear DNA's. The proper choice of pH in a buoyant gradient (Fig. 6) should also lead to useful separations.

The renaturation of completely titrated closed DNA is expected to occur at optimal rates under conditions of approximately 0.2 M salt, neutral pH, and 80°C, a temperature 25°C below the melting temperature, by analogy with linear DNA. In such experiments, however, the strand scission rate is significant. It has been reported that the fully denatured RF form of ϕX174 DNA may be reannealed in the pH interval 10.9–11.8 at varying rates in the temperature range between 40° and 60°C [25], in contrast to an earlier report in which this form, prepared by neutralization of the DNA after alkaline denaturation, could not be reannealed [26].

VI. Reactions with Intercalating Substances

Any ligand which upon binding affects the duplex winding of DNA will alter the degree of supercoiling in a closed DNA. If the reaction causes an unwinding of the duplex with a loss of superhelical turns in the process, the net free energy of the binding reaction becomes more negative and the extent of binding is increased relative to the nicked form. The net reaction free energy in this case includes contributions from the intrinsic binding and from the loss of supercoiling. Conversely, if supercoiling is increased by the binding of a duplex unwinding reagent, the net reaction free energy becomes less negative and the extent of binding is reduced relative to the nicked form.

The principal quantitative investigations of the above effects have been performed with the trypanocide ethidium bromide (EthBr) (Fig. 9), which intercalates between the base pairs of DNA and unwinds the duplex. Native

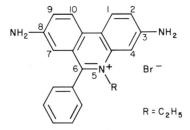

Fig. 9. Structure of ethidium bromide; 3,8-diamino-6-phenyl-5-ethylphenanthridium bromide.

SV40 DNA is approximately 4% supercoiled and the superhelical turns are of negative sign. The binding of ethidium bromide to this molecule at low dye concentrations is enhanced relative to the nicked form (Fig. 10a), and is diminished at higher dye concentrations (Fig. 10b). These effects are presented diagrammatically in Fig. 11. At a critical free dye concentration, in which the closed molecule is depleted of supercoils, the binding to the two forms is the same. In the presence of this dye concentration, which corresponds to the intersection of the binding isotherms in Fig. 10a, the nicked and closed DNA molecules differ only by the presence or absence of a chain scission and may be reversibly interconverted. Further increase in the dye concentration after the closed, relaxed form is attained requires the accompanying formation of positive superhelical turns. This process becomes increasingly unfavorable and accounts for the separation between the isotherms in Fig. 10b. There is an effective cessation of binding to the closed form within the dye solubility limits when only about two-thirds of the potential binding sites are occupied.

The properties of closed circular DNA may be understood in terms of two general concepts: the topological constraint and the free energy of super-coiling. If the latter quantity were negligible, nicking a supercoiled molecule would have none of the effects on the hydrodynamic or chemical properties that have been described above. The free energy as a function of the degree of supercoiling has been estimated from an analysis of the data for the binding of ethidium bromide to closed and nicked circular SV40 DNA's, molecular weight 3.1×10^6 daltons [6]. This analysis was performed over a region in the binding isotherm in which the degree of supercoiling varied from -3.9 to $+11.7\%$. The analysis of the binding isotherms was carried out with the aid of two different models for the noncooperative binding reaction, and essentially the same results were obtained in both cases. The models used in the calculations were the independent site model, in which the ligands are assumed to bind without interaction to a fixed number of sites (spaces between base pairs), and the excluded site model, in which the ligands bind without interaction to all sites but upon binding convert the adjacent spaces to non-sites. These models may be used to describe the binding isotherms to open circular DNA. The independent site model yields the relationship

$$\nu = \frac{\nu_m kc}{1 + kc} \tag{3a}$$

whereas the excluded site model results in the relation

$$\nu = \tfrac{1}{4}[1 - (1 + 4kc)^{-1/2}] \tag{3b}$$

In the above equations the quantity ν is the molar ratio of bound dye to

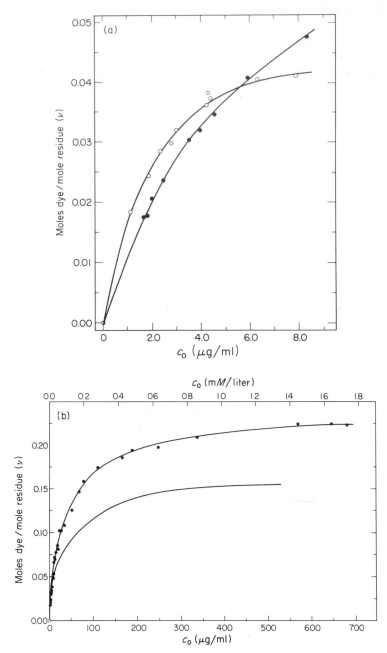

Fig. 10. The molar binding ratio of ethidium bromide to open (●) and closed (○) circular SV40 DNA's in buoyant CsCl at 25°C. (a) Low dye concentrations. (b) High dye concentrations. (From Bauer and Vinograd [2]; reproduced by permission of *J. Mol. Biol.*)

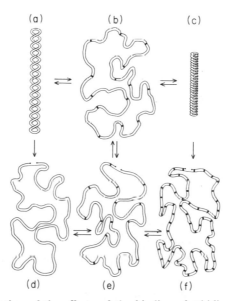

Fig. 11. Representation of the effects of the binding of ethidium bromide to SV40 DNA I and II. The upper part of the diagram presents three stages in the reversible binding of dye to SV40 DNA I. (a) The dye-free molecule with 14 superhelical turns, $\tau_0 = -14$. The addition of 420 molecules of ethidium bromide completely unwinds the superhelical turns to form the relaxed molecule, (b). The addition of a further 720 dye molecules, which occurs at a free dye concentration of 100 μg/ml, leads to the formation of a positive superhelical molecule, $\tau = 24$, shown in (c). The lower part of the diagram shows three stages in the reversible binding of dye to SV40 DNA II, which remains relaxed throughout. (e) represents the relaxed, nicked molecule with the same number of dye molecules bound as to the relaxed, intact molecule in (b). The arrows joining (b) and (e) indicate that the nick may be introduced or repaired without change of the 420 molecules of ethidium bromide bound. The nicked molecule is nearly saturated, as shown in (f), with 1860 molecules of ethidium bromide bound. Introduction of a single-strand scission into I in the presence of a high dye concentration results in an irreversible unwinding of the superhelix accompanied by an approximately 46% increase of the amount of dye bound. In this diagram each block (■) represents 30 molecules of dye. All of the dye molecules are not shown in (c). (From Bauer and Vinograd [2]; reproduced by permission of *J. Mol. Biol.*)

nucleotide; ν_m, the fixed maximum site ratio; c, the molar free dye concentration; and k, the intrinsic association constant. The excluded site model requires that the value of ν_m be fixed at 0.25, whereas ν_m is an experimentally determined parameter in the independent site model.

The binding of ethidium bromide to closed circular DNA is not adequately described by either of the above relationships, which take no account of the free energy of supercoiling. This free energy varies with the number of supercoils, a quantity which in turn depends upon the superhelix density in the absence of dye and upon the amount of dye bound.

$$\sigma = \sigma_0 + \frac{10\phi}{\pi} \nu \tag{4}$$

The constant ϕ represents the angle in radians by which the DNA duplex winding is changed upon the binding of one molecule of ethidium bromide. This value is taken to be $-\pi/15$, or $-12°$, based upon an estimate made from model-building studies [27] and substantially confirmed by the results of the alkaline buoyant titration as described below. It is generally assumed, but has not been established, that ϕ is independent of the amount of dye bound.

The models described above were shown to adequately represent the binding isotherm for closed SV40 DNA provided that the binding constant, k, is modified to take into account the changing contribution of the free energy of supercoiling to the total reaction energy as the binding ratio is varied. The result is

$$k_{\mathrm{I}}(\sigma) = k_{\mathrm{II}} \exp\left(-a\sigma - b\sigma^2\right) \tag{5}$$

where σ may be calculated from Eq. (4). The constants a and b were determined from a best least-squares fit of the binding data for both closed and nicked circular SV40 DNA's, using the combined Eqs. (4), (5), and (3) as appropriate.

Based upon the above analysis, the free energy of supercoiling at 25°C in buoyant CsCl may be written explicitly as

$$\Delta G_\sigma = 0.039 \, M \, \sigma^2(1 - 1.6\sigma) \tag{6}$$

where M is the molecular weight of the sodium salt of the DNA. This formulation reveals the untested assumption that $\Delta G_\sigma/M$ is an intensive quantity, independent of the molecular weight [28–30].

The values of ΔG_σ for SV40 DNA vary from zero for the relaxed, closed molecule to approximately 100 kcal/mole DNA for the native molecule in the absence of dye, and to 180 kcal/mole for the positively supercoiled DNA molecule at the highest dye concentration studied. While these values are very large, the free energy per base pair is, of course, smaller and, for example, amounts to approximately 25 cal/mole base pair in the native molecule. The combined Eq. (3) and (5), using the coefficients determined with native SV40 DNA, were successfully used to calculate both the amount of dye bound and the superhelix densities for a series of enzymically prepared SV40 DNA's, as described in Section VII.

The differential ligand occupancy upon adding ethidium bromide to a mixture of closed and nicked circular DNA's often leads to the appearance of new and useful differences in physical properties. Among the differentially altered properties which have been experimentally investigated are the buoyant density, sedimentation velocity, intrinsic viscosity, fluorescence enhancement, and appearance in electron micrographs.

The closed and nicked forms of circular DNA do not differ in buoyant density under ordinary solution conditions. In the presence of a small amount of ethidium bromide, however, the binding advantage of the closed DNA molecule dictates that a relatively greater number of dye molecules are bound to it [2]. Since the ethidium bromide is light relative to neutral CsDNA, the closed molecule diminishes in density more than does its nicked counterpart and the bands move apart. In this early binding region the total amount of dye bound to either component is small, however, and poor resolution between components is obtained for SV40 DNA, the only species for which detailed experimental measurements are available. This early separation is obviated by the progressive removal of superhelical turns by dye and by the eventual attainment of the closed, relaxed form. Relaxation occurs at the critical dye concentration and the closed and nicked circular components are again indistinguishable in buoyant density.

The continued addition of ethidium bromide again separates closed and nicked components on the basis of buoyant density, but the nicked molecule now possesses a progressively increasing binding advantage over the closed molecule. The results of experimental measurements of buoyant densities for SV40 DNA and varying amounts of ethidium bromide in the analytical ultracentrifuge are shown in Fig. 12a. The two components are completely physically resolved above approximately 75 μg/ml ethidium bromide, and the density separation above 100 μg/ml EthBr is nearly constant and equal to 40 mg/ml. The failure of this final buoyant density separation to further decrease at the highest dye concentrations attainable in the buoyant solvent demonstrates the energetic undesirability of introducing large numbers of supercoils into the closed molecule.

The large buoyant separation between closed and nicked components at high ethidium bromide concentrations has been used to advantage as a method for physically separating closed circular DNA from a mixture. The results of such an experiment are shown in Fig. 12b [30a]. The two bands consist of nicked (upper band) and closed (lower band) mitochondrial DNA's. The preferred conditions for the experiment include an ethidium bromide concentration of 330 μg/ml and aqueous CsCl at a density of 1.56 gm/ml. The fluorescence enhancement upon the binding of ethidium bromide to DNA permits direct visualization of moderate molecular weight DNA in amounts as low as 0.1 μg.

The sedimentation coefficient of closed circular DNA changes as ethidium bromide is added [2,8,31,32] in a manner which is reminiscent of the varia-

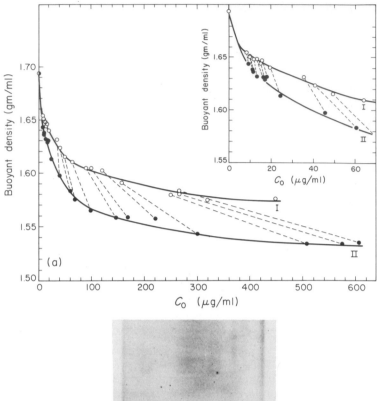

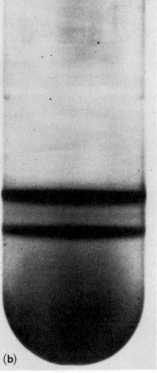

(b)

286

tions in s which are observed in a homologous series of closed DNA's of varying superhelix densities [5,8]. Figure 13 presents the variation of s with free dye concentration for an SV40 DNA of high superhelix density. The shape of the curve is consistent with the concept that the negative superhelical turns which are present in native closed DNA are removed by the intercalation of ethidium bromide. The curve shown in Fig. 13 differs from that of Fig. 3 in two respects. First, dye binding does not cease with the attainment of the closed, relaxed DNA molecule. It has not yet proved possible to prepare dye-free closed duplexes with significant positive superhelix density. In addition, the general shape of the curve in Fig. 13 appears distorted compared to that of Fig. 3. This distortion is due to the nonlinear increase in the amount of dye bound as c increases, as well as to the changes in s which result from the buoyant effects of increasing amounts of bound dye.

The actual amount of dye bound at a given free dye concentration is not the same for open and for closed DNA, and in the latter case depends upon the superhelix density of the closed DNA in the absence of dye. Binding isotherms for ethidium bromide have been determined for a variety of linear and nicked circular DNA's, by spectrophotometric, fluorometric, and dialysis equilibrium methods. The complete binding isotherm for ethidium bromide and closed circular SV40 DNA has been measured at 25°C and 5 M CsCl by the buoyant shift method. Wang [5] has measured by the fluorometric method the binding of ethidium bromide to closed $\lambda b_2 b_5 c$ DNA with a superhelix density of -0.029, up to a dye concentration adequate to raise the superhelix density to $+0.008$.

The known effect of supercoiling upon the binding constant, Eq. (5), may be used to calculate values of ν as a function of c for a closed DNA, provided the intrinsic binding constant to the nicked form, k_{II}, and the superhelix density in the absence of dye, σ_0, are known. Upholt *et al.* [8] used this procedure for a family of enzymically closed SV40 DNA's with different superhelix densities. A distortion in the ordinate axis of Fig. 13 results from the negative buoyant effect of the bound ethidium on the sedimentation of the DNA–ethidium bromide complex in dense salt solutions. This effect is evaluated from a knowledge of ν and the extent of the buoyant shift per added ethidium.

Fig. 12. (a) The buoyant density of SV40 DNA I (○) and II (●), in CsCl containing ethidium bromide. The dashed lines connect the data points obtained in a single analytical cell containing both forms at equilibrium. (From Bauer and Vinograd [2]; reproduced by permission of *J. Mol. Biol.*) (b) A CsCl–EthBr density gradient showing two fluorescent bands of closed (lower) and nicked (upper) mitochondrial DNA isolated from human leukocytes. The tube was illuminated with light at 365 nm and photographed through appropriate filters on red-sensitive film [30a].

The application of both of the above corrections to a family of SV40 DNA's is shown in Fig. 14. These curves are approximately superimposable when the corrected sedimentation coefficients are plotted as a function of superhelix density. In addition, the curves are similar to the corresponding curve obtained with dye-free molecules (Fig. 3a). This agreement indicates that the perturbations in the frictional coefficient associated with the ethidium bromide-induced changes in stiffness and the extension of the DNA molecule

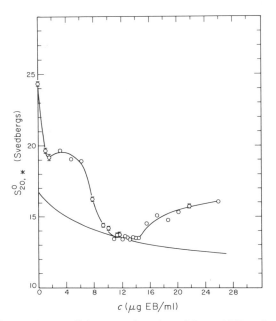

Fig. 13. Sedimentation coefficient as a function of free ethidium bromide concentration for SV40 L3 (○) and SV40 II (solid line with no data points) DNA's. SV40 L3 has a superhelix density of −0.071. (From Gray *et al.* [37]; reproduced by permission of *J. Mol. Biol.*)

are small as compared with the effects of supercoiling. The agreement also demonstrates the validity of the free energy coefficients as determined from ethidium binding data on one SV40 DNA.

Closed circular DNA is a useful tool in the examination of the mode of binding of small molecules to DNA. The presence of the minimum in a sedimentation velocity titration or of the maximum in a viscometric titration [33,34] with such materials is a good indication that binding occurs and induces unwinding of the duplex. The existence of such an extremum does not, however, provide support for any particular unwinding mechanism, including intercalation. A variety of dyes and antibiotics have been investigated by the velocity procedure [35]. The various acridines, actinomycin, and several antibiotics were found to unwind duplex DNA. Several other substances, such as spermine, streptomycin, chloromycin, mithramycin, LSD, and chloropromazine, did not induce a dip in the sedimentation velocity of the RF form of ϕX174 DNA under the conditions of the experiment.

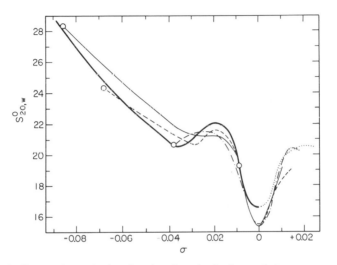

Fig. 14. Sedimentation velocity–dye titrations in the form of $s^0_{20,w}$ vs σ curves superimposed on the $s^0_{20,w}$ vs σ curve (———) of Fig. 3. The open circles indicate the starting points of the dye titrations of materials with initial superhelix densities of -0.009 (---); -0.039 (—·—); -0.068 (———); and -0.085 (···). (From Upholt *et al.* [8]; reproduced by permission of *J. Mol. Biol.*)

VII. The Measurement of Superhelix Density

The superhelix density is the characteristic new structural property of closed circular DNA, and considerable attention has therefore been devoted to the measurement of this quantity. The several techniques which have been developed for this purpose are all derived from one of two distinct chemical reactions of DNA. The first of these reactions is the deprotonation of the duplex at high pH, and the second is the binding of intercalating dyes.

The alkali-induced duplex instability was used to advantage in the alkaline buoyant titration of polyoma DNA, as described in Section V. The early region of the alkaline buoyant titration, between pH 11.4 and 11.7, corresponds to the descending region of the dip in the sedimentation velocity–pH titration and represents a change in buoyant density of 3.2% of the 57 mg/ml shift for the complete titration of open circular polyoma DNA. Assuming that the fractional buoyant shift is directly proportional to the degree of titration, it was calculated that 3.2% of the base pairs had been disrupted to form unwound loop(s), resulting in a completely relaxed molecule free of supercoils. From this it is concluded that the superhelix density of native polyoma DNA is -0.032 (3.2% supercoiled).

The binding of the intercalating dye ethidium bromide to closed circular DNA has likewise been the basis for the development of several techniques for the measurement of the superhelix density. These include the sedimentation velocity–dye titration, the viscometric dye titration, the buoyant density–dye titration, and the buoyant separation method.

The sedimentation velocity–dye titration [2,31] requires determination of the critical free dye concentration, c', at which all superhelical turns are removed, corresponding to the principal minimum in Fig. 13. Knowledge of the binding isotherm to the open form of the DNA in the same solvent is also required. Equations (3) and (4) are then combined for the calculation of σ_0 with $\sigma = 0$, $c = c'$, and $\nu = \nu_c$. This has so far been the most extensively used method. The experiments are most conveniently performed by band sedimentation in aqueous CsCl of density 1.35 gm/ml (2.85 M) at 20°C. It has been proposed that the values so obtained be used as standard superhelix densities [8].

In the direct viscometric dye titration [33,34] the flow time in a capillary viscometer is measured as a function of the total dye concentration. These titrations exhibit a maximum flow time at the critical dye concentration, c'. The slope of a plot of c' vs DNA concentration is in turn directly equal to ν_c, the critical binding ratio. This method does not require the supplementary determination of either the intrinsic binding constant or of any region of the binding isotherm.

The buoyant density–dye titration [2] takes advantage of the shifts in buoyant density which occur upon the addition of dye. The critical buoyant density decrement, $\Delta\theta_c$, has been directly related to the amount of dye bound, and its determination is sufficient for calculation of the superhelix density. This method requires measurement of neither the critical dye concentration nor of the critical binding ratio. The method, however, does require sufficient data points to adequately define the critical point in the early buoyant shift region and it has so far been used only for SV40 DNA.

The buoyant separation method [36,37] represents the most rapid and convenient technique so far developed for the determination of superhelix density. This method takes advantage of the separation between a closed molecule and its open derivative in a buoyant density gradient containing high dye concentrations. The magnitude of this separation depends linearly upon the superhelix density in the absence of dye, σ_0. Figure 15 illustrates the principle of the method for three closed DNA's differing only in superhelix density. The addition of dye to the three molecules is considered to proceed in two stages. Stage I consists of the addition of exactly enough dye to each molecule to bring about complete relaxation. The amount of dye bound at each of these three critical dye concentrations increases linearly with superhelix density according to the relationship

$$\nu_c = -0.67\sigma_0 \tag{7}$$

The number of sites remaining available for binding in each molecule is decreased to the value $(\nu_m - \nu_c)$, where ν_m is the maximum number of sites. Further dye is added in stage II to a final free dye concentration of 330 μg/ml for all three species. Dye-binding in this process is restricted relative to the nicked molecule in two ways, which contribute in different proportions to the final difference indicated by the solid area in the right-hand bar graphs of Fig. 15. The dominant restriction factor for the DNA of greatest initial supercoiling is the reduced number of sites which remain at the conclusion

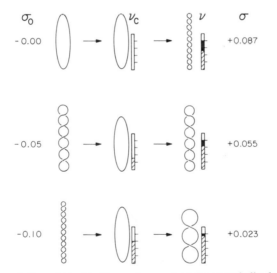

Fig. 15. The variation of the binding ratios, and of the superhelix density, σ, at three selected values of the native superhelix density, σ_0. The values of ν and ν_c are shown by the bar graphs adjacent to the corresponding molecules. The full extension of each bar graph corresponds to the theoretical maximum binding ratio, 0.25 mole dye/mole nucleotide. The cross hatched region in the bar graph corresponds to ν_c in the center column and to ν_I in the right column. The solid region in the bar graph corresponds to $(\nu_{II} - \nu_I)$ and represents the binding restriction. Each cross-over in the superhelical molecules represents a degree of supercoiling of approximately 1%. The values of ν and σ shown on the right-hand side of the diagram were evaluated at 330 μg EthBr/ml in buoyant CsCl at 25°C. (Modified from Bauer and Vinograd [36]; reproduced by permission of *J. Mol. Biol.*)

of stage I. The number of positive supercoils introduced in stage II for this DNA is small and the resultant positive superhelix free energy is expected to make a relatively small contribution to the final restriction. The restriction on the DNA of zero initial superhelix density, on the other hand, is due entirely to the high degree of positive supercoiling as dye is added.

The relationship between σ_0 and the buoyant separation was calculated [36] from the known values of the free energy of supercoiling and the measured variations of buoyant density with dye occupancy for native SV40 DNA. This relationship was further investigated experimentally [37] for a series of enzymically closed SV40 DNA's. The sedimentation velocity–dye titration was used to provide an independent measure of σ_0 in each case. In order to minimize the effects of possible variations in centrifugal field, temperature, dye concentration, and solution density in the buoyant experiments, a reference sample of known superhelix density is always included in a second tube in the same rotor or, when radioactive, in the same tube. The expected relationship between σ_0 and the distance separation, Δr, was obtained:

$$\sigma_0 = k_1 + k_2 \bar{r} \, \Delta r, \tag{8a}$$

where $\bar{r}$ is the radial distance between the bands. The parameters k_1 and k_2 depend upon θ_0, the buoyant density in the absence of dye, upon the reference DNA, and upon the binding properties of the dye. For ethidium bromide at 330 μg/ml and 20°C, the values are

$$k_1 = \sigma_0^* - 0.115 \tag{8b}$$

$$k_2 = \frac{0.115}{\bar{r}^* \, \Delta r^*} \left(\frac{\theta_0 - 1}{\theta_0^* - 1} \right)^2 \tag{8c}$$

where the asterisk refers to the reference DNA. The corresponding values for propidium diiodide (PDI) under the same conditions are

$$k_1 = \sigma_0^* - 0.095 \tag{8d}$$

$$k_2 = \frac{0.095}{\bar{r}^* \, \Delta r^*} \left(\frac{\theta_0 - 1}{\theta_0^* - 1} \right)^2 \tag{8e}$$

The above equations may be written in simpler forms which are applicable when $\theta_0 \simeq \theta_0^*$ and $\bar{r}^* \simeq \bar{r}$.

$$\text{EthBr:} \quad \Delta\sigma_0 = 0.12 \left(\frac{\Delta r}{\Delta r^*} - 1 \right) \tag{8f}$$

$$\text{PDI:} \quad \Delta\sigma_0 = 0.10 \left(\frac{\Delta r}{\Delta r^*} - 1 \right) \tag{8g}$$

where $\Delta\sigma_0 = \sigma_0 - \sigma_0^*$.

The distance separations in PDI are typically 80% larger than in EthBr. Measurements with higher molecular weight DNA's, including $\lambda b_2 b_5 c$, 25×10^6 daltons, and an F-factor with a molecular weight of 60×10^6 daltons, gave rise to a similar relationship with the same slope but with the intercept k_{I} in Eq. (8b) changed to $(\sigma_0^* - 0.102)$. Equations (8) have been experimentally confirmed with a mouse mitochondrial DNA having a superhelix density of -0.11.

It is to be emphasized that the above relationships were determined with purified DNA's which were entirely in the duplex form, both closed and open. The presence of significant extents of single-stranded regions in the upper band DNA, or of nonrestricted duplex regions in the lower band DNA, lead to significant changes in buoyant density in the presence of dye. The attachment of proteins or RNA also affects the buoyant position.

VIII. Other Factors Which Affect the Superhelix Density

The sedimentation coefficient in the range of low superhelix density is sensitive to very small changes in superhelix density and, as shown in Eq. (2), to very small changes in the duplex winding number, β. The duplex winding number is equal to the number of duplex turns per 10 base pairs and is taken to be unity for the B form of DNA. The average duplex rotation angle, ϕ, is proportional to β and in a closed DNA is given by

$$\phi = 36(A - \sigma) \text{ degrees} \qquad (9)$$

The invariance of the topological winding density, A, then leads to the result that changes in ν are accompanied by proportional changes in σ,

$$\Delta\phi = -36\Delta\sigma \text{ degrees} \qquad (10)$$

Wang [5] has investigated the effects of temperature, counterion, and ionic strength upon ϕ with closed double-stranded $\lambda b_2 b_5 c$ DNA. It was found that the superhelix density varies with temperature in the range of $5°-35°C$ with a coefficient of $1.3 \times 10^{-4}/°C$, corresponding to a change in the average rotation angle per base pair of -4.7×10^{-3} degrees/°C, or $-1.3 \times 10^{-2}\%/°C$. The substitution of Cs^+ ion ($3\ M$) for Na^+ ion ($2\ M$) as counterion made the superhelix density more negative by 2×10^{-3}. Similarly, changing the solvent from $0.1\ M$ CsCl to $3\ M$ CsCl made the superhelix density more negative by 4×10^{-3}. These effects have been confirmed semiquantitatively with SV40 DNA [8]. The addition of perchlorate salts to PM2 DNA makes the superhelix density more positive by $5.4 \times 10^{-3}/$equiv liter^{-1} for the sodium salt [23] and by $15 \times 10^{-3}/$equiv liter^{-1} for the magnesium salt [38].

IX. The Enzymic Preparation of Closed DNA of Preselected Superhelix Density

The enzymic conversion of open to closed circular DNA was first demonstrated [39] in crude extracts from *E. coli* containing tRNA as an inhibitor of endonucleases. It was found that hydrogen-bonded λ circles were converted to species sedimenting rapidly in alkali. These covalently closed circles, however, sedimented more slowly than the *in vivo* closed λ circles in neutral salt solutions. The sedimentation coefficient of the enzymically prepared material could be increased by 30% upon the addition of 3 μM actinomycin D. These results are now interpreted as indicating that the enzymically closed DNA has a low superhelix density. In more detailed experiments it has been shown that an enzyme, polynucleotide ligase, present in *E. coli*, is capable of joining an adjacent 3'-OH and 5'-phosphoryl group into a covalent linkage in duplex DNA [40,41]. This enzyme requires Mg^{2+} and DPN as a cofactor [42,43]. A similar enzyme formed upon infection of *E. coli* with a variety of bacteriophages requires ATP [44–46]. A comparable enzymic activity in animal cells requires ATP [47].

The superhelix density of the enzymically prepared closed DNA is substantially zero under the reaction conditions. The reaction can also be conducted under conditions in which the duplex is underwound at the time of closure as, for example, by the addition of ethidium bromide. The principle of the procedure is illustrated in Fig. 16 for closure in the absence and in the presence of ethidium bromide. The addition of ethidium to a nicked DNA decreases the duplex winding number, β. Upon closure the resultant DNA has a reduced value of the topological winding number, α. Subsequent removal of ethidium restores β to its normal value and then gives rise to supercoiling to an extent determined by the topological constraint Eq. (1).

Other unwinding reagents may also be used in this procedure. Wang [48] has closed λ DNA in the presence of known amounts of actinomycin D under conditions in which the drug was almost quantitatively bound to the DNA. Subsequent determination of the superhelix density of the actinomycin-free DNA by the sedimentation velocity–ethidium bromide titration led to the result that the unwinding angle for actinomycin binding was indistinguishable from that of ethidium.

X. Heterogeneity in Superhelix Density

The superhelix density is a macromolecular parameter and is, therefore, subject to variation within a population of molecules. Such variation can occur as a distribution around a mean or, alternatively, can result from the admixture of homogeneous or heterogeneous populations with different means.

The buoyant separation method provides a sensitive assay for the presence of such heterogeneity, especially in double-label experiments. It was found, for example, that intracellular SV40 DNA has a lower degree of supercoiling than does the SV40 DNA isolated from the virion [49]. A similar result has been found for PM2 bacteriophage DNA [50]. The intracellular and virion DNA's from these systems are therefore distinguishable on the basis of mean superhelix density.

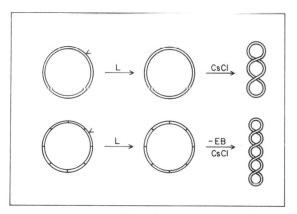

Fig. 16. The enzymic preparation of closed circular DNA with a preselected number of supercoils. The closure of a nicked circular molecule in the absence of dye results in a nonsupercoiled closed molecule in the reaction medium. For a hypothetical molecule with a molecular weight of 4.3×10^6 daltons, transfer to 3 M CsCl results in the formation of -2 superhelical turns. Alternatively, incubation of the nicked DNA in a reaction medium in which 120 molecules of EthBr are bound results in the formation of two additional superhelical turns upon transfer to CsCl. Each bar in the lower circular form represents 15 molecules of EthBr.

The intracellular SV40 DNA was found to be itself heterogeneous by means of experiments in which two differently labeled DNA samples are prepared and fractionated in a dye–CsCl density gradient. Subsequent rebanding of mixtures of the opposing halves of the bands gives rise to overlapping, but clearly displaced, bands. No displacements were observed with the virion DNA. The lower band formed from the mitochondrial DNA extracted from mouse cells grown in tissue culture contains three identified species with superhelix densities which range from -0.01 to -0.03. One of these proved to be a replicative intermediate, D-loop DNA. Such experiments are best performed with propidium diiodide, a reagent which approximately doubles the resolving power of the dye–CsCl buoyant system.

Heterogeneity in superhelix density should in principle be observable in the broadening of sedimenting bands of closed DNA in dye titrations in the region in which the sedimentation coefficient depends sensitively upon the degree of supercoiling.

XI. Complex Forms of Closed Circular DNA

In the first application of the dye–buoyant density procedure [51], with extracts of the DNA from HeLa cells, an electron microscope examination of the material from the lower (closed) band revealed the presence of circular DNA molecules of double the length of the mitochondrial DNA. These molecules proved to consist of two interlocked closed submolecules. When one submolecule is nicked while the other remains intact, a species is formed which is intermediate in buoyant density in an ethidium bromide–CsCl gradient (Fig. 17). Such molecules are called *catenanes* and have been found in preparations of nonencapsulated DNA's isolated from a variety of sources [52–54]. The level of catenanes ranges from 1 to 2% by mass in intracellular SV40 DNA to 40% in HeLa mitochondrial DNA, and is as large as 50% in the mitochondrial DNA of SV40-transformed mouse cells. Higher catenated oligomers have also been reported and, when present in adequate concentrations, form identifiable discrete intermediate bands in ethidium bromide–CsCl gradients.

Catenated dimers and trimers have also been prepared synthetically by cyclizing 5-Bu-λ DNA via hydrogen-bonding in solutions containing previously closed circles of 186 DNA molecules [55]. In a buoyant CsCl gradient these catenanes formed bands at an intermediate density. The synthetic procedure has also been used to prepare closed and singly open catenanes.

Sedimentation properties of catenated dimers, when both submolecules are closed (doubly closed), appear to be similar to those of the corresponding simple double-length circular molecules [52,56]. When both submolecules are nicked (doubly open), the sedimentation coefficient is similar to that of the nicked double-length circle in the case of λ DNA [56], but appears to be somewhat reduced in the case of human mitochondrial DNA [57]. The singly open catenated dimers have sedimentation coefficients intermediate between the doubly closed and doubly open species.

XII. Closed Circular Replicative Intermediates

Other types of closed circular duplexes which appear to be in the act of replication have been isolated from various fractions of the ethidium bromide–CsCl density gradient. A very large fraction, up to 50%, of the closed mito-

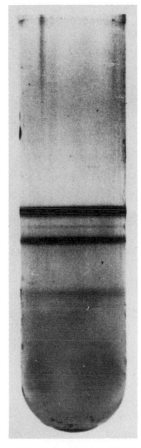

Fig. 17. Three species of HeLa mitochondrial DNA are resolved in an ethidium bromide–CsCl density gradient. The upper dark fluorescent band contains nicked (open) DNA, the narrow middle band contains singly open catenanes, and the bottom band, closed DNA. The pale band in the lower part of the tube contains an unidentified material.

chondrial DNA molecules isolated from mouse cells in tissue culture consists of a species termed *displacement DNA* or *D-loop DNA* [29]. These molecules contain a three-stranded region 400–500 nucleotides long in which a new single strand appears inserted into the duplex with displacement of a corresponding segment of the still closed DNA (Fig. 18). These molecules band at a position in the dye–buoyant density gradient corresponding to a superhelix density of -0.012. The band position is expected to be affected, however, by the unrestricted binding of ethidium bromide to both the duplex and single-stranded regions in the D-loop, which together account for 4.6% of the mass of the DNA. When the displacing strand is removed, the superhelix density becomes

−0.028. The D-DNA cosediments with mitochondrial DNA having a super-helix density of −0.027 in 2.85 M CsCl at 20°C, but sediments significantly slower in 0.5 M NaCl at 20°C.

Replicating forms of SV40 DNA from primary monkey kidney cell cultures appear to have a structure similar to that of displacement DNA but in which the displacement region is largely duplex and variable in size up to nearly the full length of the genome [58,59]. These molecules band heterogeneously in the dye–CsCl density gradient, with the more extensively replicated mole-

Fig. 18. Diagrammatic representation of closed circular D-loop mitochondrial DNA containing a displacement loop. The displacing strand is represented by a heavy line and the displaced strand by a curved line with attached bars.

cules banding in the region of open DNA. The bands were observed to rise in the gradient as the extent of replication increased. A similar situation appears to apply in larger mitochondrial DNA replicative intermediates [60].

XIII. Circular DNA as a Substrate in Nucleic Acid Enzymology

The size homogeneity of circular DNA and the absence of ends renders these materials important substrates for the study of enzymes which act on nucleic acids. The first endonuclease hit is readily followed as a result of the discrete change in the sedimentation velocity and banding of EthBr. Further random nucleolytic events have very little effect on these properties. The enzymic activity may be monitored by observing the gain in fluorescence which accompanies the increased dye binding associated with the loss of the topological restriction [61,62]. Correspondingly, the activity of the polynucleotide ligase can be followed by determining the relative reduction in fluorescence which is associated with the establishment of the topological restriction.

Closed DNA's provide test substrates for investigation of the requirement for a chain end in DNA or RNA synthesis. It has been shown by this means that *E. coli* polymerase I requires an end, whereas the *E. coli* RNA polymerase can transcribe RF ϕX174 in the absence of a strand scission.

With single-stranded circular ϕX DNA as a template and an oligonucleotide as a primer, Goulian *et al.* [63] obtained the enzymic synthesis of a closed duplex circle with polymerase I and an excess of polynucleotide ligase. The closed duplex was used as an intermediate in the first demonstration of the *in vitro* synthesis of an infective DNA.

A protein has been extracted and highly purified from crude extracts of *E. coli*, which interacts with negatively supercoiled DNA and appears to reduce the degree of supercoiling. This molecule, referred to by Wang [64] as the ω protein, appears to act in the absence of detectable amounts of the known polynucleotide ligase and endonuclease activities and does not require a cofactor. Nicked circular DNA could not be isolated as an intermediate in the relaxation process, nor was the nicked circle closed in the presence of the protein. Positive superhelical DNA, generated by the addition of ethidium bromide, did not serve as a substrate. The mechanism for the change in α is at the present time obscure. It is to be noted that the detection of this reaction as presently understood is possible only with the aid of the chemistry of closed circular DNA. The action of the ω protein has not yet been shown to be catalytic. A similar activity has been purified from mouse cells [65]. In this case closed circular DNA containing positive superhelical turns could also be relaxed.

Helinski and collaborators [66,67] have found that the covalently closed colicinogenic and F-factors, when isolated by gentle lysis with nonionic detergents, form a *relaxation complex* with a tightly bound protein. This protein introduces a nick into a unique strand of the DNA molecule when the protein–DNA complex is exposed to the action of a variety of reagents, such as strong ionic detergents or alkali or heat. This work is fully discussed in the recent review by Helinski and Clewell [1].

XIV. The Occurrence of Closed Circular DNA

At the present time more than 50 different closed circular DNA's have been isolated from both prokaryotic and eukaryotic sources, as well as a viral and intracellular forms of viruses. These DNA's include the mitochondrial DNA's, the chloroplast DNA in euglena, the kinetoplast DNA in protozoa, the plasmid and episomal DNA in prokaryotes, the DNA in the papova group of animal viruses, and one bacteriophage DNA and in several insect viruses [68]. More detailed, recent summaries of the presently known distribution of closed circular DNA are available [1,28].

Glossary of Terms Used in the Description of Closed Circular DNA

The description of the properties of closed circular DNA has made necessary the introduction of a variety of new terms, some of which are redundant. These terms are listed here with the authors' preferences indicated. The glossary does not include some new terms necessary to describe replicative closed circular DNA.

Preferred terms	Definition	Alternative terms
Closed double-stranded DNA	A double-stranded DNA molecule containing two circularly continuous single strands	Covalently closed (circular) DNA
Closed duplex DNA	A closed double-stranded DNA having the native secondary structure	DNA I; form I.
Supercoiled DNA	A closed duplex DNA containing tertiary turns	Superhelical DNA; twisted DNA; compact DNA; fast neutral form
Closed denatured DNA at neutral pH	A closed double-stranded DNA in which the nucleotide pairs are substantially out of register	Form IV; den I
Closed denatured DNA at alkaline pH	A fully alkali-titrated closed double-stranded DNA	Double-stranded cyclic coil; fast alkaline form
Relaxed closed DNA	A closed duplex DNA containing no tertiary turns	Form I'
Tertiary turns	The net winding in space of the duplex helix axis (refer to *superhelix density* for sign convention)	Superhelical turns; supercoils
Duplex turns	The winding of the single strands about the duplex axis	
Superhelix winding number	The quantity τ, given by $\tau = \alpha - \beta$, is the number of tertiary turns (see also, *superhelix density*)	Tertiary winding number; writhing number
Duplex winding number	The quantity β, equal to the number of duplex turns	Directional writhing number
Topological winding number	The number of turns of one strand about the other when the duplex axis is constrained to a plane; designated by the symbol α	Linking number
Superhelix density	The number of superhelical turns per ten base pairs, σ; this quantity has a negative sign in closed DNA molecules in which superhelical turns are removed by the removal of duplex turns	Superhelical density

Preferred terms	Definition	Alternative terms
Standard superhelix density	The superhelix density in 2.85 M CsCl at 20°C, density 1.35 gm/ml	
Degree of supercoiling	The absolute value of the superhelix density multiplied by 100 and expressed as percent	
Superhelix axis	The axis about which tertiary winding may be considered to occur	
Open circular DNA	A circular duplex DNA containing at least one swivel point and free of the topological restraint	Form II; DNA II; open form; nicked or gapped circular DNA
Complex closed DNA	Catenated and concatenated circular oligomers	
Catenane	Two or more interlocked circular submolecules	Interlocked DNA; catenated dimer; catenated oligomer
Submolecule	One of the constituent circular duplexes in a catenane	
Singly open dimeric catenane	Catenated dimer containing one open submolecule	Singly nicked catenane
Doubly open dimeric catenane	Catenated dimer containing two open submolecules	Doubly nicked catenane
Closed catenane	Catenane with closed submolecules	
Circular concatamer	Two of more more DNA's connected end-for-end into a circle	Circular oligomer
Concatenated dimer	Circular concatamer containing two DNA's	Circular dimer
Early melting	That portion of the helix–coil transition accompanied by the complete loss of negative superhelical turns	Early helix–coil transition
Late melting	That portion of the helix–coil transition which begins with the melting of relaxed closed DNA	Late helix–coil transition
Rapid renaturation	The rapid reannealing process that occurs when a partially melted closed or linear duplex is allowed to renature	Snap-back renaturation
Nick	The introduction of a chain scission in the backbone chain leading to the formation of a swivel	Single-strand scission
Swivel	A site for free rotation in the complementary strand opposite a nick or a gap	

References

1. D. R. Helinski and D. B. Clewell, *Annu. Rev. Biochem.* **40**, 899 (1971).
2. W. Bauer and J. Vinograd, *J. Mol. Biol.* **33**, 141 (1968).
3. J. Vinograd and J. Lebowitz, *J. Gen. Physiol.* **49**, 103 (1966).
4. F. B. Fuller, *Proc. Nat. Acad. Sci. U.S.* **68**, 815 (1971).
5. J. C. Wang, *J. Mol. Biol.* **43**, 25 (1969).
6. W. Bauer and J. Vinograd, *J. Mol. Biol.* **47**, 419 (1969).
7. D. Glaubiger and J. E. Hearst, *Biopolymers* **8**, 692 (1967).
8. W. B. Upholt, H. B. Gray, Jr., and J. Vinograd, *J. Mol. Biol.* **62**, 21 (1971).
8a. M. F. Maestre and J. C. Wang, *Biopolymers* **10**, 1021 (1971).
8b. W. W. Dean and J. Lebowitz, *Nature (London), New Biol.* **231**, 5 (1971).
8c. J. Paoletti and J.-B. LePecq, *J. Mol. Biol.* **59**, 43 (1971).
9. H. Jacobson, *Macromolecules* **2**, 650 (1969).
10. H. Jacobson, *Macromolecules* **3**, 702 (1970).
11. E. P. Geiduschek and T. T. Herskovits, *Arch. Biochem. Biophys.* **95**, 114 (1961).
12. T. T. Herskovits, *Arch. Biochem. Biophys.* **97**, 474 (1962).
13. W. F. Dove and N. Davidson, *J. Mol. Biol.* **5**, 467 (1962).
14. C. Schildkraut and S. Lifson, *Biopolymers* **3**, 195 (1965).
15. Yu. S. Lazurkin, M. D. Frank-Kamenetskii, and E. N. Trifonov, *Biopolymers* **9**, 1253 (1970).
16. H. R. Mahler, B. D. Mehrotra, and C. W. Sharp, *Biochem. Biophys. Res. Commun.* **4**, 79 (1961).
17. G. Felsenfeld, G. Sandeen, and P. H. von Hippel, *Proc. Nat. Acad. Sci. U.S.* **50**, 644 (1963).
18. K. Hamaguchi and E. P. Geiduschek, *J. Amer. Chem. Soc.* **84**, 1329 (1962).
19. R. Haselkorn and P. Doty, *J. Biol. Chem.* **236**, 2738 (1961).
20. P. Doty, *J. Cell. Comp. Physiol.* **49**, Suppl. 1, 27 (1957).
21. J. Vinograd, J. Lebowitz, R. Radloff, R. Watson, and P. Laipis, *Proc. Nat. Acad. Sci. U.S.* **53**, 1104 (1965).
22. J. Vinograd, J. Lebowitz, and R. Watson, *J. Mol. Biol.* **33**, 173 (1968).
23. W. Bauer, *J. Mol. Biol.* **67**, 183 (1972).
24. J. G. Wetmur and N. Davidson, *J. Mol. Biol.* **31**, 349 (1968).
25. W. Strider and R. C. Warner, *Fed. Proc., Fed. Amer. Soc. Exp. Biol.* **30**, 1053 (1971).
26. P. H. Pouwels, C. M. Knijnenburg, J. van Rotterdam, J. A. Cohen, and H. S. Jansz, *J. Mol. Biol.* **32**, 169 (1968).
27. W. Fuller and M. J. Waring, *Ber. Bunsenges. Phys. Chem.* **68**, 805 (1964).
28. W. Bauer and J. Vinograd, *Prog Mol. Subcell. Biol.* **2**, 181 (1971).
29. H. Kasamatsu, D. Robberson, and J. Vinograd, *Proc. Nat. Acad. Sci. U.S.* **68**, 2252 (1971).
30. N. Davidson, *J. Mol. Biol.* **66**, 307 (1972).
30a. R. Watson, W. Bauer, and J. Vinograd, *Anal. Biochem.* **44**, 200 (1971).
31. L. V. Crawford and M. J. Waring, *J. Mol. Biol.* **25**, 23 (1967).
32. J. C. Wang, *J. Mol. Biol.* **43**, 263 (1969).
33. B. M. J. Révet, M. Schmir, and J. Vinograd, *Nature (London)* **229**, 10 (1971).
34. E. M. Smit and P. Borst, *FEBS Lett.* **14**, 125 (1971).
35. M. J. Waring, *J. Mol. Biol.* **54**, 247 (1970).
36. W. Bauer and J. Vinograd, *J. Mol. Biol.* **54**, 281 (1970).
37. H. B. Gray, W. B. Upholt, and J. Vinograd, *J. Mol. Biol.* **62**, 1 (1971).
38. W. Bauer, *Biochemistry* **11**, 2915 (1972).

39. M. Gellert, *Proc. Nat. Acad. Sci. U.S.* **57**, 148 (1967).

40. B. M. Olivera and I. R. Lehman, *Proc. Nat. Acad. Sci. U.S.* **57**, 1426 (1967).

41. M. L. Gefter, A. Becker, and J. Hurwitz, *Proc. Nat. Acad. Sci. U.S.* **58**, 240 (1967).

42. B. M. Olivera and I. R. Lehman, *Proc. Nat. Acad. Sci. U.S.* **57**, 1700 (1967).

43. S. B. Zimmerman, J. W. Little, C. K. Oshinsky, and M. Gellert, *Proc. Nat. Acad. Sci. U.S.* **57**, 1841 (1967).

44. B. Weiss and C. C. Richardson, *Proc. Nat. Acad. Sci. U.S.* **57**, 1021 (1967).

45. A. Becker, G. Lyn, M. Gefter, and J. Hurwitz, *Proc. Nat. Acad. Sci. U.S.* **58**, 1996 (1967).

46. T. Ando and T. Kosawa, *Biochim. Biophys. Acta* **204**, 257 (1970).

47. T. Lindahl and G. M. Edelman, *Proc. Nat. Acad. Sci. U.S.* **61**, 680 (1968).

48. J. C. Wang, *Biochim. Biophys. Acta* **232**, 246 (1971).

49. R. Eason and J. Vinograd, *J. Virol.* **7**, 1 (1971).

50. R. T. Espejo, E. Espejo-Canelo, and R. L. Sinsheimer, *J. Mol. Biol.* **56**, 623 (1971).

51. R. Radloff, W. Bauer, and J. Vinograd, *Proc. Nat. Acad. Sci. U.S.* **57**, 1514 (1967).

52. B. Hudson and J. Vinograd, *Nature (London)* **216**, 647 (1967).

53. D. A. Clayton, C. A. Smith, J. M. Jordan, M. Teplitz, and J. Vinograd, *Nature (London)* **220**, 976 (1968).

54. L. Pikó, D. G. Blair, A. Tyler, and J. Vinograd, *Proc. Nat. Acad. Sci. U.S.* **59**, 838 (1968).

55. J. C. Wang and H. Schwartz, *Biopolymers* **5**, 953 (1967).

56. J. C. Wang, *Biopolymers* **9**, 489 (1970).

57. I. H. Brown and J. Vinograd, *Biopolymers* **10**, 2015 (1971).

58. R. Jaenisch, A. Mayer, and A. Levine, *Nature (London), New Biol.* **233**, 72 (1971).

59. E. D. Sebring, T. J. Kelly, Jr., M. M. Thoren, and N. P. Salzman, *J. Virol.* **8**, 478 (1971).

60. D. L. Robberson and D. A. Clayton, *Proc. Nat. Acad. Sci. U.S.* **69**, 3810 (1972).

61. J.-B. LePecq, *Methods Biochem. Anal.* **20**, 41 (1971).

62. C. Paoletti, J.-B. LePecq, and I. R. Lehman, *J. Mol. Biol.* **55**, 75 (1971).

63. M. Goulian, A. Kornberg, and R. L. Sinsheimer, *Proc. Nat. Acad. Sci. U.S.* **58**, 2321 (1961).

64. J. C. Wang, *J. Mol. Biol.* **55**, 523 (1971).

65. J. J. Champoux and R. Dulbecco, *Proc. Nat. Acad. Sci. U.S.* **69**, 143 (1972).

66. D. B. Clewell, D. G. Blair, D. J. Sherratt, and D. R. Helinski, *Fed. Proc., Fed. Amer. Soc. Exp. Biol.* **29**, 725 (1970).

67. D. B. Clewell and D. R. Helinski, *Biochemistry* **9**, 4428 (1970).

68. M. D. Summers and D. L. Anderson, *J. Virol.* **9**, 710 (1972).

69. D. Denhart, *J. Mol. Biol.* (in press).

70. M. Schmir, B. M. J. Révet, and J. Vinograd, *J. Mol. Biol.* (in press).

5

DINUCLEOSIDE MONOPHOSPHATES, DINUCLEOTIDES, AND OLIGONUCLEOTIDES

PAUL O. P. TS'O

In Chapter 6, Volume I the intrinsic properties and the interactions of the monomeric units of nucleic acid were described. In nucleic acids, the monomeric nucleotidyl units are linked together through the 3'-5' phosphodiester bonds. Having such linkages, dimers, oligomers, and polymers possess some properties not found in the monomers. In this chapter, we shall examine the intrinsic properties and the interaction of the dimers and oligomers, with special emphasis on the influence of the phosphodiester linkages on the conformation and interaction of these short segments of nucleic acids.

I. Intrinsic Properties

A. Stereochemical Considerations

The 3'-5' phosphodiester linkage of the nucleic acid backbone consists of six chemical bonds as shown in Fig. 1 [1]: $P_i \rightarrow$ O-5'(ω); O-5' $\rightarrow$ C-5'(ϕ); C-5' $\rightarrow$ C-4'(ψ); C-4' $\rightarrow$ C-3'(ψ'); C-3' $\rightarrow$ O-3'(ϕ'); and O-3' $\rightarrow P_{i+1}$(ω'). The distances and the torsion angles of these bonds completely define the conformation of the nucleic acids backbones. Among these six bonds, five of them are, in principle, freely rotatable, while the C-4' to C-3' bond has a very limited degree of rotational freedom and is related to the furanose conformation.

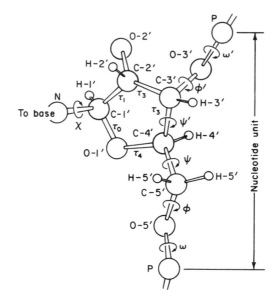

Fig. 1. The chemical bonds along the phosphodiester linkage of the nucleic acid backbone. (From Sundaralingam [1].) See Table I for values of various torsion angles.

1. *X-ray Diffraction Studies on Solid State*

Only by the x-ray diffraction study on crystals or oriented fibers can the parameters (distances and angles) of these bonds be defined in great detail. Currently, certain informative reports on the x-ray study of the crystals of dinucleoside monophosphates have appeared and these will be briefly discussed later in this section. Most of the information, therefore, has to rely on the crystallographical data of the monomeric nucleosides and nucleotides, as well as the computer-analyzed data of the helical nucleic acids in oriented fibers. Two recent reviews [1,2] are most valuable in this area and should be consulted for more details.

In Table I, the pertinent data on these six bonds are summarized [1–5]. The bond distances cited were considered probably good to ± 0.02 Å [4]. The torsion angle is defined by the clockwise rotation of the far bond relative to the near bond when looking along any bond, and the angle is always positive (0° to 360°) as shown in Fig. 2. The backbone is fully folded at 0° (or 360°) and fully extended at 180°. A more detailed definition for each angle is given in Table I. The mid-range values in the major zone of these torsion angles compiled from over 15 compounds are listed together with the conclusion by Sundaralingam [1] on the preferred conformations for the backbone of right-handed double-helical nucleic acids. In addition, the torsion angles for the backbone of helical DNA and ribosyl polynucleotide complexes obtained by Arnott and co-workers [2–5] through computer refinement of the fiber diffraction patterns are also listed in Table I. DNA in low humidity and low salt exists in the A form, an 11-fold double helix with 28.15 Å pitch. At high humidity, DNA exists in B form which is a 10-fold double helix with a 33.7 Å pitch. In B form, the helix axis passes centrally through the base pairs that are roughly perpendicular to it with the rise per residue along the

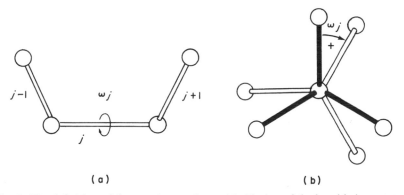

(a)   (b)

Fig. 2. The definition of the rotation angle ω_j; (a) sideview of the bond being rotated; (b) front view of the bond being rotated. (From Sundaralingam [1].)

TABLE I

The Rotatable Bonds along the Phosphodiester Linkage of the Nucleic Acid Backbone

Bonds [a]	$P_i \to O_{5'}$	$O_{5'} \to C_{5'}$	$C_{5'} \to C_{4'}$	$C_{4'} \to C_{3'}$	$C_{3'} \to O_{3'}$	$O_{3'} \to P_{i+1}$
Nomenclature [b] (Sundaralingam)	ω	ϕ	ψ	ψ'	ϕ'	ω'
Bond distances [c] (Å)	1.60	1.44	1.51	1.52	1.43	1.60
0° Position [d]	$O_{5'}$—$C_{5'}$ cis planar to P_i—$O(3')_{i-1}$	$C_{5'}$—$C_{4'}$ cis planar to P_i—$O_{5'}$	$C_{4'}$—$C_{3'}$ cis planar to $O_{5'}$—$C_{5'}$	$C_{3'}$—$O_{3'}$ cis planar to $C_{5'}$—$C_{4'}$	$O_{3'}$—P_{i+1} cis planar to $C_{3'}$—$C_{4'}$	P_{i+1}—$O(5')_{i+1}$ cis planar to $C_{3'}$—$O_{3'}$
Mid-range of torsion angles [d] (major zone)	$290° \pm 10°$	$170° \pm 30°$	$55° \pm 15°$	$120° \pm 40°$	$220° \pm 50°$	$200° \pm 100°$
Preferred conformations for right-handed double helix [d] ($\pm 20°$)	285°	170°	60°, 175°	80°, 150°	210°	280°, 210°
DNA-A form torsion angle [e]	291°	191.4°	49.5°	79.2°	199.8°	297.1°
DNA-B form torsion angle [e]	319°	194°	33.4°	145°	170°	258.1°
$(rI)_n \cdot (rC)_n$ helix torsion angle [e]	291.6°	194.7°	50.0°	79.2°	193.2°	301.7°
$r(A)_n \cdot r(U)_n$ torsion angle [e]	288.7°	186.2°	58.5°	79.2°	203.7°	292.8°
GpC torsion angle [f]	284°	186°	51°	—	209°	291°
ApU (1 and 2) torsion angle [f]	289°, 296°	177°, 169°	57°, 57°		213°, 220°	294°, 285°
RNA 11 torsion angle [f]	294°	186°	49°	—	202°	294°
RNA 10 torsion angle [f]	261°	180°	84°	—	212°	287°

[a] The graphic presentation of these bonds is shown in Fig. 1.

[b] From Sundaralingam [1]. The nomenclature for these angles adopted by Arnott [2] is ψ, θ, ξ, σ, ω, and ϕ, respectively; by Lakshminarayanan and Sasisekharan [3] is ψ, θ_1, θ_2, (not defined for C-4' to C-3' bond), θ_3, and ϕ, respectively; by Olson and Flory [3a] is ψ'', ϕ', ϕ'', ω', ω'', and ψ' respectively; and by Sussman et al. [9] is ω, ϕ, ψ, ρ, ϕ', and ω', respectively.

[c] From Arnott et al. [4].

[d] From Sundaralingam [1]. The definition of the torsional angle is, "When looking along any bond, the far bond rotates clockwise relative to the near bond" [1]. The angle is always positive (0°–360°), and has a right-handed rotation (see Fig. 2). The chain is fully folded at 0° and fully extended at 180°. At 180°, the first bond described is trans planar to the second bond described. For example, $\omega = 180°$, when O-5' to C-5' is trans planar to P_i—$O(3')_{i-1}$.

[e] From Arnott [2]. The angles for helical reovirus RNA have almost the same values as helical $(rA)_n \cdot (rU)_n$. The angles were obtained after computerized analyses of the x-ray diffraction data on fiber samples.

[f] From Day et al. [10a].

helix of 3.37 Å; while in the A form, the bases are nearly 5 Å further away from the axis and make an angle of only 70° to it with the rise per residue of 2.56 Å. As for the ribosyl polynucleotide helices, $(rI)_n \cdot (rC)_n$ exists in an A′ form [2] which is a 12-fold double helix with a 36.2 Å pitch (or a 3.01 Å rise per residue), and with bases 4.7 Å away from the helix axis and making an angle of 81° to it. The $(rA)_n \cdot (rU)_n$ complex exists in a B–A form [2] which is an 11-fold double helix with a 30.0 Å pitch (or a 2.81 Å rise per residue), and with bases 4.3 Å away from helix axis and making an angle of 76° to it. There occurs an A → A′ transformation in RNA helix just like the well-known A → B transformation in DNA helix. The torsion angles of these helical nucleic acids reported by Arnott and co-workers [2–5] are in satisfactory accord with those recommended by Sundaralingam [1] as the preferred angles for right-handed double helix (Table I).

The torsion angles of the nucleic acid helical complexes are also in good agreement with the mid-range values of the torsion angles compiled from all these compounds including nucleosides and nucleotides. This agreement suggests the existence of physicochemical factors that determine the rotation of these bonds, regardless of whether they are in monomers or polymers. From a simple stereochemical consideration of the bonds of the backbone of a dinucleoside monophosphate unit as shown in Fig. 1 the rotation of the two P—O bonds (ω and ω') must have the least amount of restriction as compared to the other three bonds (ϕ, ϕ', and ψ, while the rotation of ψ' in the furanose ring is even more restricted). These two angles, ω and ω' ($\sim 300°$), in fact, assume a semifolded (fully folded is 360°), *gauche* conformation. Sundaralingam [1] proposed that this situation reflects a significant conformational rule: "In molecular systems such as

$$
\begin{array}{ccc}
& \text{O} & \\
& \| & \\
\text{—C—O—P—O—C—} & \quad \text{and} \quad & \text{—C—O—} \overset{\displaystyle \text{H}}{\underset{\displaystyle \text{H}}{\text{C}}} \text{—O—C—} \\
& | & \\
& \text{O} &
\end{array}
$$

where heteroatoms with lone-pair electrons are attached to a central atom, the *gauche* conformations about one or both of the O—P or O—C bonds represent the preferred conformations." This was explained on the basis of the electrostatic repulsion of the lone pair of electrons of the oxygen atom as shown in Fig. 3a. Similarly, Sasisekharan and Lakshminarayanan [6] have calculated the potential energy of dimethylphosphate and concluded that the most stable conformations are the *gauche–gauche* conformations about these bonds. They indicated (60°, 60°) or (300°, 300°) would be most favorable for the phosphodiester unit. As for the C—O bonds (ϕ or ϕ'), they are in an extended position ($\sim 200°$; 180° is the fully extended form) presumably to reduce steric hindrance. For the O-5′ to C-5′ bond, the angle ϕ is such that the P_i to O-5' bond is *trans* to the C-5' to C-4' bond and is *gauche* to both the C-5' to H-5' and C-5' to H-5'' bonds (bisecting them). For the C-3' to O-3' bond, the angle ϕ' is such that the C-3' to H-3' is *cis* to O-3' to P_{i+1} bond, and allows both C-4' and C-2' substituted on C-3' to be away from the P atom. As for the C-5' to C-4' bond, the angle ψ is such that the C-5' to O-5' bond is *gauche* to both C-4' to C-3' and C-4' to O-1' and the C-4' to H-4' is *gauche* to both C-5' to H-5' and C-5' to H-5'' bonds (bisecting them). It should be noted that while these angles may be the preferred ones as observed by the x-ray diffraction studies, according to the potential energy calculation by Lakshminarayanan and Sasisekharan [3], the differences in the energy for various rotational minima for ϕ, ψ, and ϕ' are small, less than 1 kcal/mole.

In Fig. 3b are shown two ω–ω' maps for 3'- and 5'-nucleotides, dinucleoside monophosphates, and polynucleotides, from the recent work of Sundaralingam [6a]. As illustrated, most of the ω–ω' angles of the polynucleotides are in the *gauche–gauche* (*gg*) area, especially those with the right-handed turn (g^-g^-). In the cases of the monomer and the dimer, again a majority are in the *gauche–gauche* configuration, even though a considerable number have the left-handed turn (g^+g^+).

The first x-ray crystallographic study [7,8] on dinucleoside monophosphate was made on adenosine-2',5' uridine phosphoric acid ($A_{2'}p_{5'}U$). In this molecule, the adenine residue is protonated at N-1 by proton transfer from the

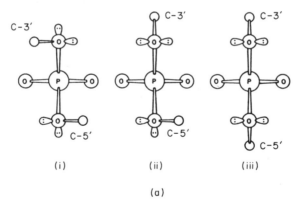

(a)

Fig. 3 (a) The phosphodiester conformation. (From Sundaralingam [1].) The electrostatic repulsion of the lone-pair electrons of the two oxygen atoms causes the *gauche–gauche* conformation (i), or *gauche–anti* conformation (ii), to be preferred conformation. The *anti–anti* conformation (iii) is not preferred due to this repulsion effect; even though this extended conformation has less steric hindrance.   (b) $\omega(P_i \rightarrow O_{5'}) - \omega'(O_{3'} \rightarrow P_i)$ maps for 3'- and 5'-nucleotides, dinucleoside monophosphates, and polynucleotides. The symbol g^- denotes *gauche* configuration from 270° to 360°; g^+ denotes *gauche* configuration from 0° to 90°; t denotes *trans* configuration; asterisk denotes a structure which has been accurately determined [6a].

phosphate group and the uracil ring is in keto form. The ribose conformation of uridine is $C_{3'}$-*endo* and that of adenosine is $C_{2'}$-*endo*. The torsional angle of the glycosyl C—N bond, χ_{CN} ($\approx -\phi_{CN}$) is 55° for adenosine and −5° for uridine, both in *anti* range. The planes of the base residues are nearly parallel

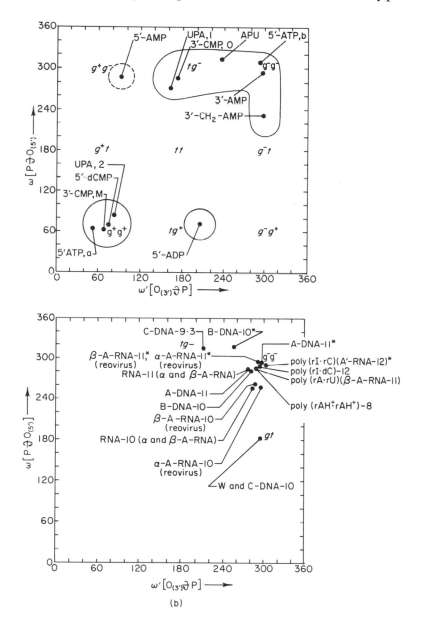

(b)

(deviating by about 15°); their closest approach is about 3.4 Å, although they actually overlap only to a small extent.

Most recently, two brief reports [9,10] have appeared independently on the studies of the crystal structure of the naturally occurring dinucleoside monophosphate-uridine-3',5'-adenosine phosphate hemihydrate ($U_{3'}p_{5'}A$). Both laboratories obtained the same monoclinic ($P2_1$) crystals, one crystallized from aqueous methanol at pH 3.5 [10] and the other from $10^{-3}\ M$ HCl aqueous solution [9]. There are 4 UpA and 2 water molecules in a unit cell,

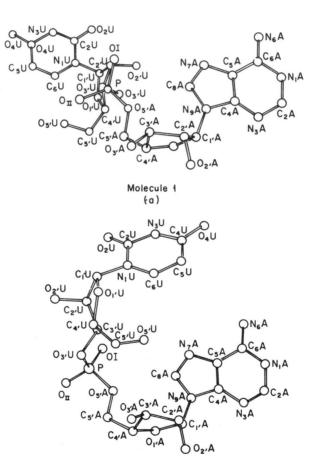

Molecule 1
(a)

Molecule 2
(b)

Fig. 4. The conformations of two independent molecules of UpA. (From Rubin *et al.* [10].) (a) molecule 1 is in open conformation; (b) molecule 2 is in the closed (inverted) conformation.

and the asymmetric portion of the unit cell contains two crystallographically independent UpA molecules. The conformations of these two independent molecules of UpA are shown in Fig. 4.

All four nucleoside moieties exhibit the *anti* conformation with respect to the glycosyl torsion angle χ_{CN}. The angles are: 48° for adenosine and 20° for uridine in molecule 1; 38° for adenosine and 9° for uridine in molecule 2 [10]. The ribofuranose rings exhibit the twist conformation of $C_{3'}$-*endo* to $C_{2'}$-*exo*, similar to that in the double-helical RNA [2,4].

The hydrogen-bonding schemes of the UpA in the crystal are shown in Fig. 5. Since adenine residues are protonated at N-1 by the phosphate protons, it is not surprising that the scheme consists of A–A and U–U, instead of A—U. In fact, all the adenines lie in one plane, the uracils in another, and the planes are parallel with a separation of 3.4 Å. The adenine of UpA(1) is hydrogen bonded to the adenine of UpA(2) in the same manner as in polyadenylic acid, displaying a pair of N_6—H⋯N_7 hydrogen bondings (Fig. 5a). The uracils form hydrogen-bonded dimers (Fig. 5b); N-3 of U in UpA(2) donates a hydrogen to O-2 of U in UpA(1) and N-3 of U in UpA(1) donates a hydrogen to O-4 of U in UpA(2). This type of uracil–uracil base-pairing, in which there is no local twofold axis perpendicular to the plane of the hydrogen bonds, has not been previously observed (Chapter 6, Volume I). Both laboratories [9,10] noted the relatively short distance (3.1 to 3.2 Å) between H of C-8 and O-5′ in all adenosines and H of C-6 and O-5′ in all uridines. These data suggest a possible hydrogen-bonding interaction between the 5′-OH of the ribose and the H-8 of adenine and the H-6 of uracil. This possibility as indicated in the crystals of monomers has been discussed in Section II,A, of Chapter 6, Volume I.

In a helical nucleic acid structure, all the sugar residues must be aligned similarly with respect to the helix axis as in Fig. 6b and the torsion angle involving O-5′ to P and P to O-3′ phosphodiester bonds is a *gauche–gauche* form. The conformation of the phosphodiester base of these two UpA molecules is quite different from a helical nucleic acid, indicating the freedom of rotation of these two P—O bonds. If the P to O-3′ bond of U in molecule 1 is rotated by about 110°, then the molecule 1 may assume a conformation similar to an 11-fold RNA helix, as shown in Fig. 6b and c.* The phosphodiester conformation of molecule 1 is *gauche-trans*. As for molecule 2, while the conformation of the phosphodiester linkage is *gauche-gauche*, it is oriented in an opposite direction to that of helical nucleic acid as shown in Fig. 6a and b. Rotation of the P to O-5′A bond by 175° of UpA(2) would

---

* UpA(1) in ref. 10 is equivalent to UpA(2) in ref. 9; UpA(2) in ref. 10 is equivalent to UpA(1) in ref. 9. In ref. 10, Rubin *et. al.* refer to UpA(1) as the molecule which differs by one P—O bond rotation from the RNA helix, while UpA(2) differs by two P—O bond rotations from the RNA helix.

allow the two fingers in Fig. 6a to point in the same direction as the helical RNA (Fig. 6b). The second rotation of the P to O-3'U bond in UpA(2) would convert it to UpA(1). The above discussion indicates that the helical RNA segment, UpA(1) and UpA(2), are all interconvertible through rotation of P—O bonds. This again emphasizes the relative freedom in rotation of the two P—O bonds; their conformation in the crystals of dinucleoside monophosphate is flexible, and subject to influence of other factors, such as hydrogen bonding.

Most recently, the MIT group [10a,b] has succeeded in crystallizing the ribosyl (3'–5') dinucleoside monophosphates GpC and ApU in *neutral* solutions. The sodium salts of GpC and ApU crystallize in a monoclinic unit

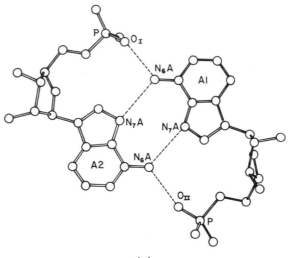

(a)

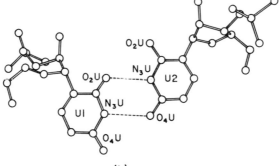

(b)

Fig. 5a. The base-pairing scheme of the UpA in crystal. (a) the self-pairing of adenylic acid residues involving the phosphate bonded to N-6; (b) the self-pairing of the uridylic acid residues. (From Rubin *et al.* [10].)

cell with one molecule in the asymmetric unit. Each molecule is related to another molecule by a 2-fold rotation axis which results in the formation of an antiparallel, right-handed double helix with complementary hydrogen bonding of Watson-Crick type between A:U and G:C. The close similarity of the right-handed double helix generated in the GpC and ApU crystals to the RNA-11 form obtained from the helical viral RNA, is shown in Fig. 6a. This similarity is also indicated by the torsional angles of the ribosyl phosphate linkage listed in Table I. The hydrogen bonding schemes and the base-base overlapping patterns for GpC and ApU in the crystals are shown in Fig. 6b and Fig. 6c. It should be noted that the helical structures of the inverted sequences UpA and CpG would have very different types of base stacking with the backbone maintained in the same conformation.

In both GpC and ApU, all the ribose residues are in 3'-*endo* conformation, similar to that found in the protonated UpA. The nucleoside glycosyl torsion (χ_{CN}) is 13° for G and 25° for C in GpC and 7° and 2° for A and 29° or 30° for U in ApU, while this angle is 14° for A and 15° for U in RNA-11 form. Both crystals are highly hydrated, with 36 water molecules in GpC and 22 water molecules in ApU per unit cell. A sodium ion is found in the minor groove of the ApU helix complexed to two free O(2) atoms of two uracils. The knowledge of the precise structure of these dinucleoside monophosphates allows us to construct the helical RNA molecular model with much greater certainty.

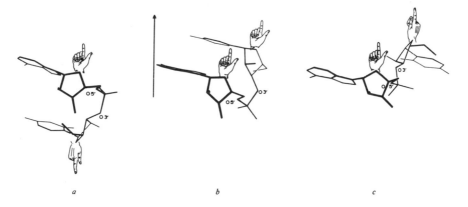

a   b   c

Fig. 5b. View of (a) UpA (molecule 2), (b) UpA in an 11-fold helical RNA conformation, and (c) UpA (molecule 1) so that the bases of the adenosine portions are similarly aligned as indicated by the hand sign at the ribose ring-oxygen. (After Seeman *et al.* [9].)

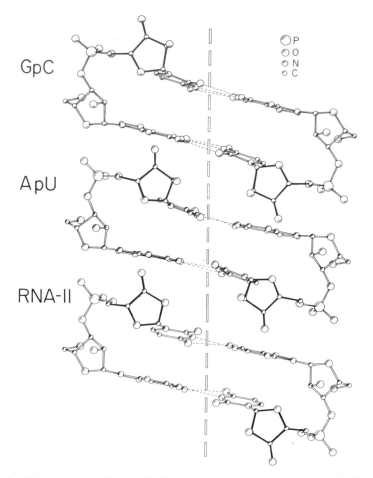

Fig. 6a. The view of ApU and GpC in crystals normal to the vertical helix axis in comparison to that of RNA-11 form from viral RNA. The separation of the bases in the stack is 3.4 Å. (From Day *et al.* [10a].)

2. *Nuclear Magnetic Resonance Studies on Solution State*

Considerable discussion has been given in Chapter 6, Volume I and else-where (for example, Ts'o [11]) to the problems and uncertainties in the extra-polation of information collected in the solid state to the solution state, especially concerning the conformation of the rotatable bonds. At present, only NMR studies can, in principle, provide stereochemical information about these bonds in solution in some detail through the measurement of spin–spin coupling constants (Chapter 6, Volume I).

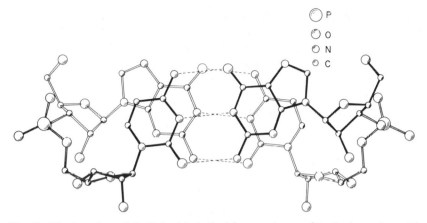

Fig. 6b. The top view of GpC double-helical fragment normal to the base plane. The shaded nucleosides are nearest to the readers. The distances of N4(C) to O6 (G) and N3 (C) to N1 (G) are both 2.90 Å; the distance of O2 (C) to N2 (G) is 2.86 Å; the distance of Cl′ of G to Cl′ of C is 10.67 Å. (From Day *et al.* [10a].)

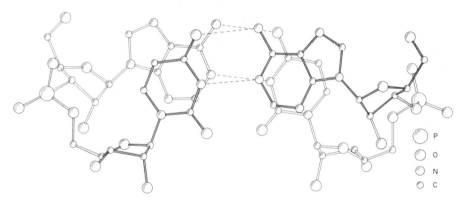

Fig. 6c. The top view of ApU double-helical fragment normal to the base plane. The shaded nucleosides are nearest to the readers. (From Rosenberg *et al.* [10b].)

The conformation of the C-4′ to C-5′ bond (ψ) of β-pseudouridine and uridine in aqueous solution was first studied [12,13] using a 220 MHz PMR spectrometer with the aid of a computer analysis. The designation of the protons and the schematic representation of the three possible rotational isomers around the C-4′ to C-5′ bond is shown in Fig. 7. At 28°–30°C, the coupling constants (H_2) for these protons are: $J_{4'-5'B}$ (3.0), $J_{4'-5'C}$ (4.4), and $J_{5'B-5'C}$ (-12.7) for uridine; $J_{4'-5'B}$ (3.2), $J_{4'-5'C}$ (4.6), and $J_{5'B-5'C}$ (-12.7) for β-pseudouridine. (It should be noted that the assignment for the H-5′ and

H-5″ is not certain; therefore, the assigned $J_{4'-5'B}$ and $J_{4'-5'C}$ values can be reversed, which will affect the calculation to some extent.) The estimation of the relative mole fraction of the rotational isomers (Fig. 7) P_I (*gauche–gauche*), P_{II} (*gauche–trans*) and P_{III} (*trans–gauche*) are based on the assumption that there are three energy minima as designated, and the lifetime of a molecule in any of the three energy minima is long compared to the time spent rotating between energy minima. For a sufficiently rapid interconversion, the H-4′ to H-5′ coupling constants will be a weighted average (weighted according to the relative populations) of the coupling constants in the three rotamers. Algebraically, this can be expressed as Eq. (1a) and (1b):

$$J_{4'-5'B} = P_I J_{IB} + P_{II} J_{IIB} + P_{III} J_{IIIB} \qquad (1a)$$

$$J_{4'-5'C} = P_I J_{IC} + P_{II} J_{IIC} + P_{III} J_{IIIC} \qquad (1b)$$

where J_{IB} is the coupling constant between H-4′ and H-5′B in rotamer I position (Fig. 7) and J_{IIB} is the same coupling constant in rotamer II position, etc. $J_{4'-5'B}$ and $J_{4'-5'C}$ were obtained from the PMR data and the J_{IB}, J_{IIB}, etc. are obtained from the Karplus equation (see Chapters on monomers and nuclear magnetic resonance) with a proper choice of J_0 value for the equation. In addition, there is the uncertainty about the precise angle of the rotation at the energy minima, which can deviate from the classical 60° stagger conformers [13]. Mindful of these uncertainties and assumptions, the calculation clearly indicates that the *gauche–gauche* form (I in Fig. 7) is the predominant conformation, representing about 55% ± 10% of the population. This conclusion is in accord with that derived from the x-ray diffraction study on crystals described above for angle ψ. It should be noted that the close similarity of the values at 28° and 78°C for $J_{4'-5'B}$ and $J_{4'-5'C}$ suggests

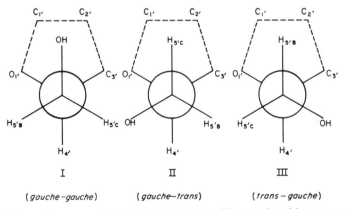

Fig. 7. Schematic representation of the three possible rotational isomers around the C-4′ to C-5′ bond of uridine. (After Blackburn *et al.* [13].)

the difference in enthalpy between these conformational states of these rota-mers is small. Another indication of the small differences in energy between these rotamers is the observation that in β-cyanuric acid riboside which has three oxo groups (at C-2, C-4, and C-6) the rotamer I (*gauche–gauche* form) no longer is the predominant form [14]. This is explained on the basis of the repulsion between the 5'-hydroxyl group and one of the two α-keto groups (C-2 or C-6) of the base. Also in the case of α-pseudouridine, the preference for the *gauche–gauche* rotamer is much less than in β-pseudouridine [15]. Recently, it was proposed that the rotation of the C-4' to C-5' bond may be related to the puckering of the furanose ring; the PMR data suggests that the *gauche–gauche* conformation is favored if the furanose ring is in C-3'-*endo* conformation and less favored if the furanose is in C-2'-*endo* conformation [16]. This information was obtained by plotting the existing data of $J_{4'-5'B}$ + $J_{4'-5'C}$ vs $J_{3'-4'}$, and vs $J_{1'-2'}$ of about 12 β-pyrimidine nucleoside derivatives. It was found that the sum of $J_{4'-5'B}$ + $J_{4'-5'C}$ *decreases* as $J_{3'-4'}$ increases, while the sum of $J_{4'-5'B}$ + $J_{4'-5'C}$ *increases* as $J_{1'-2'}$ increases. Hruska *et al.* [16] reasoned that a decrease in the proportion of the *gauche–gauche* rotamer should be manifest in an increase in the observed sum of $J_{4'-5'B}$ + $J_{4'-5'C}$; and the increasing proportion of the $C_{3'}$-*endo* (and/or $C_{2'}$-*exo*) should be manifest in an increase of $J_{3'-4'}$ and a decrease of $J_{1'-2'}$. Another recent comprehensive study [17] indicates that the *gauche–gauche* form is also predominant for the 3'-UMP and 3'-β-pseudouridine monophosphate, but not for deoxyuridine ($J_{4'-5'}$ = 3.4; $J_{4'-5''}$ = 5.1); for the deoxyuridine, the *gauche–gauche* and the *gauche–trans* rotamers have approximately equal probabilities. These results are not affected by addition of 2 M NaClO$_4$ or by change of temperature from 10° to 80°C.

As for the rotation of the $C_{5'}$—O bond (ϕ), $C_{3'}$—O bonds (ϕ'), the hydroxyl protons for four common ribonucleosides and four deoxynucleosides were studied in mixtures of dry DMSO-d$_6$ and C$_6$D$_6$ [18]. Use of this organic solvent system permitted the study of the exchange-free spin-coupled multi-plets for the hydroxyl protons. The $J_{\text{H 5' to OH-5'}}$ was found to be about 5.0 Hz for all eight nucleosides, and the $J_{\text{H-3' to OH-3'}}$ to be about 3.8–4.2 Hz for all nucleosides except uridine (4.7 Hz). The $J_{\text{H-2' to OH-2'}}$ for the ribonucleosides are 5.5–6.0 Hz. From these data, Davies and Danyluk [18] concluded that the $C_{5'}$—O bond (ϕ) rotates freely in solution, since the observed H-5' to OH-5' coupling constant is the same as that (5.3 ± 0.3 Hz) of the hydroxyl coupling constants of the aliphatic alcohols (methanol, ethanol, etc.) in similar solvent systems. As for the $C_{3'}$—O bond, based on the empirical relation between the vicinal coupling constant J_{HCOH} and the dihedral angle, as well as on the assumption about the existence of the three classical staggered, rotational states (two *gauche* and one *trans*), the computation [18] indicates that the O—H$_{3'}$ group exists for 80% of its time in a *gauche* conformation with respect to the C-3' to H-3' bond. In the solid state with the phosphate derivatives, the $H_{3'}$-$C_{3'}$—$O_{3'}$—P assumes more or less a *cis* conformation, however (Table I).

Stereochemical information about the dihedral angle (α) in a H—C—O—P system can be provided by the coupling constant, J_{H-P}, in a manner analogous to the constant $J_{H-H'}$ in an H—C—C—H system [19-21].

From the stereochemistry of the H—C—O—P system, the stereochemistry of the C—C—O—P system can be defined through the geometrical relation

of the

$$\begin{matrix} H & & H \\ \diagdown & & \diagdown \\ & C- & in the & C—O—P \ system. \\ \diagup & & \diagup \\ C & & C \end{matrix}$$

The J_{H-P} of an H—C—O—PO_2 system in aqueous solution is 1.5–4.5 Hz when C—H is *gauche* to O—P ($\alpha = 60°$), and about 22 Hz when C—H is *trans* to O—P ($\alpha = 180°$). The J_{H-P} of diethyl phosphate anion $(CH_3CH_2O)_2PO_2{}^-$ was found to be about 7.0 Hz in D_2O; and the J_{H-P} of the 5'-inosine monophosphate and 5'-guanosine monophosphate was found to be about 4.5 Hz, (the $J_{H(5')-P}$ and $J_{H(5'')-P}$ are practically equivalent) [20]. This low J_{H-P} value suggests that P is *trans* to C-4' (or *gauche* to H-5' and H-5''). If the value of J_t is adopted to be 22.1 Hz and J_g to be 2.3 Hz, together with the assumption of the existence of the three classical staggered conformations, the population of *trans* to C-4' conformation was estimated to be 78% [20]. In solid state, C-4' is indeed usually *trans* to P-5' ($\phi = 170° \pm 30°$, Table II) for the nucleotides and polynucleotides. The $J_{H(3')-P}$ of 3'-UMP and 3'-pseudouridylic acid was found to be 7.7 $\pm$ 0.1 Hz in aqueous solution [17]. From these data, it has been suggested that the time averaged dihedral angle of the H—$C_{3'}$—O—P system would be about 50°, especially with the arrangement of C-3' to C-2' *gauche* to the $O_{3'}$—P and C-3' to C-4' *trans* to $O_{3'}$—P [17]. This suggestion is again supported by the x-ray diffraction data on ϕ' ($\phi' \approx 200°$, Table I).

The J_{H-P} values of two (3'-5') dinucleoside monophosphates, ApA and UpU have been reported by Tsuboi *et al.* [21]. The J_{H-P} values obtained at room temperature by the first order approximation were found to be practically the same for both dimers. The values are 3.4, 6.5, and 8.1 Hz for ApA; 3.4, 6.7, and 8.2 Hz for UpU. The 3.4 Hz and 6.6 $\pm$ 0.1 Hz were assigned to be the J_{H-P} of the two $H_{5'}$—C—O—P and $H_{5''}$—C—O—P systems with the 5'-CH_2—; and the corresponding dihedral angles are estimated to be about 64° and 56°, a situation similar to that of the 5'-GMP and 5'IMP, as described above. Therefore, the $C_{4'}$—$C_{5'}$—O—P group has nearly the *trans* conformation. The remaining J_{H-P} (8.1 Hz) was assigned to the $H_{3'}$—C—O—P system with the dihedral angles of either 50° or 130°. After the consideration of other stereochemical factors, including the stacking requirement [21], it was suggested that the most likely conformation of the C_3—O bond is that of the $C_{4'}$—$C_{3'}$—O—P in nearly *trans* conformation, and the $C_{2'}$—$C_{3'}$—O—P in nearly *gauche* conformation, as described above for the 3'-UMP and 3'-pseudo-UMP [17]. While the preceding discussion on the conformation of the dimers as revealed by the J_{H-P} measurements may be subject to various uncertainties, two definite conclusions can be reached. The first one is that in all the H—C—O—P systems studied so far, none of them is in a predominant *trans* conformation on a time-average basis. From

the $\overset{\displaystyle H}{\underset{\displaystyle R}{\diagdown}}$C—O—P relationship this suggests, therefore, that the R—C—O—P

system is more likely to be in a *trans* conformation, especially in the case of the $C_{4'}$—CH_2—O—P system. The second conclusion is even more far-reaching. As shown in later sections, the extent of stacking of ApA is much larger than that of UpU, and yet these two dimers have practically identical J_{H-P} values. In fact, the J_{H-P} values of the monomers ($J_{H(5')-P}$ of 5'-IMP acid = 4.5 Hz; $J_{H(3')-P}$ of 3'-UMP = 7.7 Hz) are very similar to the corresponding J_{H-P} values of the dimers (3.4 and 6.7 for the $J_{H5'-P}$ and 8.2 for the $J_{H(3')-P}$). It can be concluded, therefore, that the rotation of the two C—O bonds (ϕ and ϕ', Fig. 1) is not mainly responsible for the degree of stacking or destacking [21]. This important conclusion supports the notion discussed in Section I that among the five rotatable bonds (Fig. 1) the rotation of the two O—P bonds (ω and ω') is probably most responsible for the flexibility and rotation of the nucleic acid backbone.

B. Experimental Methods for the Investigation on the Conformation of Dinucleoside Monophosphates

A discussion on the methodology is often a necessary prerequisite for the discussion of the scientific results. Our knowledge and ignorance about the system are directly related to the powers and limitations of our techniques.

This simple truth will become self-evident in later sections of this chapter, particularly in comparing in detail the results derived from several different experimental approaches. Other chapters in Volumes I and II have already described the theoretical basis and the experimental application of some of these techniques. Therefore, only a very brief account will be given here concerning the aspects most pertinent to the studies of dinucleoside monophosphates. Emphasis will be placed on spectroscopic methods through which the conformation of dimers can be investigated from the stereochemical or geometrical viewpoint in detail. Other thermodynamic methods, such as titration (21a) which has been employed to investigate the free energy of stacking (for instance, about 1 kcal for ApA at 20°C), will not be described here.

1. *Ultraviolet Spectroscopy*

Hypochromicity (or hypochromism) in UV absorbence at $\lambda_{\max}$ has been used most often for the studies on the conformations of dinucleotides, oligonucleotides, and nucleic acids. This approach is one of the topics in Chapter 2. Hypochromicity in percentage at a given wavelength (usually at $\lambda_{\max}$) is defined as $h\% = (1 - \varepsilon_D/\varepsilon_M) \times 100$, where ε_D and ε_M are the extinction coefficients of dimer and monomer, respectively. Similarly, hypochromism in percentage is defined as $H\% = (1 - f_D/f_M) \times 100$, where f_D and f_M are the oscillator strength which can be obtained by the measurement of the absorbence at the entire absorption band. From the theoretical point of view, hypochromism (fractional decrease in integrated intensity) is more appropriate than hypochromicity (or point hypochromism) as a parameter of base–base interaction. However, the difference in absorbence between monomer and dimer is relatively small. Measurement of hypochromism involves the separation of the overlapping absorption bands, a step which can introduce additional error in calculation for a complex spectrum. While experimentally this measurement is not difficult to make, caution should be taken to obtain accurate results, since the percent hypochromicity for dimers and small oligomers is small.

Table II lists the percent hypochromism and percent hypochromicity of 16 dinucleoside monophosphates at three pH (1.0, 7.0, and 11.5) and room temperature, as reported by Warshaw and Tinoco [22]. Large values of hypochromism or hypochromicity are indicative of strong base–base interaction by stacking (Chapter 2). Therefore, ApA has been classified as having a stacked conformation, while UpU has an unstacked conformation [22]. When both bases are charged or ionized, such as ApA and CpC in pH 1.0, or GpG and UpG at pH 11.5, the $H\%$ or $h\%$ is considerably lower than the corresponding value at pH 7.0, the neutral state. This observation is explained by reason of charge repulsion which causes the unstacking of the dimer. Also, sequence effect on hypochromism or hypochromicity has been noted, such as the comparison between the sequence isomers of ApG vs GpA, and UpG vs GpU at pH 7.0 [22].

At the present level of knowledge, especially limited by the information we have about the excited states of the base chromophores, it is not possible to relate the hypochromicity (or H) data to the interaction between the bases in a dimer by specific, geometrical terms. However, for a homodimer, the situation is simplified; the interaction of the dipoles of the transition moments

TABLE II[a]

HYPOCHROMISM ($H\%$)[b] AND HYPOCHROMICITY ($h\%$)[b] OF THE DINUCLEOSIDE
MONOPHOSPHATES AT THESE pH, ROOM TEMPERATURE, AND
0.1 IONIC STRENGTH[c]

Dimer	pH 1.0		pH 7.0		pH 11.5	
	$H(\%)$	$h(\%)$	$H(\%)$	$h(\%)$	$H(\%)$	$h(\%)$
ApA	−0.5	0.3	6.8	9.4	7.4	9.4
GpG	−1.0	0.5	9.1	6.9	4.4	−0.8
CpC	−5.5	0.1	4.9	7.2	5.7	7.2
UpU	−2.0	3.0	−3.6	1.7	0.8	2.5
ApG	1.4	3.0	2.6	5.8	1.3	4.0
GpA	2.7	3.7	6.0	7.6	2.9	5.8
ApC	0	1.8	7.3	7.6	9.7	9.0
CpA	−2.3	−1.0	5.2	7.8	6.3	6.5
ApU	2.7	3.0	1.6	5.0	0	3.8
UpA	3.0	3.3	1.4	3.0	1.7	1.5
GpC	6.8	10.7	7.2	8.7	4.4	6.0
CpG	6.8	9.8	6.2	3.2	5.4	9.2
GpU	4.9	6.2	−1.2	2.4	0	4.3
UpG	6.5	5.6	−3.6	1.7	3.0	4.2
CpU	1.5	2.7	4.2	6.3	4.2	4.0
UpC	−0.3	3.2	0.7	2.4	0.8	2.5

[a] From Warshaw and Tinoco [22].
[b] Hypochromism (H) was calculated from oscillator strength. The integration was from a cutoff wavelength in the vicinity of the absorption minimum. Hypochromicity (h) is calculated from extinction coefficient at λ_{max}. See text for definition.
[c] Additional UV spectral data on dinucleotides can be obtained from Section G, "Handbook of Biochemistry" (H. A. Sober, ed.), Chem. Rubber Publ. Co., Cleveland, Ohio, 1970.

of two identical bases in the dimer can be related by the intersection of a given set of geometrical axes. Glaubiger *et al.* [23] have established some useful relationships for the analysis of the optical properties of the homodimer based on dipole approximation. In their model, the two identical base planes of the homodimers are assumed to be parallel to each other. As the first approximation, the relative motion between base planes in their model is restricted to rotation without stretching or lateral displacement. The optical property is analyzed on the basis that each chromophore has an isolated, optically inactive, electronically allowed transition at frequency ν_a. In the analysis of hypochromism, Glaubiger *et al.* [23] had made the following additional simplification: The effect of permanent dipoles is negligible, and the symmetry of the polarizability term $\alpha(\nu_a)$ is that of an elliptical cylinder with its axis perpendicular to the plane of base. From this approximation, the following expression for hypochromism, H, was derived:

$$H(\theta) = \frac{2}{|R_{12}|^3} [\alpha_1 \cos^2 (\theta + \phi) + \alpha_2 \sin^2 (\theta + \phi)] \tag{2}$$

where θ is the angle between μ_{10a} and μ_{20a} which are the transition dipoles in planes of base 1 and base 2, respectively, α_1 and α_2 are the polarizabilities at frequency, ν_a, along the principal axes e_1 and e_2 which lie in the plane of the base, and ϕ is the angle between μ_{20a} and e_1. From Eq. (2) and other considerations, we obtain the following results in regard to the dependence of hypochromicity on the geometrical relationship of two bases in the dimer [24]. (1) If the base (such as adenine) is isotropic with respect to the polarizabilities (i.e., $\alpha_1 = \alpha_2$), then Eq. (2) is reduced to

$$H(\theta) = \frac{2}{|R_{12}|^3} [\alpha] \tag{3}$$

Under this condition, the hypochromism is not sensitive to θ and therefore not sensitive to rotation of the base planes relative to each other. Recently, Takashima [25] has calculated the polarizability of adenine along two axes (α_x and α_y) by both Hückel and SCF methods. The results indicated that the ratio of $\alpha_x : \alpha_y$ is less than 1.4. Though the calculation of Takashima is for the ground-state polarizabilities and the terms in Eq. (2) are the polarizabilities of the excited state (at ν_a), the ratio of $\alpha_y : \alpha_x$ (at ν_a) is unlikely to be much larger than the ratio at ground state. This consideration suggests that for certain bases, at least, hypochromism is rather insensitive to the value of θ. In addition, since the angle term (θ) in Eq. (2) is a square term, the sign of the angle (θ) will not be important. Hypochromicity will be the same regardless of whether the screw axis of the dimer is right-handed or left-handed. (2) Hypochromism is dependent on the $|R_{12}|^{-3}$. (3) In the dimer model of Glaubiger *et al.* [23], no lateral movement of the two base planes relative to each other is allowed. The CPK models of these dimers clearly show that lateral movement must be involved upon rotation of the phosphodiester bonds. Therefore, we shall consider the term $G_{ia,jb}$ used by DeVoe and Tinoco (Eq. 5 in DeVoe and Tinoco [26]) instead of the T_{12} term used by Glaubiger *et al.* ([23], Eq. 9), where $G_{ia,jb}$ (or T_{12}) represents the spatial relationship for the interaction term $V_{ia,jb}$ (or $V_{ia,2b}$). For our present parallel model, $G_{ia,jb}$ can be simplified to

$$\left(\frac{1 - 3 \cos^2 a}{R_{ij}^3} \right)$$

where a is the angle between R_{ij} (the distance between the center of two bases) and the transition moment μ_{ia}, and H is directly proportional to this term as

$$H \propto \frac{1}{R_{ij}^3} (1 - 3 \cos^2 a) \qquad (4)$$

According to Eq. (4), when the bases are directly on top of each other, (i.e., $a = 90°$, $\cos a = 0$), then the H value is the highest; when a is near $55°$ ($3 \cos^2 55° \approx 1$), H is near zero, and when a is from $50°$ to $0°$ (at $0°$ the bases are lying horizontally side by side), H will have a negative value (or the interaction leads to hyperchromism instead). Therefore, it is clear that hypochromism is sensitive to lateral movement of the two bases relative to each other.

2. *Circular Dichroism* (Optical Rotatory Dispersion)

The CD(ORD) technique has been a favorite and useful method for the study of dimers and oligomers. Again, since this is the topic of Chapter 2 the description here will be very brief. As in the presentation in Chapter 2, only the CD data will be discussed because of the simplification of the CD spectra and the greater ease of interpreting CD results, even though a considerable amount of early work was done by the ORD technique.

A comprehensive and valuable study on the CD of deoxydimers has been reported by Cantor *et al.* [27]. In Table III and Fig. 8, their results are presented. Two simple conclusions can be drawn from these results. First, the CD bands of the dimers are very different from their constituent monomers and usually much larger. Second, the CD bands of the sequence isomers such as dApdG vs dGpdA, etc., are very different. Both conclusions indicate that the CD data contain valuable information about the interaction between the bases in the dimers. Similar comprehensive CD study on ribosyl dimers (including two 2'-5' dimers) has also been reported by Warshaw and Cantor [28], as presented in Table IV and Fig. 9. The two conclusions concerning the deoxydimers can be applied equally well to the ribosyl dimers. Brahms *et al.* [29] have reported previously the CD spectra of 13 ribosyl dimers and the temperature effects ranging from $-20°$ to $80°C$ on these spectra in a solution of $4.7\ M$ KF–$0.01\ M$ tris, pH 7.5. Their earlier results [29] are not entirely in agreement with those reported by Warshaw and Cantor [28]. Some of these differences may be due to partial unstacking or change of conformation resulting from high-salt concentration used as solvent by Brahms *et al.* Davis and Tinoco found as much as 35% reduction in ORD magnitude upon going from low salt to 25% LiCl [30]. Warshaw and Cantor [28] also suggested that high salt and low temperature conditions used by Brahms *et al.* promoted aggregation of GpC.

TABLE III

CIRCULAR DICHROISM OF DEOXY DIMERS [a,b]

Compound	λ (nm)	$[\theta]$ ($\times 10^{-4}$)	λ (nm)	$[\theta]$ ($\times 10^{-4}$)	λ (nm)	$[\theta]$ ($\times 10^{-4}$)
			Cotton effects in order of decreasing wavelength			
d(ApAp)	269	1.66	250	−2.26	218	3.90
d(ApT)	272.5	1.62	252	−1.43	219	1.18
d(ApCp)	273	1.66	241	−1.25	217.5	0.71
d(ApG)	281	−0.64	272.5	−0.63	260	−0.56
d(TpAp)	270	0.85	250	−1.07	217.5	0.65
d(TpT)	279	0.72	250	−0.62	215	−0.08
d(TpC)	278.5	0.85	239	−0.32	215.5	−0.54
d(TpGp)	285	0.35	260	−0.73	212.5	−0.03
d(CpA)	272.5	1.17	212.5	−1.47	—	—
d(CpT)	280	1.55	235	−0.36	213	−0.92
d(CpC)	277	1.42	215	−0.84	—	—
d(CpG)	278	0.69	247.5	−0.88	207.5	−2.09
d(GpA)	270	−0.41	247.5	0.88	215	0.76
d(GpT)	283.5	0.56	265	−0.35	246	0.20
d(GpC)	278	0.60	249	−0.25	228.5	−0.15
d(GpGp)	280	−0.46	255	0.38	234	−0.12

[a] From Cantor *et al.* [27].

[b] In 0.01 M sodium phosphate buffer, pH 7.2 and 0.1 M sodium perchlorate, at 26°C. Except for TpT, all data were calculated from the assumption that the extinction coefficients of the deoxydimers are identical to those of the ribosyl dimers. This assumption may introduce errors of a few percent.

To relate the CD data to the interaction between the bases in a dimer by specific, geometrical terms is a formidable task (see Chapter 2). Considerable progress has been made, especially that revealed by the paper of Johnson and Tinoco [31], which will be discussed further in later paragraphs. However, the complex computation based on monopole–monopole interaction does not provide the investigator the basic notion about the relative importance of various geometrical factors in determining the intensity of the CD band derived from the exciton interaction. In order to achieve this purpose, we [24] again go back to the homodimer model developed by Glaubiger *et al.* [23] based on the dipole approximation as described above for the analysis of hypochromism [Eq. (2) and (3)]. The analysis is formulated on the basis that each chromophore has an isolated, optically inactive, electronically allowed transition at frequency ν_{0a}. According to their analysis, the circular dichroism (the original derivation was for optical rotatory dispersion, but

TABLE IV

CIRCULAR DICHROISM OF RIBO MONOMERS AND DIMERS [a,b]

Compound	Cotton effects in order of decreasing wavelength					
	λ (nm)	$[\theta]$ $(\times 10^{-4})$	λ (nm)	$[\theta]$ $(\times 10^{-4})$	λ (nm)	$[\theta]$ $(\times 10^{-4})$
A	263.5	−0.38	227	0.11	—	—
U	267.5	0.92	237.5	−0.51	217.5	−0.56
C	272.5	1.04	217.5	−0.94	—	—
G	252	−0.29	215	0.81	—	—
ApA	271	2.03	252	−2.51	220	2.00
2′ → 5′ApA	271.5	1.97	252	−1.65	218.5	3.45
ApU	269	1.35	248	−0.68	217.5	0.82
ApC	275	2.35	237.5	−1.15	220	0.74
2′ → 5′ApC	278.5	1.68	255	−1.36	220	2.35
ApG	280	0.75	260	−1.28	216.5	0.43
UpA	267.5	0.76	244	−0.42	—	—
UpU	271	1.61	242.5	−0.75	210	−0.32
UpC	275	1.94	237.5	−0.40	214	−0.84
UpG	280	0.59	257.5	−0.30	235	−0.20
CpA	274	1.88	232.5	−1.09	207.5	−1.56
CpU	275	2.72	235	−0.90	210	−0.93
CpC	280	3.08	230	−0.74	215	−0.99
$C_{2'-OMe}pC$	280	3.34	230	−0.81	213.5	−0.84
CpG	282.5	1.28	229	−0.60	212.5	−0.95
GpA	285	−0.24	265	0.84	247.5	−0.16
2′ → 5′GpA	295	−0.04	262.5	0.61	215	0.69
GpU	280	0.29	251	0.46	217	0.34
GpC	269	0.94	227.5	−0.60	212.5	−0.79
GpG	280	−0.13	259	0.79	240	−0.31

[a] From Warshaw and Cantor [28].
[b] In 0.01 M sodium phosphate buffer, pH 7.2, and 0.1 M sodium perchlorate, at 26°C.

it is the same for circular dichroism) is proportional to the $V_{12}R_{0a}$, where V_{12} is the interaction energy term arisen from the splitting of the degenerate transition into two energy levels which have the rotational strength, R_{0a}, of equal magnitude but opposite sign. In the dipole approximation, V_{12} and R_{0a} can be written as follows [23]:

$$V_{12} = \frac{1}{|R_{12}|^3}\left[\mu_{10a}\cdot\mu_{20a} - 3\frac{(R_{12}\cdot\mu_{10a})(R_{12}\cdot\mu_{20a})}{R_{12}2}\right] \quad (5)$$

$$R_{0a} = \frac{\pi\nu_{0a}}{2c}R_{12}\cdot(\mu_{10a}\times\mu_{20a}) \quad (6)$$

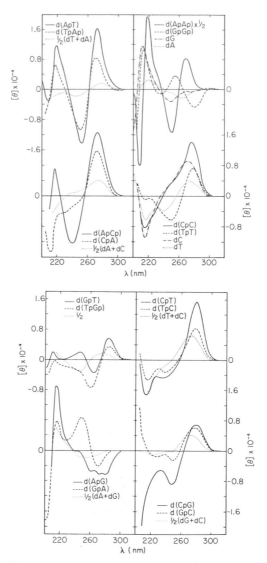

Fig. 8. Circular dichroism of 16 deoxy dimers and their constituent monomers at 26°, in 0.01 *M* sodium phosphate buffer, pH 7.2, and 0.1 *M* sodium perchlorate. All spectra are given in units of molar ellipticity per residue. Note that scale used for d(ApAp) is twice as large as that of other compounds. (From Cantor *et al.* [27].)

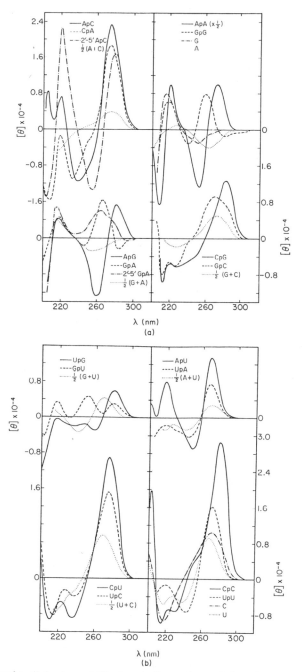

Fig. 9. Circular dichroism of 18 ribosyl dimers (two of them have 2′-5′ linkage) and their constituent monomers at 26°, in 0.01 *M* sodium phosphate buffer, pH 7.2 and 0.1 *M* sodium perchlorate. All spectra are given in units of molar ellipticity per residue. Note that scale used for ApA is twice as large as that of other compounds. (From Warshaw and Cantor [28].)

where R_{12} is the distance from the transition dipole in base 1 to the transition dipole in base 2 of the dimer, and μ_{10a} and μ_{20a} are the transition dipole moment in the planes of base 1 and base 2, respectively, as in Eq. (2). Since these two planes are always parallel, R_{12} is perpendicular to both μ_{10a} and μ_{20a}. Under this condition, the second term in V_{12} vanishes and one obtains

$$[\theta]_{\mathrm{CD}} \propto R_{0a} V_{12} \propto \frac{|\mu_{10a}|^2 |\mu_{20a}|^2}{|R_{12}|^2} \sin\theta \cos\theta \tag{7}$$

where θ (not $[\theta]_{\mathrm{CD}}$) is the angle between μ_{10a} and μ_{20a}. From expressions (6) and (7), we obtain the following results in regard to the dependence of circular dichroism measurement on the geometrical relationship of two bases in the dimer. (1) The circular dichroism value is very sensitive to the angle θ between the two transition dipoles. It becomes zero when θ is at $0°$, $90°$, and $180°$ and reaches a maximum when θ is near $45°$. The circular dichroism value is dependent on the sign of θ; therefore it is dependent on the handedness of the screw axis. A detailed calculation on the dependence of optical activity of $A_{3'}p_{5'}A$ on the angle θ has been reported [32]. (2) The circular dichroism value is dependent on the $|R_{12}|^{-2}$. Although Eq. (7) was developed for a model without relative lateral displacement, the circular dichroism value is affected by the lateral displacement between the bases when allowed, only to the same extent that this displacement affects $|R_{12}|^{-2}$. This is because the lateral component of the interaction terms $[R_{\parallel} \cdot (\mu_{10a} \times \mu_{20a})]$ becomes zero as substituted into expression (6).

In summarizing Eqs. (2)–(7), which describe the dependence of circular dichroism and hypochromism in ultraviolet absorbence (H) measurements on the geometrical relationship between two parallel bases in a homodimer (especially adenine), we can make the following comparison. (1) In terms of rotation or angle θ, circular dichroism is very sensitive to both the sign and the magnitude of this parameter, while H is not. (2) In terms of the distance R_{12}, circular dichroism is dependent on $|R_{12}|^{-2}$ and H is dependent on $|R_{12}|^{-3}$. (3) In terms of the lateral movement or $\cos a$, circular dichroism is not sensitive to this term other than the effect on $|R_{12}|^{-2}$ while H is dependent $(1 - 3\cos^2 a)$ in addition to the effect on $|R_{12}|^{-3}$. The overall evaluation is that the circular dichroism value is more dependent on the angle between the bases, and the H value is more dependent on the distance between the two bases. In utilizing these relationships, the assumptions and the approximations should be clearly kept in mind. For instance, this analysis is for the in-plane π–π^* transition and not for out-of-plane n–π^* transition, etc. These approaches are also limited by the inherent problems in the dipole approximation.

3. *Proton Magnetic Resonance*

PMR has become, perhaps, the most powerful method for the study of the conformation of dimers. The capabilities of this approach have been greatly increased by the improvement in sensitivity through the introduction of computer average transient technique and the Fourier transform technique, and by the improvement in resolution through the availability of high frequency (220 and 300 MHz) spectrometers.

Experimentally certain precautions are necessary for obtaining proper spectral data. At the concentration range of 0.01–0.05 M usually adopted for the study of dinucleoside monophosphates, the chemical shifts of the base protons can be strongly concentration-dependent due to self-association, especially at low temperature [33,34]. Therefore, the chemical shift data often have to be extrapolated to infinite dilution to avoid the intermolecular self-association, so that the data can be properly interpreted for intramolecular base–base interaction in the dimer. Another problem is that the dinucleotides and oligonucleotides all have strong affinity for metal ions. Contaminating paramagnetic ions would broaden the linewidths and may lead to erroneous interpretation [34]; therefore, these ions should be removed, for instance, by passing through a chelating column. It is also not certain whether or not some dinucleotides and oligonucleotides would interact with the internal standard in solution owing to the hydrophobic properties of these compounds. For this reason, an external standard contained in a capillary is usually employed instead. With the use of an external standard in a capillary, change of magnetic susceptibility of the solvent has to be considered when there is a change of temperature or solvent (including addition of large amounts of salt).

The first task in the PMR study is the spectral assignment, which is usually not a difficult problem for a heterodimer, but can be challenging for a homodimer and a homooligomer, especially the purine compounds. Fortunately, the H-8 of the purine bases are exchangeable with the deuterium in D_2O to about 50% when heated at 90°C for 2 hr [34,35]. This procedure provides a simple means by which to distinguish the H-8 protons from the H-2 protons. Another very useful procedure exists for separating the H-8 and H-1' protons of the 5'- residues (such as -pA) from the H-8 and H-1' protons of the 3'- residue (such as Ap-). This approach was first developed by Chan and Nelson [33] for the assignment of the two H-8 protons in rAprA. It is based on the observation that certain divalent metal ions, such as Mn^{2+}, are bound mainly to the phosphate group of the nucleotides at low concentrations of the ions and nucleotides. If the divalent ion is a paramagnetic ion, such as Mn^{2+}, it will tend to broaden the resonances of the nearby protons through the dipole–dipole interaction. This interaction is extremely distance dependent (proportional to $(\mu^2/r^6)\tau_c$, where r is the distance between the paramagnetic ion and the proton, μ is the magnetic moment, and τ_c is the correlation time). Since the distance of the H-8 of prA residue to the phosphate group where the Mn^{2+} ion would be located is closer than the H-8 of rAp residue when the nucleotidyl units are in the *anti* conformation, the H-8 of prA residue will be broadened preferentially over the H-8 of rAp residue as observed in the presence of dilute Mn^{2+} solution [33,36]. The assignment of the H-1' protons is based on the same procedure, except now the effect is reversed. The H-1' of 3'-dAMP (or 3'-rAMP) is now closer to the PO_4—Mn^{2+} group than the H-1' of 5'-dAMP (or 5'-rAMP). Therefore, in the presence of increasing concentrations of Mn^{2+}, the line width of H-1' of 3'-dAMP and 3'-rAMP is being broadened to a greater extent than the linewidth of H-1' of 5'-dAMP and 5'-rAMP [36]. In this assignment procedure for the H-8 protons, it is implicitly assumed that the -pA residue spends a considerable fraction of its time in an *anti* conformation. Therefore, the broadening effect of H-8 observed in the presence of Mn^{2+} not only serves the purpose of assignment but also indicates the *anti* conformation of the -pA residue in the dimer or oligomer [33,34]. The above two procedures not only are very useful, but can also exemplify the approaches in solving the assignment problem, i.e., specific deuteration and specific effect on relaxation (or shifts) by an added agent (Mn^{2+}, in this case).

In the analysis of the chemical shift data for the study of conformation or base–base interaction, one useful procedure is the calculation of the "dimerization shifts ($\Delta\delta_D^{t^\circ}$)." This term is defined as the difference between the chemical shift values of the 5'-nucleotidyl unit or the 2'- or 3'- nucleotidyl unit in the dimer and the chemical shift values of the corresponding 5'-mononucleotides or 2'- or 3'-mononucleotides. ($\Delta\delta_D^{t^\circ} = \delta_{Np}^t - \delta_{NpN'}^t$; $= \delta_{pN'}^t - \delta_{NpN'}^t$). Most of the $\Delta\delta_D^{t^\circ}$ originates from the nonbonded interaction between the two neighboring units in the dimer as indicated by the reduction of the $\Delta\delta_D^{30^\circ}$ values to a very low level when the dimers are dissolved in dimethyl sulfoxide-d_6. The residual $\Delta\delta_D$ values are reduced further to practically zero when the $A_{3'}p_{5'}A$ dissolved in dimethyl sulfoxide-d_6 is measured above 60° [34]. From this study and our previous work [37], it can be concluded that under this condition, the individual nucleotidyl unit in ApA becomes totally free from the influence of the neighboring unit in the dimer [34]. There is one special case where a contribution to the dimerization shift is made indirectly by the formation of the phosphodiester linkage. The deshielding of the H-8 proton of the 5'-nucleotidyl unit (or the H-6 proton of the 5'-pyrimidine nucleotidyl unit) by the 5' phosphate (Chapter 6, Volume I; Section I,D) is reduced by the conversion of the phosphomonoester to the phosphodiester in the formation of the dimer from the 5'-mononucleotides [34,38]. Other than this interesting exception, all the contribution to the dimerization shift is apparently from the nonbonded interaction. This is because the phosphate group exerts little direct influence on the chemical shift of the base and H-1' protons in a nucleotide other than the deshielding effect of the 5'-phosphate group to the H-8 of the purinyl unit or H-6 of the pyrimidinyl unit [38]. Most of the dimerization shift apparently originates from the shielding effect of the ring-current magnetic anisotropy of the neighboring base as indicated by a series of extensive research on the effect of concentration of the chemical shifts of the monomers (Chapter 6, Volume I; Section II,C).

In Table V, the chemical shifts (relative to TMS capillary in ppm, $28°-30°C$, pD 5.9) and dimerization shifts ($\Delta\delta_D^{30°}$) of 16 ribosyl dimers and TpT are given. The GpG spectrum could not be obtained at $30°C$, even at a concentration level of 0.004 M due to extensive broadening, but could be observed at $60°C$ [34]. These data will be discussed in the following section concerning the conformational model for these dimers [39]. Ribosyl dimers of ApA, ApC, and CpA, TpT, TpdU, dUpT, etc., have also been extensively studied by Chan and co-workers ([33,40,41] and are described in the chapter on NMR).

Both theoretical and experimental studies indicate that the purine bases have a much higher ring-current magnetic anisotropy than the pyrimidine bases (especially uracil), a subject which will be further discussed in later sections. Suffice it to state here that for this reason, the PMR studies on pyrimidine dimers yield less information about their conformation than the PMR studies on purine dimers.

4. *Fluorescence*

The properties of the "excited states" of the nucleic acid bases are the topic of Chapter 3. Therefore, only a brief account of the study on dinucleoside monophosphates will be given here. Förster and Kasper [42] showed that neighboring aromatic molecules in their excited singlet state may form an excited dimer or *excimer*. The energy of this complex is lower than that of the excited monomer, as a result of stabilizing electron–electron interaction. In the ground state, the two neighboring molecules normally repel each other at distances shorter than the sum of van der Waal's radii (~ 3.4 Å for aromatic compounds). For molecules in the excited state, the potential minimum may exist at a shorter intermolecular distance. Since the excited-state energy is lowered by the dimer interaction and since the ground-state energy is raised by the intermolecular repulsion at a smaller internuclear separation, the emission from this excited-state dimer (excimer) will invariably be at longer wavelengths than the emission from the monomer. Moreover, since the ground state of the dimer is unbound, it has no discrete vibrational levels, thus excimer emission is always broad and featureless [43]; and since no dimer is formed in the ground state, the absorption spectrum will be unchanged from

TABLE V

COMPARATIVE STUDY ON THE EFFECT OF SEQUENCE ON THE CHEMICAL SHIFTS (δ) AND DIMERIZATION SHIFTS (ΔδD) OF THE BASE AND H-1' PROTONS OF THE DINUCLEOSIDE MONOPHOSPHATES[a],[b]

	A or G						C, U, or T					
	H-8		H-2		H-1'		H-6		H-5		H-1'	
	δ	$\Delta\delta_D$	δ	$\Delta\delta_D$	δ	$\Delta\delta_D$	δ	$\Delta\delta_{D'}$	δ	$\Delta\delta_D$	δ	$\Delta\delta_D$
ApC [c]	8.79	−0.015	8.61	0.08	6.49	0.06	8.15	0.28	6.12	0.41	6.18	0.24
CpA	8.82	0.11	8.63	0.06	6.51	0.05	8.11	0.18	6.24	0.25	6.13	0.24
ApU	8.75	0.03	8.60	0.09	6.47	0.08	8.18	0.26	6.06	0.33	6.18	0.23
UpA	8.83	0.10	8.63	0.06	6.51	0.05	8.15	0.16	6.18	0.14	6.13	0.24
GpC	8.44	+0.01	—	—	6.29	0.10	8.23	0.20	6.13	0.40	6.31	0.10
CpG	8.43	0.12	—	—	6.31	0.04	8.16	0.13	6.31	0.18	6.19	0.17
GpU [d]	8.42	0.01	—	—	6.33	0.06	8.26	0.18	6.17	0.22	6.33	0.08
UpG [d]	8.45	0.10	—	—	6.32	0.03	8.18	0.13	6.21	0.11	6.20	0.17
CpU (C)	—	—	—	—	—	—	8.31	+0.03	—[e]	—[e]	—[e]	—[e]
(U)							8.34	+0.10	—[e]	—[e]	—[e]	—[e]
UpC (C)	—	—	—	—	—	—	8.35	0.08	6.45	0.08	6.37	0.04
(U)							8.33	0.02	6.28	0.04	6.26	0.11
ApG [d] (A)	8.67	0.11	8.52	0.17	6.34	0.21						
(G)	8.35	0.20			6.24	0.11						
GpA [d] (A)	8.76	0.17	8.62	0.07	6.51	0.05						
(G)	8.32	0.11			6.12	0.27						
ApA [c] (5')	8.69	0.24	8.61	0.07	6.41	0.15						
(3')	8.66	0.12	8.47	0.22	6.29	0.26						
CpC (5')							8.34	0.09	—[e]	—[e]	—[e]	—[e]
(3')							8.29	0				
UpU (5')							8.34	0.10	6.32	0.07	6.37	0.04
(3')							8.33	+0.02	6.29	0.03	6.31	0.06
									(methyl)			
TpT (5')							8.12	0.08	2.32	0.02	6.75	0.02
(3')							8.08	0.02	2.32	0 [f]	6.66	0.05 [f]

[a] 28°–30°C, pD 5.9; in ppm; all negative unless otherwise indicated, TMS capillary. For 5'-nucleotidyl unit $\Delta\delta_D = \delta_{pN'} - \delta_{NpN'}$; for 3'-nucleotidyl unit $\Delta\delta_D = \delta_{Np} - \delta_{NpN'}$.

[b] Unless specified, the dimer concentration was 0.02 *M*. The accuracy of $\Delta\delta_D$ is about ±0.01 ppm. Data from Ts'o *et al.* [34] and Kondo *et al.* [24].

[c] At infinite dilution.

[d] At 0.004 *M* concentration.

[e] Spectral peaks indiscernible.

[f] The δ values of thymidine are substituted for the δ values of 3'-TMP.

that of the monomer. In view of the fact that the energy minimum of the excimer is created by a covalent interaction between the π-electron systems, the stacking of the aromatic molecules increases the probability of excimer formation. Therefore, the presence of a fluorescence spectrum of the excimer type is considered as an indication of stacking phenomenon.

Figure 10 shows the examples of fluorescence spectra of four dinucleoside monophosphates in ethylene glycol:water (1:1) glass, measured at 85°K.

The broad and red-shifted fluorescence spectra of the dimers at pH 7 versus those of the corresponding monomers are the evidence of the excimer excited state of the stacked dimer. At pH 7, CpC, UpA, ApC, and to a lesser extent TpT, all give the excimer fluorescence, indicating that these dimers are stacked under this condition; while at pH 2 for CpC (protonated) and at

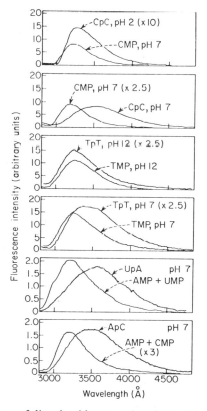

Fig. 10. Fluorescence of dinucleoside monophosphates. The total base concentrations are $3 \times 10^{-3} M$ in ethylene glycol:water (1:1) glass, measured at 85°K. (From Eisinger *et al.* [44].)

pH 12 for TpT (ionized), these dimers have the same fluorescence as their respective monomers, indicating that the charged dimers do not stack as expected. More dimer fluorescence spectra (such as UpA, GpU, etc.) and the comparison between the dimer fluorescence and the corresponding polynucleotide fluorescence can be found in Chapter 3 and other references [43,44]. In addition, by using this fluorescence technique in conjunction with

UV hypochromicity measurements, a study of the base–base interaction in a series of dinucleotide analogs has been made [45–47]. In these analogs, B—$(CH_2)_3$—B', where B and B' are 9-substituted adenine, guanine, purine derivatives (such as dimethyl guanine) or 1-substituted cytosine, thymine, and uracil residue, the two bases are connected by a trimethylene chain [—$(CH_2)_3$—] instead of a pentosyl phosphodiester bond. This study [45] concludes that in the series B—$(CH_2)_3$—B', the order of interaction in neutral aqueous solution is purine–purine > purine-pyrimidine > pyrimidine-pyrimidine, a conclusion identical to that derived from the study of monomer–monomer interaction (Chapter 6, Volume I; Section II,C). For example, the fluorescence emission of 9,9-trimethylene–bisadenine [Ad—$(CH_2)_3$—Ad] was characteristic of an excimer state. The emission of adenine–$(CH_2)_3$–cytosine and guanine–$(CH_2)_3$–thymine occurred from both excimer states and from excited singlet states, and the fluorescence emission of thymine–$(CH_2)_3$–thymine, cytosine–$(CH_2)_3$–cytosine, and cytosine–$(CH_2)_3$–thymine occurred only from excited singlet states similar to those found for the isolated chromophores of the molecules. While the origin of this excimer fluorescence phenomenon is now well understood (Chapter 4, Volume I), this approach is unlikely to provide information about the dimer conformation in detail. Also, the fluorescence measurement can be carried out only under limited conditions, such as in a frozen glass.

C. A General Conformational Model for Dinucleoside Monophosphates

While the UV hypochromicity and fluorescence studies can only indicate an extensive stacking interaction between the bases in the dimer, the CD and the PMR studies are able to provide more detailed information about the conformation of the dimer. As indicated in the discussions in Chapter 6, Volume I, Section I, and in Section I,A of this chapter, the main conformational feature of the dimer is governed by the rotation of glycosyl bond (ϕ_{CN}) between the furanose and the base, and by the rotation of the two P—O bonds in the phosphodiester linkage. Therefore, the general conformational model for a dinucleoside monophosphate can be described by three principal stereochemical considerations: (1) the conformation of the nucleosidyl unit (*syn* vs *anti* conformation); (2) the extent of base–base overlap; (3) the screw axis of the stack (right-handedness vs left-handedness).

The concept of chirality deserves additional discussion, since this consideration is derived from the joining of the monomeric units into an oligomer or polymer. Strictly speaking, the handedness of the helical conformation of the oligo- or polynucleotide chain is defined by the configuration of the furanose–phosphodiester backbone only, and is not defined by the arrangement of the bases. In the case of dinucleoside monophosphate, it contains only one unit (or one link) of the furanose–phosphodiester backbone; therefore, it does not contain sufficient information to define the handedness of the helical turn based solely on the configuration of the backbone. Thus, in this situation, the arrangement of the bases is used for the definition of the handedness of the screw axis. The plane of the rotation is defined by the plane of the base (base 1) which is justifiably taken to be parallel to the plane of the neighboring base (base 2); and the angle in rotation is the one formed between the geometrical principal axis of the two bases. If the axis of the stack is advancing *upward* from the plane of the base (as base 1 in Fig. 11a; base 1 lies in the same plane as the paper while base 2 lies above the paper) and is simultaneously rotating *counterclockwise* in following the direction of the backbone, then the stack is termed right-handed, as shown schematically in Fig. 11a. Conversely, if the axis of the stack is advancing *downward* from the plane of the base (base 1) and is simultaneously rotating counterclockwise in following the direction of the backbone, then the stack is termed left-handed, as shown in Fig. 11b. However, the handedness of the stack can also be changed from right-handed to left-handed by maintaining the same direction of advancement of the axis (say, upward), but changing the direction of rotation (say, from counterclockwise to clockwise). This change is illustrated in the sequence shown in Fig. 11c and d. In Fig. 11c, the bases are rotating from a stacked position shown in Fig. 11a to a parallel position. This is the situation of a straight stack without turn, since the principal axes of the bases do not intersect each other. As the bases are rotating farther away from each other and assuming the positions shown in Fig. 11d, the rotation angle from the axis of base 1 to the axis of base 2 changes from counterclockwise to clockwise. By definition, therefore, this conformation can be designated as left-handed, the same term used to describe the conformation shown in Fig. 11b. However, these two conformations are fundamentally different in two ways: (1) the bases are overlapped extensively in the conformation of Fig. 11b, but are far apart in the conformation of Fig. 11d; (2) the extension of the dimer conformation in Fig. 11d to the trimer, oligomer, etc., in a repetitive fashion, will lead to the construction of a polymer in which the hydrophobic bases are exposed to the outside of the helix, and the hydrophilic, charged, phosphate groups are congregated on the inside of the helix (Fig. 12g).

This concept can be described by the presentations shown in Fig. 12. In Fig. 12a–d the dimer in straight stack (a); in right-handed stack (b); in left-handed (unstacked) conformation (c); and in left-handed stack (d) is shown for the purpose of comparison to the trimer. This series also illustrates the point that with only one link (one phosphodiester unit), it is not possible to define the helical sense from the backbone alone without the aid from the orientation of the bases. In Fig. 12e–h, the trimer in straight stack (e); in right-handed stack (f); in left-handed (unstacked) conformation (g); and in left-handed stack (h) is illustrated. With two linkages, the helical sense of the backbone can now be identified even in the absence of the bases. The conformation shown in 12g is left-handed. However, this helical conformation

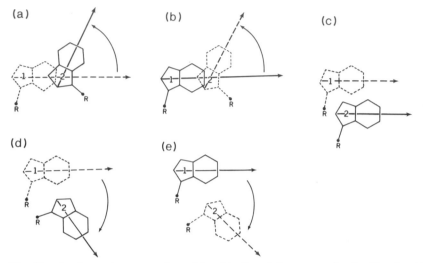

Fig. 11. A schematic presentation of the chirality of the screw axis of a dinucleoside monophosphate containing two purine bases. The two planes of the bases are both parallel to the plane of the paper and the advancement of the axis starts from base 1 to base 2. The geometrical principal axes and their intersectional angles are shown; the heavy-line base is above the dotted-line base. (a) The advance of the screw axis is *upward* (base 2 above base 1) and the rotation is counterclockwise in this right-handed conformation; (b) the advance of the screw axis is downward (base 2 below base 1) and the rotation is still counterclockwise in this left-handed conformation; (c) the bases are stacked in a parallel fashion, no rotation and no helical turn; (d) the advance of the screw axis is upward and the rotation is clockwise. By definition, this conformation can be designated as left-handed. However, for reasons stated in the text, this conformation is designated simply as an unstacked conformation; (e) this conformation is analogous to the conformation shown in (d), with the downward advancement instead. The relationship between the conformations in (b) and (e) is the same as that between the conformations in (a) and (d). Therefore, it is also designated as an unstacked conformation. Finally, the change of conformation from (a) to (d), to (e), and then to (b), represents the sequence of transformation from a right-handed dimer to a left-handed dimer. The reverse sequence is, therefore, the transformation of a left-handed dimer to a right-handed dimer.

will lead to the exposure of the hydrophobic bases to the outside of the helix, and the congregation of the charged phosphate on the inside of the helix. Such an arrangement is, energetically, highly unfavorable. In this conformation (Fig. 12g), the bases are not overlapped. For these reasons, the conformation shown in Fig. 12c for the dimer and in Fig. 12g, will be simply termed as unstacked; and the term left-handed stack is reserved for the conformation of the dimer shown in Fig 11b and 12d, and of the trimer shown in Fig. 12h.

The transformation of a right-handed stack of dinucleoside monophosphate to a left-handed stack can be followed in the sequence shown in Fig. 11. Upon rotation of the backbone (mainly the two P—O bonds), the bases become

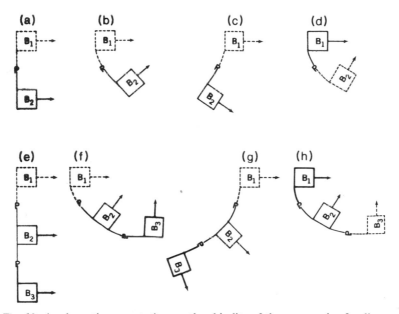

Fig. 12. A schematic presentation on the chirality of the screw axis of a dimer and a trimer. All of the base planes are parallel to each other and to the plane of the paper, and the advancement of the axis starts from base 1 to base 2 and then to base 3. The geometrical principal axes and their intersectional angles are shown; the heavy-lined base is above the normal-lined base, and the dotted-line base is below the normal-lined base. (a) Dimer in a straight stack, same as Fig. 11c; (b) dimer in a right-handed stack, same as Fig. 11a; (c) dimer in a left-handed unstacked conformation, same as Fig. 11d; (d) dimer in a left-handed stack; (e) trimer in a straight stack; (f) trimer in a right-handed stack; (g) trimer in a left-handed unstacked conformation; (h) trimer in a left-handed stack.

separated and the right-handed conformer (Fig. 11a) is transformed to an unstacked conformation (Fig. 11c or d; these two conformations differ only slightly), with base 1 still below base 2. Further rotation changes the unstacked conformer (Fig. 11d) to the unstacked conformer (Fig. 11e) which has the base 1 located above the base 2. Upon stacking of the bases, the unstacked conformer (Fig. 11e) can assure the conformation of a left-handed stack (Fig. 11b). Reversal of this process will transform the left-handed conformer

(Fig. 11b) back to the right-handed conformer (Fig. 11a). Examination of the CPK models indicates that the right-handed to left-handed transformation (or vice-versa) involves mostly rotation of the two P—O bonds (ω and ω', Fig. 1 and Table I), and perhaps with a slight adjustment (10°–30°) of the O-5' to C-5' bond, C-4' to C-5' bond, and glycosyl bond [ϕ, ψ, and ϕ_{CN} (or χ), respectively; Fig. 1 and Table I]. From this and other considerations, the general conformational model discussed in this section is on a time-averaged basis of a dynamic equilibrium between various conformers.

As described in Chapter 2 significant progress has been made in calculating the CD–ORD spectra of nucleic acids according to the exciton theory based on a given set of geometrical considerations. For dinucleoside monophosphates, Bush and Tinoco [32] were first to calculate the ORD patterns of 16 ribosyl dimers and to compare the calculation with the experimental results. A monopole-monopole interaction method was adopted and the geometry of the dimers was assumed to be that of the single strand in the B form of DNA. This means that the nucleosidyl units are in *anti* conformation and the bases are in parallel planes spaced 3.4 Å apart with 10 bases per right-handed turn (or the angle in the helical turn between the two successive bases is 36°). The agreement between the calculated spectra and the ORD spectra is good for ApA, UpU, ApU, UpA, and less satisfactory with dimers containing C and G because of the lack of knowledge of their excited state properties. Nevertheless, the calculation is sufficiently successful to warrant the main features of the theory and of the conformation adopted. Moreover, their calculation clearly indicates the dependence of the optical activity on the angle between the bases in the dimer as discussed in Eq. (7) in Section I,B. Their calculation on the rotation of ApA vs the angle between the two parallel stacked bases in ApA is shown in Fig. 13, which shows that when the geometrical axes of the two bases are parallel or nearly perpendicular to each other, the rotations from the exciton effect will be very low.

Another major attempt has been made more recently by Johnson and Tinoco [31] to calculate the CD spectra in comparison to the observed spectra measured at very low temperature. Both coordinates of DNA in B form and the coordinates of helical RNA were employed. The RNA helix was chosen to have 11 bases per turn and an 11° tilt of the bases with respect to the helix axis. The new theoretical approach, however, was found to produce less improvement than expected, since the errors in the previous method of calculation were found to cancel each other. The best agreement between the calculated and the observed spectra is shown in Fig. 14 for ApA, CpC, and UpU. In general, the calculation based on RNA conformation offers no improvement. In summary, we can conclude from this comparison between the experimental and calculated CD spectra that the general conformational model of the dimers in solution is similar to that in the helical nucleic acids,

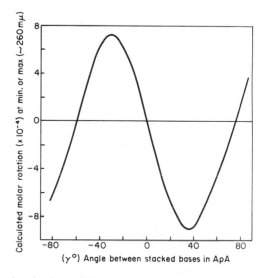

Fig. 13. Calculated values of the maximum (or minimum) optical rotation of ApA vs the angle between the bases (angle = 36° for 10 bases per helical turn). (From Bush and Tinoco [32].)

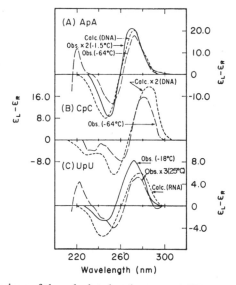

Fig. 14. Comparison of the calculated and measured CD spectra: (a) ApA, calculated for DNA conformation; (b) CpC calculated for DNA conformation; (c) UpU for RNA conformation. (From Johnson and Tinoco [31].)

i.e., nucleosidyl unit in *anti* conformation and the bases are stacked with a right-handed screw axis.

There is another strong argument which indicates that D-rA$_{3'}$p$_{5'}$rA must exist in aqueous solution predominantly in a right-handed stack. As described in the following section, L-rA$_{3'}$p$_{5'}$rA has been synthesized from L-adenosine [48]. This dimer must have a conformation opposite to the naturally occurring D-ApA, and its CD spectrum is opposite in sign but same in magnitude as that of D-ApA. The L-ApA forms a complex with D-poly U having the same stoichiometry (1A:2U) as the complex of D-ApA with D-poly U. The CD spectra of the L-ApA · 2-D-poly U complex, D-ApA · 2-D-poly U complex, and finally the D-poly A · 2-D-poly U complex are all nearly identical, indicating that the overall conformations of these helical complexes are similar. X-ray diffraction studies indicate that the chirality of the D-poly A · 2-D-poly U complex is right-handed [1,2,48]; therefore, the conformation of both L-ApA and D-ApA in the triple-stranded complex must also be right-handed. The T_m of the D-ApA · 2D-poly U complex (13.7°C) is significantly higher than the T_m of L-ApA · 2-D-poly U complex (5.5°C) under the same conditions. The thermodynamic analysis of these data indicates that this difference in T_m reflects the difference in conformation between the free L-ApA and the free D-ApA, the conformation of the free D-ApA being closer to the conformation of the ApA inside the helix. Since the ApA in the helix has a right-handed conformation, this study concludes that the conformation of the free D-ApA in solution is predominantly right-handed also, while the conformation of the free L-ApA is therefore mainly left-handed.

PMR studies on the dimers can provide even greater detail on their conformation than the optical methods. Concerning the conformation of the nucleosidyl units in the dimer, PMR results definitely indicate the dominance of the *anti* conformation. This conclusion is supported by the following observations: (1) The specific broadening effect of Mn^{2+} on the H-8 of -rpA residue in rAprA [33], of -pdA residue in dApdA [36], and of -prI in rIprI [49] has demonstrated that all these nucleosidyl units in the 5'-residue of these dimers are in *anti* conformation. As discussed above, only when the nucleosidyl unit is in *anti* conformation, would the H-8 proton be near enough to the phosphate–Mn^{2+} complex for this specific broadening effect to occur. (2) The specific deshielding effect of the phosphate group on the H-8 of prAp-residue in prAprA [34] and of prIp- residue in prIprI [49], as well as the H-6 of prUp- residue in pUpU, has indicated that the nucleosidyl units in the 3'- residue of these dimers are also in *anti* conformation. This conclusion was obtained by the comparison between the chemical shifts of H-8 of pPupPu vs that of PupPu (Pu = purine nucleosides), or by the comparison between the chemical shifts of H-6 of pPypPy vs that of PypPy (Py = pyrimidine nucleosides). As discussed in above paragraphs and in Chapter 6, Volume I, the negatively charged phosphate group exerts a deshielding effect on the H 8 proton of purine or H-6 proton of pyrimidine in proximity when the nucleosidyl unit is in *anti* conformation. (3) This specific phosphate deshielding effect on H-8 (purine) or H-6 (pyrimidine) of the 5′-residue (-pN) can also be detected in the dimer NpN, especially when N is a pyrimidine base which has little or no ring current magnetic anisotropy, such as uracil or thymine [34]. This effect has been confirmed after the synthesis of alkyl phosphotriester of dApdA and dTpdT [50]. Comparison of the chemical shifts of H-8 or H-6 of -pdN in dNpdN, which has a charged phosphate group, to those of -p(R)dN in dNpRdN (R = methyl, or ethyl), in which the phosphate group has been neutralized to become an alkyl triester, clearly indicates the deshielding effect of the negatively charged phosphate group [50]. Thus, this observation provides important evidence that the -pN residues in dApdA and dTpdT are in the *anti* conformation. Examination of the stacking patterns of many dimers by the aid of molecular models (such as CPK) indicates that only an *anti, anti* conformation (both nucleosidyl units in *anti* conformation) can explain all the chemical shifts data or the dimerization shifts data [24,33,34, 36,40]. Therefore, the four observations cited above firmly established that the nucleosidyl units in the dimer are predominantly in *anti* conformation.

As to the extent of the base–base overlap in the dimer, the quantitative interpretation of the dimerization shifts data ($\Delta\delta_D^t$) is based on the assumption that this parameter is originated entirely from the influence of diamagnetic anisotropy of the neighboring base. Johnson and Bovey [51], following the approach of Waugh and Fessenden [52], have applied the ring current concept to account quantitatively for the additional shielding observed for the aromatic protons (such as those in benzene) over the closely related olefinic protons (such as those in 1,3-cyclohexadiene). In this approach, the electrons are assumed to circulate in two loops of radius a (radius of the benzene ring) separated by $+P$ and $-P$ above and below from the plane of the carbon. Satisfactory agreement between the calculation and the observed data was obtained.

The shielding zones for the purines and pyrimidines have been calculated in a plane 3.4 Å distant from the molecular surface by Giessner-Prettre and Pullman [53]. The shielding zones of adenine and cytosine at distance of 2, 3, 4, and 5 Å from the plane have been computed (B. Pullman and C. Giessner-Prettre, unpublished data). Examination of these calculations indicates that models cannot be built to correlate the observed data with the patterns computed at distances equal to 2 and 5 Å [39]. This study suggests that patterns of shielding zones computed at a distance of 3–4 Å are most suitable for model construction; owing to the x-ray diffraction data on the dimer crystals and oriented fibers (see Section I,A), the distance of 3.4 Å appears to be most appropriate. In Fig. 15, the schematic presentation of the front view of the conformational model for rAprA is shown. The bases are taken to be parallel with a vertical distance of 3.4 Å. This model is constructed on the basis of $\Delta\delta_D$ values measured at 4° [54] and of the isoshielding curves [53]. The isoshielding zones of the 3'- residue (rAp-) at the bottom are shown in Fig. 15a, while the isoshielding zones of the 5'- residue (-prA) at the top are shown in Fig. 15b. The arrangement of the two residues is not only governed by the consideration of matching the $\Delta\delta_D$ values of six protons to two appropriate sets of shielding, but also by the constraint of the backbone as indicated by both CPK and Kendrew models. The location of the H-1'

rAprA 4°

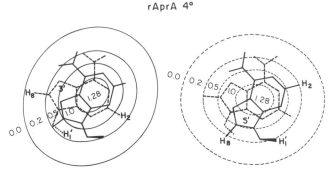

Fig. 15. Schematic presentation of the conformational model of rAprA as viewed in the 5' to 3' direction. The dimensions of the bases and the magnetic isoshielding zones are from the work of Giessner-Prettre and Pullman [53]. The bases are parallel to each other with a distance of 3.4 Å (bases with solid lines are above bases with dotted lines) and to the plane of the paper with the H-1' protons out of the plane. The relative orientation between the two bases was determined by consideration of the PMR data at 4°; the isoshielding zone values of Giessner-Prettre and Pullman; and backbone constraint determined by building CPK and Kendrew models. The nucleotidyl units have an *anti* conformation and the screw axis is right-handed, i.e., the axis advances upward from the plane of the paper in a counterclockwise manner. (From Kondo *et al.* [54].)

is less certain since they are not in the same plane as the base protons. The same approach is adopted to construct the schematic presentation of dApdA in Fig. 16. It should be noted that these static models are only the representation of the time-average results of a dynamic phenomenon. The differences between the conformational model of rAprA and dApdA will be discussed in Section I,D.

The dimerization shifts of 16 ribosyl dimers and TpT at 28°–70°C are tabulated in Table V. These data can be used therefore to construct conformational models in the estimation of the extent of base–base overlap of these dimers. The pyrimidine dimers have small $\Delta\delta_D^{30°C}$ values owing to their small ring-current effect and their low tendency in stacking (Chapter 6,

Volume I). It is interesting to note the differences in $\Delta\delta_D$ values between the sequence isomers such as CpA vs ApC (Table V). This difference has been very helpful previously in the construction of conformational models [34,40]. The above discussion, therefore, demonstrates that the extent of base–base overlap in a dimer can be quantitatively estimated from the PMR data.

The dimerization shifts data of the base protons in the dimer, however, provide little information about the handedness of the screw axis of the stack, especially of the homodimer. Because of the symmetry of the ring-current magnetic anisotropy, in reference to the plane of the bases, the same dimerization shifts of the base protons can be obtained from either the right-handed stack or the left-handed stack. Therefore, the chirality of the stack can be derived mainly from the $\Delta\delta_D^t$ data of the furanose protons. The

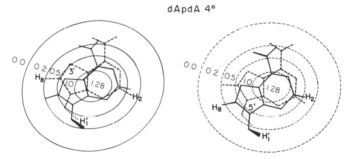

Fig. 16. Schematic presentation of conformational model of dApdA. Explanation is the same as that for Fig. 15. [From Kondo *et al.* (54).]

measurement, assignment, and analyses of the furanose protons in a dimer are a formidable task. For the ribosyl dimers, the H-2' and H-3' protons are too close to the HDO peak for adequate resolution. For the deoxyribosyl dimers, the spectral patterns are complex due to multiple spin-spin coupling. Fortunately, the analyses of the furanoses of the dApdA have been accomplished [36]. The success of this study is due to the availability of all the necessary reference compounds (such as 3'-AMP, 5'-dAMP, rApdA, dAprA, and rAprA), and to the availability of both 220 MHz spectra and 100 MHz spectra for comparison. Through the help of these reference compounds and the Mn^{2+} broadening procedure discussed above, definite assignments of all the base protons and the H-1', H-2', and H-2" protons were made [36]. The dimerization shifts (in ppm) at 28° in D_2O for the furanose protons are as follows: For -pdA residue, H-1' (0.20), H-2' (*endo*, −0.06), and H-2" (*exo*, −0.03); for dAp- residue, H-1' (0.36), H-2' (*endo*, 0.61), and H-2" (*exo*, 0.21); where H-2' (*endo*) proton is defined as *trans* to H-1' proton and H-2" (*exo*) proton is defined as *cis* to the H-1′ proton. These data indicate that the H-2′ (*endo*) of the dAp- residue is located inside the stack, very near to and strongly shielded by the 5-membered ring of the adenine in the -pdA residue; in fact, this is the most shielded proton in the dimer. This observation that the H-2′ (*endo*) is much more shielded than the H-2″ (*exo*) clearly reveals that the dApdA is predominantly a right-handed stack. In this model, the H-2″ (*exo*) is located outside the stack. For a left-handed stack, the H-2′ (*exo*) would be more shielded than the H-2″ (*endo*); in this case, the H-2″ (*exo*) would be located inside and the H-2′ (*endo*) outside the stack. Similarly, from a right-handed conformational model, the H-1′ of dAp- residue can be expected to be more shielded than the H-1′ of -pdA residue as observed. Since the H-1′ protons of rAp- residue [34,54] and of rIp- residue [49] are more shielded

TABLE VI

The Coupling Constants of Mono- and Dideoxynucleoside
Monophosphates [a,b,c]

	$J_{1'-2''(cis)}$		$J_{1'-2'(trans)}$		$J_{2'-3'(cis)}$		$J_{2''-3'(trans)}$	
	(3′)	(5′)	(3′)	(5′)	(3′)	(5′)	(3′)	(5′)
dApdA	5.6	6.6	8.8	6.7	5.5	6.7	2.2	4.0
dAprA	5.7		8.7		5.5		2.0	
3′-dAMP	6.0		8.1		6.3		2.9	
5′-dAMP		6.6		6.9		5.8		3.8

[a] Measured at 220 MHz, in D_2O, 0.05 M, pD 6–7, and 20°C.
[b] The coupling constants are given in Hz.
[c] From Fang *et al.* [36].

than the H-1′ of -prA and of -prI, respectively in rAprA and rIprI, these two ribosyl dimers are also in a right-handed stack. As for the heterodimers, this approach becomes less certain. The differences in the shielding patterns of the base protons and the H-1′ proton between the two sequence isomers, such as ApC vs CpA have been used as supportive evidence to indicate that these dimers are in a right-handed stack [34].

The coupling constants of the furanose protons provide valuable information about the furanose conformation of the dimer. In Table VI, the four coupling constants of the 3′- residue and the 5′- residue of dApdA are shown together with their respective monomers and one analog, dAprA. The data show that the coupling constants between the dAp and pdA residues of dApdA are similar to those of 3′-dAMP and 5′-dAMP, while the coupling constants of 3′-dAMP and 5′-dAMP are not the same. Through the application of the Karplus equation (Chapter 6), four dihedral angles, $\phi_{1'-2'}$ (*cis*),

$\phi_{1'-2'}$ (*trans*), $\phi_{2'-3'}$ (*cis*), and $\phi_{2'-3'}$ (*trans*) were determined [36]. The results are shown in Table VII together with the dihedral angles for the four standard, plausible furanose conformations [55]. These analyses suggest that the furanose conformation for 3'-dAMP and dAp- in dApdA is that of $C_{2'}$-*endo* (envelope) or $C_{2'}$-*endo*–$C_{3'}$-*exo* (twisted form), while the furanose conformation for 5'-dAMP and pdA in dApdA is that of a rapid equilibrium between $C_{2'}$ *endo* and $C_{3'}$ *endo*. As discussed in Section I,D, furanose conformations of the residues in rAprA are quite different from those of their respective monomers, contrary to the observations here concerning dApdA and dAMP.

TABLE VII

Calculated Dihedral Angles for dApdA,[a] 5'-dAMP,[a] and 3'-dAMP,[a] and for the Four Plausible Furanose Conformation [b]

	$\theta_{1'-2''(cis)}$	$\theta_{1'-2'(trans)}$	$\theta_{2'-3'(cis)}$	$\theta_{2''-3'(trans)}$
dAp-(3')	40°	160°	40°	60° (120°)
-pdA(5')	30°	145°	30°	130° (50°)
5'-dAMP	35°	145°	40°	130° (50°)
3'-dAMP	35°	155°	35°	55° (125°)
C-2'-*endo*-C-3'-*exo*	45°	165°	60°	60°
C-2'-*endo*	45°	165°	45°	75°
C-3'-*exo*	15°	135°	45°	75°
C-3'-*endo*	15°	105°	45°	165°

[a] The coupling constants used for the calculation of the dihedral angles are from Table VI [36].
[b] The data are from Smith and Jardetzky [55].

In summary, we may conclude from all these studies that generally the dinucleoside monophosphates form right-handed stacks and have both their nucleosidyl units in an *anti* conformation. The base planes are usually parallel to each other, separated by a distance of 3.4 Å. The extent of base–base overlap can be semiquantitatively estimated by optical methods with the UV hypochromicity measurement most sensitive to the vector distance between the bases and the CD amplitude measurement most sensitive to the intersection angle formed between the axes of the bases. PMR studies can provide more detailed information about the geometry of the stacking and stereochemistry of the furanoses.

D. The Influence of the Backbone on the Conformation of the
Dinucleoside Monophosphates

In the preceding section, we established a general conformational model
for the dinucleoside monophosphates in solution. The extent of base–base
stacking of these dimers is dependent on the kinds of bases in the dimer.
Besides the studies on dimers with the usual bases cited in previous sections,
investigation has been made into dimers containing unusual bases [56],
such as adenylyl(3′ → 5′)N⁶-(Δ²-isopentenyl)adenosine, N⁶-(Δ²-isopentenyl)-
adenylyl(3′ → 5′)adenosine and adenylyl(3′ → 5′)pseudouridine, as a model
for studies on the anticodon loop of tRNA. However, the properties of the
bases in the dimer can be predicted from their properties as monomers, as
described in Chapter 6, Volume I. For instance, as monomers, the tendency
of association of purine with purine by stacking is considerably stronger
than that of pyrimidine with pyrimidine. Therefore, in general, base–base
stacking of purine dimers is more extensive than that of pyrimidine dimers.
More interesting will be the investigation on the influence on the properties of
the dimers containing the same bases, but with different types of backbone in
linking these bases. Through this approach, we hope to learn about the contribution of the backbone to the conformation of the dimers. After all, the
difference between the monomers and dimers lies in the existence of a
linkage in the dimers. Most of the studies discussed in this section concern
the adenine dimers joined by different types of backbones.

1. *Ribosyl Dimers Joined by Phosphodiester linkage at Different Positions of
   the Furanose*

The interactions of two adenines in dinucleoside monophosphates joined
at different positions by a phosphodiester linkage (2′-5′, 3′-5′, and 5′-5′) have
been studied by proton magnetic resonance, circular dichroism, and ultraviolet absorbence over the temperature range of 4°–60°C [39]. Similarly, the
interactions of adenine and cytosine in $A_{2'}p_{5'}C$ and $A_{3'}p_{5'}C$ were also
investigated. While these five dimers possess certain basic features in common
in their conformations, the geometrical relationship between the two con-
stitutive units, as well as the temperature effect on this geometrical relationship, is quite different for each dimer when examined in sufficient detail.
These results indicate that the conformation of the dimers is strongly influenced
by the position of the phosphodiester linkage. Studies on the interaction of
adenine–adenine (or adenine-cytosine) having three types of geometrical
relationships by three physicochemical methods offer a unique comparison
among different types of dependence of these three methods (PMR, CD, and
UV absorbence) on the geometrical relationships between two bases. Since
the main conclusion in this study can be well illustrated by the work on the

adenine homodimer, the results on the adenine–cytosine dimer will not be described here.

The extinction coefficients and hypochromicity values in UV absorbence of the three adenine dimers, $A_3'p_5'A$, $A_2'p_5'A$, and $A_5'p_5'A$, are shown in Table VIII. The temperature dependence of the UV absorbence at λ_{max} over the 0°–70°C range has also been studied. Within this temperature range, the absorbence is linearly proportional to the temperature. This constant relationship between the variation in absorbence vs temperature change can be expressed by $\varepsilon_{max(t)} = \varepsilon_{max\,25°C}[1 + \theta(t - 25°C)]$, where $\varepsilon_{max\,25°C}$ is the maximal molar extinction coefficient of the individual dimer at 25°C (Table VIII), t is the temperature (°C) within the range of 0°–70°C, and the $\theta(1/°C)$ is the proportionality constant between the temperature change and the absorbence

TABLE VIII

Hypochromicity and Molar Extinction Coefficients
of Adenine–Adenine Dinucleoside Monophosphates
at Neutral pH and Room Temperature [a]

Dimer	λ (nm)	$\varepsilon(10^3)$	Hypochromicity (%)
$A_3'p_5'A$	258	13.6	11.9
$A_2'p_5'A$	258	12.9	15.9
$A_5'p_5'A$	259	12.0	22.1

[a] Date from Kondo *et al.* [24].

variation after normalization to the ε_{max} at 25°C. The value of θ is an indication of the extent of temperature effect on the absorbence. The θ values (1/°C) for the three dimers were found to be as follows: 8×10^{-4} for $A_3'p_5'A$; 10×10^{-4} for $A_2'p_5'A$; 11.7×10^{-4} for $A_5'p_5'A$. We may conclude that at room temperature, $A_5'p_5'A$ has the largest value of hypochromicity ($\sim 22\%$), $A_2'p_5'A$ has the next largest value ($\sim 16\%$), and the $A_3'p_5'A$ has the smallest value ($\sim 12\%$). As for the temperature effect, the same order also holds; that is, the absorbence of $A_5'p_5'A$ is slightly more sensitive than that of $A_2'p_5'A$, and the absorbence of $A_2'p_5'A$ in turn is slightly more sensitive than that of $A_3'p_5'A$.

The CD spectra of $A_3'p_5'A$, $A_2'p_5'A$, and $A_5'p_5'A$ in 0.05 M NaClO$_4$ and at four temperatures are shown in Fig. 17. These spectra all have similar peak positions (273 ± 2 nm) and trough positions ($\sim 253 \pm 2$ nm). Within the temperature range of 5°–60°C, the spectral positions of the peak and trough of these three dimers remain about the same and the changes of the $[\theta]$ values for both the peak and the trough are all linearly dependent on the temperature

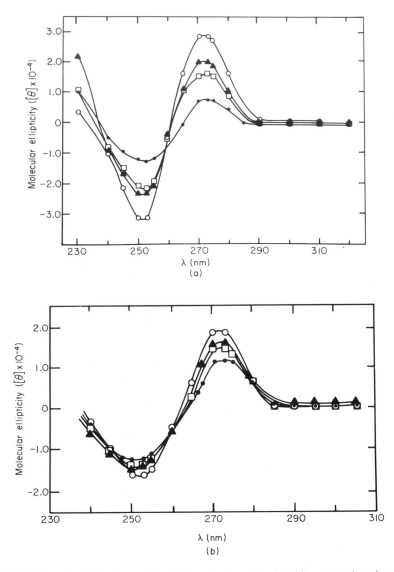

Fig. 17, Circular dichroism of adenine–adenine dinucleoside monophosphates in 0.05 *M* NaClO$_4$ (pH 7.3) and at the following four temperatures: 4.5°C (○), 23°C (▲), 30°C (□), and 60°C (●). (a) A$_3$·p$_5$·A; (b) A$_2$·p$_5$·A; (c) A$_5$·p$_5$·A. (Kondo *et al.* [24].)

(see Kondo *et al.* [24] for more details). The [θ] values of A$_{3'}$p$_{5'}$A are most sensitive to temperature, with the [θ] values of A$_{5'}$p$_{5'}$A next most sensitive, and the [θ] values of A$_{2'}$p$_{5'}$A least sensitive to temperature change. Similarly, as shown in Fig. 17, at room temperature or below, the amplitude of [θ] values between the peak and trough is largest for A$_{3'}$p$_{5'}$A, next largest for A$_{5'}$p$_{5'}$A and least for A$_{2'}$p$_{5'}$A.

The dimerization shifts of these three adenine dimers derived from PMR studies are shown in Table IX. It should be noted that the PMR spectrum of A$_{5'}$p$_{5'}$A consists of only one set of proton resonances, such as one H-8, one H-2, etc., instead of two sets of resonances as in A$_{3'}$p$_{5'}$A, such as two H-8, two H-2, etc. This magnetic equivalence of the same protons in both residues requires a symmetry consideration in building the conformational model. Naturally, the conformation of A$_{5'}$p$_{5'}$A would be different from those of A$_{3'}$p$_{5'}$A and A$_{2'}$p$_{3'}$A. In addition, the $\Delta\delta_D$ values of A$_{5'}$p$_{5'}$A are large in comparison to those of A$_{2'}$p$_{5'}$A and especially to A$_{3'}$p$_{5'}$A (Table IX). These results indicate extensive base–base overlapping in A$_{5'}$p$_{5'}$A. The dimerization shifts in Table IX also show that the stacking patterns of A$_{2'}$p$_{5'}$A and A$_{3'}$p$_{5'}$A are not the same. Comparing the six protons in two residues, A$_{2'}$p$_{5'}$A has higher $\Delta\delta_D$ values in the other two protons. Based on these dimerization shifts and the isoshielding zones computed by Giessner-Prettre and Pullman [53] discussed previously, as well as the aid of CPK models, the conformational models (Fig. 18) of A$_{3'}$p$_{5'}$A, A$_{2'}$p$_{5'}$A, and A$_{5'}$p$_{5'}$A have been proposed for comparison [39]. All of them have an *anti, anti*, right-handed conformation. For A$_{5'}$p$_{5'}$A, the model is constructed with a twofold axis of symmetry, C_{2v}, parallel to the two base planes and bisecting the phosphorus atom of the 5'-5' dimer. This twofold axis is generated by the intersection of the two

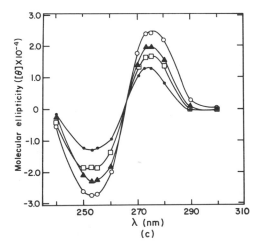

λ (nm)

(c)

TABLE IX

Dimerization Shifts ($\Delta\delta_D$) of the Base and H-1' Protons of $A_{3'}p_{5'}A$,
$A_{2'}p_{5'}A$, and $A_{5'}p_{5'}A$ at Various Temperatures [a,b]

		3' or 2'			5'		
	Temp. (°C)	H-8	H-2	H-1'	H-8	H-2	H-1'
$A_{3'}p_{5'}A$	4	0.17	0.345	0.285	0.30	0.125	0.19
	30	0.11	0.215	0.26	0.235	0.075	0.15
	60	0.12	0.145	0.19	0.175	0.105	0.11
	% change [c]	—[d]	58%	33%	42%	—[d]	42%
$A_{2'}p_{5'}A$	4	0.195	0.53	+0.01	0.53	0.07	0.33
	30	0.155	0.46	0.02	0.485	0.07	0.315
	60	0.155	0.385	0.03	0.445	0.04	0.29
	% change [c]	—[d]	28%	—[d]	16%	—[d]	12%
$A_{5'}p_{5'}A$	4				0.43	0.185	0.21
	30				0.40	0.185	0.18
	60				0.35	0.16	0.18
	% change [c]				18%	—[d]	—[d]

[a] From Kondo *et al.* [24].
[b] pD = 5.9; in parts per million, all negative unless otherwise indicated. For 5'-nucleotidyl unit $\Delta\delta_D = \delta_p A - \delta A_p A$ and for 3' or 2' nucleotidyl unit $\Delta\delta_D = \delta A_p - \delta A_p A$.
[c] Percentage change $= [1 - (\Delta\delta_D^{60°}/\Delta\delta_D^{4°})] \times 100$.
[d] The magnitude of change is too small to be accurate or significant.

perpendicular planes formed by the four oxygen atoms (one plane contains the two ester oxygens and the other plane contains the two ionized oxygens) in the tetrahedron of the phosphate group. In this conformation (Fig. 18), the hydrogen bonding sites (NH_2 and N-1 groups) of each of the two adenines in $A_{5'}p_{5'}A$ are facing in different directions from each other. Attempts to build a 2:1 complex between poly U and $A_{5'}p_{5'}A$ with this conformation were unsuccessful. This is in accord with the experimental findings that $A_{5'}p_{5'}A$ does not form a complex with poly U under the normal conditions which allow the formation of $A_{3'}p_{5'}A \cdot 2$ poly U complex [24].

From the dimerization shifts obtained from the PMR studies and the conformational models (Fig. 18) derived from these data, the order for the extent of the overlap of the two adenines in the dimers is as follows: $A_{5'}p_{5'}A >$

$A_{2'}p_{5'}A > A_{3'}p_{5'}A$. This order agrees with the order of the hypochromicity data (Table VIII). This agreement is in accord with the conclusion on the relationship between hypochromicity and the geometry of the dimer, particularly the homodimer, discussed in Section I,B of this chapter. It was shown in Eqs. (3) and (4) that the hypochromicity of the homodimer is strongly dependent on the distance between the two bases. The correlation of the PMR results and the models with the CD data is less certain. Superficially, the amplitudes between the peak and the trough of CD spectra give the following order for the extent of the base–base interaction: $A_{3'}p_{5'}A > A_{5'}p_{5'}A > A_{2'}p_{5'}A$. This order from CD studies is not in agreement with those indicated from UV hypochromicity data and from PMR data. As shown by

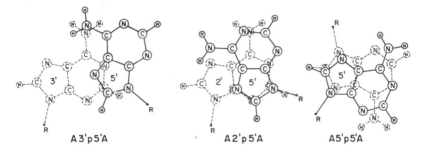

A3'p5'A   A2'p5'A   A5'p5'A

Fig. 18. Schematic presentation of the front view of the conformational models for $A_{3'}p_{5'}A$, $A_{2'}p_{5'}A$, and $A_{5'}p_{5'}A$. The dimensions of the bases are from x-ray studies and the base planes are parallel to each other (bases drawn by the dotted line are located at the bottom) and to the plane of the paper. The relative orientation between the two bases is positioned from the consideration of the PMR data at 4° and the constraint of the backbone as indicated by the CPK model. As shown, the nucleotidyl units have an *anti* conformation and the screw axis is right-handed. (Kondo *et al.* [24].)

Eq. (7) in Section I,B the [θ] values are sensitive to the sin θ cos θ, where θ is the angle between the transition dipole moment in the planes of bases (μ_{10a} and μ_{20a}). It was pointed out [24] that the angle θ between the transition moments (which is the same as the angle between the geometrical principal axes) of $A_{3'}p_{5'}A$ is about 60° ± 10°, and that θ for $A_{2'}p_{5'}A$ is about 80° ± 10°. In other words, the angle between the bases in $A_{3'}p_{5'}A$ is more oblique while the angle in $A_{2'}p_{5'}A$ is more perpendicular (Fig. 18). Figure 13 shows that when the angle between the stacked bases in ApA is close to 80°, the optical activity derived from the exciton interaction vanishes. It was proposed [24] that this may be the reason why the amplitude of the CD spectrum of $A_{2'}p_{5'}A$ is small, even though there is an extensive overlap of bases in this dimer as indicated by UV data and PMR data. Similar reasoning was offered as the explanation of the smaller amplitude of $A_{5'}p_{5'}A$ vs that of $A_{3'}p_{5'}A$, even though the overlap of bases in $A_{5'}p_{5'}A$ is clearly more extensive [39]. It should be noted, therefore, that the dependences of these three methods (PMR, CD, and UV hypochromicity) in their measurement of the base–base interaction on the geometrical relationships between the two bases are quite different. Caution should be exercised in the interpretation of the results. A concurrent study by all these three methods on a series of comparable dimers is a powerful procedure.

There is an additional problem concerning these studies, particularly the CD study. This problem is the possible existence in the population of some dimer molecules which have a left-handed turn for their screw axis in the stack, even though the CD and the PMR studies indicate that a major portion of the population should be right-handed conformers. The existence of a certain percentage of the left-handed conformers in solution in a dynamic equilibrium would have a very large effect on the amplitude of the CD spectrum; it would have a negligible effect on the $\Delta\delta_D$ for the base protons, but some effect on the $\Delta\delta_D$ for the H-1′ proton, which, however, is difficult to evaluate quantitatively. Molecular model studies (CPK and Kendrew) clearly show that the steric hindrance for the formation of the left-handed stack may be much smaller for the 2′-5′ dimer (such as $A_{2'}p_{5'}A$ or $A_{2'}p_{5'}C$) than for the 3′-5′ dimer (such as $A_{3'}p_{5'}A$ or $A_{3'}p_{5'}C$). This situation may be partially responsible for the small amplitudes of the CD spectra observed for the 2′-5′ dimer.

The above investigation reveals that the 5′-5′ dimer is not a good model for nucleic acids since this type of dimer cannot be elongated by homogeneous linkage to give a polymer, and it does not form a complex with complementary polynucleotides. The 2′-5′ dimers (both $A_{2'}p_{5'}A$ and $A_{2'}p_{5'}C$) have more extensive base–base overlap in their stack as compared to the respective 3′-5′ dimers ($A_{3'}p_{5'}A$ and $A_{3'}p_{5'}C$). One may raise the question of why the nucleic acids, especially RNA, contain exclusively the 3′-5′ phosphodiester linkage. What kind of disadvantage will the polynucleotides of 2′-5′ linkage have as informational macromolecules in comparison to the polynucleotides of 3′-5′ linkage? The following section may provide some information relevant to this question.

Recently, a comparative study on $G_{2'}p_{5'}G$ vs $G_{3'}p_{5'}G$ by UV absorption and ORD technique was reported [57]. The $G_{3'}p_{5'}G$ has a slightly larger extinction coefficient than the $G_{2'}p_{5'}G$. The ORD patterns of these two isomers are also significantly different from each other; the magnitude of rotation of the long-wavelength peak of $G_{2'}p_{5'}G$ is larger than that of $G_{3'}p_{5'}G$. However, no simple and definitive interpretation of these observations can be offered at present.

2. Comparison between the Adenine Dimers Containing D- Ribose and those containing L- Ribose

In the preceding section, it was recognized that the left-handed stack may also exist in the population, especially for the 2'-5' dimers. However, there is no simple method to study quantitatively this question of the chirality of the screw axis of the stack in solution. For this reason, L-adenylyl-(3'-5')-L-adenosine(L-A$_{3'}$p$_{5'}$A) and L-adenylyl-(2'-5')-L-adenosine(L-A$_{2'}$p$_{5'}$A) were synthesized, starting from L-adenosine [48]. Comparative study of these enantiomeric pairs, i.e., D-ApA(s) vs L-ApA(s), may provide more pertinent information about the chirality of the dimers.

As expected, the L-ApA(s) have the same PMR spectra, the same UV spectra, the same electrophoretic mobilities and the same chromatographic properties in silica gel as their corresponding D-ApA(s). The L-ApA(s), however, are distinguishable from the D-ApA(s) in two intrinsic properties. First, L-ApA(s) are either completely or extremely resistant to both spleen and venom phosphodiesterases, while D-ApA(s) are very sensitive to these enzymes. Second, the CD spectra of the L-ApA(s) are opposite in sign but the same (or nearly the same within experimental error) in magnitude as those of the corresponding D-ApA(s), as illustrated in Fig. 19. All these studies confirm the expectation that the conformation of L-ApA(s) is the mirror image of the corresponding D-ApA(s). Therefore, while the conformation of the D dimers is predominantly an *anti, anti* right-handed stack as discussed above, the conformation of the L dimers is predominantly an *anti, anti* left-handed stack [48].

D-ApA(s) form a complex with D-poly U with a 1:2 stoichiometry. (Oligomer–polymer interaction is the topic of Section II of this chapter; therefore only a brief discussion is given here.) From the mixing curves, L-ApA(s) were found also to form a complex with D-poly U having the same stoichiometry (1A:2U). In addition, the CD spectra of the four ApA-2 poly U complexes (Fig. 20) are closely similar to each other especially the CD spectra of D-A$_{3'}$p$_{5'}$A·poly U and L-A$_{3'}$p$_{5'}$A·2 D-poly U, which are nearly identical (Fig. 20). This observation suggests that the overall conformations of these helical complexes are very similar. The thermal stability of these complexes, however, is not the same. In tris (pH 7.5)-Mg^{2+} (0.01 M), with 1×10^{-4} M poly U and 5×10^{-5} M ApA, the T_m values for these complexes are: D-A$_3$p$_{5'}$A·2 D-poly U (13.7°C), L-A$_{3'}$p$_{5'}$A·2 D-poly U (5.5°C), D-A$_{2'}$p$_{5'}$A·2 D-poly U (11.2°C), and L-A$_{2'}$p$_{5'}$A·2 D-poly U (13.3°C).

The thermodynamics of the complex formation between ApA and poly U is analyzed in accordance with the following formulation and equations:

$$\text{ApA (free)} + 2 \text{ poly U} \rightleftharpoons \text{ApA (stacked)} \cdot 2 \text{ poly U} \qquad \text{(I)}$$

$$\text{ApA (free)} \;\rightleftharpoons\; \text{ApA (stacked)} \qquad\qquad\qquad \text{(II)}$$

$$\text{ApA (stacked)} + 2 \text{ poly U} \;\rightleftharpoons\; \text{ApA (stacked)}\cdot 2 \text{ poly U} \qquad \text{(III)}$$

Process I represents the overall process of the formation of the ApA·2 poly U complex. For the present purpose, the change of states of ApA in the process of complex formation is divided into twō steps. In the first step (process II) the ApA in the "free" state is changed into a "stacked" state. In the free state, ApA is not complexed and assumes a conformation determined by the energies of the isolated molecule in solution, while in the stacked state, ApA is still not complexed but assumes a conformation identical with that in the ApA·2 poly U complex. Thus, the ApA in the stacked state is ready to combine with the poly U (process III) to form the helical complex without any further change in conformation. Therefore, process I is the sum of process II and process III and the standard free-energy change in process I, ΔF_{I}, is the sum of ΔF_{II} and ΔF_{III}, the standard free-energy change in processes II and III, respectively:

$$\Delta F_{\mathrm{I}} = \Delta F_{\mathrm{II}} + \Delta F_{\mathrm{III}}$$

We assign $\Delta F_{\mathrm{I}}^{\mathrm{D}}$ to be the standard free-energy change in process I for the formation of D-$A_{3'}p_{5'}A$·2 D-poly U complex, $\Delta F_{\mathrm{I}}^{\mathrm{L}}$ to be the standard free-energy change in process I for the formation of L-$A_{3'}p_{5'}A$·2 D-poly U complex, and $\Delta F_{\mathrm{I}}^{\mathrm{D-L}}$ to be the difference between $\Delta F_{\mathrm{I}}^{\mathrm{D}}$ and $\Delta F_{\mathrm{I}}^{\mathrm{L}}$ as

$$\Delta F_{\mathrm{I}}^{\mathrm{D}} = \Delta F_{\mathrm{I}}^{\mathrm{L}} + \Delta F_{\mathrm{I}}^{\mathrm{D-L}}$$

It can be shown [48] that $\Delta F_{\mathrm{I}}^{\mathrm{D-L}}{}_{(T_{\mathrm{m}}^{\mathrm{L}})} = (T_m^{\mathrm{D}} - T_m^{\mathrm{L}}) \, \Delta S_{\mathrm{I}}^{\mathrm{D}}$. The T_m values were reported in the preceding paragraphs, and the $\Delta S_{\mathrm{I}}^{\mathrm{D}}$ of $-(40 \pm 2)$ is adopted here on the basis of the ΔS value of the monomer–polymer complex (-42 cal/mole-deg [58]) and of the ΔS value of the polymer–polymer complex (-34 to -36 cal/mole deg; see Tazawa *et al.* [48] for details). Thus, $\Delta F_{\mathrm{I(5°C)}}^{\mathrm{D-L}} = (8.5°\mathrm{C})(-40 \text{ cal/mole-deg}) = -340 \pm 20 \text{ cal/mole}$. In other words, the difference in free energy of the complex formation between D-$A_{3'}p_{5'}A$·2 D-poly U complex and that of L-$A_{3'}p_{5'}A$·2 D-poly U complex and that of L-$A_{3'}p_{5'}A$·2 D-poly U at 5°C is about 350 cal/mole in favor of D–D complex, as reflected by the difference in T_m of about 8.5°C.

Since the CD spectrum of the D-$A_{3'}p_{5'}A$·2 D-poly U complex is almost identical with that of the poly A·2 poly U complex [59], the chirality of the helix of the dimer·polymer complex is expected to be that of the polymer· polymer complex. From the x-ray diffraction study (see Tazawa *et al.* [48]), the chirality of poly A·2 poly U can be shown to be right-handed; therefore, the chirality of D-$A_{3'}p_{5'}A$·2 D-poly U complex is also right-handed. Since the circular dichroism spectrum of L-$A_{3'}p_{5'}{}'A$·2 D-poly U is almost the same as that of D-$A_{3'}p_{5'}A$·2 D-poly U, we suggest that the chirality of L-$A_{3'}p_{5'}A$·2

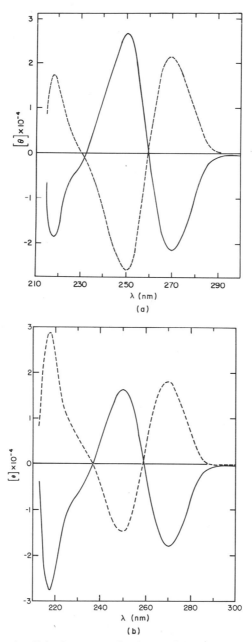

Fig. 19. (a) Circular dichroic spectra of L-A$_3$·p$_5$·A (—) and D-A$_3$·p$_5$·A (---) in 0.05 M NaClO$_4$ (pH 7.0) at 21.5°C. (b) Circular dichroic spectra of L-A$_2$·p$_5$·A (—) and D-A$_2$·p$_5$·A (---) in 0.01 M tris (pH 7.5)–0.01 M MgCl$_2$ at 20.5°C. (From Tazawa *et al.* [48].)

357

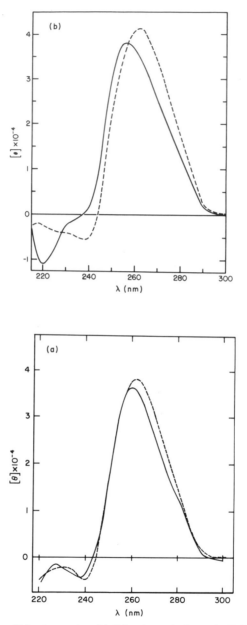

Fig. 20. Circular dichroic spectra. (a) Of D-A$_{3'}$p$_{5'}$A · 2 D-poly U (- - -) and L-A$_{3'}$p$_{5'}$A · 2 D-poly U (—) at $-0.5°C$ in 0.01 M tris (pH 7.5)–0.01 M MgCl$_2$. (b) Of D-A$_{2'}$p$_{5'}$A · D-poly U (- - -) and L-A$_{2'}$p$_{5'}$A · 2 D-poly U (—) at $-0.5°C$ in 0.01 M tris (pH 7.5)– 0.01 M MgCl$_2$. (From Tazawa *et al.* [48].)

358

D-poly U is right-handed as well. From this consideration, we can further define process II(D) and process II(L) as

$$\text{ApA}^{\text{D}} \text{ (free)} \rightleftharpoons \text{ApA}^{\text{D}} \text{ (stacked)} \qquad \text{II(D)}$$

$$\text{ApA}^{\text{L}} \text{ (free)} \rightleftharpoons \text{ApA}^{\text{L}} \text{ (stacked)} \qquad \text{II(L)}$$

From this analysis, we have good reason to assign most of the difference in the standard free-energy change for the overall process (process I) between these two complexes, $\Delta F_{\text{I}}^{\text{D-L}}$, to the difference in standard free-energy change of process II, i.e.,

$$\Delta F_{\text{I}}^{\text{D-L}} = \Delta F_{\text{II}}^{\text{D-L}}$$

We view that once the ApA exists in a correct stacked state, the standard free-energy change of process III is not related to the backbone of the ApA dimer [48]. From this consideration, we can further define process II(D) and process II(L) as

$$\text{ApA}^{\text{D}} \text{ (free)} \rightleftharpoons \text{ApA}^{\text{D}} \text{ (right-handed stack)} \qquad \text{II(D)}$$

$$\text{ApA}^{\text{L}} \text{ (free)} \rightleftharpoons \text{ApA}^{\text{L}} \text{ (right-handed stack)} \qquad \text{II(L)}$$

From the consideration that the enantiomers are mirror images of each other, the standard free energy, F, for the conformational states of the enantiomers will be the same as

$$F(\text{ApA}^{\text{D}} \text{ (right-handed stack))} = F(\text{ApA}^{\text{L}} \text{ (left-handed stack))}$$

$$F(\text{ApA}^{\text{D}} \text{ (left-handed stack))} = F(\text{ApA}^{\text{L}} \text{ (right-handed stack))}$$

$$F(\text{ApA}^{\text{D}} \text{ (free))} = F(\text{ApA}^{\text{L}} \text{ (free))]}$$

Thus

$$\begin{aligned}
\Delta F_{\text{II}}^{\text{D-L}} &= [F(\text{ApA}^{\text{D}} \text{ (right-handed stack))} - F(\text{ApA}^{\text{D}} \text{ (free))]} \\
&\quad - [F(\text{ApA}^{\text{L}} \text{ (right-handed stack))} - F(\text{ApA}^{\text{L}}\text{(free))]} \\
&= - [F(\text{ApA}^{\text{D}} \text{ (left-handed stack))} - F(\text{ApA}^{\text{D}} \text{ (right-handed stack))]} \\
&= -\Delta F_{\text{IV}}
\end{aligned}$$

where ΔF_{IV} describes the standard free-energy change of process IV which is

$$\text{ApA}^{\text{D}} \text{ (right-handed stack)} \rightleftharpoons \text{ApA}^{\text{D}} \text{ (left-handed stack)} \qquad \text{(IV)}$$

Therefore, the $-\Delta F_{\text{I(5°C)}}^{\text{D-L}}(=340 \text{ cal/mole})$ can be considered as the standard free-energy change of process IV, i.e., the equilibrium between the stacked conformation of D-$A_{3'}p_{5'}A$ having a right-handed turn to the stacked conformation having a left-handed turn. The orientation of the bases in the stacked conformation is that of the $A_{3'}p_{5'}A \cdot 2$ poly U helix. From the relationship of $-RT \ln K = 340 \text{ cal/mole}$ [Eq.(7)], the K_{IV} value (5°C) of process IV is calculated to be 0.54, or the distribution of the population at 5°C is calculated to have 34% of the D-$A_3'p_5'$A in the left-handed stack. Since the entropy of ApA^D (right-handed stack) should be identical with the entropy of ApA^D (left-handed stack), therefore the ΔS_{IV} (ΔS_{II}^{D-L} or ΔS_I^{D-L}) is zero. It follows that ΔF_{IV} (ΔF_{II}^{D-L} or ΔF_I^{D-L}) equals ΔH_{IV} (ΔH_{II}^{D-L} or ΔH_I^{D-L}). From this consideration, we can calculate the K value of process IV at 25° by the relationship of $\ln K = [(-340 \text{ cal/mole})/R \cdot 298°\text{K}]$. K_{IV} is found to be 0.57, or the distribution is calculated to be about 36% in left-handed stacks at 25°. From the x-ray diffraction we learn that poly A · poly U helix has a geometry of 11 bases per turn. Following the previous arguments about the similarities of the circular dichroism spectra among poly A · poly U complex, poly A · 2 poly U complex and $A_3'p_5'$A · 2 poly U complex, we shall assume that the $A_3'p_5'$A · 2 poly U complex also has a geometry of 11 bases/turn. In this conformation, the intersection of the principal axes of the neighboring adenines will have an angle of about 33°. Therefore, the conformation of D-$A_3'p_5'$A in the right-handed stack will also have an intersection of the principal axes of the two adenines at an angle of about 33°. Process IV can be redefined as

$$A_3'p_5'\cdot A^D \text{ [right-handed stack (33°)]} \rightleftharpoons A_3'p_5'\cdot A \text{ [left-handed stack (33°)]}$$

The above thermodynamic analyses show that the difference in free energy (or enthalpy) between the conformation of D-$A_3'p_5'$A in a 33° right-handed stack to the conformation of D-$A_3'p_5'$A in a 33° left-handed stack is about 340 cal mole. At present, it is the only known thermodynamic data in the estimation of the free energy difference between the right-handed conformation and the left-handed conformation for a ribosyl linkage in an 11-fold helix. While this value for an individual linkage is not large, for a polynucleotide with 100 linkages, this value can become a respectable value of 34 kcal in favor of the right-handed form for this polymer. On the other hand, the dimer or polymer can assume a left-handed conformation if other forces can overcome this difference of 340 cal/linkage. Some examples are given in the following sections.

The above analysis is for the conformation of a 33° stack; the distribution of D-$A_3'p_5'$A between the 33° right-handed stack and the 33° left-handed stack would be about 2:1. However, the conformational model proposed for the free D-$A_3'p_5'$A in solution discussed above is the right-handed stack with the intersection angle of about 60° for the principal axes of the two adenines. At present, the standard free-energy change of process V cannot be estimated.

$$A_3'p_5'\cdot A^D \text{ (right-handed stack (60°))} \rightleftharpoons A_3'p_5'\cdot A^D \text{ [left-handed stack (60°)]} \quad \text{(V)}$$

Since the CD and PMR data both indicate the predominance of the right-handed form, the ΔF_V is likely to be larger than 340 cal/mole.

While the above thermodynamic analyses can also be applied to the study of $A_2'p_5'$A · 2 poly U complex, the quantitative aspect is less certain. The stereochemistry of the $A_{2'}p_{5'}A \cdot 2$ poly U complex is not clearly known; the resemblance between the CD spectrum of the L-$A_{2'}p_{5'}A \cdot 2$ poly U and that of D-$A_{2'}p_{5'}A \cdot 2$ D-poly U is not as close; and interestingly, the T_m of L-$A_{2'}p_{5'}A \cdot$ 2 D-poly U is 2°C higher than that of D-$A_{2'}p_{5'}A \cdot 2$ D-poly U. Regardless of the quantitative aspect of this estimation, the experimental results clearly indicate that L-$A_{2'}p_{5'}A$ and D-poly U can form a complex which has a stability against temperature perturbation even slightly greater than that of the D-$A_{2'}p_{5'}A \cdot 2$ D-poly U complex. Since the stack of L-$A_{2'}p_{5'}A$ is most likely right-handed in the complex because of the similarity between the circular dichroism spectra of the D–D complex and the L–D complex, the T_m results suggest that L-$A_{2'}p_{5'}A$ can form a right-handed stack easily (in fact, the apparent suggestion is that L-$A_{2'}p_{5'}A$ can form a right-handed stack more easily than D-$A_{2'}p_{5'}A$). From the symmetry consideration of the enantiomers, therefore, the comparison between the T_m results on both $A_{3'}p_{5'}A$ and $A_{2'}p_{5'}A$ suggest that D-$A_{2'}p_{5'}A$ can form a left-handed stack with greater ease than D-$A_{3'}p_{5'}A$. This conclusion is supported by the model-building studies described in the preceding section. Thus, from this comparative study on D-$A_{3'}p_{5'}A$ and L-$A_{3'}p_{5'}A$ vs D-$A_{2'}p_{5'}A$ and L-$A_{2'}p_{5'}A$, we may safely conclude that the D-2'-5' dimer can form a left-handed stack more readily than the D-3'-5' dimer. In other words, the selectivity in chirality of the screw axis of the stack of the 2'-5' dimer is less than the selectivity of the 3'-5' dimers.

As students of the structure and conformation of nucleic acids, we often wonder why in nature the nucleic acid is built with a 3'-5' phosphodiester linkage instead of a 2'-5' phosphodiester linkage, especially for RNA. Since the selectivity in chirality of the 3'-5' dimer stack is higher than that of the 2'-5' dimer stack, it is reasonable to predict that the selectivity in the conformation of the 3'-5' polymer is higher than that of the 2'-5' polymer, or the entropy state of the 3'-5' polymer is lower than the entropy state of the 2'-5' polymer. If we consider that the existence of a biological system depends on a high degree of selectivity (or low state of entropy), then the nucleic acid built with a 3'-5' linkage may be preferred over the nucleic acid built with a 2'-5' linkage in the process of evolution. Furthermore, these two types of nucleic acid are unlikely to coexist, since the stability of the double-stranded helix consisting of both 3'-5' strands will be higher than the stability of the double-stranded helix consisting of one 3'-5' strand and one 2'-5' strand.

3. Dinucleoside Monophosphates Containing Additional Linkage (or Constraint) in the Molecule

In this section, we shall discuss two dimers which have additional linkage (therefore, constraint) in the molecule. As the result of this constraint, these dimers apparently assume a left-handed conformation.

1. The first dimer is the dinucleoside monophosphate (A^{s}pA^s) of 8,2′-anhydro-8-mercapto-9-(β-D)arabinofuranosyl adenine [60,61]. The nucleosidyl units of this dimer, 8,2′-S-cycloadenosine, have a linkage between the base (C-8) and the furanose (C-2′). Therefore, the torsion angle between the base and the furanose (ϕ_{CN}) of this nucleoside is rigidly fixed at $-108°$ instead of about $-10°$ for adenosine in crystal form (Table V; Chapter 6, Volume I). The optical properties of pA^s, A^{s}pA^s as compared to those of pA and ApA are shown in Fig. 21. The hypochromicity of A^{s}pA^s is about 15% (Fig. 21); this value is definitely higher than that of the ApA (12%, Table VIII). This

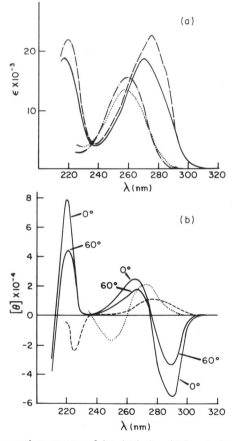

Fig. 21. **(a)** UV absorption spectra of A^{s}pA^s (—), pA^s (---), ApA ($\cdots$) and pA (—·) in 0.01 *M* phosphate buffer -0.1 *M* KF; pH 7.5, room temperature. (b) CD spectra of A^{s}pA^s at 0°C (—) and at 60°C (—), pA^s at room temperature (---), and ApA at 25°C ($\cdots$). Buffer: 0.01 *M* phosphate, pH 7.0. ε and [θ] are calculated per nucleoside residue. (From Uesugi *et al.* [61].)

UV data certainly indicates a strong interaction between the two bases in A^{s}pA^s. The pA^s CD spectrum (Fig. 21) is quite different from the pA or other naturally occurring purine nucleotides, which normally have a dominant negative CD region around 260 nm, and a small positive CD band at 280 nm. The CD spectrum of A^{s}pA^s at neutral pH is also very different from that of ApA (Fig. 21); the CD spectrum of A^{s}pA^s is more or less opposite to that of ApA and the cotton effect of the CD spectrum of A^{s}pA^s is much larger than in those of pA^s and ApA. The CD property of A^{s}pA^s indicates a substantial interaction between the base chromophores and suggests that the conformation of A^{s}pA^s is a left-handed stack.

Some PMR studies on pA^s and A^{s}pA^s have been made [61]. At 30°, the dimerization shift of H-1' is 0.59 ppm and 0.17 ppm for the A^{s}p- and -pA^s residues, respectively; and the dimerization shift of H-2 is 0 ppm and 0.37 ppm for A^{s}p- and -pA^s residues, respectively. Based on these dimerization shifts,

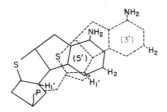

Fig. 22. Schematic presentation of A^{s}pA^s. Solid line is 5'-linked nucleoside and dotted line shows 3'-linked nucleoside, or conformation of a left-handed stack based on PMR data. (From Uesugi *et al.* [61].)

a left-handed stack has been constructed as the conformational model for A^{s}pA^s (Fig. 22). This left-handed model is supported by the observation that poly U does not form a complex with A^{s}pA^s. Also, A^{s}pA^s is extremely resistant to spleen and venom phosphodiesterase. In addition, the left-handed stack of A^{s}pA^s is relatively stable against thermal perturbation.

Extension of the dimer model in Fig. 22 to a polymer in a repetitive fashion will lead to the formation of a left-handed helix with bases located inside and the phosphate group outside the helix. This difference of handedness between the conformation of A^{s}pA^s and that of ApA is attributable to the difference of ϕ_{CN} ($-180°$ vs $-10°$). In order to maintain a maximal base–base overlap and a minimal stereochemical hindrance in the backbone, the 5'-nucleosidyl unit in A^{s}pA^s rotates upward in a clockwise fashion instead of a counterclockwise fashion. Examination of the CPK models reveals that the change of the conformation of the 3'-OH-phosphate-5'-OH linkage of the stack of ApA to the stack of A^{s}pA^s is rather small: the exact reason for the thermal stability of the conformation of A^{s}pA^s is still open to conjecture. The absence of the oscillation around the torsional angle could be one reason that a degree of freedom and flexibility has been removed from this cyclonucleoside.

2. The second dimer is the cyclic deoxythymidine dinucleotide, designated as dp͡Tp͡T [62]. In this molecule, the terminal 5′-phosphate is joined to the terminal 3′-hydroxyl end [63]; thus, this dimer has an additional linkage between the two units. In Fig. 23, the CD spectra of dpT, dpTpT, and dp͡Tp͡T

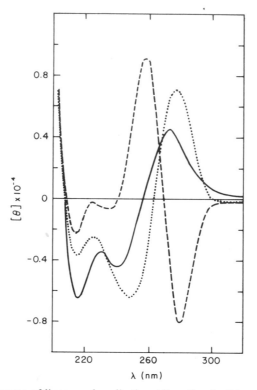

Fig. 23. CD spectra of linear and cyclic thymidine dinucleotides and thymidylic acid in neutral aqueous buffer at 26°C (——) dpT; (---) dp͡Tp͡T; (···) dpTpT. (From Cantor *et al.* [62].)

are shown. In aqueous solution, while the UV absorption spectra of dpTpT and dp͡Tp͡T are nearly identical [62], the CD spectrum of dp͡Tp͡T is opposite to that of dpTpT at the region of 230–300 nm. The PMR studies indicate that the two bases in dp͡Tp͡T are magnetically equivalent, a situation very different from that of dpTpT where a clear magnetic inequivalence of the two bases can be observed. Based on these studies, a left-handed stack with C_2 axis of symmetry for dp͡Tp͡T was proposed [62], as shown in Fig. 24. The conformation of dp͡TpTp͡T was also shown to be different from that of dpTpTpT by the CD studies.

4. *Neutral Dimers—Adenine and Thymine Dinucleoside Monophosphate Alkyl Phosphotriesters*

The methyl and ethyl phosphotriester derivatives of TpT and dApdA [50] and the ethyl phosphotriester of dIpdI [63a] were synthesized and were obtained as a pair of diastereoisomers in each case, due to the asymmetry the phosphorus atom after alkylation (Fig. 25a) [64]. These neutral dimers are useful for two important purposes: First, the phosphate–alkyl groups serve as monitors for the interaction between the backbone and the bases. The PMR

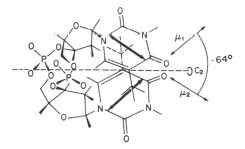

Fig. 24. Schematic drawing of a possible model for the conformation of dp͡Tp͡T. The arrangement of atoms in the phosphate–ribose backbone has been distorted to allow all of the atoms to be visible. The thick double arrows in the planes of the bases show the known orientation of the transition moment of the 266-nm absorption band of thymine. (From Cantor *et al.* [62].)

resonances of these groups provide valuable information about the dynamics of the dimer conformation. Second, these dimers provide the material for the evaluation of the negative influence of the charged phosphate group on the stability of the polynucleotide–oligonucleotide complexes.

As shown by UV hypochromicity, CD, and PMR measurements, the conformations of the triesters in solution are quite similar to those of the parent dimer–diesters, although there is less base-stacking in the triesters [50]. For instance, the percent of hypochromicity of dAp(CH$_3$)dA (6.5%), and of dAp(C$_2$H$_5$)dA (11%) is smaller than that of dApdA (17%). At 1.5°, the CD spectra of dApdA, dAp(CH$_3$)dA, and dAp(C$_2$H$_5$) all have the same peak position (~270 nm), same trough position (~250 nm) and same crossover position (~260 nm); except that the spectra of both triesters have only about 60% of the amplitude between the peak and the trough that dApdA does. Also, the resonances of the H-2, H-1', and H-8 (3'- residue) of the dApdA

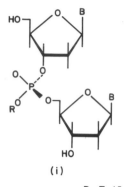
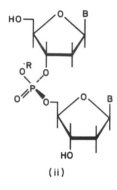

(i)   (ii)

B = T, dI, dA
R = CH₂CH₃

(a)

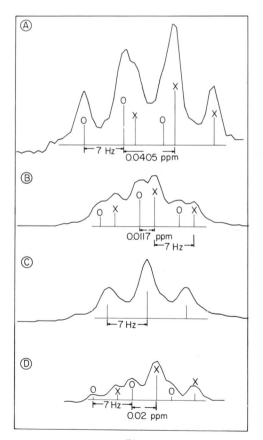

(b)

366

triesters all are located about 0.05 ppm downfield as compared to the corresponding protons of dApdA, as an indication of reduction in shielding by the neighboring adenine. The resonance of the H-8 of 5'- residue of the dApdA triester, however, is shifted upfield by about ~ 0.08 ppm vs that of dApdA; this is due to the alkylation of the phosphate group which removes the charged effect of the phosphate group on this particular H-8 proton in the triester. As discussed above, this observation is part of the evidence in support of the conclusion that the nucleosidyl unit in dApdA is in the *anti*

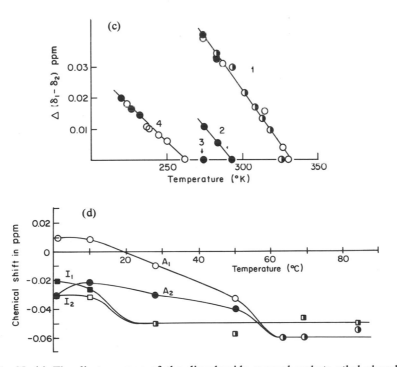

Fig. 25. (a) The diastereomers of the dinucleoside monophosphate ethyl phosphotriesters. (From Kan *et al.* [64].) (b) PMR spectra of the methyl resonances of the ethyl phosphate group at various temperatures. (A) dAp(Et)dA at 1°C in D_2O; (B) dIp(Et)dI at 1°C in D_2O; (C) Tp(Et)T at 1°C in D_2O; (D) dAp(Et)dA at 220°K in CD_3OD. (From Kan *et al.* [64].) (c) Difference in chemical shifts $(\delta_1-\delta_2)$ of the phosphate–methyl resonances of the two diastereomers of Np(Et)N in D_2O and CD_3OD vs temperature. (1) dAp(Et)dA in D_2O; (2) dIp(Et)dI in D_2O; (3) Tp(Et)T in D_2O; (4) dAp(Et)dA in CD_3OD. (○) data from HA-100; (●) data from HR-220; (◑) data from Miller *et al.* [50]. (d) The difference between chemical shifts of the phosphate methyl resonance of dAp(Et)dA and dIp(Et)dI with respect to that of Tp(Et)T in D_2O. (○) and (□) indicate the δ_1 values of A and I, respectively. (●) and (■) indicate the δ_2 values of A and I, respectively. The half-shaded symbols represent the simultaneous presence of both empty and full symbols. (From Kan *et al.* [64].)

conformation. Apparently, the alkyl group on the phosphate of the triester does hinder the overlapping of the bases to some extent.

Another interesting observation is that the methyl and ethyl triesters of TpT and dApdA are completely resistant to snake venom and spleen phosphodiesterase and to micrococcal nuclease, under conditions in which the parent diesters, TpT and dApdA, were completely cleaved. Thus, it would appear that the negatively charged phosphate in the dimer is an important recognition site for these nucleases.

Interaction of the dinucleoside alkyl phosphotriesters with their complementary polynucleotides is of considerable physicochemical and biological interest. Neutralization of the negatively charged phosphate should increase the stability of the triester–polymer complex by decreasing electrostatic repulsion. On the other hand, the phosphate alkyl substituent may sterically hinder the formation of the complex.

As a model system, the interaction of polyuridylic acid with dApdA and with the alkyl phosphotriesters was studied. Poly U and dApdA form a complex in $0.01\ M\ Mg^{2+}$ with a stoichiometry of 2U:1A. Likewise, the phosphotriesters, $dAp(CH_3)dA$ and $dAp(C_2H_5)dA$, form complexes with a 2U:1A stoichiometry, as illustrated by the mixing curves [50]. In addition, the CD spectra of all these three complexes are virtually identical. Thus the overall conformation of each complex appears to be the same. The thermal stabilities of the complexes are quite different as shown by the melting experiments. The 2 poly U–dApdA complex melts at $7.6°C$ in $0.01\ M\ Mg^{2+}$. Under the same conditions, the 2 poly $U \cdot dAp(C_2H_5)dA$ complex melts at $12.0°C$, while the 2 poly $U \cdot dAp(CH_3)dA$ complex melts at $13.2°C$. It should be noted that these experiments were conducted in $0.01\ M\ Mg^{2+}$. Under this condition, all the binding sites of the phosphate groups are fully occupied by the Mg^{2+} ions. Apparently under this condition, charge repulsion between the phosphate groups still takes place. The increase in the T_m of the 2 poly U–dAp(r)dA complex evidently reflects a decrease in electrostatic repulsion. It is interesting to note that the stability of the 2 poly $U \cdot dAp(C_2H_5)dA$ complex is slightly lower than that of the 2 poly $U \cdot dAp(CH_3)dA$ complex. This small difference may represent the slight increase in the steric hindrance by the ethyl group of the triester for complex formation.

As for the studies of the interaction between the alkyl group and the bases by PMR, this discussion will be deferred to Section I,E.

5. *Comparison among Dimers with Deoxyribose, Ribose, 2'-O-Methyl Ribose, and Arabinose—Influence of the Substituents on the C-2' of the Furanose on Conformation*

The difference in DNA and RNA conformation is of great significance in biology. 2'-*O*-Methyl nucleotides are natural constituents of tRNA and arabinosyl nucleosides of cytosine and adenine are powerful antiviral and antitumor agents [65,66]. Therefore, comparison of the conformation among dimers with different furanose backbones is of considerable interest, and the research becomes a systematic study on the influence of various substituents in the C-2′ of the furanose on the nucleic acid conformation. This study at the polynucleotide level will be discussed in a later volume.

The most comprehensive study on the conformational differences between deoxyribosyl and ribosyl dimers by CD method was done by Warshaw and Cantor [28]. The CD spectra of these deoxyribosyl and ribosyl dimers are reported in Fig. 8 and 9 [27,28]. After a quantitative comparison, it was concluded [28] that in no case is the CD spectrum for the deoxyribosyl dimer

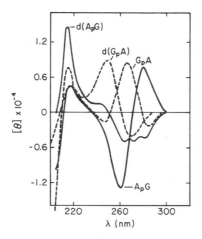

Fig. 26. Comparison of CD spectra of dApdG vs rAprG, and of dGpdA vs rGprA.

a simple multiple of that for the ribosyl compound. (The closest to this situation of difference in proportionality is that between dCpdC and rCprC, which will be discussed in later paragraphs.) Some of the drastic differences between the CD spectra of ribosyl and deoxyribosyl dimers containing the same bases can be shown in Fig. 26. On the other hand, the UV absorption spectra of these four dimers in Fig. 26 are very similar [28]. This study established that the geometry of the average stacked conformation of the ribosyl dimer is often different from that of the deoxyribosyl dimer. Therefore, this important conclusion does not support the study [67] which has been based on implicit assumptions that the mode of stacking of the ribosyl dimer and the corresponding deoxyribosyl dimer is the same, and that the differences in CD lie in the degree of stack formation with the same geometrical mode. In the following paragraphs, the PMR studies amply support this conclusion based on optical studies and provide a more detailed description of the differences in the geometrical mode of stacking and in the backbone between the ribosyl and the deoxyribosyl dimers. We shall concentrate our discussion on the two homodimers, adenine dimers and cytosine dimers, which have been studied most extensively.

a. *Adenine Homodimers* [54].   The conformation of six adenine dinucleoside monophosphates were studied and compared concurrently by UV absorbence, CD, and PMR methods. Four of them are 3'-5' dimers, which represent all possible combinations of the ribosyl and deoxyribosyl nucleotide units, i.e., rAprA, rApdA, dAprA, and dApdA. Two of them are 5'-5' dimers which have been selected for the following special reason. In the 5'-5' ribosyl dimer, the 2'-OH group can no longer reach the phosphate group, since the phosphate linkage has been moved from the 3'-position to the 5'-position. Therefore, the 2'-OH group in the 5'-5' dimer can only interact with the base, if indeed this group can interact with any other component of the molecule at all. In addition, according to the conformational model [39], the two 2'-OH groups are located at the outside of the stack of $rA_{5'}p_{5'}rA$, and should not provide steric hindrance to the stacking. Therefore, it appears that comparison of $dA_{5'}p_{5'}dA$ and $rA_{5'}p_{5'}rA$ may yield valuable information about the interaction of the 2'-OH group.

The UV absorbence studies on the four 3'-5' adenine dimers reveal that the percent hypochromicity of dApdA (16.5%) and dAprA (17.6%) is higher than those of rAprA (11.7%) and rApdA (13.7%). The extent of hypochromicity, and therefore the extent of stacking, appears to be determined by the furanose of the 3'- residue of the dimers; the dimer which has a ribose nucleotide at the 3'- residue (rAp-) has less hypochromicity and hence less stacking. As described in Section I,D and Table VIII, 5'-5' dimers have larger hypochromicity and larger base–base overlap than the 3'-5' dimers. The percent hypochromicity of $rA_{5'}p_{5'}rA$ (22.1%) is virtually the same as that of $dA_{5'}p_{5'}dA$ (24.3%), indicating that the extent of stacking between these two 5'-5' ribosyl and deoxyribosyl dimers is very similar.

The dimerization shifts, $\Delta\delta_D$, of the base and the H-1' protons of six adenine dimers from PMR study at 4°–60° are shown in Table X. Within the experimental error (± 0.02–0.03 ppm from extrapolation to infinite dilution) and the temperature range of 4°–60°C, the $\Delta\delta_D$ values for $rA_{5'}p_{5'}rA$ are the same as those for $dA_{5'}p_{5'}dA$; and the $\Delta\delta_D$ values for dApdA are the same as those for dAprA. The $\Delta\delta_D$ values for rAprA at 30° and 60°C are also the same as that for rApdA; though at 4°C, there appeared to be a small but real difference (0.09 ppm). These data suggest that the conformation of $rA_{5'}p_{5'}rA$ is very close to that of $dA_{5'}p_{5'}dA$; similarly, the conformation of dApdA is close to that of dAprA; and to a certain extent, rAprA is similar to rApdA. On the other hand, the dimerization shift data clearly indicate that the conformation of dApdA is substantially variant from that of rAprA. The $\Delta\delta_D$ values for four protons of dApdA (H-8 and H-1' from 3'- residue; H-2 and H-1' from 5'- residue) are much larger than those from the corresponding protons of rAprA. While the $\Delta\delta_D$ of H-8 from the 5'- residue of rAprA is significantly larger than that of dApdA, the $\Delta\delta_D$ of H-2 from the 3'- residue

TABLE X

DIMERIZATION SHIFTS, $\Delta\delta_D$,[a] OF THE BASE AND H-1' PROTONS OF ADENINE DINUCLEOSIDE MONOPHOSPHATES AT 4°, 30°, AND 60°C (D_2O, pD 7.4)

Dinucleoside	Temp. (°C)	($A_{3'}p-$) residue			($-p_{5'}A$) residue		
		H-8	H-2	H-1'	H-8	H-2	H-1'
rAprA	4	0.155	0.315	0.285	0.285	0.11	0.19
rApdA		0.15	0.225	0.325	0.23	0.15	0.225
dApdA		0.325	0.31	0.42	0.16	0.175	0.275
dAprA		0.32	0.28	0.41	0.16	0.21	0.25
$rA_{5'}p_5rA$					0.425	0.18	0.205
$dA_{5'}p_5 dA$					0.41	0.185	0.22
rAprA	30	0.12	0.215	0.26	0.235	0.075	0.15
rApdA		0.11	0.19	0.29	0.19	0.13	0.16
dApdA		0.295	0.27	0.36	0.125	0.15	0.20
dAprA		0.295	0.235	0.35	0.155	0.17	0.18
$rA_{5'}p_5rA$					0.40	0.185	0.18
$dA_{5'}p_5·dA$					0.345	0.20	0.22
rAprA	60	0.12	0.135	0.19	0.17	0.08	0.10
rApdA		0.14	0.165	0.21	0.17	0.07	0.12
dApdA		0.25	0.23	0.32	0.12	0.15	0.16
dAprA		0.24	0.18	0.32	0.155	0.175	0.17
$rA_5 p_5·rA$					0.335	0.165	0.18
$dA_{5'}p_5·dA$					0.28	0.16	0.17

[a] $\Delta\delta_D = \delta Ap- - \delta ApA$; or $\delta-pA - \delta ApA$ [54].

of rAprA is about the same as that of dApdA at 4°, and became smaller than that of dApdA at 30° and 60°. These data not only indicate that the extent of base–base stacking is different between dApdA and rAprA, but also that the mode of the base–base stacking is different. Based on these dimerization shift values, conformational models for rAprA and dApdA are constructed as shown in Fig. 15 and 16, respectively.

A comparison between the model of rAprA (Fig. 15) and that of dApdA (Fig. 16) reveals interesting differences. First, the extent of the base–base overlap is significantly larger in dApdA than in rAprA. Second, the bases are almost parallel to each other in dApdA, while the bases are definitely oblique to each other in rAprA. For instance, the angle between the two C-4 to C-5 bonds in dApdA is about 15° (0° would indicate a parallel orientation between two adenines), while this angle is 45°–50° in rAprA. These differences in conformation can account for the difference in $\Delta\delta_D$ values between these two dimers. Since the $\Delta\delta_D$ values of dAprA are practically identical to those of dApdA, the conformational model of dAprA should be the same as that of dApdA. Similarly, the model of rApdA is probably close to that of rAprA. Again, the $\Delta\delta_D$ values of $dA_5 \cdot p_5 \cdot dA$ are the same as those of $rA_5 \cdot p_5 \cdot rA$; therefore the model of $dA_5 \cdot p_5 \cdot dA$ should be the same as the model of $rA_5 \cdot p_5 \cdot rA$, which has been constructed by the same approach (Fig. 18). The PMR data

TABLE XI

TEMPERATURE AND SOLVENT EFFECT ON THE COUPLING CONSTANTS OF THE H-1′
PROTONS OF ADENINE DINUCLEOSIDE MONOPHOSPHATES (D_2O, pD 7.4).
(PARTS PER MILLION FROM TETRAMETHYLSILANE CAPILLARY) [a]

Temp.	rAprA		rApdA		dAprA		dApdA		$rA_5 \cdot p_5 \cdot rA$	$dA_5 \cdot p_5 \cdot dA$
(°C)	$J(3')$	$J(5')$	$J(3')$	$J(5')$	$J(3')$	$J(5')$	$J(3')$	$J(5')$	$J(5')$	$J(5')$
5°	2.5	2.0	3.5	5.6	8.8, 4.8	6.0	8.8, 5.6	6.5	4.3	6.5
30°	3.2	3.5	4.0	6.0	8.0, 5.5	5.7	8.5, 5.5	6.4	4.6	6.5
60°	4.5	4.1	4.5	6.3	8.1, 6.0	5.3	7.9, 6.0	6.7	4.7	7.0
DMSO-d_6 (30°)	6.8	5.2			7.5, 5.2	5.2	7.9, 5.7	6.1	5.7	7.1

[a] From Kondo *et al.* [54].

and the models constructed from the PMR data are in good agreement with the results from UV absorbence studies described above.

The coupling constants $J_{1'-2'}$ for the ribosyl dimers and $J_{1'-2'}$, $J_{1'-2''}$, or $J_{1/2(1'-2'+1'-2'')}$ for the deoxyribosyl dimers are presented in Table XI. As described in Section I,C and Table VI and VII, the furanose conformations of 5′-dAMP, 3′-dAMP and dApdA have been extensively analyzed. The J values of dApdA were found to be (1) fairly temperature- and solvent (DMSO-d_6)-independent, and (2) quite similar to the values of the constitutent monomers which were also temperature- and solvent-independent. The result of this work suggested that the furanose conformation of 3′-dAMP and the dAp- portion of dApdA is $C_{2'}$-*endo* (envelope) or $C_{2'}$-*endo*–$C_{3'}$-*exo* (twisted form), while that of 5′-dAMP and the -dpA portion of dApdA is a rapid equilibrium between $C_{2'}$-*endo* and $C_{3'}$-*endo*. Unlike the deoxyribosyl dimer case, the H_1–H_2' coupling constants of rAprA are much smaller than those of the constituent monomers. As shown in Table XI, the (H-1')–(H-2') coupling constants of rAprA increase in the temperature range 5°–60°C. In DMSO-d_6 the J values are large and similar to those of the corresponding monomers. Thus in aqueous solution, the furanose conformation of the monomer and the conformation of the furanose in the dimer rAprA are not the same. Furthermore, the results indicate that the furanose conformation of rAprA is influenced by the stacking of the bases, while the furanose conformation of dApdA is not [36,54]. The J values of both the deoxyribosyl and ribosyl portions of the mixed dimer dAprA are essentially temperature invariant. On the other hand, the coupling constants of rAp- residue in the mixed dimer rApdA increase with increasing temperature, although this increase is smaller than that observed for the parent dimer rAprA. These results on the 3'-5' dimers suggest that the dependency of the furanose conformation of the rA residue on temperature or on base-stacking in the dimer can be listed in the following increasing order: dAprA < rApdA < rAprA. Similarly, the dependency of the furanose conformation of dA residue on temperature or on base-stacking in the dimer can be listed as: dApdA ≈ dAprA < rApdA. This summary indicates that the furanose conformation in a dimer will be sensitive to temperature and base-stacking if this dimer has a ribosyl nucleotide at the 3'- residue as rAp-. The coupling constant of the 5'-5' deoxydimer $dA_{5'}p_{5'}dA$ is not sensitive to temperature and is very similar to that of dApdA (Table XI). The coupling constant of the 5'-5' ribosyl dimer $rA_{5'}p_{5'}rA$ is slightly temperature-dependent.

The circular dichroism spectra of rAprA, dApdA, rApdA, and dAprA at 3° in the wavelength range 230–300 nm are shown in Fig. 27. The shapes of all four curves are quite similar, with a peak at 272 nm, a trough at about 250 nm, and a crossover point at approximately 248 nm. Although qualitatively similar, the spectra of the four 3'-5' dimers differ quantitatively. At 3 ± 1°, the magnitude of [θ] from peak to trough of the four 3'-5' dimers can be arranged in the following decreasing order : rAprA > dAprA ≈ dApdA > rApdA. The CD spectra of $rA_{5'}p_{5'}rA$ and $dA_{5'}p_{5'}dA$ are similar to each other (peak, 272 nm; trough, 254 nm; and crossover, 264 nm); however, the magnitude of [θ] of $rA_{5'}p_{5'}rA$ is about twice that of $dA_{5'}p_{5'}dA$. A quantitative interpretation of the CD results is more difficult. Not only do the amplitudes in [θ] between the peak and the trough of these six dimers not follow the order of hypochromicity or the order of base–base overlap in the conformational models built from PMR data, but also the pair members in two pairs (i.e., rAprA vs rApdA; $rA_{5'}p_{5'}rA$ vs $dA_{5'}p_{5'}dA$) do not have similar amplitudes. According to the discussions in Sections I,B and C [Eq. (7), for instance], the differences in amplitude among these six adenine homodimers can be attributed to the following three factors: (1) The averaged

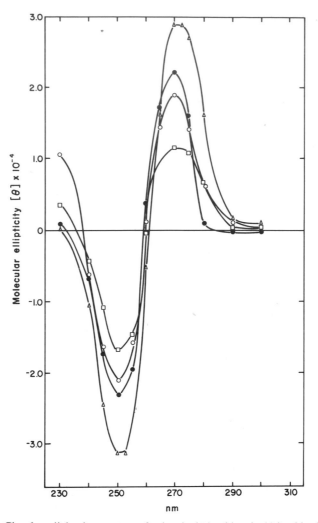

Fig. 27. Circular dichroic spectra of rAprA (△), dAprA (●), dApdA (○), and rApdA (□) in 0.05 *M* NaClO₄ (pH 7.3) at 3° ± 1°C. (From Kondo *et al.* [54].)

distance between the two transition dipoles of the two adenines; (2) the averaged angle θ (not $[\theta]_{CD}$) between the two transition dipoles; (3) the averaged proportion of the right-handed conformer vs the left-handed conformer in the population. In the preceding paragraphs, it was pointed out that the bases are almost parallel to each other in dApdA, while the bases are definitely oblique to each other in rAprA. As a consequence, the $\theta_{transition}$ (same as the angle between the two C-4 to C-5 bonds) is only about 15° in dApdA, but about 45°–50° in rAprA. This substantial difference in $\theta_{\text{transition}}$ most likely is the factor which causes the amplitude of rAprA to be larger than that of dApdA even though dApdA has a larger base–base overlap. The contribution of the third factor, i.e., the proportion of the right-handed conformer vs that of the left-handed conformer in solution, cannot be evaluated at present. However, this factor may be the major reason in the explanation for the difference in CD amplitude between rAprA vs rApdA, as well as between $rA_5'p_5'rA$ vs $dA_5'p_5'dA$, since as far as UV hypochromicity and PMR data are concerned, the pair members in these two pairs have the same extent of base–base interaction within experimental error. Therefore, the consideration of the proportion between the right-handed conformer and the left-handed conformer in solution is likely to be the important factor in explaining the CD difference between the pair members of these two pairs.

Our present study also indicates that the strong influence of the 2′-OH group on the conformation of the dimer can be observed only when the 2′-OH group is located next to a 3′-phosphodiester linkage. In the case of the 5′-5′ dimer, the 2′-OH groups are *not* located next to a 3′-phosphate linkage; thus, the stacks of $dA_5'p_5'dA$ and $rA_5'p_5'rA$ have very similar conformations and stabilities, indicated by the UV and PMR data. The comparison between these two ribosyl and deoxyribosyl 5′-5′ dimers reveals that the possible hydrogen bonding of the 2′-OH group to the bases, if it occurs, has little influence on the conformation or stability. Thus, among these four 3′-5′ dimers, rAprA and rApdA belong to one group which has conformational properties of ribosyl compounds, while dApdA and dAprA belong to another group which has properties of deoxyribosyl compounds. A likely stereochemical reason for these observations is that the nonbonded repulsion between the 2′-OH group in the rNp- portion of the ribosyl dimer and the base and the ether oxygen of the furanose in the -prN portion of the dimer may provide a steric hindrance to stacking in ribosyl dinucleoside monophosphate; such a hindrance, if it does in fact exist, should be considerably reduced in the situation of deoxyribosyl dinucleoside monophosphate.

The $J_{1'-2'(2'')}$ values of the six dimers (Table XI) relate a revealing story. The data strongly suggest that the decrease in $J_{1'-2'(2'')}$ values of the dimer vs that of the monomer represents the compression of the furanose ring due to stacking. This compression can be released at elevated temperature or in a destacking solvent such as DMSO-d_6. In dApdA and dAprA, the $J_{1'-2'(2'')}$ values of all four furanoses are larger and similar to those of the mononucleotides, as well as temperature-independent and solvent-independent. This observation implies that the furanoses in these dimers are not compressed in stacking; perhaps this is the reason why the bases in dApdA and dAprA can have a parallel, extensive overlap. In rAprA and rApdA, the $J_{1'-2'(2'')}$ values of the three riboses are significantly smaller than those of the mononucleotides and are temperature-dependent and solvent-dependent. This observation implies that the riboses in these dimers are being compressed in stacking. Thus, the 2'-OH group in these riboses may provide a steric hindrance in forcing the bases to have an oblique and less extensive mode of stacking in rAprA and rApdA. It is interesting to note that the $J_{1'-2'}$ of the -prA residue in rAprA indicates that the ribose in this 5'- portion is being compressed. Since the C-2' atom and the 2'-OH group of the 5'-nucleosidyl unit are located outside the stack (Fig. 15 and 18), these groups should not provide any stereochemical hindrance to stacking. We have proposed, therefore, that the change of $J_{1'-2'}$ of the -prA may reflect not the change of the C-2' position, but rather the C-1' atom instead [24]. From this point of view, this change of $J_{1'\,2'}$ of the -prA residue may be taken as evidence that the O-1' to C-1' region is being compressed by the 2'-OH group of the rAp-residue in the dimer. The average value of $J_{1'-2'}$ and $J_{1'-2''}$ of -pdA residue in rApdA is slightly smaller than that of the mononucleotides (6.6–6.7 Hz) or that of the dApdA (6.5 Hz), especially at 5° (5.6 Hz; Table XI). This J value is also somewhat temperature-dependent. Again, we propose that this phenomenon may be due to the compression of the O-1' to C-1' region of the deoxyribose in -pdA residue by the 2'-OH group of the ribose in rAp- residue.

In summary, the UV, PMR and possibly the CD data on these six adenine dinucleoside monophosphates indicate that the influence of the 2'-OH group of the ribose on the conformation of the ribose-containing dimers is exerted through the steric hindrance of this group (particularly the 2' oxygen atom) and not through its hydrogen-bonding properties. Upon stacking, the ribose of the 3'- residue is being compressed by the furanose of the adjacent 5'-residue (possibly by the base of the 5'- residue as well). This steric interference between the 3'-furanose group and the immediately following 5'- residue prevents an extensive overlap of the adenines in a parallel fashion. Such an interference is absent in the dimers containing a deoxyribosyl 3'- residue, such as dApdA and dAprA.

This important conclusion about the difference between ribosyl and deoxyribosyl dimers is also supported by the study on the properties of the monomers. In dilute solution, the $J_{1'-2'}$ values of 5'-rAMP and 5'-dAMP are insensitive to temperature variation as well as insensitive to change of solvent from D_2O to DMSO-d_6. However, the $J_{1'-2'}$ of 5'-rAMP in aqueous solution is sensitive to the concentration of nucleotides or to addition of purine [34], while under identical conditions, the $J_{1'-2'}$ of 5'-dAMP is *insensitive* to both nucleotide concentration and the presence of purine [36], This phenomenon was attributed to the influence of the neighboring molecules in the stack formation; apparently, the formation of stacks has a large effect on the furanose conformation (as monitored by $J_{1'-2'}$) of 5'-rAMP, but has *no* effect on 5'-dAMP. Study on the polynucleotides also completely supports this conclusion from studies on monomers and dimers.

This conclusion is also supported by the study on the 2'-*O*-methyl derivatives in which the hydrogen-bonding properties of the 2'-OH group are now blocked. The chemical shifts of protons in 2'-*O*-methyl Ap 2'-*O*-methyl A have been compared with those in ApA at 37°C and were found to be closely similar [68]. This study on the polynucleotide yielded even more conclusive information. The chemical shifts and polymerization shifts of poly rA, poly 2'-*O*-methyl A, and poly dA from 20° to 75°C clearly indicate that the conformation of poly 2'-*O*-methyl A closely resembles that of poly rA, but differs distinctly from that of poly dA [68a]. The amplitude of the CD curve of 2'-*O*-methyl Ap 2'-*O*-methyl A is smaller than that of rAprA in 4.7 *M* KF at temperatures below 40°–50°C [68]. Also, at 23°C and 0.01 *M* Tris, the amplitude of the CD curve of 2'-*O*-methyl AprA is smaller than that of rAprA [69]. Since dApdA also has a smaller amplitude than ArprA, and in view of the problem in the interpretation of CD, this observation does not lead to a simple conclusion. At the level of polynucleotides, however, the interpretation is unambiguous. The conformation of poly 2'-*O*-methyl A resembles that of poly rA, but differs from that of poly dA [68a]. The study on the 2'-*O*-methyl derivatives, therefore, supports the conclusion that the difference between the deoxyribosyl compounds and the ribosyl compounds is not due to the hydrogen-bonding properties of 2'-OH group but due to the steric properties of the 2'-OH group vs those of the H atom. Similar comparisons between poly dI, poly 2'-*O*-methyl I and poly rI have been made [68a,b]. Therefore, we may conclude that this is a general property of the purine derivatives.

b. *Cytosine Homodimers.* From the preliminary studies, it appears that the difference between the ribosyl pyrimidine compounds (especially cytosine) and the deoxyribosyl pyrimidine compounds is not the same as that between the ribosyl and deoxyribosyl compounds of purine derivatives discussed above. The CD spectra of rCprC and 2'-*O*-methyl CprC are nearly identical ([28], Fig. 9); while the CD spectrum of dCpdC [27,28], Fig. 8) is similar to that of rCprC in shape but has a much smaller peak (277 nm, $[\theta] = 1.42 \times 10^4$, 26°) than that of rCprC (280 nm, $[\theta] = 3.08 \times 10^4$, 26°). It should also be noted that the CD spectra of cytosine homodimers are highly nonconservative in nature. The situation of these CD spectra of cytosine dimers in some respects is similar to the situation of the CD spectra of the adenine dimers. However, there is one important difference. The CD spectrum of single-stranded poly rA is vastly different from that of poly dA [68c], while the CD spectrum of poly rC is similar to that of poly dC but with a much smaller magnitude of $[\theta]$ at peak position [70]. This observation has led to the suggestion that single-stranded poly rC has more secondary structure (stacking) than single-stranded poly dC in solution at moderate temperature. This notion is supported by UV hypochromicity and PMR data. It should be noted that this conclusion is in contrast to that about the adenine derivatives, i.e., that the single-stranded poly dA has more secondary structure (stacking) than the poly rA.

At present, there is no quantitative data on the comparison of hypochromicity between dCpdC and rCprC. As for the PMR study on rCprC vs dCpdC [71], the H-5 and H-1′ protons of rCprC are significantly more shielded (from 0.05 to 0.17 ppm) than the corresponding protons of dCpdC; while the H-6 proton of rCprC is about the same or slightly less shielded (≈ 0.02 ppm) than the H-6 of dCpdC. The same pattern holds true from 4° to 60°. It may be concluded tentatively that the degree of base–base overlap of rCprC is larger than that of dCpdC, but data also suggest that the mode of stacking may be different from rCprC and dCpdC. It should be noted (Section I,B) that the ring-current magnetic anisotropy effect is less in pyrimidine than in purine; therefore, less information can be derived from the PMR results on pyrimidine molecules.

CD spectra of homodimers of cytosine with arabinose backbone, i.e., arabinosyl C, have been reported [70,72]. The CD spectra of these compounds grossly resemble (but are not the same as) those of rCprC and dCpdC. The magnitudes of [θ] at peak positions of these compounds are much smaller than that of rCprC, but are larger than that of dCpdC at room temperature. However, the difference between the CD spectrum of araCparaC and that of the monomer, 5′-araCMP, is very small, even smaller than the difference between the CD spectrum of dCpdC and that of dCMP. Since the CD spectrum of araCparaC is practically identical to its monomer, it was concluded that the araCparaC has little or no secondary structure [70,72]. Discussion in the past [70,72] about the reason for this apparent lack of secondary structure in araCparaC vs rCprC has been concerned mostly with the availability of the hydrogen-bonding of the hydroxyl group substituted at C-2′ position. It has been argued that in an arabinosyl backbone such as araCparaC since the 2′-OH group is now at a *trans* position vs 3′-OH (or 3′-phosphoryl group), the original hydrogen-bonding capacity of the 2′-OH group of the ribosyl has now been abolished. However, since 2′-*O*-methyl CpC has the same CD spectrum as rCprC [28], this argument therefore is not valid. Upon model building, it is rather clear that the *trans* 2′-OH group of the arabinosyl backbone provides a substantial steric hindrance to the overlapping of bases in the dimer. This obstruction to stacking is most likely the reason why the araCparaC has little secondary structure. The same explanation may be applicable for the lack of stacking in other arabinosyl dimers, such as araCpA [72].

E. CONFORMATION DYNAMICS

It has been emphasized that the conformational models discussed in Sections I,C and D are on a time-averaged basis in terms of an entire population; the dimers in solution do assume various conformational states at a given time and are continuously in motion. Characterizing this dynamic process sufficiently to obtain useful information is a formidable challenge. Generally, upon increase in the temperature, the UV absorbence of the dimer increases (or decreases in percent hypochromicity, Section I,D,1); the amplitude of the CD spectrum decreases (for example, Fig. 17); and the chemical shifts of the protons are moved downfield (or reduction in dimerization shift, Table X). All these observations indicate that the degree of stacking between bases is reduced at high temperature. Can these data be analyzed on a more quantitative basis?

One early approach to this problem [29] was to assume that the unstacking process is a two-state process,

$$D_{(s)} \; \rightleftharpoons \; D_{(u)} \tag{8}$$

where $D_{(s)}$ represents the conformational state of a completely stacked form at low temperature, and $D_{(u)}$ represents the conformational state of a completely unstacked form at high temperature. At intermediate temperature, t, this approach assumes that the population consists of a mixture of dimers in $D_{(s)}$ and in $D_{(u)}$ states. If the equilibrium constant (K_t) of Eq. (8) and its temperature dependence can be derived from the data, then the thermodynamic function of this process can be evaluated. Experimentally, this approach requires an unambiguous determination of the properties of the dimer at $D_{(s)}$ and at $D_{(u)}$, which presumably are the properties of the entire dimer population in solution at a sufficiently low temperature and at a sufficiently high temperature, respectively.

The limiting conditions for the two temperature levels are supposedly reached when the measured properties become insensitive to temperature variation at the two extremes. However, for most of the natural dimers, especially the purine dimers, these limiting conditions could not be achieved in aqueous solution. At temperatures near 85° and 0°C, where experimental measurement became difficult, the temperature dependence of the optical and magnetic properties of the dimer usually has not been significantly diminished as compared to that at room temperature. This problem is especially serious for reaching the low temperature limit, and attempts have been made to use 4.7 M KF [29] and 25.2% LiCl [30] solutions as solvents, so that measurements can be made at $-20°$ and $-65°C$, respectively. In 4.7 M KF at $-20°C$, the optical properties of many dimers are still highly sensitive to temperature variation [29]; in 25.2% LiCl, certain dimers such as rAprA and rCprC appear to reach the temperature insensitive state at about $-60°C$. However, the optical properties of the dimers appear to be altered in the presence of the strong salts [28,30]. Nevertheless, $\Delta H°$ values for stacking of rAprA and rCprC in 25.2% LiCl have been computed by this procedure on ORD data to be about 5.3 and 4.9 kcal/mole, respectively [30]. $\Delta H°$ for stacking of rAprA in dilute buffer ($\sim 0.1\ M$ salt) was computed to be 5.3 kcal/mole by Warshaw and Tinoco [22], 8.0 kcal/mole by Brahms, Michelson, and Van Holde [73], and 5.3 kcal/mole in 4.7 M KF by Brahms, Maurizot, and Pilet [72]. $\Delta H°$ for stacking of rCprC in dilute buffer was computed to be 6.9 kcal/mole by Davis and Tinoco [30] and to be 7.5 kcal/mole in 4.7 M KF by Brahms, Maurizot, and Michelson [29]. Though all these computations from different laboratories were based on CD or ORD data, there appears to be little consistency among them.

In fact, Davis and Tinoco [30] pointed out that the $\Delta H°$ of rAprA computed on ORD and CD data based on this two-stage model varies from 5.3 to 8 kcal/mole, and the $\Delta H°$ computed on UV absorbence–hypochromicity data varies from 8.5 to 10 kcal/mole. This large variation (5–10 kcal/mole) of $\Delta H°$ values, on the one hand, may reflect the experimental difficulty in obtaining the boundary conditions ($D_{(s)}$ and $D_{(u)}$), but, on the other hand, it also may reveal the inconsistency between the theoretical expectation of a two-state model and the actual results. If unstacking is indeed a two-state process then the same thermodynamic parameter should be obtainable, using any experimental method which is capable of detecting the difference between the stacked and the unstacked form. In addition, the basic postulate of the two-state model is that the energy difference between these two major states is much larger than the small energy differences among the microstates within each major state. In considering the helix-coil transition of double stranded DNA, such a postulate is reasonable. However, in the case of the dimer, there can be a range of stacked conformations and a range of unstacked conformations with no sharp division between them. Under this condition, a two-state model would not be applicable.

Most recently, Powell *et al.* [39] reexamined this question on the basis of the UV absorption, CD and ORD data of rAprA and rCprC at various temperatures. They provided evidence to indicate that previous experiments done in concentrated LiCl solution in order to reach the low-temperature plateau for the optical data have created undesirable complications due to the interaction of the salt directly with the dimers. They proposed, therefore, to obtain the four variables $\Delta S°$, $\Delta H°$, $D_{(s)}$ (data describing the conformation state of the completely stacked form at the low-temperature plateau) and $D_{(u)}$ (the data describing the conformational state of the completely unstacked form at high temperature), by an iterative process through computer analysis of the optical data at various temperatures. The resulting families of curves were examined to see if they were linearly related to each other and whether they were expressible in terms of two linearly independent components. For instance, experimentally a well-defined isosbestic point for the series of UV absorption and CD spectra at various temperatures is a good indication of a two-variable process. They concluded that these optical data obtained in the range 1°–80°C could be analyzed by thermodynamics formulated on the basis of a two-state model. However, they carefully pointed out "the possibility that the parameters used to observe the equilibrium are not sufficiently sensitive to discriminate between the microscopic substates" [39]. To amplify this point further, they state:

> We emphasize that our results do not rule out multistate models, such as that of Tinoco and co-workers, but merely show that intermediates between the stacked and unstacked forms do not have recognizable identities in terms of spectroscopic properties. The methods at present available are responsive only to these two extreme states, each of which can clearly contain a population of substates, which again are not spectroscopically distinguishable.

The $\Delta H°$ values obtained for ApA and CpC were 8.1–8.9 kcal/mole and the $\Delta S°$ values were 28–31 cal/deg-mole.

In the above analysis, the stacked state, $D_{(s)}$, and the unstacked state, $D_{(u)}$, have been defined only operationally, but not in absolute geometrical terms, which is probably an impossible task for optical studies. It has been clearly noted in our PMR studies on the temperature dependence of the chemical shifts of various protons in the adenine dimers, that the results do not support the two-state model [24,54]. The data from seven adenine dimers indicated that the percent change from 4°–60°C is different for each proton in the dimer.

As shown in Table IX, for instance, the change of $\Delta \delta_D$ 4°–60°C in $rA_{3'}p_{5'}rA$ is 58% for H-2 of 3'- residue, 42% for both H-8 and H-1' of 5'- residue, 33% for H-1' of 3'- residue, and less than 20% for H-2 of 5'- residue. A similar situation can be observed for all the adenine dimers (Tables IX and X). It is difficult to explain this differential temperature effect on dimerization shifts by a two-state model, since this model would predict an equal temperature dependence (normalized as percent change) for all protons within a given dimer. This conclusion is derived from the fact that the dimerization shift values ($\Delta \delta_D$) for all the protons of a dimer at the totally unstacked state are zero [34]. Thus, the PMR data indicate that at moderate temperature (such as room temperature), the dimer population *cannot* be classified into two major groups: the completely unstacked dimers which have no dimerization shifts for all the protons measured, and the completely stacked dimers, which have the maximal dimerization shifts. On the contrary, the dimerization shift results show that the dimers exist in a variety of conformations with varying degrees of base–base overlap, and a population-averaged conformation can be constructed from these data as shown in Figs. 15, 16, and 18.

The study on the dinucleoside monophosphate alkyltriesters described in Section I,D provides additional information about the dynamic states of the dimer conformation, since the alkyl group (especially the ethyl group) on the phosphate moiety serves as an indicator of the interaction between the backbone and the two bases [50,64]. As shown in Fig. 25a, the population of the dimer–triesters consists of a pair of diastereoisomers due to the asymmetry of the phosphate group after substitution. The difference in chemical shifts of the methyl protons, $\Delta(\delta_1 - \delta_2)$, of the diastereomeric phosphate–ethyl groups of Tp(CH$_2$CH$_3$)T is very small from 1° to 84°C (Fig. 25b), indicating that these groups are in very similar magnetic environments. On the other hand, the phosphate–ethyl groups of the diastereomeric dAp (CH$_2$CH$_3$)dA's and diastereomeric dIp(CH$_2$CH$_3$)dI's must be in different magnetic environments, since their chemical shifts (methyl protons) are not the same (Fig. 25b). This may result from a difference in the degree of shielding from the adjacent adenine rings experienced by the two ethyl groups. The effect of temperature on the difference in chemical shifts, $(\delta_1 - \delta_2)$, of the two diastereomeric methyl resonances of dAp(Et)dA and dIp(Et)dI in D$_2$O are shown in Fig. 25c. From about 84° to about 60°C, the values of $(\delta_1 - \delta_2)$ for dAp(Et)dA and dIp(Et)dI are zero, indicating that the two diastereomeric methyl resonances are practically equivalent in this temperature range. At temperatures below 60°C, $(\delta_1 - \delta_2)$ from dAp(Et)dA is measurable, and increases linearly to 0.04 ppm at 1°C (Fig. 25b,c). For dIp(Et)dI, the $(\delta_1 - \delta_2)$ value becomes detectable at 18°C and increases to 0.01 ppm at 1°C (Fig. 25b,c). It is important to note that the increase in the value of $\delta_1 - \delta_2$ vs lowering in temperature per degree is about the same for both dAp(Et)dA and dIp(Et)dI, except that for dAp(Et)dA the increase starts from zero at about 60°C, and for dIp(Et)dI the increase starts from zero at about 18°C (Fig. 25c).

The dAp(Et)dA was also studied in CD$_3$OD; the spectrum of the two methyl resonances at -53°C (220°K) is shown in Fig. 25b(D) with a $(\delta_1 - \delta_2)$ value of about 0.02 ppm. The $(\delta_1 - \delta_2)$ value starts from zero at about -12°C (261°K) and again increases linearly upon lowering the temperature (Fig. 25c). Also the slope of the line in the $(\delta_1 - \delta_2)$ vs temperature plot of dAp(Et)dA in CD$_3$OD is similar to that in D$_2$O (Fig. 25c). However, the starting point of the $\delta_1 - \delta_2$ value from zero is at 261°K in CD$_3$OD vs 333°K in D$_2$O. The lack of base-stacking of dAp(Et)dA in CD$_3$OD at room temperature is also indicated by the base protons spectrum which only shows four resonance lines; at -43°C, eight resonance lines can now be observed for the base protons. Only in a stacked conformation is the arrangement of the base planes influenced by the ethyl group; therefore, under this condition the four base protons of the two diastereomers become nonequivalent and afford an eight-line spectrum. In summary, under conditions favorable for base stacking (such as low temperature in D$_2$O), the differences in the magnetic environment of the two methyl resonances from the diastereomers are clearly measurable for dAp(Et)dA and dIp(Et)dI. These differences diminish upon increasing temperature and vanish completely at a temperature characteristic for each dimer. Clearly, the phenomenon of magnetic nonequivalence of the two methyl groups of the diastereomeric phosphate ethyl substituents is related to the degree of base stacking and the ring-current magnetic anisotropy of the purine bases.

The comparison among the methyl resonances from the three dinucleoside phosphotriesters in D_2O solution provides additional valuable information on their conformational dynamics. Above room temperature, the methyl resonances of Tp(Et)T were found to be located at a field lower than those of dAp(Et)dA and dIp(Et)dI. This is because the chemical shifts of the phosphate methyl protons in Tp(Et)T reflect the intrinsic field position of these resonances without the effect of the diamagnetic anisotropy of the bases [50,64]. This notion is supported by the observation that the two methyl resonances of Tp(Et)T are practically equivalent, an indication of the absence of shielding from the base moieties. Therefore the difference in chemical shifts ($\delta_A - \delta_T$ or $\delta_I - \delta_T$) between the phosphate methyl resonances of Tp(Et)T (low field) and those of dAp(Et)dA and dIp(Et)dI (high field) is a reliable index of the shielding effect of ring-current magnetic anisotropy of the bases on the methyl protons. The differences ($\delta_A - \delta_T$ or $\delta_I - \delta_T$) in chemical shifts among the methyl resonances of these three phosphotriesters at varying temperature are shown in Fig. 25d. The $\delta_A - \delta_T$ value reaches the maximal plateau at about 60°C, the $\delta_I - \delta_T$ values at about 28°C; in both cases, no increase was found up to 84°C. It should be noted that the maximal values for $\delta_A - \delta_T$ and $\delta_I - \delta_T$ are nearly the same. As the temperature is lowered, the $\delta_A - \delta_T$ and the $\delta_I - \delta_T$ values begin to diminish. Unfortunately, in D_2O, the experiment could not be carried out below 0°C. It thus appears, however, that these values diminish toward zero. As the temperature is reduced the two methyl resonances begin to separate for dAp(Et)dA and dIp(Et)dI (Fig. 25d), a phenomenon described in the preceding section.

The following model for the conformational changes of these dinucleoside phosphotriesters appears to afford the best explanation for the two basic phenomena observed above. In D_2O and at low temperature, the two planar bases of most dAp(Et)dA are in an overlapping position with each other, thus leaving the phosphate ethyl moieties outside and at a fixed distance from the two bases. For this reason, the methyl protons in dAp(Et)dA are not shielded by the adenines at low temperature and become similar to those of Tp(Et)T. Since each diastereomeric phosphate ethyl substituent occupies a unique spatial position about the phosphate group of the stack dimer, the distance of the methyl group to the two bases in one diastereomer is different from that in the other diastereomer, creating two different magnetic environments for the two groups of methyl protons as observed. This explanation is fully in accord with the molecular models (CPK models) which in fact suggest that

structure i (Fig. 25a) is the diastereomer which has the methyl group away
from the two bases, and structure ii (Fig. 25a) is the diastereomer which has
the methyl group close to the two bases. It is likely, therefore, that the methyl
resonances of diastereomer i will be located at a lower field than the methyl
resonance of diastereomer ii. The observation of the large shielding effect
(recorded as the dimerization shifts) which the base protons and the H-1'
protons experiences in dAp(Et)dA under these conditions [50] supports the
conclusion that most of the dimers in solution of D_2O solution at low temperature are in a stacked form. As a consequence, the population of dAp(Et)dA
at low temperature has a narrow distribution in terms of various conformers.

At elevated temperature and in organic solvent, the planar bases of most
dAp(Et)dA are away from each other as indicated by the reduction in
dimerization shifts of the base and H-1' protons [50], but the methyl protons
of the ethyl substituents are now close to the bases of the unstacked dimer as
indicated by the increased shielding of the ring current of the adenines. Yet
under this condition, the magnetic nonequivalence of the methyl resonances
of the two diastereomers vanishes. The only logical explanation is that at
high temperature, there exists a wide range of geometrical positions of the
two planar bases relative to each other. Thus, the average distances of the
two diastereomeric methyl groups to the bases of the unstacked dimers become
very similar and the methyl groups of these two diastereomers become magnetically indistinguishable. On the contrary as prescribed by the "two-state
model," if there exists a unique unstacked conformation of the dimer with
the bases close to the phosphate ethyl substituent (as required by the chemical
shift data), then the magnetic nonequivalence of the methyl resonances should
be preserved at high temperature.

The above discussion of the PMR data is on the basis of an average en-
semble in a population of dimers measured at a given time. The data clearly
established that in the population of stacked dimers at low temperature, the
relative position of the two base planes in the dimer is unique and the
distribution of the conformers is narrow. On the other hand, in the population
of unstacked dimers at high temperature, the two planar bases assume a
large variety of positions relative to each other and thus the distribution of
the conformers is very wide. These findings indicate that the "stacked"
state may be unique but the "unstacked" state of the dimers is a collection
of many conformers. Such a finding together with the former PMR results
[24,54] does not support the application of a simple "two-state model" for
the analysis of the conformational changes.

When the model developed above on the basis of a population of dAp(Et)dA
is adopted to explain the properties of a typical dimer under thermal perturbation, this explanation is more in accord with an "oscillation model." In
D_2O and at low temperature, the two planar bases of a dAp(Et)dA dimer

spend most of their time in a unique, overlapping position with respect to each other and the ethyl substituent is kept outside of the stack. When the temperature is elevated, the bases begin to move away from each other and to assume various relative positions, probably in an oscillation mode at moderate temperature. Finally, under sufficient perturbation at high temperature, the planar bases constantly move away from each other probably in a rotation mode. Constant rotation at the backbone allows the methyl protons to be fully exposed to the bases and therefore subjected to maximum shielding by the ring current of the bases. Also, the relative movement of the bases and the rapid rotation of the backbone greatly diminish the difference in distances to the bases between the two methyl protons from the two diastereomers. As a result, the methyl resonances from both diastereomers are shifted upfield and beome magnetically equivalent (within the resolution limit of 0.002 ppm) at high temperature. This description is in agreement with the observed deshielding effect of the base protons and H-1' protons of dApdA due to elevation of temperature [54].

When the solvent for dAp(Et)dA is changed from D_2O to CD_3OD, the same process occurs except at a temperature $\sim 70°$ lower than the process in D_2O (Fig. 25c). This reflects the relative ease of destacking of the two adenine bases in dAp(Et)dA in methanol vs that in water. An attempt at a more quantitative evaluation of this process is presented in a following paragraph.

Comparison between dAp(Et)dA and dIp(Et)dI reveals additional information about the stereochemistry of the interaction between the ethyl groups and the purine bases. Theoretical calculation [53,63a] indicates that the ring-current magnetic anisotropy of the six-membered ring in guanine and hypoxanthine is much less than that in adenine, while the ring-current magnetic anisotropy of the five-membered ring in guanine and hypoxanthine is nearly equivalent to that in adenine. It is of interest to note that at the plateau region of high temperature, the extent of the shielding of the methyl groups from dIp(Et)dI is about the same as that for dAp(Et)dA (Fig. 25d). In addition, the slope of the line in the plot of $\delta_1 - \delta_2$ vs temperature (Fig. 25c) appears also to be the same for both dIp(Et)dI and dAp(Et)dA, though the starting temperature for the separation ($\delta_1 - \delta_2 > 0$) begins at different levels for these two dimers. These observations strongly suggest that the shielding effect exerted by the purine bases on the methyl protons is through the ring-current magnetic anistropy of the five-membered ring of the bases. This suggestion is supported by the molecular models, since the five-membered ring of the base is closer to the backbone region of the dinucleoside monophosphate than the six-membered ring when the nucleosidyl units are in an *anti* conformation.

The model described above concerning the conformation dynamics of the dinucleoside monophosphate triester as revealed by the PMR studies can be more quantitatively described by the diagram in Fig. 28. At a sufficiently low temperature, T_1, where the conformation of the dimer is no longer affected by a further lowering of temperature (the low-temperature plateau), the base planes will have a maximal extent of overlapping. This maximal extent of overlapping can be characterized by the angle, θ, formed by the two principal axes of the two bases—the value of θ should be small, and if the one base is exactly on top of the other, the value of θ is zero. Under these conditions, the distribution of the conformers in the population is very narrow and these

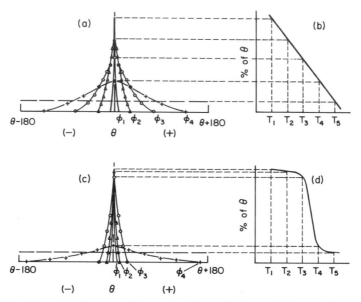

Fig. 28. Schematic diagram of two possible oscillating-rotating models for dimer conformation at varying temperatures. $T_1(—)$; T_2 ($\triangle$); T_3 ($\bigcirc$), T_4 ($+$), and T_5 (— —). For the definition of θ and ϕ_{1-5}, see text. (From Kan *et al.* [64].)

conformers can be characterized by having the angle within the range of $\theta + \phi_1$ to $\theta - \phi_1$. At a higher temperature, T_2, the two base planes begin to move away from each other probably by oscillation; these conformers are described by the angle $\theta + \phi_2$ to $\theta - \phi_2$ where ϕ_2 is larger than ϕ_1. Thus the population has a wider distribution at T_2, though the main fraction of the population is still centered at the conformation characterized by θ. Similar situations occur at the higher temperatures, T_3 and T_4, with $\phi_4 > \phi_3 > \phi_2$; though the distribution of conformers is wider at higher temperature, the conformer having angle θ still comprises the largest fraction. When the temperature is sufficiently high, T_5, the conformation of the dimer is no longer affected by a further elevation of temperature (at the high-temperature plateau). Then the distribution of conformers in the population becomes random and even, as shown in the diagram. Under such a situation the model of relative movement of the base planes is likely to be a continuous, rapid rotation.

Thus, our previous PMR studies on the base–base interaction and the present PMR studies on the base–backbone interaction all support the diagram described in Fig. 28 as the model for the unstacking process of the dimers. These PMR results are *not* in accord with a model which divides the population into two main fractions: one completely stacked fraction consisting of conformers having angle θ, and the other a completely unstacked fraction having angle $\theta + \phi$ (and/or $\theta - \phi$) with ϕ having a large value (say 90°–180°). In this model, the population distribution between these two main fractions (each one can have its own standard deviation) is then governed by the temperature—low temperature favoring the fraction of θ and high temperature favoring the fraction of $\theta + \phi$ (and/or $\theta - \phi$). The PMR data certainly cannot be described by such a two-state model, when the two states are defined by such explicit conformational terms. On the other hand, there can be various types of temperature effects on the distribution of conformers among the population as described by the diagram in Fig. 28. If the fraction θ (or the remaining portion of the population $1 - \theta$) is plotted against the variation of temperature (i.e., T_1–T_5), in principle there can exist at least two types of temperature-transition profiles. For the first type (Fig. 28b) the transition is gradual, and the effect of temperature on the two fractions per degree change is more or less constant over the entire temperature range. For the second type (Fig. 28b), the transition is abrupt, the effect of temperature on the two fractions per degree change is negligible at first, then rises to a maximal value, and finally decays to zero again. In other words, the transition is gradual and continuous for the first type, but abrupt and discontinuous for the second type.

This new insight may bring forth a resolution of the apparent conflicts between the results from the optical data from Powell *et al.* [39] and those from the PMR studies. If the optical measurements in relation to the conformation of the dimers can indeed be separated into two such classes—those of the fraction within θ pertaining to the stacked form and those of the remaining portion $(1 - \theta)$ pertaining to the unstacked form, then the optical measurements vs temperature data can be analyzed in terms of two variables. Such a possibility does exist in view of the complexities in the distance dependence, angle dependence, etc., of the interaction between the chromophores, especially of a homodimer (see Section I,B). If both CD measurement and UV absorption measurement are dependent on the conformation in a similar manner, then these two measurements can be linearly related. However, the meaning of the thermodynamic analysis of such data based on a two-state model is not certain, since there is no discrete, large energy barrier between the fraction of θ and the remaining portion, $1 - \theta$; these fractions were arbitrarily determined by the optical measurements. This is not true for the transition of type II discussed above (Fig. 28c,d). As shown in the diagram of this hypothetical case, there exists a definite and discrete energy barrier between the distribution of T_3 and that at T_4. Clearly the transition from the distribution pattern at T_3 to that at T_4 can be described as the transition between two states, and thus can be analyzed by a two-state theory. In terms of molecular geometry, it is likely that oscillation is the predominant model of the relative movement between the two base planes at temperature below T_3, while full rotation becomes the major model of the relative movement at temperatures above T_4. Thermodynamic analyses of such a system based on a two-state model would provide information about difference in energy between the distribution of conformers caused by an oscillation mode of movement vs that caused by a rotation mode of movement in response to perturbation. The challenging question is still what type of physical measurement can accurately reflect the distribution of the conformers in the population? If the optical data can accurately gauge the fraction of θ, vs the remaining fraction of $1 - \theta$ in the population of conformers, then the thermodynamic analysis of such data does provide very useful information. So far, neither optical data [24,29,30,39,54] nor PMR data [24,33,40,54] of adenine dimers obtained at various temperatures show an abrupt and discontinuous transition profile as described in Fig. 28d. Therefore, this statement and the diagram of Fig. 28 c and d remain a hypothetical case. On the other hand, the gradual and continuous transition profile observed for both optical data and the PMR data is more in agreement with the diagram described in Fig. 28 a and b.

In brief, the present studies on the conformation dynamics of the dimer suggest a model of oscillation at low temperature and a model of continuous, full rotation at high temperature. In aqueous solution, and at low temperature, the bases of a typical dimer tend to spend a long time in a stacked conformation with the backbone outside the stack. Thus, from the standpoint of the entire population at a given time, most of the dimers exist in the stacked conformation at low temperature. At high temperature or in organic solvents, the bases tend to spend much less time in a stacked conformation, and the dimer rotates in such a manner that the backbone (as indicated by the triester–alkyl group) atoms are now in proximity to the base planes. Under these conditions, the population of the unstacked dimers consists of many conformers with bases having different positions relative to each other.

F. OLIGONUCLEOTIDES

Though the extensive discussion on the dinucleoside monophosphates in the above sections has prepared the ground for our inquiry into the properties of oligonucleotides, the study of oligonucleotides is still in its infancy. We must distinguish two effects of chain length on the physicochemical properties from an extension of dimer to oligomer series: (1) Effect of chain length on the measurable parameters—an effect related to the problem of measurement. For example, even if dimers and oligomers had identical structures, the observed properties might be different due to the existence of distant neighbor effect; and (2) Effect of chain length in altering the actual structures assumed by oligonucleotides—an effect related to the intrinsic properties of the oligomers. The separation of these two effects presents a formidable challenge to our present knowledge; in addition, preparation and proper identification of oligomers in series is technically demanding.

From the standpoint of conformational analyses, we can envisage at least three distinct changes when the dimer is lengthened to trimer or oligomer. (1) The presence of more than one phosphate in the trimer (NpNpN, or even dinucleotide, pNpN, etc.) introduces the factor of electrostatic interaction between the negatively charged groups. (2) In the trimer, the interior residue is generated for the first time, and is sandwiched between the exterior residues and therefore is less accessible by the solvent. (3) Additional restriction to the conformational equilibrium is imposed on the trimer; for instance, the right-handed stack ⇌ left handed stack interconversion should be much reduced in the trimer as compared to the dimer; the conformational mobility of the interior residue also should be considerably suppressed. These considerations should be kept in mind when the properties of the dimer are compared with those of the trimer and oligomer.

Preparation and identification of the oligomer series requires a considerable amount of effort. As an example, Fig. 29 shows the elution profile of the oligoinosinate (oligo I) from the DEAE cellulose column with a linear gradient with 0–1 M triethylammonium acetate buffer (pH 6.5) in 7 M urea [49]. The oligomers were well separated and the chain length of the oligomers in this series was unambiguously identified up to the octamer ($n = 8$). Figure 30 shows a series of sharply fractionated dAT oligomers which has been prepared by digestion of the dAT copolymer with pancreatic DNase, followed by preparative electrophoresis in concentrated acrylamide gel [74]. All these oligomers were obtained from the degradation of the polymer synthesized by enzymic process.

1. *Ultraviolet Absorbance Studies*

A phenomenological description of the nearest-neighbor interaction can be readily developed for a variety of physicochemical measurements; an example is cited here for hypochromicity [75]. If the hypochromic effect is derived only from the interaction of the nearest-neighboring chromophores and if the

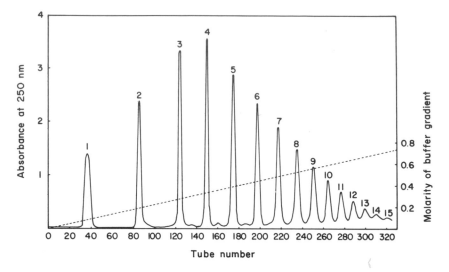

Fig. 29. DEAE cellulose elution profile of poly I degradation by alkaline hydrolysis, followed by acid treatment and alkaline phosphatase treatment. The linear gradient elution was carried out with 0–1 *M* triethylammonium acetate buffer (pH 6.5) in 7 *M* urea. (From Tazawa *et al.* [49].)

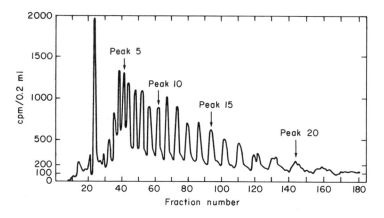

Fig. 30 Elution pattern of ^{32}P-labeled dAT oligomers fractionated by preparative polyacrylamide gel electrophoresis. (After Elson and Jovin [74].)

chain length of the oligomer does not enhance base-stacking in a dimeric segment, we might expect to find a decrease in absorptivity ($\Delta\varepsilon$) proportional to the number of pairs of residues, as indicated in the following equations:

$$\varepsilon_2 = \frac{2\varepsilon_0 - \Delta\varepsilon}{2}$$

$$\varepsilon_3 = \frac{3\varepsilon_0 - 2\Delta\varepsilon}{3} \qquad (9)$$

$$\varepsilon_n = \frac{n\varepsilon_0 - (n-1)\,\Delta\varepsilon}{n}$$

where ε_0 is the absorbence of the isolated residue (or the monomer), ε_2, ε_3, and ε_n are the observed, mean residue absorbences of dimer, trimer, and oligomer, respectively, and n is the chain length.

Upon rearrangement of Eq. (9), the following relation can be obtained:

$$\varepsilon_n = (\varepsilon_0 - \Delta\varepsilon) + \Delta\varepsilon/n \qquad (10)$$

A plot of ε_n vs $1/n$ would be linear, with the slope equal to $\Delta\varepsilon$ and with the intercept at $1/n = 0$ of $(\varepsilon_0 - \Delta\varepsilon)$, corresponding to the absorbence of the polymer.

The hypochromicity data of the oligonucleotides can be analyzed in another manner. We shall divide the bases in the oligomer into two types; those located externally (at the ends), and those located internally (at the center). Based on the assumptions for the derivation of Eq. (9) or Eq. (10), we can formulate that the extinction coefficient of the external bases at the end is the same as that of the dimer ($\varepsilon_{\text{dimer}}$). Thus,

$$\varepsilon_n = f_{\text{ext}}\varepsilon_{\text{dimer}} + f_{\text{int}}\varepsilon_{\text{int}}$$
$$= (2/n)\varepsilon_{\text{dimer}} + (1 - 2/n)\varepsilon_{\text{int}} \qquad (11)$$

where f_{ext} is the fraction of the external bases of the oligomer ($f_{\text{ext}} = 2/n$), and f_{int} is the fraction of the internal bases ($f_{\text{int}} = 1 - 2/n$). When the physicochemical measurements of the oligomer series are describable by the assumptions involved in Eq. (10), then these two equations [Eq. (10) and (11)] are basically the same. However, when the data cannot be fitted in a straight line into an ε vs $1/n$ plot demanded by Eq. (10), then the analysis from Eq. (11) graphically illustrates the dependence of the properties of internal bases as a function of chain length. The usefulness of Eq. (11) can be demonstrated by a comparison between Fig. 31 and Fig. 32.

The hypochromicity data ($H_n = \varepsilon_n/\varepsilon_0$) of seven oligonucleotide series are plotted in accordance with Eq. (10) in Fig. 31, and with Eq. (11) in Fig. 32. The data of these seven oligomer series can be divided into two classes. The (Up)$_{2-11}$ series [75], the (pdT)$_{2-10}$ series [76], and the (pdA)$_{2-10}$ series [76] all

belong to the first class; the data of these three oligomer series can be described by Eq. (10). Within experimental error, the hypochromicity data of
these oligomer series can be fitted to a straight line (Fig. 31) in the H_n vs $1/n$
plot; and the hypochromicity of the first internal base in the trimer is the
same as the internal bases of the polymers (Fig. 32). This apparent compliance
of the data of the oligo U series, oligo T series, and oligo dA series with the
two basic assumptions needed for the derivation of Eq. (10) is significant.
The data—especially the data of $(pdA)_{2-10}$—reveal that the hypochromic

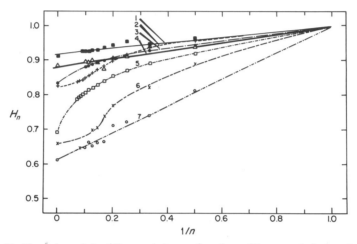

Fig. 31. Hypochromicity ($H_n = \varepsilon_n/\varepsilon_0$) as a function of inverse chain length. (1) rU;
(2) rI; (3) dC; (4) dT; (5) rC; (6) rA; (7) dA. Data were obtained from the following
sources. $(Ip)_{n-1}I$ (+); Tazawa *et al.* [49]; conditions: 0.005 M NaOAc, pH 6.0. $(Up)_n$
(■); Simpkins and Richards [75]; conditions: 0.01 M NaCl + 0.02 M tris, pH 7.0.
$(pdA)_n$ (○) and $(pdT)_n$ (△): Cassani and Bollum [76]; conditions: 1 mM tris, pH 8.0.
$(Cp)_{n-1}C$ (□) and $(dCp)_{n-1}dC$ (●): Adler *et al.* [77]; conditions: 0.08 M NaCl +
0.02 M tris, pH 8.5. $(Ap)_{n-1}A$ (×): Brahms *et al.* [73]; conditions; 0.1 M NaCl, 0.01 M
tris, pH 7.4. (From Tazawa *et al.* [49].)

effect in the oligonucleotides rests only on the interaction of the nearest-
neighboring chromophores, and that the interactions among chromophores
farther away than the two nearest neighbors are of little consequence. This
is the requirement for the first assumption. As for the second assumption,
there are two ways to satisfy its requirement. First, when the bases of a particular oligomer have very little tendency to stack, then the increase in chain
length of the oligomer would have no effect on the stacking of the bases in the
oligomer. Thus, owing to the minimal tendency in stacking, the extent of base-
stacking in the trimer under this condition can be essentially the same as that

of the polymer. This is likely the explanation for the compliance of the data from the oligo U series and the oligo T series with the requirement of the second assumption. Second, when the bases of other oligomers have a great tendency to stack, then the increase in chain length of the oligomer would have little effect on the stacking of the bases in the oligomers, if a *maximal* degree of stacking has also been reached. Thus, owing to the maximal tendency in stacking, the extent of base-stacking in the trimer under this condition can be the same as that of the polymer. This could be the reason for the compliance of the data of the oligo dA series with the requirement of the second assumption. PMR studies [36] indeed indicate the extensive stacking of the dimer, dApdA.

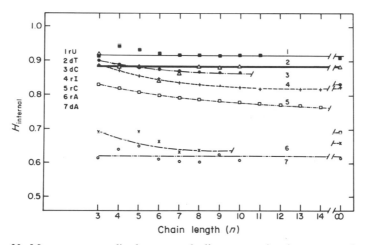

Fig. 32. Monomer normalized $\varepsilon_{\text{internal}}$ of oligomers and polymers as a function of chain length. Data and conditions are the same as in Fig. 31. (From Tazawa *et al.* [49].)

The data of the oligo I [49], oligo rA [73], oligo rC [77], and oligo dC [77] series belong to the second class. The data in this class do not give a linear line in H_n vs $1/n$ plot (Fig. 31) and therefore cannot be described by Eq. (10). This failure can be due to the noncompliance with the requirements of either or both of the two assumptions. However, since it has been concluded in the preceding paragraph that the phenomenon of UV hypochromicity in oligo-nucleotides rests on the interaction of the nearest-neighboring chromophores only, the inadequacy in Eq. (10) must be due to the requirement in the second assumption. In other words, the chain length of these oligomers does exert an influence on the degree of stacking of the bases in this class of oligomers. It is interesting to note that while the maximal extent of hypochromicity of oligo I

(~ 0.82) is less than that of oligo rA (~ 0.64), the dependence of the hypo-chromicity of the internal bases becomes equivalent to that of the polymer at *n* about 7–8 for the oligo rA series and at *n* about 8–9 for the oligo I series. The situation for the oligo rC series and oligo dC series is apparently different. The hypochromicity of the polymers is still much larger than that of the internal base at a chain length of fourteen in the oligo rC series. Apparently, the effect of chain length on the degree of base-stacking in the cytosine oligomer series is more gradual than that of the oligo I or oligo rA series.

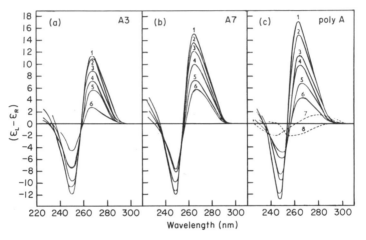

Fig. 33. Circular dichroism spectra of adenylate oligonucleotides at pH 7.4 at various temperatures [73]. (a) The trimer in 0.1 *M* NaCl, 0.01 *M* tris at: (1) $-2°C$; (2) $+0.5°C$; (3) $4.5°C$; (4) $18°C$; (5) $25°C$; (6) $47°C$. (b) The heptamer at: (1) $-2°C$; (2) $+0.5°C$; (3) $8°C$; (4) $18.5°C$; (5) $32°C$; (6) $40°C$. (c) Poly A at: (1) temperature range from $-2°$ to $+6°C$; (2) $17°C$; (3) $34°C$; (4) $42°C$; (5) $57°C$; (7) poly A quaternary ammonium salt in 98% ethanol at $0°C$; (8) AMP at $0°C$.

2. Circular Dichroism Studies

The success of the theoretical studies on the CD spectral properties of rAprA has been described in Chapter 2 and in Section I,B of this chapter. Serious attempts have been made to calculate the CD spectra of rAprA and several other dimers as a function of conformation based on this theoretical approach [78]. Also, CD spectra of oligo rA have been measured and their properties explained with a certain degree of satisfaction on the basis of coupled oscillator splitting of the strong $\pi \rightarrow \pi^*$ transition [73]. The CD spectra of trimer and heptamer of adenylates in comparison with AMP and poly A are depicted in Fig. 33. As shown, the positive band and the negative band in the spectra increase with the degree of stacking or chain length of the oligomer more or less proportionally, as anticipated by the theory, and this phenomenon was described as "conservative Cotton effect." Results in other oligonucleotide series, however, indicate that the situation is much more complex.

The most unusual case is that of the oligoinosinate series [49]. The dependence of CD spectra on the chain length of oligo I is displayed in Fig. 34.

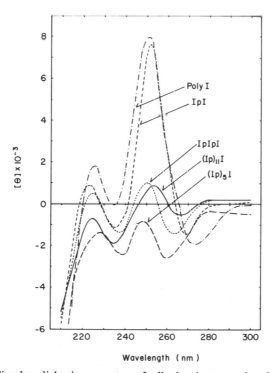

Fig. 34. Circular dichroism spectra of oligoinosinates and poly I at 2°C, 0.05 *M* NaClO₄ (pH 6.5). (From Tazawa *et al.* [49].)

This property of the oligo I series was also observed by Pochon and Michelson [79]; their proposed explanation concerning the change of nucleosidyl conformation involving *syn–anti* equilibrium will be discussed in later paragraphs. The unusual dependence of the CD spectra of the oligo I's on chain length is summarily presented in Fig. 35 [49]. After subtracting the values for 5'-IMP, the *remaining* intensities of the peak (248–252 nm) and the trough (270 nm) of the dimer, trimer, hexamer, dodecamer, and polymer at 2°C, 0.05 *M* NaClO₄ (pH 6.5) are plotted as a function of chain length (Fig. 35). While the details of these curves cannot be certain, the general tendency is clear. The differential intensity of the peak in CD decreases initially and significantly from the dimer to the trimer stage; then at some stage between the hexamer and dodecamer, it begins to increase again. At the polymer level, the differential peak intensity becomes about the same or slightly larger than that of the dimer. The differential intensity of the trough, on the other hand,

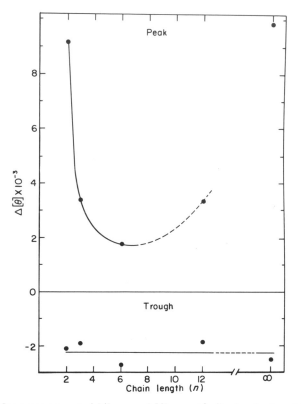

Fig. 35. Monomer corrected $[\theta]_{peak}$ and $[\theta]_{trough}$ of oligoinosinates at 2°C in 0.05 *M* NaClO$_4$ (pII 6.5). (From Tazawa *et al.* [49].)

remains practically the same from dimer to polymer. The CD spectral data of five oligonucleotide series, rI, rU, rA, dA, and rC, are shown in Fig. 36 for comparison. In this figure, the intensity values at the peak and at the trough of the first CD band of the five series are all normalized with respect to those of the corresponding polymers except in one instance. The results of the oligo U series in Fig. 36 (in dash lines) are actually derived from the ORD measurements which showed that UpU, oligo U, and poly U all have practically the same ORD spectra [75]; therefore, there is no chain length effect on the CD spectra of UpU through poly U, reflecting a lack of base–base interaction in this rU series. The peak intensity of the first CD band in the oligo rA series and oligo rC series increases with longer chain length, while the peak intensity in the oligo dA series decreases with longer chain length, and the peak intensity in the oligo rI series first decreases and then increases

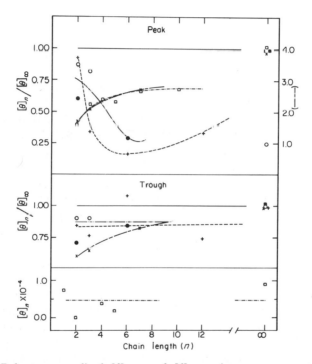

Fig. 36. Polymer normalized $[\theta]_{peak}$ and $[\theta]_{trough}$ (monomer corrected) of oligonucleotides as a function of chain length. The oligonucleotides are the same as in Fig. 31: $(Ip)_{n-1}I$ (+); $(Up)_n$ (■); $(dA)_{n-1}dA$ (○); $(pdA)_n$ (●); $(Cp)_{n-1}C$ (□); $(Ap)_{n-1}A$ (×). The raw values for $[\theta]_{trough}$ of rC are plotted in the bottom frame. (From Tazawa *et al.* [49].)

with increasing chain length. As for the trough intensity of the first CD band, only that of the oligo rA series increases with longer chain length; in all the other three series (rI, dA, and rC), the trough intensity shows little dependence on chain length. Therefore, only the CD spectra of the oligo rA series fulfill the expectation of a "conservative Cotton effect" derived from coupled oscillator splitting of the strong $\pi \rightarrow \pi^*$ transition, which gives positive and negative bands of nearly equal magnitude. This survey reveals no general rule in governing the CD dependence on chain length for these oligomer series, and the "conservative Cotton effect" appears to be an exceptional case rather than a common example. The geometry of the stacking pattern is known to be different between rAprA vs dApdA and between poly rA vs poly dA (Section I,D). The adenine bases are arranged in a more oblique manner in the ribosyl derivatives than the adenine bases in the deoxyribosyl derivatives, which are arranged in a more parallel manner. This consideration has been proposed previously as the explanation for the reduction of the "conservative Cotton effect" due to the reduction in the angle between the transition moments in a dipole approximation [39,54]. Bush and Scheraga [69] proposed that a $n \rightarrow \pi^*$ band at the 280 nm region is responsible for the optical activity of oligo- and poly dA at this wavelength; this band could be shifted to a shorter wavelength in the monomer due to solvation and reappears in the polymer spectrum when the bases become interior and are kept from contacting the solvent. A small difference between the MCD spectrum of adenine at pH 2 vs that at pH 7 did suggest the possibility of a $n \rightarrow \pi^*$ transition in this region ([80]; Chapter 6, Volume I). As for the cytosine chromophores, the broad band at the 280 nm region consists of both B_{1u} and B_{2u} transitions; and the 230 nm absorption band is an $n \rightarrow \pi^*$ transition ([80]; Chapter 6, Volume I). Though the assignments of these bands provide us with more insights about the optical properties of the isolated chromophores, a considerable amount of both theoretical and experimental work is needed for a better understanding of the optical activities derived from the interaction of these chromophores in the oligo-nucleotides.

As noted earlier in this section, the presence of more than one phosphate in the trimer (NpNpN) or even dinucleotide (pNpN) introduces the factor of electrostatic interaction between the charged groups. This consideration suggests that the conformation of these oligomers are more sensitive to ionic strength than the dinucleoside monophosphates (NpN). However, this situation is not often distinctly demonstrable. For instance, the PMR properties of the oligo I series are only moderately affected by ionic strength; on the other hand, the CD properties of this series are unusually sensitive to the influence of electrostatic interaction [49]. Therefore, the CD properties of oligo I are discussed further as an example of this interesting phenomenon.

The negative band at 245 nm of the CD spectra of inosine, 3'-IMP and 5'-IMP is practically the same, indicating that the addition of a phosphate to the nucleoside has little effect on the chromophore. In 0.05 M NaClO$_4$, the spectrum of pIpI at 2° (Fig. 37b) closely resembles the spectrum of (Ip)$_2$I between 24° and 40° (Fig. 38b) and is not similar at all to the spectrum of IpI at 2° (Fig. 37a). At 2°–4°, the spectrum of pIpI (Fig. 37b) now is nearly equivalent to that of 5'-IMP. In a similar fashion, addition of a 5'-phosphate

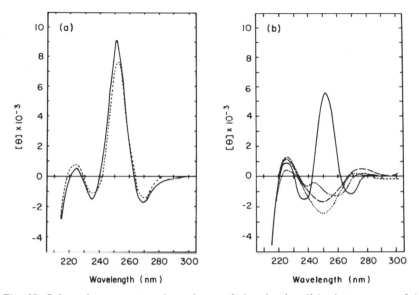

Fig. 37. Salt and temperature dependence of the circular dichroism spectra of (a) IpI and (b) pIpI in 0.05 *M* NaClO₄ (pH 6.5). Conditions: 2°C (---); 24°C (——); 40°C (· · ·); 2°C in 1 *M* NaCl (——). (From Tazawa *et al.* [49].)

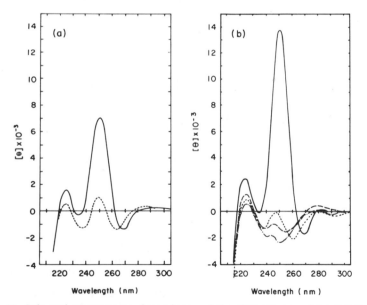

Fig. 38. Salt and temperature dependence of the CD spectra of (a) IpIpI and (b) pIpIpI in 0.05 *M* NaClO₄ (pH 6.5). Conditions 2°C (---), 24°C (——), 60°C (—·), and 2°C in 1 *M* NaCl (——). (From Tazawa *et al.* [49].)

group to $(Ip)_2I$ also has a substantial effect on the CD spectrum. At 0.05 M NaClO$_4$, the CD spectrum of p(Ip)$_2$I at 2° (Fig. 38b) is about the same as that of $(Ip)_2I$ at 24° [49], but not that of $(Ip)_2I$ at 2°. At about 60°, the CD spectrum of p(Ip)$_2$I resembles that of 5'-IMP. Since the effect of the addition of the terminal phosphate on the CD spectra is electrostatic in nature, such a phenomenon should be highly dependent on salt concentrations. As shown in Fig. 37a and b, while the change of 0.05 M NaClO$_4$ to 1 M NaCl solution

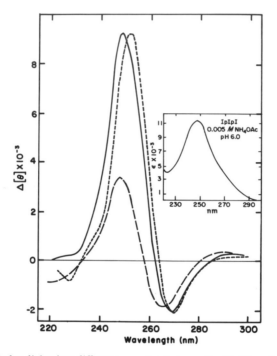

Fig. 39. Circular dichroism difference spectra (minus pI) of IpI (---) and IpIpI (— —) at 2°C in 0.05 M NaClO$_4$ (pH 6.5) and of IpIpI (——) at 2°C in 1 M NaCl. Inset is UV absorbence spectrum. (From Tazawa *et al.* [49].)

has little effect on the CD spectrum of IpI at 2°, the change to 1 M NaCl has an immense effect on the CD spectrum of pIpI. Similarly, the change to 1 M NaCl has a very large effect on the CD spectra of $(Ip)_2$ and p(Ip)$_2$I (Fig. 38a,b). In 1 M NaCl, and at 2°, the CD spectra of IpI, pIpI, $(Ip)_2I$, and p(Ip)$_2$I all become very similar to each other; the $[\theta]$ value of the peak at ~250 nm, however, is largest for p(Ip)$_2$I (about 14×10^3) and smallest for pIpI (about 6×10^3). In Fig. 39, the CD spectrum of $(Ip)_2I$ in high salt (1 M NaCl), the spectrum of $(Ip)_2I$ in low salt (0.05 M NaClO$_4$) and the spectrum of IpI in low salt arc compared, all after *subtraction* of the CD spectrum of 5'-IMP. In comparison of these *different* spectra, that of $(Ip)_2I$ in high salt is practically identical to that of IpI in low salt, except a slight blue shift of the CD band from 252 to 248 nm; while the intensity of the CD band at 248 nm of the $(Ip)_2I$ in low salt is definitely much smaller. It should be clearly noted that there is little change in the negative CD band at 265–270 nm, which may correspond to the shoulder of the UV absorption spectrum of $(Ip)_2I$ (shown in the inset of Fig. 39); all the change takes place only at the 248–252 nm CD band, corresponding to the main UV absorption band of $(Ip)_2I$ at 249 nm.

Carefully reviewing the information in Figs. 36–39, we are led to two important conclusions: (1) The effect on the CD spectra of these oligo I's by change in temperature, in phosphate to base ratio, and in salt concentration all takes place exclusively in the 245–255 nm CD bands without major variation in other spectral areas. The CD band at 245–255 nm can be changed from a negative value (monomer, -3×10^3; pIpI in 0.05 M salt at 40°C, -2×10^3) to a positive value (IpI at 2°, 8×10^3; $p(Ip)_2I$ at 2° in 1 M NaCl, 0.05 M NaClO$_4$, 14×10^3). In the UV absorption spectrum of the inosine derivative (see insert, Fig. 39), the 245–255 nm region is the location of the main absorption band and has been identified as the B_{1u} band of the $\pi \rightarrow \pi^*$ transition by magnetic CD studies [80]. The CD spectra of the dimer and trimer also show a negative CD band at the ~ 270 nm region, corresponding to the shoulder of the UV absorption spectrum which has been classified as the B_{2u} band of the $\pi \rightarrow \pi^*$ transition by the MCD studies. This CD band is little affected by all these perturbations (temperature, phosphate/base ratio, salt, etc.) directly. This observation, together with the *different* spectra shown in Fig. 39, indicates convincingly that the optical activity derived from the interaction of the chromophores in IpI and $(Ip)_1I$ is *not* the "conservative" type due to the coupled oscillator splitting of the $\pi \rightarrow \pi^*$ transition as in the case of ApA. The optical activity of IpI due to interaction is rather small and does not possess positive and negative bands of nearly equal magnitude.

The influence of the electrostatic interaction among the charged phosphate groups in the oligomer conformation is now well documented; in fact, the effect on the CD spectra by change in temperature, the introduction of phosphate group and the nucleoside residue into the chain (IpI vs pIpI vs IpIpi vs pIpIpI), and addition of salt (1 M NaCl) are all directly interrelated. Starting from the monomeric inosine, addition of a phosphate for the formation of nucleotide has little effect on its CD spectrum. Introduction of an inosine residue to pI for the formation of IpI has a significant effect on the CD of pI (or Ip), indicating the interaction of two linked inosine residues. In dilute salt solution (0.05 M), addition of a 5'-phosphate group to IpI and $(Ip)_2I$ to form pIpI and pIpIpI, respectively, has an effect on the CD spectra of IpI and $(Ip)_2I$ equivalent to a temperature elevation of 55° and 22°C, respectively [49]. Change from dilute salt solution to 1 M NaCl solution can counteract both effects of the introduction of the phosphate group and of the elevation of temperature, while the change to 1 M NaCl has little effect on the CD spectra of IpI (Fig. 37a). In strong salt condition, the spectra of IpI, pIpIp, IpIpI, and pIpIpI now all closely resemble each other. In dilute salt solution and at elevated temperature, the spectrum of pIpI and to some extent that of pIpIpI can be converted to one which is very similar to that of 5'-IMP. These studies conclusively demonstrate the electrostatic effect of the phosphate group on the conformation of oligo I as revealed by their CD spectra; the intricate, counterbalancing interaction between the phosphate and the base, as related to their influence on the oligo I conformation is well demonstrated also in these studies.

The effects of the addition of the phosphate group on the optical activities of the dinucleoside monophosphates have been previously noticed. As reported by Bush and Scheraga [69], the peak and the trough of the CD spectrum of ApAp ($[\theta]_{peak} = 1.00 \times 10^4$; $[\theta]_{trough} = -1.81 \times 10^4$) is substantially smaller than that of ApA ($[\theta]_{peak} = 2.48 \times 10^4$; $[\theta]_{trough} = -2.4 \times 10^4$). The same conclusion was reached by Inoue and Satoh [80a] in their comparative ORD study on the phosphate effect on ApA, CpC, etc. They found that the presence of a 3'-phosphate group (and to a lesser extent, also a 2'-phosphate group) at ApAp(3') reduces the first Cotton effect by about 42%, as compared to that of ApA, and that this reduction effect can be practically removed by the methylation of the 3'-terminal phosphate. Similarly, the rotation of CpCp(3') appears to be smaller than CpC at neutral pH. Inoue and Satoh [80a] had also observed a noticeable change of ORD spectrum of IpIp(3') from solution of 0.1 ionic strength to solution of 1.0 ionic strength. Comparison of the CD curve on dApdA [50] and that on pdApdA [69] again reveals a significant decrease in amplitude (33%). On the other hand, Cantor *et al.* [27] reported that dTpdT, pdTpdT, and pdTpdTp all have similar CD spectra which are insensitive to salt concentration. This could be due to the low level of base-stacking of the oligo T; however, they also reported that the CD spectra of dApdG and dApdGp are also similar.

3. *Proton Magnetic Resonance Studies*

At the present stage, PMR study of oligomers is still in its infancy, although rapid growth is anticipated in the coming year. Various technical problems, such as resolution and sensitivity, have been resolved by advancements in instrumentation such as the availability of the 220 and 300 MHz spectrometers and the Fourier transform data system. Nevertheless, the problem of assignment of the many resonance lines in a spectrum remains a formidable one, especially in the case of short-chain homooligonucleotides.

A comparative study of 3'-IMP, 5'-IMP, IpI, pIpI, IpIpI, and pIpIpI is briefly described here as examples [49]. This study is to serve as a supplement to the optical studies mentioned above; the main objective of this PMR investigation is to ascertain the conformation of nucleosidyl units in the dimers and the trimers, with respect to the torsional angle of the glycosyl bond. The chemical shifts at infinite dilution and the assignments of these resonance lines at 5°, 30°, and 60° from 3'-IMP, 5'-IMP, IpI, pIpI, (Ip)$_2$I, and p(Ip)$_2$I are reported in Table XII. The assignments of the resonances from the dimers and trimers have been established by deuteron exchange of H-8 at elevated temperature with D$_2$O, specific line-broadening of H-8 and H-1' due to Mn^{2+} binding, and consideration of the conformational model, as discussed above for the dimer case. While the assignments for the dimer resonances can be made with certainty, the assignments for the resonances of the trimers, though most likely to be correct, should be verified by a more direct approach in the future [49].

The observed specific line broadening of H-8 resonances of pI, -pI, and -pIp-, in the presence of Mn^{2+} is a clear indication that these residues are in an *anti* conformation (see Sections I,B,C,D, and Tazawa *et al.* [49]). This conclusion is supported by the comparison of the chemical shifts of the H-8 resonances of the pIp- residues in pIpI and (pI)$_3$, versus those from the Ip-residues in IpI and IpIpI. The H-8 resonances of the pIp- residues are located at a lower field position than those of the Ip- residues due to the specific deshielding effect of a 5'-phosphate group exerted on the H-8 of a purine nucleotidyl unit when it has an *anti* conformation. Furthermore, these two H-8 resonances are the only resonances (Table XII) which are shifted downfield upon a change of pD from 6.0 to 7.0. This change of pD should increase the ionization of the secondary phosphate group in the pIp- residue, and it has been demonstrated that the increase in 5'-phosphate ionization enhances the specific deshielding effect of the phosphate group when the nucleosidyl unit is in *anti* conformation [38,81]. These studies, in fact, reveal that all the residues in dimer and trimer are predominantly in an *anti* conformation.

The coupling constants, $J_{H-1' \text{ to } H-2'}$, of IMP, dimer, and trimer, are shown in Table XIII. These J values from the inosinyl dimers and trimers are significantly lower than the respective values in monomers at low temperature; they become larger, approaching the values of the monomers, at higher temperature. In this respect, the ribosyl inosinyl dimers and trimers are similar to the ribosyl–adenylyl dimers and trimers, and not to the deoxyadenylyl dimers and trimers [54]. These considerations indicate that upon base stacking in IpI and (Ip)$_2$I, the conformation of the ribose residue in the dimers and trimers is being compressed (Section I,D).

Dimerization and trimerization shifts ($\Delta\delta$, Table XIV) are the parameters derived from PMR studies which are the direct indication of the neighboring

TABLE XII

Chemical Shifts of Base and H-1' Protons of Monomers, Dimers, and Trimers of Inosinate.[a] (Parts per Million from Tetramethylsilane Capillary)

Compound	pD	Temp. (°C)		Chemical shifts δ (negative)		
				H-8	H-2	H-1'
3'-IMP	6.0	5		8.77	8.62	6.54
		30		8.83	8.695	6.60
		60		8.82	8.70	6.60
5'-IMP	6.0	5		8.87	8.60	6.545
		30		8.92	8.68	6.61
		60		8.90	8.69	6.61
IpI	6.0	5	Ip-	8.64	8.53	6.34
			-pI	8.73	8.56	6.43
	6.0	30	Ip-	8.71	8.62	6.43
			-pI	8.81	8.64	6.53
	6.0	60	Ip-	8.73	8.64	6.47
			-pI	8.82	8.65	6.55
pIpI [b]	6.0	5	pIp-	8.80	8.53	6.38
			-pI	8.75	8.555	6.46
	6.0	30	pIp-	8.85	8.63	6.48
			-pI	8.82	8.64	6.54
	7.0	30	pIp-	8.91	8.63	6.505
			-pI	8.83	8.64	6.55
IpIpI	6.0	5	Ip-	8.59	8.50	6.39
			-pIp-	8.69	8.49	6.34
			-pI	8.69	8.53	6.40
	6.0	30	Ip-	8.68	8.59	6.48
			-pIp-	8.78	8.59	6.42
			-pI	8.79	8.62	6.51
	6.0	60	Ip-	8.69	8.62	6.51
			-pIp-	8.80	8.62	6.44
			-pI	8.81	8.63	6.54
pIpIpI [c]	6.0	5	pIp-	8.71	8.49	6.41
			-pIp-	8.685	8.49	6.34
			-pI	8.685	8.53	6.42
	6.0	30	pIp-	8.80	8.59	6.49
			-pIp-	8.80	8.59	6.43
			-pI	8.80	8.61	6.51
	7.0	30	pIp-	8.85	8.59	6.49
			-pIp-	8.79	8.58	6.43
			-pI	8.80	8.61	6.515

[a] All extrapolated to infinite dilution unless specified [49].
[b] Concentration at 0.014 M (dimer) for experiments at 30°C.
[c] Concentration at 0.014 M (trimer) for all experiments.

interaction (mostly base–base interaction) of the residues in the dimer and the trimer. Three major conclusions concerning the $\Delta\delta$ data in Table XIV are discussed below: (1) Within experimental error (± 0.02 ppm), the $\Delta\delta$ values of IpI and pIpI are essentially the same, and similarly, the $\Delta\delta$ values of IpIpI and pIpIpI are nearly identical. Thus, the PMR data do not indicate a *significant* change of conformation upon addition of a 5'-phosphate group to the dimer or trimer; for instance, the change of an *anti* to *syn* conformation of one or all nucleosidyl units can be safely excluded. This observation is in

TABLE XIII

THE COUPLING CONSTANT, $J_{H-1',H-2'}$,[a] OF MONOMERS, DIMERS, AND TRIMERS OF INOSINATE

Compound	Temp. (°C)	Hz		
		(p)Ip-	-pIp-	-pI
3'-IMP [b]		5.6	—	—
5'-IMP [b]		—	—	5.6
IpI	5	4.7	—	4.0
	30	5.0	—	4.7
	60	5.3	—	4.9
IPI, 1 *M* NaCl	5	4.6	—	3.8
pIpI	5	4.1	—	4.1
pIpI, 1 *M* NaCl	5	4.0	—	4.0
IpIpI	5	4.2 [c]	4.6	3.8 [c]
	30	5.0	5.2	4.8
	60	5.2	5.2	5.0

[a] Error in measurement is on the order of 0.2 Hz [49].
[b] Value is temperature-independent at concentrations studied (<0.02 *M*).
[c] Due to peak overlap, error somewhat larger, ~ 0.4 Hz.

contrast to the data from CD studies described above. (2) A change to 1 *M* NaCl solution from dilute salt also does not noticeably change the $\Delta\delta$ values of IpI, except that of the H-8 in -pI residues to the extent of 0.05 ppm, barely above the level of experimental error. This result can be attributed to a variety of factors: a small increase in base stacking (since $\Delta\delta$ of the H-2 of Ip- is also slightly increased, 0.03 ppm); a slight change in the nucleosidyl conformation (to a conformation closer to *syn* form); or a minute reduction in charge of the phosphate group in IpI due to Na^+ binding, etc. A change to 1 *M* NaCl does cause a change of $\Delta\delta$ values of H-8 of -pI (0.1 ppm), H-8 of

TABLE XIV

DIMERIZATION AND TRIMERIZATION SHIFTS ($\Delta\delta$) [a] OF
IpI, pIpI, IpIpI, and pIpIpI [b]

Compound	Temp. (°C)		H-8	H-2	H-1'
IpI	5	Ip-	0.13	0.09	0.20
		-pI	0.14	0.05	0.11
	30	Ip-	0.11	0.07	0.17
		-pI	0.11	0.04	0.08
	60	Ip-	0.09	0.05	0.13
		-pI	0.08	0.04	0.05
IpI, 1 M NaCl	5	Ip-	0.12	0.11	0.195
		-pI	0.19	0.05	0.12
pIpI	5	pIp-	0.10	0.08	0.20
		-pI	0.115	0.06	0.09
	30 [c]	pIp-	0.105	0.07	0.16
		-pI	0.10	0.04	0.06
pIpI, 1 M NaCl	5	pIp-	0.14	0.14	0.205
		-pI	0.21	0.055	0.11
IpIpI	5	Ip-	0.18	0.12	0.15
		-pIp-	0.21	0.13	0.24
		-pI	0.18	0.07	0.14
	30	Ip-	0.15	0.10	0.12
		-pIp-	0.17	0.10	0.22
		-pI	0.13	0.06	0.10
	60	Ip-	0.13	0.08	0.09
		-pIp-	0.14	0.08	0.20
		-pI	0.095	0.06	0.07
pIpIpI [c]	5	Ip-	0.185	0.13	0.175
		-pIp-	0.21	0.13	0.24
		-pI	0.18	0.08	0.13
	30	Ip-	0.15	0.11	0.155
		-pIp-	0.15	0.11	0.21
		-pI	0.12	0.07	0.095

[a] In ppm, $\Delta\delta_{\text{D or T}} = \delta_{\text{dimer}}$ (or δ_{trimer}) $- \delta_{\text{monomer}}$ [49].

[b] All values obtained from chemical shifts extrapolated to infinite dilution unless otherwise specified.

[c] Concentration at 0.014 M (dimer or trimer).

pIp- (0.04 ppm, barely significant) and H-2 in pIp- (0.06 ppm) of pIpI. This observation suggests that a small increase in base stacking probably does take place for pIpI in 1 M NaCl; the increase in $\Delta\delta$ of H-2 in pIp- is important in this respect, since the $\Delta\delta$ value of this proton is mainly sensitive to stacking in a dimer with both nucleotidyl units in *anti* conformation. (3) As described previously relating to the problem of assignment of resonances, the $\Delta\delta$ values of IpI and pIpI are in complete agreement for a conformation of a right-handed, *anti*, *anti* stack.

The comparison between the dimerization shifts of IpI versus the trimerization shifts of IpIpI allows us to examine the dependence of chemical shifts on chain length. The simplest expectation in this comparison is: The observed values for the Ip- residues and for the -pI residues are the same in both dimer and trimer, and the observed value for the -pIp- residue is the summation of the values for Ip- residues and for -pI residues. This expectation is based on two assumptions: (1) The conformation of the trimer is constructed from an exact extension of the dimer—the time-averaged geometrical relationship between the first and the second residues in the trimer is exactly the same as that between the second and third residue, as well as exactly the same as that in the dimer. (2) The magnetic field effect of the ring-current anisotropy or any other through-space-field-effects do not extend beyond the nearest neighboring residue—the chemical shifts of the first residue are not influenced by the third residue and vice versa. Various considerations do suggest that these assumptions are unlikely to hold. For instance, the dynamic process of right-handed stack to left-handed stack interconversion occurs to a significant extent in the dimer [48], but probably takes place hardly at all in a trimer. This situation will affect the chemical shifts of the H-1' protons. The rotational freedom of the nucleosidyl unit of the interior residue (the middle residue) around the glycosyl linkage (or ϕ_{CN}) is likely to be more restricted, since this interior base is sandwiched between the two outside bases. If the H-8 proton of the interior base is kept at a close distance to the 5'-phosphate group of the -pIp- residue, this proton can be more deshielded. Finally, besides the possible increase in the extent of base stacking, the possible magnetic field effects between the first and the third residues cannot be discounted on a priori grounds.

The comparison between the dimerization shifts ($\Delta\delta_D$) of IpI and the trimerization shifts ($\Delta\delta_T$) of IpIpI is shown in Table XV. With one exception, all the $\Delta\delta_D$ values of the Ip- and -pI residues are smaller than the corresponding $\Delta\delta_T$ values. This could reflect an increase in base stacking in the trimer, or in the magnetic field effect between the first and the third residues (simply termed as distant-neighboring effect), or both. The exception noted above concerns a negative $\Delta\delta_{Dif}$ value (Table XV), which is that from the H-1' of Ip- residue. The $\Delta\delta_D$ of this H-1' in the dimer is larger than the $\Delta\delta_T$ of this

TABLE XV

COMPARISON BETWEEN THE DIMERIZATION SHIFTS ($\Delta\delta_D$) OF IpI AND THE
TRIMERIZATION SHIFTS ($\Delta\delta_T$) OF IpIpI (IN PPM)

			$\Delta\delta_T{}^a$	$\Delta\delta_D{}^a$	Difference ($\Delta\delta_{Dif}$) [b]
5°	H-1'	Ip-	0.15	0.20	−0.05
		-pIp-	0.24 (0.29)	(0.31)	−0.07
		-pI	0.14	0.11	+0.03
	H-2	Ip-	0.12	0.09	+0.03
		-pIp-	0.13 (0.19)	(0.14)	−0.01
		-pI	0.07	0.05	+0.02
	H-8	Ip-	0.18	0.13	+0.05
		-pIp-	0.21 (0.36)	(0.27)	−0.06
		-pI	0.18	0.14	+0.04
30°	H-1'	Ip-	0.12	0.17	−0.05
		-pIp-	0.22 (0.22)	(0.25)	−0.03
		-pI	0.10	0.08	+0.02
	H-2	Ip-	0.10	0.07	+0.03
		-pIp-	0.10 (0.16)	(0.11)	−0.01
		-pI	0.06	0.04	+0.02
	H-8	Ip-	0.15	0.11	+0.04
		-pIp-	0.17 (0.28)	(0.22)	−0.05
		-pI	0.13	0.11	+0.02
60°	H-1'	Ip-	0.09	0.13	−0.04
		-pIp-	0.20 (0.16)	(0.18)	+0.02
		-pI	0.07	0.05	+0.02
	H-2	Ip-	0.08	0.05	+0.03
		-pIp-	0.08 (0.14)	(0.09)	−0.01
		-pI	0.06	0.04	+0.02
	H-8	Ip-	0.13	0.09	+0.04
		-pIp-	0.14 (0.23)	(0.17)	−0.03
		-pI	0.09	0.08	+0.01

[a] The values in parentheses are obtained from the summation of $\Delta\delta$ values from Ip- and -pI residues [49].

[b] $\Delta\delta_{Dif} = \Delta\delta_T - \Delta\delta_D$.

same proton in the trimer to about the same extent from 5° to 60°. While no definitive explanation can be offered at present why this H-1' of the Ip-residue is more shielded in the dimer than in the trimer, one reasonable suggestion is that this finding may be related to the absence of right-handedness to left-handedness interconversion in the trimer. From the consideration of the conformational model, a left-handed stack would cause a significant increase in shielding of H-1' of the Ip- residue.

The $\Delta\delta_T$ values of the protons in the interior residue (-pIp-) deserve a more thorough discussion. Except in two cases to be discussed later, the $\Delta\delta_T$ values of the protons in the -pIp- are *smaller* than those derived from the summation (the bracketed values in Table XV) of the $\Delta\delta_T$ values of the Ip- residue and those of the -pI residue. This observation could be an indication of the distant-neighboring effect which the Ip- and -pI residues receive. If so, together with the assumption that the conformations of the dimer and trimer are essentially similar, the observed $\Delta\delta_T$ values for the -pIp- residue and the summed $\Delta\delta_D$ values (the bracketed values under the column of $\Delta\delta_D$ in Table XV) from the $\Delta\delta_D$ values of Ip- and -pI can be compared. Such a procedure cannot be used for the H-1′ protons, since the $\Delta\delta_D$ of H-1′ in the Ip- residue is larger than the $\Delta\delta_D$ of the same proton for the reason already discussed, i.e., possibly due to the absence of right-handedness to left-handedness interaction in the trimer. Such a procedure was found to be applicable for the H-2 protons. Indeed, within the experimental error, the $\Delta\delta_T$ values of H-2 protons in the -pIp- residue of the trimer are the same as the summation of the $\Delta\delta_D$ values from Ip- and -pI residues in the dimer. This is an observation which supports the existence of the distant-neighboring effect. However, for the H-8, the $\Delta\delta_D$ values of the -pIp- residue in the trimer are consistently smaller (0.03–0.06 ppm less) than the summed values from the dimer. We propose that the H-8 proton of the -pIp- residue in the trimer is held at an averaged distance closer to the 5′-phosphate group than the H-8 of the -pI residue in the dimer owing to the fact that the nucleosidyl unit in the -pIp- is sandwiched between the two outside bases.

The H-1′ protons present another interesting situation. In comparing the observed $\Delta\delta_T$ values of H-1′ in the -pIp- residue with the summed values (in brackets, Table XV) from the $\Delta\delta_T$ values of H-1′ in Ip- and -pI residues of the trimer, an interesting trend was observed. At 5°C, the observed $\Delta\delta_T$ is smaller than the summed value as may be expected due to the existence of distant-neighboring effects at the two terminal residues; at 30°C, the observed $\Delta\delta_T$ value is equal to the summed value, and at 60°C the observed $\Delta\delta_T$ value is larger than the summed value. The data reveal that the observed $\Delta\delta_T$ of the H-1′ in -pIp- is relatively insensitive to temperature change (a reduction by about 20% from 5° to 60°C), while the summed value is decreased by about 45% from 5° to 60°C. Obviously, the H-1′ proton in the -pIp- residue is receiving a larger shielding effect from two terminal residues than that exerted by the -pIp- residue in return to the two H-1′ protons in the two terminal residues. One possible explanation for such a phenomenon is that the dynamic geometrical relationship among the three residues in the trimer is not symmetrical, i.e., the relationship between the first residue and the second residue is not identical to that between the second and the third residues.

In summary, the comparison between the dimerization shifts of IpI and the trimerization shifts of IpIpI reveals the following information: (1) a possible existence of the distant-neighboring field effect between the two terminal residues in the trimer; (2) the absence of right-handedness to left-handedness interconversion in the trimer; (3) a reduction of the rotational freedom of the nucleosidyl unit in the interior residue in the trimer (more rigidly fixed in an *anti* conformation); (4) a possible asymmetry in the geometrical relationship among the residues in the trimer. It should be noted that the polymerization shifts ($\delta_{\text{polymer}} - \delta_{\text{monomer}}$) of poly rI at 30°C are 0.24 ppm for H-1', 0.15 ppm for H-2, and 0.21 ppm for H-8 [82]. The corresponding observed trimerization shifts of the interior residue (Table VII) are 0.22 ppm for H-1', 0.10 ppm for H-2, and 0.17 ppm for H-8. Therefore, the trimerization shift of the H-1' is almost that of the polymerization shift already and the others are about 60–80%. This analysis suggests that the distant-neighboring effect cannot be very extensive along the chain, and that the level of shielding of the interior residue in the trimer is close but not equal to that in the polymer. In a recent comparative study on TpT, TpA, TpTpA, ApT, ApTpT, and poly T, it is concluded that a distant shielding effect is exerted by the 5'-terminal A residue (or by the 3'-terminal A residue) on the other terminal residue in TpTpA or ApTpT [82a].

4. *The Correlation among the Studies on UV Hypochroism, CD and PMR*

In a comparison between the data in Fig. 32 and those in Fig. 36, there is no apparent correlation between the chain length dependence on UV hypochromicity and the chain length dependence on CD properties. For instance, in the rI series, the UV hypochromicity of the interior bases in I_{8-9} is about the same as that in the poly I, but the CD spectrum (either the original spectrum, Fig. 34, or the subtracted spectral properties, Fig. 35, or the normalized spectral properties, Fig. 36) of I_{11} is still very different from that of poly I. In the rA series, again the hypochromicity of the interior bases in rA_{7-8} is about the same as that in poly A (Fig. 32), but the CD amplitude of rA_{10} is still much smaller than that of the poly rA (Fig. 36). In the dA series, the interior base of the trimer already has the same degree of hypochromicity as the poly dA; as for the CD properties, not until the level of hexamer (Fig. 36; [69]) does the CD spectrum of poly dA resemble that of the oligomer $(pdA)_6$. The series of rC cannot be evaluated because of the lack of data on the oligomer with n larger than 15. As for the series of rU and dT, the optical properties of these compounds are not chain length-dependent above the dimer stage. We may conclude that in general, the requirement in chain length for the oligomer to attain the same degree of residue–residue interaction as the polymer is much shorter for the UV hypochromicity than for the CD. The chain length effect on UV hypochromicity is more abrupt, while that on CD is more gradual.

The present study on the influence of chain length on the PMR properties of oligonucleotides is only the beginning stage. Besides the comparison between IpI and IpIpI, another important aspect of the PMR study is the comparison between IpI vs pIpI, as well as the comparison between IpIpI vs pIpIpI, in the absence or the presence of 1 M NaCl. All these alterations (dimer to trimer, addition of 5'-phosphate group to the terminal residue, and addition of 1 M NaCl) cause a dramatic change of the CD spectra. However, these alterations do not change the basic conformational model as far as the PMR measurements are concerned. The most outstanding example is the comparison between the CD spectra of IpI vs pIpI (Fig. 37) on the one hand, and the comparison between the dimerization shifts of IpI vs pIpI (Table XIV) on the other. While the difference in CD is large, the difference in dimerization shifts is minimal. Similarly, the effect of a change to 1 M NaCl solution on the CD spectra of pIpI is great, while the effect of strong salt on the dimerization shifts of pIpI is small. The PMR data do not indicate that the nucleosidyl units in these oligomers have been changed from the *anti* conformation to the *syn* conformation—both residues of pIpI clearly are in *anti* conformation. Therefore, further studies are needed in order to understand the real nature of conformational changes which have such a significant effect on the CD spectra of the oligo I. This comparison indicates that each measurement—the UV absorbence (hypochromicity), CD spectrum, and PMR spectrum—has a different degree of dependence or sensitivity to various aspects of the conformation of the dimer and oligomers. Only when all these measurements (and perhaps others as well) can be correlated and unified in a satisfactory manner, can we have a full understanding about the conformation of these oligonucleotides as well as the measurements themselves.

II. Interaction between Dinucleotides and Oligonucleotides

Interactions between the well-defined, shorter segments of nucleic acids can be characterized much more thoroughly owing to the relative simplicity of these systems in comparison to polymer–polymer interaction. For this reason, much effort has been devoted to this study in the past few years, and certainly more will be forthcoming.

A. Interaction of the Dimers in Organic Solvents

The formation of hydrogen-bonded base pairs of the monomers in organic solvents such as chloroform has been fully described in Chapter 6, Volume I. In this nonaqueous medium, the specificity of the complementary base pair formation of these monomers is essentially the same as that in the nucleic acid duplex.

In the past, such study was confined to the level of monomers because of the low solubility of oligomers in organic solvents such as chloroform. When the alkyl phosphotriester of the dinucleoside monophosphate became available [50], experiments in organic solvents became possible. These neutral dimers can be dissolved in chloroform, especially when the remaining 3'- and 5'-hydroxyl groups are covered by protecting groups. For this purpose, the neutral dimers of structure I and II shown in Fig. 40 were prepared [83]. Compound I is the dithymidine monophosphate ethyl phosphotriester with 5'-monomethoxytrityl group and 3'-acetyl group [designated as mTA–Tp(Et)T], and compound II is the dideoxyladenosine monophosphate ethyl phosphotriester with 5'-dimethoxytrityl group and 3'-benzoyl group [designated as dT,B–dAp(Et)dA]. The association of these compounds through hydrogen bonding in $CDCl_3$ was followed by the absorbence of NH

TABLE XVI

INTRINSIC ASSOCIATION CONSTANTS OF MONOMERS AND PROTECTED ALKYL
PHOSPHOTRIESTER OF DINUCLEOSIDE MONOPHOSPHATES IN $CDCl_3$ AT
ROOM TEMPERATURE[a]

Compound [b]		$K_i (M^{-1})$	Published values
	Self Association		
T		3.6	3.2 [c]
mT,A–Tp(Et)T		4.2	
A	Pair formation	3.1	3.1 [d]
	Chain formation	2.2	1.4 [e], 4.8 [e]
dT,B–dAp(Et)dA	Pair formation	6.5	
	Chain formation	4.0	
	Cross Association		
T + A		92	130 [c]
mT,A–Tp(Et)T + A		113	
T + dT,B–dAp(Et)dA		72	
mT,A–Tp(Et)T + dT,B–dAp(Et)dA		192	

[a] From DeBoer *et al.* [83]. Measurements were made at 22°–24°C. The intrinsic association constants were calculated on the basis of residue concentrations.

[b] Compounds: T = 1-cyclohexylthymine; A = 9-ethyladenine; mT,A-Tp(Et)T = 5'-monomethoxyltrityl, 3'-acetyldithymidine monophosphate ethyl phosphotriester (compound I in Fig. 40); dT,B–dAp(Et)dA = 5'-dimethoxyltrityl, 3'-benzoyl dideoxyadenosine monophosphate ethyl phosphotriester (compound II in Fig. 40).

[c] From Kyoyoku *et al.* [84]; at 25°C.

[d] From Kyogoku *et al.* [85]; at 24°C.

[e] From Nogel and Hanlon [86]. The first value is for pair formation and the second for higher associations, all taking place in a chain formation reaction.

and NH_2 stretching bonds (frequency from 3397 yo 3530 cm^{-1}) using infra-red spectroscopy [83]. The results of this IR study on the monomer–monomer association, dimer–monomer association, and dimer–dimer association are shown in Table XVI.

The intrinsic association constants (K_i) determined in these experiments are in good agreement with those published earlier [84–86], serving as a check on the reliabilities of these studies.* These constants (Table XVI) calculated on the basis of residue concentration further show that the cross-association constants of monomer to dimer [Tp(Et)T + A; T + dAp(Et)dA] are similar to that of the monomer to monomer (A + T). This is not unexpected,

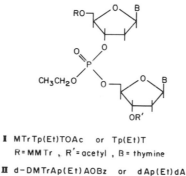

I MTrTp(Et)TOAc or Tp(Et)T
R = MMTr , R' = acetyl , B = thymine

II d – DMTrAp(Et)AOBz or dAp(Et)dA
R = DMTr , R' = benzoyl , B = adenine

Fig. 40. Structural formulas of the two protected neutral alkyl phosphotriesters of dinucleoside monophosphates. Structure I: protected Tp(Et)T; structure **II**: protected dAp(Et)dA. (From De Boer *et al.* [83].)

since the hydrophobic-stacking interaction of the bases is abolished in chloroform. The most important observation is that the intrinsic cross association constant for the dimer–dimer association [Tp(Et)T + dAp(Et)dA] is only twice as large as the monomer–monomer association (A + T). Similarly, the constants for self-association of the dimer [dAp(Et)dA, and Tp(Et)T] are only 1.5 to 2 times larger than those of the self-association of the monomer (A and T). This observation implies that the binding of one residue of a particular dimer to the residue of another dimer has only a small influence on the further association of the remaining residues in these two dimers. In other words, the association of the two residues in the dimer is separate and unrelated.

* To calculate the actual molecular association constants, the K_i value should be multiplied by a statistical factor of two for monomer–dimer association and a factor of four for dimer–dimer association.

The above conclusion reveals that the linkage between the two residues must be rather flexible. If the backbone is rigid, and fixed in a favorable position for the pairing of the two dimers, then the association constant of the dimer–dimer interaction should be manifold greater than that of the corresponding monomer–monomer interaction—approaching a square term of the constant of the monomer–monomer association because of the doubling of the enthalpy term in the association. On the other hand, if the backbone is rigid but fixed in an unfavorable position for pairing of both residues of the dimer, then the association constant of the dimer–dimer interaction should be significantly diminished in comparison to that of the monomer–monomer interaction.

In aqueous solution, dAp(Et)dA has been shown to interact with poly U to form a triple helix of 2U:1A with a stability higher than that of the dApdA·2 poly U complex [50]. All these observations indicate that the backbone of the neutral dimers, dAp(Et)dA and Tp(Et)T in chloroform, is flexible and allows free rotation along its axis; there is little intrinsic, stereochemical interference with the rotation along the axis of the backbone when the base–base stacking is abolished and when the charged phosphate group is neutralized by an alkyl substituent. This result will be further discussed in the later chapter on polynucleotides, in which the origin of the observed rigidity of the backbone of the polynucleotides in aqueous solution will be considered. It will suffice to state that the findings reported here reinforce the conclusion from the polynucleotide studies that the restriction of the rotation of the backbone of the polynucleotide in solution cannot be due to steric hindrance alone; the data support the suggestion that the electrostatic interaction between the charged groups of the polynucleotide backbone and between the charged groups and water could be the cause of the rigidity of the polynucleotide in a random coil conformation observed in aqueous solution.

B. ASSOCIATION OF HOMOOLIGONUCLEOTIDES IN NEUTRAL AQUEOUS SOLUTION

1. *Oligouridylic Acid*

The formation of the helical structure of polyuridylic acid under the conditions of low temperature (0°–5°C) and high Mg^{2+} concentration was first reported by Lipsett [87]. It was found that the T_m of such a helical structure can be much enhanced upon addition of diamines such as spermine [88]. A recent comprehensive investigation [89] shows that the ordered conformation of poly U is a double-helical structure resulting from the folding of the molecule upon itself to form a single hairpin or multihairpins, depending on salt concentration and temperature.

The study by Dourlent *et al.* [90] on oligo U was to confirm and extend the above study on poly U. The oligo U series was prepared by brief alkaline hydrolysis of poly U and fractionated by DEAE cellulose column. These oligomers were investigated in 0.5 M CsCl and in 2.5 × 10⁻⁵ M spermine at neutral pH. The absorbence of the oligomer solution is not lowered even at −5°C until the chain length of the oligomer is above 12; i.e., $(U)_{12}$. $(U)_{15}$ was reported to have a T_m of −4.5°C in CsCl and a T_m of 2°C in spermine solution, while $(U)_{31}$ and poly U have a T_m of 5.6° and 10.5°C, respectively in CsCl, and a T_m of 15° and 31°C, respectively in spermine solution. The molar enthalpy (ΔH) for the formation of the ordered structure based on a two-state model was calculated to be ranging from −22 to −35 kcal, depending on the chain length (15–31) and the medium (CsCl or spermine solution). The most important observation on the oligomers in support of the hairpin model is that the T_m of the helix–coil transition is insensitive to concentration of oligomer (a 10-fold change in concentration of oligomer without affecting the T_m)—an indication that this is a unimolecular process. The preliminary temperature-jump study on the kinetic aspect of the transition also yielded results in accord with this model of hairpin formation.

2. Oligoguanylic Acid

The unusually strong association tendency of oligoguanylic acid was noted quite early (see brief review in Lipsett [91]). The gel formation of 5′-GMP and 3′-GMP was discussed in Chapter 6, Volume I (Sections II,A and C). The x-ray diffraction patterns from fibers drawn from the GMP gels indicated the existence of the helical structure with the hydrogen bonding scheme shown in Fig. 10 of Chapter 6, Volume I.

Quantitative study on the aggregation of guanine oligonucleotides began with the report by Lipsett [91]. The aggregation of the oligo G was measured by the hypochromicity of the UV absorbence at about 252 and 275 nm; the loss of absorbence was as large as 28 and 45%, respectively, at these two wavelengths for GpG (0.378 mM, pH 7.1) changing from 32° to 3°C. The aggregation was also clearly demonstrated by chromatography in a Sephadex G-75 column which separates the oligomers according to their sizes. When stored (1 mM concentration of G in 0.2 M NaCl, 0.002 M Na cacodylate, pH 6.2) and eluted at 37°C, GpGpG migrated in the same manner as pUpUpU in the gel titration experiment, indicating that the GpGpG was not aggregated under this condition. On the other hand, when stored and eluted at 3°C, 80% of the GpGpG migrated in a manner similar to poly U, indicating that a large extent of aggregation did take place under this condition [91].

These results indicate that the rate of aggregation and disaggregation of oligo G is unusually slow; the time required for equilibrium is measured in terms of days and months for aggregation, and in hours for disaggregation. The slow rate of the disaggregation process is presumably the cause of a large equilibrium constant highly favorable to aggregation, while the slow rate of aggregation suggests that a strong intramolecular interaction exists and that the aggregation process may involve a change of the intramolecular interaction and the conformation of the individual oligomer. Upon elevation of temperature, a somewhat abrupt increase of absorbence can be observed over a few degrees change in temperature—the exact interpretation of this "T_m" is difficult for such a slow and multistaged process. In contrast to the situation of oligo U (Section II,A,1), the "T_m" values are sensitive to concentration of oligo G, indicating a multimolecular process in aggregation. In 0.2 M salt, pH 6–7, GpG has a "T_m" of 17°–18°C at $\sim$0.4 mM and a "T_m" lower than 3°C at 0.04 mM. For GpGpG, at pH 7.4, the "T_m" is 21.8°C at 0.09 mM and 24.3°C at 0.42 mM [91]. Upon interaction with Hg^{2+} (stoichiometrical ratio of 1 Hg to 2 G), the GpGpG + Hg complex behaved mostly like monomeric (at most dimeric) GpGpG in the gel filtration experiment, despite the marked hypochromicity of the solution. This observation, coupled with the knowledge that Hg^{2+} usually are attached to ring nitrogen of purine and pyrimidine, suggests that these nitrogen atoms are involved in the aggregation.

The aggregations of GpGpG, GpGpU, GpGpGpU, etc., have also been investigated by ORD and equilibrium sedimentation [92]. The aggregation of GpGpG brought about a 2- to 3-fold enhancement of optical rotation at 273 ± 2 nm (positive branch) and at 248 ± 1 nm (negative branch) with little change at the crossover point (near 260 nm). Again, the aggregation and disaggregation were found to be slow, using the optical rotation at 275 nm ($\phi_{275 \text{ nm}}$) as an indicator. In a log plot of $\phi_{275 \text{ nm}}$ vs time for the disaggregation of GpGpGpU (near 1×10^{-5} M) at 63°C, 0.5 M NaCl, pH 7.0 a linear line was observed, indicating a first-order process with a drop of $\phi_{275 \text{ nm}}$ of 7×10^4 at 0 time, to $\phi_{275 \text{ nm}}$ of 1×10^4 in 35 min. After the thermal disaggregation, the reaggregation process takes days to recoup, but can be accelerated by quick freezing. Sedimentation equilibrium studies indicated that the number-averaged molecular weight of GpGpGpU at 5×10^{-5} M, 0.6 M salt solution, 24°C, was 5- to 6-fold larger than the formula weight when the solution was prepared freshly from the dilution of a stock solution frozen at 10^{-3} M. After equilibration of this solution at 50°–60°C for 2–3 hr, the molecular weight of GpGpGpU was found to be the same as the formula weight, indicating complete dissociation [91]. Attempts have been made to analyze the aggregation process more quantitatively [91]; however, the accuracy of the experimental data and usefulness of the treatment cannot be really evaluated at present from the literature.

C. ASSOCIATION OF HOMOOLIGONUCLEOTIDES IN ACIDIC AQUEOUS SOLUTION

1. *Oligoadenylic Acid*

The helical conformation of poly A in acidic solution is a favorite subject of polynucleotide research and therefore will be described in considerable detail in a later volume. Suffice it to state here that the stability of this helix is dependent on the electrostatic interaction of the positively charged, protonated adenine bases and the negatively charged phosphate groups of the opposite strands. Thus, while acidity usually decreases the stability of

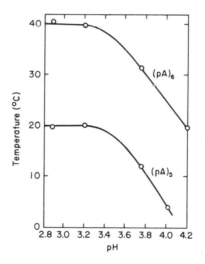

Fig. 41. T_m values of $(pA)_5$ and $(pA)_6$ as a function of pH in 0.12 M salt and 1×10^{-5} M oligomer concentration. (From Eigen and Pörschke [93].)

the double helix (such as DNA or poly A·poly U) by the disruption of the hydrogen bonding, in the case of protonated poly rA, acidity actually increases the stability of the helical structure. Similarly, a lowering of ionic strength usually decreases the stability of the double helix by enhancing the repulsion of charged phosphate groups, but reduction in ionic strength actually increases the stability of the protonated poly rA helix. In Fig. 41, the dependence of the T_m of $(pA)_5$ and $(pA)_6$ is shown to illustrate this point [93]. The T_m increases linearly with decrease in pH until the bases are fully protonated. The base-pairing scheme of the acidic oligo A helix is expected to be the same as that of the acidic poly A helix.

Most of the studies on helical oligo A are based on optical spectroscopy. Brahms *et al.* [73] reported the CD spectra of $(Ap)_6A$ and $(Ap)_{44}A$ in pH 4.5,

0.15 *M* salt, at varying temperatures (Fig. 42), which are essentially the same as that of acidic poly A. In 0.15 *M* salt, pH 4.0 and with concentration of 10^{-4} *M* total nucleotide, (Ap)₅A and (Ap)₆A were found to have very broad transitions (with midpoint about 8° and 22°C, respectively), while the (Ap)₇A was shown to have a moderately abrupt transition with a T_m at 38°C [73,94].

A comprehensive study on the thermodynamics of the association of oligo A was made by Eigen and Pörschke [93], as analyzed by an approach based

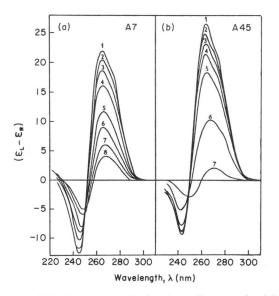

Fig. 42. Circular dichroic spectra of adenylate oligomers in 0.1 *M* NaCl–0.05 *M* acetate, pH 4.5, at various temperatures. (a) The heptamer at: (1) −2°C; (2) 2°C; (3) 6.5°C; (4) 13°C; (5) 20°C; (6) 26°C; (7) 33.5°C; (8) 45°C. (b) A₄₅ at: (1) 0 to 4°C; (2) 31°C; (3) 56°C; (4) 69°C; (5) 77.5°C; (6) 80°C; (7) 85°C. (From Brahms *et al.* [73].)

on an early attempt by Applequist and Damle [95]. In Fig. 43, the spectrum of (pA)₇ at 10°, 30°, and 55°C, pH 4.2, is displayed; Fig. 44 shows the hypochromicity at 265 nm as a function of temperature for dimer to decamer in pH 4.2 solution. The analysis of these data was based on the following model (Fig. 45). In the formation of a helical duplex, the association of the first base pair (nucleation) will be less stable than the association of the following base pairs (chain growth). An equilibrium constant, *s*, is assigned to all the steps in chain growth; and *βs* is assigned to be the equilibrium constant of the initial step in nucleation with $\beta < 1$. If β, the ratio of equilibrium constant of the first step to that of all subsequent steps, is sufficiently

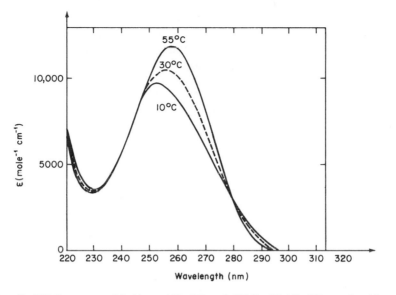

Fig. 43. UV Spectrum of $(pA)_7$ at 10°, 30° and 55°C, pH 4.2. Oligonucleotide concentration 7.0×10^{-6} M. (From Eigen and Pörschke [93].)

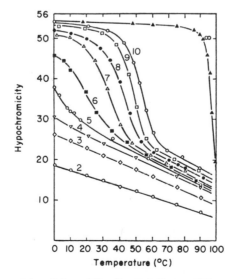

Fig. 44. Hypochromicity of the acidic oligo A helices at 265 nm, pH 4.2, as a function of temperature for different chain lengths 2 to 10. Nucleotide residue concentration 5×10^{-5} M. (From Eigen and Pörschke [93].)

small, one may assume the duplex formation between short oligonucleotides to be an all-or-none transition, i.e., the existence mainly of two states, a completely base-paired state and a completely nonpaired state. This model is applicable only when the population of the partially bonded species is small, an assumption which was verified and will be discussed in a later paragraph. From this simplified approach, the equilibrium constant (K) of the entire process is equal to βs^n, where n is the chain length of the oligomer.

The ΔH values (per base pair) of the helix–coil transition of oligo A (from $N = 6$ to $n = 10$) at pH 4.2 were analyzed by Eigen and Pörschke by three different procedures. The first procedure was based on the van't Hoff equation relating the temperature effect on the overall equilibrium constant in a $\ln K$ vs $1/T$ plot. The equilibrium constant at different temperatures was based on the estimation of the concentration of the helix species and the coil species from the hypochromicity data during the thermal transition

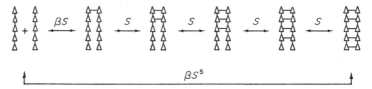

Fig. 45. The model for the analysis of the thermodynamics data on the association–dissociation of helical duplex of oligo A. (From Eigen and Pörschke [93]).

(Figs. 43 and 44). ΔH values of 5.35–8.95 kcal were obtained for $(pA)_6$ to $(pA)_{10}$. The enthalpy (per base pair) of the overall process was also evaluated from the slope of the helix–coil transition (Fig. 44) by an equation [Eq. (12)] first derived by Applequist and Damle [95]:

$$\left[\frac{\partial f}{\partial T}\right]_{T_m} = \frac{n\,\Delta H}{6RT_m^2} \tag{12}$$

where f is the fraction of bonded bases and hence $(df/dT)T_m$ is the slope of the tangent of the melting curve at T_m; n is the chain length. The enthalpy values computed from this procedure were from 5.9 to 8.6 kcal for $(pA)_6$ to $(pA)_{10}$ [93]. The enthalpy was also assessed by another equation [Eq. (13)] derived by Applequist and Damle [95] based on the assumption that β is temperature-independent:

$$\frac{1}{T} = \frac{1}{T'} + \frac{R}{n\,\Delta H}\ln\frac{C}{C'} \tag{13}$$

where T and T' are the T_m values (°K) at the concentration of C and C', respectively. ΔH values were derived from the slopes of lines in the plot (Fig. 46) of $1/T_m$ vs ln oligomer concentration* [93]. The enthalpy values so evaluated were from 6.3 to 9.1 kcal for $(pA)_6$ to $(pA)_{10}$. Therefore, the experimental ΔH values computed from these three different procedures were in agreement with each other, and were averaged to be 5.85 kcal for $(pA)_6$, 6.7 for $(pA)_7$, 7.5 for $(pA)_8$, 8.1 for $(pA)_9$, and 8.9 for $(pA)_{10}$ [93]. These values determined by Eigen and Pörschke are substantially smaller than those (14–15 kcal) reported by Brahms *et al.* [73] for $(Ap)_6A$ and $(Ap)_{11}A$ at

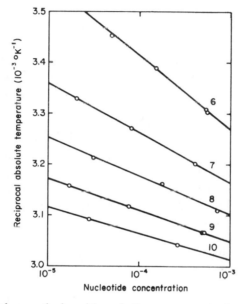

Fig. 46. $1/T_m$ values vs the logarithm of oligomer concentration (in monomer units). pH 4.2. (From Eigen and Pörschke [93].)

pH 4.5 evaluated from a van't Hoff plot; the source for this discrepancy is unknown at present.

An informative discussion on the various assumptions of this model was presented by Eigen and Pörschke [93]. They first examined the all-or-none characteristics of the transition. From a consideration of the polymer data, they assumed the β value to be about 10^{-3} mole^{-1} for the nucleation process. The equilibrium constant, s, for the successive steps in chain growth can now be estimated from the equilibrium constant, $K = \beta s^n$, for the overall process which was found to be from 6 [for $(pA)_{10}$] up to 17 [for $(pA)_6$]. By means of

* The unit of the concentration should be based on strand concentration of oligomer [95] and not on nucleotide concentration as plotted in Fig. 46.

these values, the statistical weight of partly bonded structures can be estimated by the consideration that a double helix with n' base pairs out of n
chain length can be formed in $(n - n' + 1)^2$ different ways. The distribution
of different helix species at equilibrium in the mean temperature range was
computed; the results showed that for the hexamer, 79% existed as a completely bonded helix and 18.5% existed as a duplex with five bonded pairs.
As for the decamer, 49.6% existed as a completely bonded helix, 33% as a
duplex with nine bonded pairs, 12.4% as a duplex with eight bonded pairs.
Therefore, the adoption of the all-or-none model appears to be justified,
especially for the hexanucleotides. The correction due to this statistical
consideration was estimated to be 0.3–0.7 kcal for the ΔH values of $(pA)_6$
to $(pA)_{10}$.

The question of chain length dependence of the T_m values was also examined
by Eigen and Pörschke [93] starting from the equation [Eq. (14)] of Applequist and Damle [95]:

$$\frac{1}{T_m} = \frac{R \ln \beta C_0}{n \, \Delta H} + \frac{1}{T_m\text{-polymer}} \tag{14}$$

where T_m-polymer is the melting temperature of the polymer, C_0 is the concentration of oligomer strands (n, β, and ΔH have been defined above). In
the plot of $1/T_m$ vs reciprocal chain length (Fig. 47), however, a linear line
was not obtained from the $1/T_m$ vs $1/n$ relationship; instead a linear line could
be obtained in the plot of $1/T_m$ vs $1/n - 1$, or better in the plot of $1/T_m$ vs
$1/n - 2$. A modification of Eq. (14) was proposed by Eigen and Pörschke
[93]. Instead of the assumption that β is temperature-independent (or ΔH_β

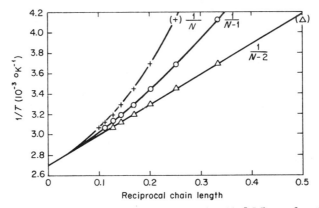

Fig. 47. $1/T_m$ values (at oligonucleotide concentration $10^{-5}\ M$) as a function of chain
length, pH 4.2; points in brackets are extrapolated from measurements at lower pH and
higher nucleotide concentrations. (From Eigen and Pörschke [93].)

is zero), it was assumed that the whole nucleation process is more or less temperature-independent, or that βs is temperature-independent. This is equivalent to the proposal that $\Delta H_\beta \approx - \Delta H_s$. At the midpoint of the transition, the equilibrium constant for the overall process, K, can be shown as

$$K = \frac{C_0/4}{(C_0/2)^2} = 1/C_0 = (\beta s)s^{n-1}$$

with

$$\ln s = - \frac{\Delta H}{TR} + \frac{\Delta S}{R}$$

at T_m,

$$\ln (\beta SC) + (n - 1)(-\frac{\Delta H}{RT_m} + \frac{\Delta S}{R}) = 0$$

At T_m-polymer, $s = 1$, thus

$$\Delta S = \frac{\Delta H}{T_m\text{-polymer}};$$

$$\frac{1}{T_m} = \frac{R \ln \beta SC_0}{(n - 1)\Delta H} + \frac{1}{T_m\text{-polymer}} \tag{15}$$

which is a modification of Eq. (14), substituting $n - 1$ for n in the denominator and substituting $R \ln \beta SC_0$ for $R \ln \beta C_0$ in the numerator. It was further proposed that the term $(n - x)$ be adopted in the denominator and $R \ln \beta S^x C_0$ in the numerator. The value of x (larger or smaller than unity) is dependent on the relative magnitude between ΔH_β and ΔH_s ($x = 0$, $\Delta H_\beta = 0$; $x = 1$, $\Delta H_\beta = - \Delta H_s$; etc.).

Other complexities concerning the thermodynamics of the helix–coil transition of protonated oligo A have also been considered by Eigen and Pörschke [93]. One consideration is the process of protonation, which has three separate aspects. The first aspect is the intrinsic enthalpy of the protonation process itself, and the difference in the ease of protonation between the helical duplex (more favorable) and the single stranded coil (less favorable). The second aspect is that the effect of protonation of the bases on the base-stacking of the single-stranded coil–protonation reduces the stacking of the neighboring bases in the single-stranded oligo A. The third aspect is the electrostatic attraction between the positively charged adenine base and the negatively charged phosphate group. Obviously, a quantitative evaluation of all these aspects for the oligomers with various chain length at different temperature is rather difficult.

Another complexity is the self-interaction of base-stacking of single-stranded oligo A, a phenomenon which was described in Section I,F. It has been recognized that the energy differential between the helical duplex state and the coil state can be strongly influenced by the self-interaction of the oligo A in the coil state. Attempts were made by Eigen and Pörschke to account for this factor [93]. However, the main question remains unanswered —whether the base-stacking in the oligo A is a positive factor for the promotion of the helix, or a negative factor in the destabilization of the helix. The answer to this question is dependent on the state of the coil form, therefore, dependent on temperature and solvent. For instance, in the comparison between the helical duplex in aqueous solution where base-stacking takes place and the noninteracting coil in organic solvents where base-stacking does not exist, the base-stacking energy is definitely a significant contribution to the stabilization of the helical duplex. On the other hand, when comparing the helical duplex and the stacking coil both in aqueous solution, in which the extent of the base–base overlap in the single strand may be larger than that in the helical duplex, the base-stacking may exert a *negative* influence on the stability of the helix.

An interesting observation made by Eigen and Pörschke [93] was that the hypochromicity of the oligomer helix formed by the $(pA)_6$, $(pA)_7$, $(pA)_8$, and $(pA)_9$ is about 92, 92, 97, and 97.5%, respectively, that of poly A ($\varepsilon_{max} = 8.8 \times 10^3$). Therefore, the chain length dependence of hypochromicity of the protonated helix appears to be small.

The kinetics of the helix–coil transition of oligo A at pH 4.2 has also been studied by Pörschke and Eigen [96] with the temperature-jump-technique in accordance with the following scheme.

$$A + A \xrightleftharpoons[k_D]{k_R} A_2$$

and

$$1/\tau = 4k_R\bar{c}_A + k_D \tag{16}$$

where $1/\tau$ is the reciprocal relaxation time, k_R is the rate of recombination, k_D is the rate of dissociation, and $\bar{c}_A$ is the concentration of the single-stranded form at equilibrium. Most of the kinetic data were obtained at constant ratio of helix to coil form—solutions were heated to temperatures corresponding to 20, 50, or 80% of the total absorbence change, which were treated as equivalent to the percent of the oligo A in coil form. The very fast process was ignored and the slow relaxation process could be analyzed. The logarithm of the absorbence change is a linear function of the time; the slope of this line yields the value of the time. Within the experimental error, only one relaxation time was found, confirming the all-or-none nature of the transition. Since $K = k_R/k_D$ and the measurement was done with a fixed fraction of the helix form converting to the coil form, a specific set of relationships can be established [96] as shown in the following tabulation. In a plot (Fig. 48a) of $1/\tau$ vs total concentration, C_t, slopes of these linear lines are proportional to the k_R, rates of recombination, which were found to be about $2\text{--}4 \times 10^6 \ M^{-1}\ \mathrm{sec}^{-1}$. These values are only slightly influenced by the degree of transition. The effect of temperature on k_R is shown in Fig. 48b. The rates of recombination, which are almost independent of chain length, were found to have negative activation enthalpics of 6–9 kcal/mole. From the value of K $(K = k_R/k_D)$ and k_R, the values of k_D, the dissociation rate constant,

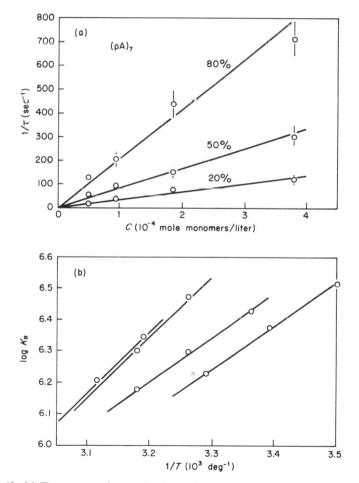

Fig. 48. (a) Temperature jump data for (pA)$_7$ at pH 4.2. Reciprocal relaxation time $(1/\tau)$ as a function of total nucleotide concentration at 20, 50, and 80% degree of transition. (b) Arrhenius plot for the rate constants of recombination (k_R). pA$_n$ at pH 4.2 with $n = 6, 7, 8,$ and 9. (From Pörschke and Eigen [96].)

were readily determined. For $(pA)_7$, 5.5, 36, and 210 sec^{-1} were found as the values of k_D at 20, 50, and 80% of dissociation.

The kinetic parameters of $(pA)_6$ were determined by another procedure as an additional check [93]. The kinetics of $(pA)_6$ were measured at four different temperatures and the rate constants were evaluated by Eq. (16). In a plot of $1/\tau$ vs $\bar{c}_A$, concentration of single-stranded form at equilibrium, the slope of the line yields k_R and the intercept of the line yields k_D. The results of this study are displayed in Table XVII. The effect of pH and salt on the kinetics has also been explored to the same extent [96].

The elegant study on the thermodynamics and the kinetics of helix–coil transition of oligo A in acidic solution by the Göttingen group [93,96] serves as an introduction to the quantitative study of the oligomer–oligomer interaction. As mentioned in the beginning, so far the study has been limited

Degree of transition	Equilibrium constant	Reciprocal relaxation
20%	$K = 10/C_t$	$1/\tau = k_R 0.9 C_t$
50%	$K = 1/C_t$	$1/\tau = k_R 3.0 C_t$
80%	$K = 1/6.4 C_t$	$1/\tau = k_R 9.6 C_t$

to optical techniques only. It would be of importance to study this problem by other physical methods as well, especially those involving measurement of molecular weight, in order to confirm the model adopted in these analyses.

2. Oligocytidylic Acid

In acidic solution, poly C also forms a helical structure. However, there are basic differences between the conformation of the protonated poly C helix

TABLE XVII

RATE CONSTANTS AND ACTIVATION ENTHALPIES FOR $(pA)_6$[a]

	7°C	18.5°C	28°C	38°C	AE (kcal/mole)
k_R	3.5×10^6	2.4×10^6	1.73×10^6	1.25×10^6	-6
k_D	1	30	185	1000	40

[a] From Pörschke and Eigen [96]. k_R = rate constant of recombination, k_D = rate constant of dissociation, AE = enthalpy of activation.

and that of the protonated poly A helix. For instance, the stability of helical poly C is greatest when bases of the polymer are only half-protonated. Further protonation weakens the helical structure. Therefore, the effect of acidity on the T_m of helical poly C goes through a maximum near the pk_a of the polynucleotide [97]. The base-pairing scheme of the acidic oligo C helix is again expected to be that of the acidic poly C helix [97].

The studies on the acidic oligo C helix have been less extensive than those on the oligo A helix. Brahms *et al.* [98] reported the T_m of oligo C in pH 4.0, 0.15 *M* salt, to be about 20°, 25°, 35°, and 42°C, respectively for $(Cp)_6C$, $(Cp)_7C$, $(Cp)_8C$, and $(Cp)_9C$; under the same conditions, the T_m of poly C is about 83°C and the T_m of $(Cp)_5C$ is below 0°C. Maurizot *et al.* [99] reported a comparative study on the deoxy oligo C helix vs the ribosyl oligo C helix. It has been noted before that the stability of the acidic ribosyl poly C helix is much less than that of the acidic deoxy poly C helix [100]. This same conclusion was reached in the comparative studies on oligo C.

In the optical studies on cytosine compounds, it should be remembered that upon protonation, there is a significant change of the spectrum of the

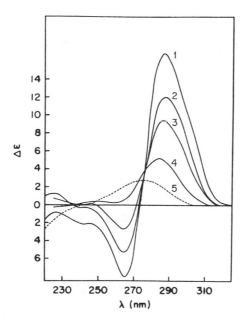

Fig. 49. CD spectra of dCMP, oligo dC and poly rC in acidic solution, 0.1 *M* acetate buffer, 0.05 *M* $(NH_4)_2SO_4$, 0°C; nucleotide residue concentration 10^{-4} *M*. (1) Poly rC, pH 4; (2) $d(pC)_7$, pH 4.75; (3) $d(pC)_5$, pH 4.75; (4) $d(pC)_3$, pH 4.75; (5) dCMP, pH 4.75. (From Maurizot *et al.* [99].)

chromophore—a red shift of the absorption maximum from 271 to 280 nm, for instance. The origin of the optical bands of the cytosine chromophore and the effect of protonation were described in Chapter 6, Volume I; (Section I,E and Table XII). It is instructive to compare the effect of protonation on the absorption and CD spectra of dimer or trimer to those of the large oligomers. Since the large oligomers can form a helix while the dimer or trimer cannot, the effect of protonation on the optical properties can then be separated from those related to helix formation.

In Fig. 49, the CD spectra of dCMP, oligo dC, and poly rC in acidic solution at 0°C are shown. In the spectra of poly rC, d(pC)$_7$, and d(pC)$_5$, compounds which can form a helical structure, the positive band at ~285 nm is greatly enhanced and a new negative band at ~268 nm appears. The extinction coefficients of (dpC)$_5$ and (dpC)$_7$ at pH 4.75, 0.15 M salt, 0°C, were found to be 8.87×10^{-3} and 8.30×10^{-3} cm^{-1} mole^{-1}, respectively. The effect of temperature on the CD band at 285 nm for various oligo dC is

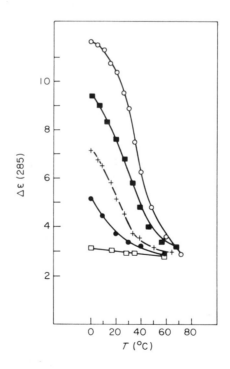

Fig. 50. Effect of temperature on the intensity of the CD band at 285 nm of the oligo dC at pH 4.75, 0.1 M acetate, 0.05 M (NH$_4$)$_2$SO$_4$, 10^{-4} M nucleotide concentration. $N = 2$ (□); 3 (●); 4 (+); 5 (■); 7 (○). (From Maurizot *et al.* [99].)

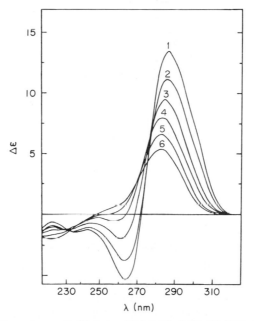

Fig. 51. The CD spectrum of r(Cp)₉ in pH 4.0, 01. *M* NaCl, 0.05 *M* acetate, 10⁻³ *M* nucleotide concentration at varying temperatures. (1) 0°C; (2) 24.5°C; (3) 30.6°C; (4) 36°C; (5) 48.7°C; (6) 71°C. (From Maurizot *et al.* [99].)

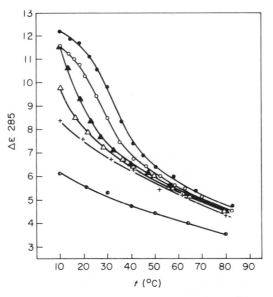

Fig. 52. Temperature effect on the intensity of the CD band at 285 nm of the oligo rC at pH 4.0, 0.1 *M* acetate, 0.05 *M* (NH₄)₂SO₄, 10⁻⁴ *M* nucleotide concentration. (◑) C₂p; (+) C₃p; (△) C₅p; (▲) C₆p; (○) C₈p; (●) C₉p. (From Maurizot *et al.* [99].)

illustrated in Fig. 50. From the slope of the transition at T_m $(df/dT)_{T_m}$, shown in Fig. 50, and by the application of Eq. (12), Maurizot *et al.* concluded that the ΔH of the transition is around 5–6 kcal/mole for the oligo dC.

As for the ribosyl oligo C series, the CD spectrum of $r(Cp)_9$ at 10^{-3} M nucleotide concentration, pH 4.00, is shown in Fig. 51. This spectrum is rather similar to that of $d(pC)_7$ at 10^{-4} M, pH 4.75. In comparing the effect of temperature on the CD band at 285 nm for the oligo rC series (Fig. 52) to that for the oligo dC series (Fig. 50), obviously the stability of the helix formed by oligo $r(Cp)_n$ is much less than that formed by oligo $d(pC)_n$.

D. SELF-ASSOCIATION OF OLIGONUCLEOTIDES CONTAINING TWO COMPLEMENTARY BASES IN ALTERNATING SEQUENCE—d(T–A) OLIGOMERS

In a series of papers [101–104], Scheffler, Elson, and Baldwin reported their interesting research on the self-association of d(T–A) oligomers. This series of deoxyoligonucleotides containing thymidylic acid and deoxyadenylic acid in alternating sequence was prepared by the cleavage of poly d(A–T) with an endonuclease–bovine pancreatic DNase [74]. This enzyme was found to be specific for the ApT bond in poly d(A–T); therefore, the resulting oligomers have equal numbers of A's and T's with T at the 5′-phosphate end of the chain and A at the 3′-hydroxyl end—$d(pTpA)_N$, or $d(T–A)_N$. Oligomers in the size range $5 \leq N \leq 25$ were prepared by electrophoresis in acrylamide gels as mentioned earlier (Fig. 30 [74]). Their molecular weights have been determined by ultracentrifugation, with the aid of an authentic sample of $d(T–A)_5$ synthesized chemically.

The research on these $d(T–A)_N$ oligomers, meant for testing the general theory describing the helix–coil transition of DNA, has two distinct advantages. First, the helix stability depends strongly on chain length in this size range; therefore, the size dependence of this oligomer series becomes an important parameter in testing the theory. Second, the heterogeneity of two base pairs ($A \cdot T$ and $G \cdot C$) has been removed since now there is only one type of base pair ($A \cdot T$) possible for these oligomers. However, precisely because of the removal of the heterogeneity in base pairing, the oligomers tend to fold back onto themselves forming one-chain, hairpin helices with a loop. Insofar as the original purpose was to study the formation of a two-stranded, straight, and fully base-paired helix, this formation of hairpin helices is a complication; on the other hand, it affords a means of investigating other properties of nucleic acid.

At low counterion concentration (<0.01 M Na^+), $d(T–A)_N$ with $N > 7$ forms one-stranded hairpin helices at room temperature or lower. This conclusion is supported by molecular weight measurements, the occurrence of only simple, monophasic, and rapidly reversible melting curves, as well as by the lack of concentration effects on the melting curves. At intermediate (<0.06 M Na$^+$) or high (<0.5 M Na$^+$) counterion concentrations, two-stranded helices are formed as shown by the biphasic melting curves. Upon elevation of temperature, the double-stranded helices are first converted into single-stranded hairpin helices. This conversion can be demonstrated by molecular weight measurements and is accompanied by a small increase in absorbence. Further heating brings about the melting of the hairpin helices into coil form and is shown by a large increase in absorbence. On cooling, after the occurrence of a rapid lowering of absorbence due to hairpin formation then a slow reaction is observed at low temperatures which appears to be the conversion of hairpins to double-stranded, straight helices by chain slippage. It was proposed [101] that the equilibrium between the conversion of open strands (C_0), the one-stranded helices (C_{1h}) and the two-stranded helices (C_{2h}) can be described by the following equations:

$$\frac{C_{1h}}{C_0} = \sigma s^n \quad \text{(one-stranded helix formation)}$$

$$\frac{C_{2h}}{C_0} = \beta C_0 s^{2n+g} \quad \text{(two-stranded helix formation)}$$

$$(17)$$

where n is the maximum number of base pairs in the one-stranded hairpin helix which has g nonbanded bases in the loop, and ($2n + g$) is the maximum number of base pairs for the two-stranded helix; s is the equilibrium constant for formation of a base pair next to a preexisting base pair, which is assumed to be the same for all base pairs; C_0, C_{1h}, and C_{2h} (about 10^{-5} M) are the molar concentrations of open strands, one-stranded hairpins and two-stranded helices, respectively. The equilibrium constants for the formation of the first base pair without a neighboring pair are σs and βs for the hairpin and duplex, respectively; σ and β are expected to be of the same order of magnitude, about 10^{-3} or 10^{-2}.

At the T_m of poly dAT, $s = 1$, the ratio of C_{1h}/C_0 in Eq. (17) is equivalent to σ (10^{-2} to 10^{-3}) and the ratio of C_{2h}/C_0 is equivalent to βC_0 (10^{-7} to 10^{-8}). Therefore, most of the oligomer is in C_0 form, little in C_{1h} form and practically none in C_{2h} form. At a lower temperature where s becomes larger (say, 10), the C_{1h} form begins to dominate, depending on the number of base pairs in the hairpin (which depends on the chain length of the oligomer); also the C_{2h} begins to emerge, depending on C_0 (the concentration of open strands) and total number of base pairs in the duplex ($2n + g$). From these two simple equations, a general picture about the composition of solution at varying temperature and different chain lengths can be formulated.

An experimental study has been made of the melting of another type of d(T–A) oligomer, the circular oligomer or d(T–A)$_{\langle N \rangle}$. To make the circles, the linear d(T–A)$_N$ were incubated first with the polynucleotide ligase of

Olivera and Lehman [105,106] and then with exonuclease for the destruction of the remaining linear forms. These circular oligomers have some rather interesting properties. Their helical form is considerably more stable than that of the linear oligomers, even that of the poly d(AT). For instance at 0.01 M Na^+, $d(T-A)_{\langle 20 \rangle}$, $d(T-A)_{\langle 31 \rangle}$, and poly d(AT) have T_m values of 60°, 55°, and 40°C, respectively. It should be noted that the T_m of $d(T-A)_{\langle 20 \rangle}$, a smaller circle, is higher than that of $d(T-A)_{\langle 31 \rangle}$, a larger circle. While the linear oligomers form open hairpin helices which melt chiefly from the open end, the circular oligomers form closed hairpin helices with a loop at each end, and melt by enlarging the loops. Although the closed hairpin helices contain two loops and the open hairpins have only one, the closed hairpins are more stable. The reason is a large difference in the conformational entropies of the two random-chain forms; the open circle can assume only a fraction of the conformations available to the linear chain.

The electrostatic effects on the helix–coil transition of these $d(T-A)_N$ oligomers were investigated. The dependence of $T_m(N)$ on log M (monovalent counterion molarity) falls off sharply with decreasing N in the range $8 \leq N \leq 22$, so that the T_m values of these oligomers at low ionic strength became very close to each other as shown in Fig. 53. At $M = 0.0028$, $d(T-A)_{20}$ has a higher T_m than poly d(AT). Also, the melting curves for all oligomers sharpen only slightly as the counterion concentration is reduced, indicating that the electrostatic interactions have a relatively small effect on the breadths of these melting curves.

A considerable amount of theoretical study has been made on this system. Only the general approach and some basic notions are introduced here. A comprehensive review [104] by Baldwin is recommended for those who are interested in the details. In the theoretical formation, it is recognized that there are two types of base pairs: the *isolated* base pair, which nucleates a new helical segment, and the *stacked* base pair, which is added at the end of an existing helix. Nucleation of the helix has a low probability [σs or βs vs s in Eq. (17)] for two reasons: (1) The *isolated* base pair lacks a stacking interaction while the *stacked* base pair gains stability via a stacking interaction with the adjacent base pair. (2) The helix may be nucleated in a variety of ways (by joining two separate chains or by closing loops of different types), all of which have unfavorable entropy ($\Delta S°$). To form a base pair, the two bases must be brought within a critical volume element, whose radius is of the order of the hydrogen bond length. For the hairpin helix, the probability of closing a loop decreases with loop size. This is expressed quantitatively by the loop-weight function [$\rho(x)$] which gives the time-average *effective concentration* of one base in the vicinity of its partner-to-be (x represents the number of internucleotide links in the loop); [$\rho(x)$] has the units of molecules/volume (such as $Å^3$).

The equilibrium constant (σs) for helix nucleation via loop closure may be correlated by the use of the loop-weighting function $\rho(x)$, with the nucleation constant (βs) for joining two chains, i.e., $\sigma s = 2\beta s [\rho(x)]^*$. The evaluation of the function $\rho(x)$ is based on the Jacobson–Stockmayer equation on the ring closure of long chain polymers on a random-walk approach and is modified by Flory and Semlyen [107] for short chains; $\rho(x)$ is dependent on $[1/\langle r_x^2 \rangle]^{3/2}$,

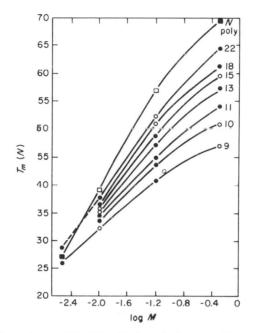

Fig. 53. The dependence of T_m (N) of hairpin helices on oligomer size ($N = 9$ to ∞) at three sodium ion concentrations (M). Filled circles ($\bullet$) are experimental, open circles ($\bigcirc$) are interpolated from plots of T_m (N) vs $1/(N - 2)$. Filled squares ($\blacksquare$) represent T_m of poly d(AT). (From Elson *et al.* [103].)

where $\langle r_x^2 \rangle$ is the mean-square end-to-end distance of a chain with x monomer residues. The value of s (or s^{app}, the apparent stability constant) should be independent of helix length and oligomer size. At the T_m of poly d(AT), s is assumed to be 1. At $M = 0.01$, $\Delta H°$ of poly d(AT) has been measured to be -7.9 kcal/mole base pairs [108] with a small ΔCp (25 cal-deg^{-1} mole^{-1} base pair). From the formulations of T_m [poly d(AT)] $= \Delta H°/\Delta S°$, and of $-RT \ln s = \Delta G° = \Delta H° - T \Delta S°$, s as a function of temperature at

* In later papers [102–104] the symbol σs was replaced by $\gamma(x)$, and $\beta(s)$ replaced by $K/2$. Equation (17) becomes $C_{1h}/C_0 = \gamma s^{n-1}$; $C_{2h}/C_0 = K/2 \, C_0 S^{2n+g-1}$.

$M = 0.01$ can be obtained. However, s is strongly dependent on the counterion concentration (M). In order to identify this dependence, ΔG_0 is divided into a chemical part (ΔG_{ch}°) and an electrostatic part (ΔG_{el}°). At a very high counterion concentration $[M_{hi}; M_{hi} = 1\text{--}2$ for natural DNA and 0.5 for poly d(AT)], where the T_m of the polymers is no longer affected by counterion concentration, ΔG_{el}° becomes small and is related to the $\Delta G_{el,M}^{\circ}$ at a lower M value by the following equation,

$$\Delta G_{el,M}^{\circ} = \Delta G_{el}^{\circ}(M_{hi}) + B \log (M/M_{hi}) \qquad (M \leqslant M_{hi})$$

From this equation, the dependence of T_m of polynucleotides on counterion concentration (M) can be formulated,

$$T_{m\text{-polymer}}(M) = T_{m\text{-polymer}}(M_{hi}) + (\frac{B}{\Delta S_{ch}^{\circ}})\log(M/M_{hi})$$

Typical values of $dT_m/d \log M$ for natural DNA's are about 18°C and a value of 22°C has been reported for poly d(AT). An extensive computation has been made by Delsi and Crothers [109] on the electrostatic contributions to the oligonucleotide transitions. In addition to the conventional approach (such as the usage of the Debye–Hückel approximation in relating the screening potential to the ionic strength), they adopt a chain model with restriction on rotational angles based on the x-ray crystallographic data, instead of the less defined models such as the linear rod and the freely jointed chain. Energies were calculated by averaging over the allowed conformations of the oligomer. For the calculation of the transition curves of d(T–A)$_N$ as a function of salt concentration and chain length, the enthalpy measurement for poly d(AT) at 0.01 M was adopted along with one adjustable parameter. Qualitative and semiquantitative agreement has been obtained between the experimental results and the computed values, especially concerning the larger oligomers such as d(AT)$_{10}$.

For the measurement of σs (or γ, see footnote on page 433) a comparison has been made between the experimental values of the dependence of $T_m(n)$ on $1/n$ and the curves generated by the computer based on various values of σs (see Scheffler et al. [102] for details). The first test of the theory is then whether one value of σs will reproduce the T_m's of all oligomers. At $M = 0.5$, $\sigma = 0.003 \pm 0.001$ was found to do this satisfactorily. The second test of the theory is to use this σ value (0.003) for the prediction of the shapes of the melting curves for different oligomers. At $M = 0.5$, again the agreement was satisfactory. However, at $M = 0.06$ and 0.01, it was not possible to compute both the shape and the T_m of even one oligomer melting curve with one value of σs. This failure is due to the effect of ΔG_{el}° at $M < M_{hi}$, which is dependent on helix length and on oligomer size. This question was the subject of the previous paragraph.

Some attempts have been made to evaluate the loop-weighting function [$\rho(x)$] from the data of the circular oligomers; $\rho[x]$ can be expressed as $\rho(x) = J/x^{3/2}$, where J is constant, dependent on chain stiffness. The predicted melting curves proved to be insensitive to the choice of exponent between $\frac{3}{2}$ and 2; but the $\rho(x)$ for small loops is characteristically different from the $\rho(x)$ for large loops; small loops are more stable, relative to large ones, than predicted by the Jacobson–Stockmayer loop-weighting function.

The above discussion demonstrates amply that the study on the helix–coil transition of the oligomers has provided valuable insight into the phenomenon of nucleic acid melting.

E. Association of Complementary Oligonucleotides in Neutral, Aqueous Solution

A duplex formed by two complementary oligonucleotides is a miniature nucleic acid helix. In comparison to the study on the polymeric helices, the study on these mini- or microhelices has the advantage of controlling the composition, the sequence, and the chain length of the helices, but the disadvantage is a certain loss of stability, specificity, etc.

Jaskunas *et al.* [110] were the first to search for the formation of a 1:1 complex by complementary di- and trinucleotide in 0.01 M MgCl$_2$, (or 0.5 M NaCl), pH 7, at 1°C and 0.01 M total residue concentration. By the ORD and the ultracentrifugation techniques, no cross-association or self-association was observed under those conditions for the following pairs: ApC + GpU, ApCpU + ApGpU, ApGpC + GpCpU (slight), ApApApA + UpUpUpU. Cross-association was observed for GpGpC + GpCpC, and for GpGpCp + GpCpC (slight) and self-association was observed for GpGpC (the self-association of oligo G has been discussed in Section II,B,2). The experimental ORD curve of GpGpC + GpCpC (1:1 mixture) is clearly different from the computed curve of the 1:1 mixture with no interaction. The continuous variation plot of optical rotation at 275 nm vs mole % composition of GpGpC in the mixture shows that the stoichiometry of the aggregate is 2 GpGpC to 1 GpCpC. The association was also confirmed by molecular weight measurements. The observed molecular weight (weight-average) of the aggregates in the solutions of 0.01 M MgCl$_2$ containing 2:1 mixtures of GpGpC and GpCpC (8700) is less than that for the GpGpC solution (15,000) but is greater than the calculated molecular weight of a complex containing 2 GpGpC and 1 GpCpC (2750). The species present in these solutions include 2:1 complexes between GpGpC and GpCpC, GpGpC aggregates, and free GpCpC oligomers. Thus, the molecular weight of the mixture is the weight-average for these different species; the weight fraction of the 2 GpGpC + GpCpC complexes in solution was estimated to be 0.75 $\pm$ 0.20. In addition, the calculation indicated that there are more than three trimers in the complex, with an average value from 5.5 to 9. The effect of temperature on the optical rotation of the 2:1 mixture of GpGpC + GpCpC is gradual with no indication of an abrupt transition. Effect of substitution of GpGpCp for GpGpC in the mixture was investigated. The presence of a 3'-phosphate on GpGpC decreases the fraction of trimers in the 2:1 complex to less than one-half of what it is with the dephosphorylated compound.

Much more defined, specific complexes between oligomers of $G(pU)_n$, $(Up)_nG$, $C(pA)_n$, $(Ap)_nC$, where $n = 5, 6$, and 7, have been investigated [111]. The uridine-containing oligomers (such as GU_6) were mixed with adenosine-containing oligomers of equal chain length (such as A_6C) in 1 M NaCl, 0.04 M phosphate (pH 7.0–7.5) and with 0.01 M oligonucleotide residues. CD measurements of these solutions at varying temperatures clearly indicated the formation of specific complexes. As expected, there is a marked dependence of the stability of the complexes on concentration. At 0.01 M, the heptamer complexes [such as $(Ap)_6G + G(pU)_6$] have T_m values around 15°–25°C; while at 0.001 M, the T_m values of these complexes drop below 0°C. Similarly, at 0.01 M, the T_m of $(Ap)_7G + G(pU)_7$ is 30.5°C; at 0.001 M, the T_m is 21°C. It is important to note that the antiparallel pair is more stable than the parallel complementary pair. For example, the T_m of the antiparallel $(Ap)_6C \cdot G(pU)_6$ is 25°C vs 18.5°C for the T_m of the parallel $(Ap)_6C \cdot (Up)_6G$; the T_m of the antiparallel $C(pA)_6 \cdot (Up)_6G$ is 19°C vs 13.5°C for the T_m of the parallel $C(pA)_6 \cdot G(pU)_6$. Similarly, the antiparallel triplex was found to be more stable than the parallel triplex. The optical activities of the oligoA $\cdot$ poly U and $(A)_6 \cdot (U)_6$ complexes are practically identical to that of the poly A $\cdot$ poly U complex [59], which has an antiparallel structure; when this is considered along with the data on these oligomer complexes, it becomes apparent that antiparallel structure is the prevalent form for the microhelices and for the oligomer–polymer complexes.

Complex formation between oligo A and oligo U has been investigated by Pörschke [112]. Oligonucleotides of chain lengths <8 always form both double helices and triple helices (measured by the decrease of extinction at 280 nm). Therefore, the double helix formation not coupled to the triple helix formation can be measured only with chain lengths higher than 7. High ionic strengths favor the formation of triple helices. As expected, an increase in nucleotide concentration leads to an increase in melting temperature; the shorter the chain length, the lower the T_m value and broader the helix–coil transition. The experimental curves of the helix–coil transition were analyzed according to a staggering zipper model with the parameters derived from measurements on the polymers.

The association of two complementary deoxypentanucleotides, d(T–T–G–T–T) and d(A–A–C–A–A), has been studied by Cross and

Crothers [113] by PMR techniques in 0.15 M NaCl, 0.1 M Na-phosphate buffer. It is interesting to note that the resonances of the base protons of both oligomers measured in separate solutions are sharpened upon lowering the temperature from 35° to 5°C, while these same resonances measured in the 1:1 mixture (0.02 M strand concentration for each pentamer) become broadened and unresolvable upon lowering the temperature to 5°C. This phenomenon of line broadening of the 1:1 mixture is an indication of complex formation. The methyl resonances of the T residues in the d(T–T–G–T–T) pentamer are slightly more shielded than the methyl resonance of 5′-TMP; this difference in δ (about 0.08 ppm) is not dependent on temperature. Upon complex formation in the mixture, the methyl resonance of the pentamer became broadened and shifted significantly upfield; this difference in δ between the methyl resonance in the complex and in the TMP is dependent on temperature. The difference is about 0.58 ppm at 0°C and only about 0.1 ppm at 35°C, indicating a nearly complete dissociation of the complexes at this temperature. From these differences of the methyl resonances, the fractions of the helical complexes (θ) at varying temperatures were estimated. A theoretical curve for a transition at $T_m = 9$°C and with a total heat of 34 kcal/mole was constructed to fit these θ values

The association of block oligomers of adenylic acid and uridylic acid in the form of $(Ap)_m(Up)_{n-1}U$ where m and n vary from 3 to 7 were studied by Martin *et al.* [114]. The oligonucleotides undergo self-association to form two-stranded helical complexes when $m \geq n$, but when $m < n$, the oligomers self-associate to form multichain aggregates containing some triple-stranded helical regions. All the melting transitions studied were instantly and completely reversible. Even with noncomplementary bases inserted between the adenylic and uridylic acid blocks [such as $(Ap)_7C(pU)_6$], no single-chain hairpin helices were observed. The T_m values of both A_7U_7 and $(Ap)_7C(pU)_6$ are strongly dependent on concentration of the oligomers; at strand concentration of 7×10^{-5} M, 1 M NaCl, 0.01 M NaPO$_4$ buffer, pH 7.0, 10^{-4} M EDTA, the complex of $(Ap)_7C(pU)_6$ has a much lower T_m (~ 20°C) than A_7U_7 ($T_m \approx 36$°C). This large difference in T_m also indicates that neither A_7U_7 nor $(Ap)_7C(pU)_6$ is forming single-stranded hairpin helices, since the hairpin helices of these two oligomers should have the same degree of stability as the unpaired base in the loop.

The dependence of T_m of $(Ap)_n(Up)_{n-1}U$ where $n = 4$–7 on oligomer strand concentration in 1.0 M salt, pH 7.0 is shown in Fig. 54. By the use of Eq. (13) (Section II,C,1), the slopes of these lines in the plot (which are equal to 2.3 R/N ΔH) yield the ΔH per mole base pair from -6.4 kcal for A_4U_4 to -7.6 kcal for A_7U_7. Krakauer and Sturtevant [115] reported a value of -8.2 kcal/mole for poly A·poly U at a lower salt concentration. Two explanations were proposed for the observed mild dependence of ΔH per mole base on the chain length. The first explanation is that the terminal base pairs do not contribute as much to the enthalpy of the complex as the internal base pairs, thereby leading to lower enthalpies for shorter oligomers. This first explanation would be correct if the enthalpy of the double helix formation were not simply proportional to the number of base pairs n, but primarily dependent on the $n - 1$ stacking interactions. The value of ΔH for each oligomer would then be increased by $n/n - 1$, bringing them closer to that obtained for the polymer and reducing the dependence on chain length. With this formulation, ΔH is 7.3 kcal/mole for A_4U_4, 8.0 kcal for A_5U_5, 7.7 kcal for A_6U_6, and 8.3 kcal for A_7U_7. It should be noted that such an argument has also been advanced by Eigen and Pörschke [93] in their modification of Eq. (14) to Eq. (15). Obviously, the end pairs may be different from the

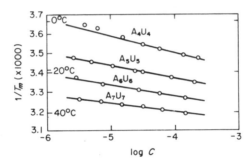

Fig. 54. Dependence of T_m of $(Ap)_n(Up)_{n-1}U$ on oligonucleotide strand concentration in 1.0 M NaCl, 0.01 M phosphate buffer (pH 7.0), 10^{-4} M EDTA. (From Martin et al. [114].)

internal pairs and certain corrections could be empirically applied. The second explanation is that the single-stranded oligomers, especially the blocks of A residues, being stacked more at lower temperatures lead to lower enthalpies of formation of complexes which dissociate at lower temperatures. It is most likely that the stacking enthalpy of the single-stranded chains is a negative contribution to the enthalpy of helix formation, a problem which will be discussed more thoroughly in a later chapter on polynucleotides. Since the amount of stacking in the single strands decreases when the temperature is raised, the observed increase in ΔH with increasing chain lengths would just result from the fact that the observations on melting of complexes with longer chain lengths were carried out at higher temperatures. Again such a problem was raised in the discussion on the oligo A microhelices (Section II,C,1).

The ΔH values have also been evaluated from the slopes at the midpoint of the melting curves by Eq. (12). From this procedure, the ΔH for A_4U_4 to A_7U_7 is 5.3 ± 0.5 kcal/mole of base; the variation is random with no dependence on chain length. This discrepancy will be further discussed in the following section on oligomer–polymer associations. The dependence of T_m of these microhelices on cation concentration was found to be less than that of the polymeric helices. Below $0.5\ M$ Na^+ concentration, the dependence of the T_m of A_5U_5 on the concentration of sodium ion is essentially linear with log $[Na^+]$. In this region a tenfold increase in $[Na^+]$ causes an increase of $12.5°C$ in T_m, an increase much lower than that of $21°C$ in T_m per factor of 10 in $[Na^+]$ observed for poly A · poly U helix.

In a following paper [116], Uhlenbeck *et al.* have investigated the insertion of a cytidylic acid or guanylic acid, or a cytidylic–guanylic pair (G·C) in the center of these A_nU_n helices as shown in Table XVIII. Insertion of a C residue in the middle of a A_nU_n oligomer significantly reduces the stability of the self-association duplex. The effect of insertion of a G residue on the other hand, is much less. Expressed in terms of the total ΔH for the transition of the oligomer complexes (derived from the plots of $1/T_m$ vs log concentration of oligomers), the values of the $(Ap)_nC(pU)_n$ were found to be about 60% of the values for the corresponding $(Ap)_{n-1}ApU(pU)_{n-1}$. Thus, when $n = 10$, the ΔH for the total complex of the A_nU_n changes from -73 to -43 kcal when the internal C residues are present as in $(Ap)_nC(pU)_n$. The corresponding change for the helix with 14 base pairs ($n = 14$) is from -107 kcal/mole

TABLE XVIII

T_m VALUES OF SELF-ASSOCIATED DUPLEXES OF ADENYLIC
ACID–URIDYLIC ACID BLOCK OLIGOMERS WITH AND
WITHOUT MISMATCHED BASES AT THE MIDDLE OF
THE OLIGOMERS [a]

Molecule	Maximum helix length	T_m (°C) at 25 μM	T_m (°C) at 100 μM
$(Ap)_3ApU(pU)_3$	8	8.1	12.4
$(Ap)_4C(pU)_4$	8	—	-5.0
$(Ap)_4G(pU)_4$	8	-1.0	5.2
$(Ap)_4ApU(pU)_4$	10	20.2	23.5
$(Ap)_5C(pU)_5$	10	5.0	9.6
$(Ap)_5G(pU)_5$	10	14.6	18.0
$(Ap)_6ApU(pU)_6$	14	37.2	39.5
$(Ap)_7C(pU)_7$	14	25.4	29.1

[a] The solvent is $1.0\ M$ NaCl–$0.01\ M$ phosphate, pH 7.0 (from Uhlenbeck *et al.* [116]).

complex to -65 kcal/mole complex. This change in total ΔH is much larger than the -15 kcal/mole, which would be lost if one A·U pair were disrupted on each side of the C residues. The most likely explanation is that the two halves of the helix are melting less cooperatively because of the interruption at the middle. As for the interruption by the G residue, it is possible that the G residue can be accommodated inside the helix through the formation of the G·U pair and contributes to the stacking energy. For instance, the difference in T_m of the self-complex at 0.1 mM strand concentration, 1.0 M NaCl–0.01 M phosphate (pH 7.0) between $(Ap)_5ApU(pU)_5$ (31.8°C) and $(Ap)_5$-$UpG(pU)_5$ (28°C) is only 3°C.

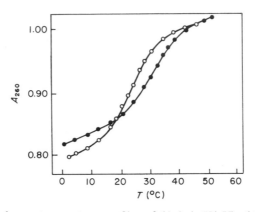

Fig. 55. Absorbence–temperature profiles of $(Ap)_4ApU(pU)_4$ (○) at 24 μM-strand concentration and $(Ap)_5G(pU)_4 + (Ap)_4C(pU)_5$ (●) at 22 μM-strand concentration normalized to $A_{260} = 1.0$ at 40°C. The solvent is 1.0 M NaCl–0.01 M sodium phosphate, pH 7.0. (From Uhlenbeck *et al.* [116].)

On the other hand, the insertion of a G·C base pair in the midst of A–U block oligomers significantly increases the stability of the complexes. In Fig. 55, the complementary complex $(Ap)_5G(pU)_4 \cdot (Ap)_3C(pU)_5$ is more stable than the self-complex of 2 $[(Ap)_4ApU(pU)_4]$; and in Fig. 56, the self-complex of 2 $[A_4CGU_4]$ which contains 2 G·C pairs in the center, is much more stable than the self-complex of A_5U_5 (by about 18°C) and even slightly more stable than the self-complex of A_7U_7. In this series of oligomers, the substitution of one G·C pair for one A·U pair increases the T_m by about the same amount as extending the pure (A, U) microhelix by two additional A·U pairs; and the effect of two G·C pairs is nearly twice that of one G·C pair. In Fig. 57, the effect of G·C pair substitution in the (A, U) microhelices on the T_m as a function of G + C content is shown. The T_m difference increases by about 0.9°C for each increase in percentage of G + C content at the

10–30% range; this observation should be compared to that of 0.5°C increase per percent of G + C observed for DNA at 40–60% range [117], or to that of 0.75°C per percent of G + C observed in double-stranded polymeric RNA [118]. Further analyses of the melting data strongly suggest that the ΔH of the G·C pair is significantly higher than that of the A·U pair.

In a recent paper by Craig, Crothers, and Doty [119], valuable data and analyses on the relaxation kinetics of the helical duplex of A·U when n is from 4 to 7 are described. Since this paper will be discussed in a later chapter on polynucleotides, only qualitative results are mentioned here. The relaxation process, which can account for most of the optical change, has a time constant

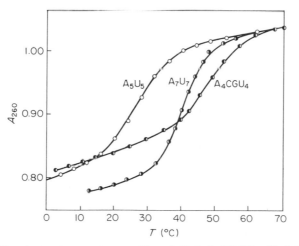

Fig. 56. Absorbence–temperature profiles of $(Ap)_4ApU(pU)_4$, $(Ap)_6ApU(pU)_6$, and $(Ap)_4CpG(pU)_4$ at strand concentration 96, 69, and 114 μM and normalized to $A_{260} = 1.0$ at 40°C. The solvent is 1.0 M NaCl–0.01 M sodium phosphate, pH 7.0. (From Uhlenbeck *et al.* [116].)

varying from 0.1 to 1 sec, depending on conditions. The two strands combine with a bimolecular rate constant of about 2×10^6 liter/mole of strands per second. The rate constant for dimerization decreases with increasing temperature. In the double helix formation, a critical intermediate having two or three pairs is formed in rapid equilibrium. Generally this critical intermediate dissociates more rapidly to single strands than it reacts to form the complete helix. When one base pair is added to this critical intermediate, a helix nucleus is found which zips up to form the complete duplex more rapidly than it dissociates to single strands. The rate for formation of the helical duplex is equal to the concentration of the critical intermediate times the rate constant for adding one base pair. The rate constant for adding a base pair to the end of an existing helical segment is estimated to be between 1×10^6 and 2×10^7 sec^{-1}. The experimental rate equation is second-order in the single strands because the concentration of the critical intermediate depends on the square of the strand concentration. The rate of helix dissociation depends on the number of pairs that must break to reach the critical intermediate, which then dissociates very rapidly. Hence, the dissociation rate constant is strongly dependent on oligomer size and on temperature, with an activation energy roughly proportionate to chain length.

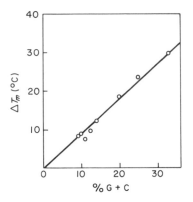

Fig. 57. The difference in T_m of $(Ap)_nCpG(pU)_n$ at 25 μM-strand concentration or $(Ap)_nC(pU)_m + (Ap)_mG(pU)_n$ at 50 μM-strand concentration and the T_m of the $(Ap)_nApU(pU)_n$ with the same chain length at 25 μM-strand concentration (T_m) plotted as a function of the percentage of G—C bonds in the oligomer. The straight line has a slope of $0.9°C/\%(G + C)$. (From Uhlenbeck *et al.* [116].)

F. FORMATION OF HAIRPIN LOOPS—$A_6C_mU_6$

Formation of hairpin loops is an important process in determining the secondary and tertiary structure of RNA, or any natural nucleic acid which does not have a complementary partner for the formation of the double helix. Some of the physicochemical factors in this looping process have been discussed in Section II,D concerning the research on d(T–A) oligomers. A significant advance has been made in this study by Uhlenbeck *et al.* [120] in their investigation of the formation of hairpin loops of ribosyl oligonucleotides of $A_6C_mU_6$ ($m = 4,5,6,8$).

The hairpin loop formation is described by the following reaction:

$$A_6C_mU_6 \text{ (single strands)} \rightleftharpoons A_6C_mU_6 \text{ (loop)}$$

with f = loop fraction, $f/(1 - f)$ = equilibrium constant, T_m = temperature at f equal to 0.5, $\Delta H°$ (total) = standard enthalpy of the reaction, $\Delta S°$

(total) = standard entropy of the reaction. The value of $\Delta H°$ can be derived from the van't Hoff equation,

$$\Delta H° \text{ (total)} = 4RT_m^2 \, (\partial f/\partial T)_{T_m} \tag{18}$$

where the slope $(\partial f/\partial T)$ is most accurate at T_m. The equilibrium can be described by

$$f/(1 - f) = \gamma_m s^{N-1} \tag{19}$$

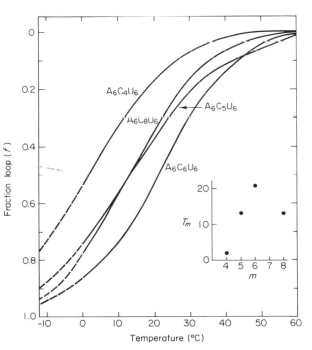

Fig. 58. Loop fraction vs temperature for $A_6C_mU_6$ ($m = 4, 5, 6,$ and 8) in 1 M NaCl, 10 mM sodium phosphate pH 7.0. (From Uhlenbeck *et al.* [120].)

where γ_m is the equilibrium constant for loop initiation for a loop of m unbonded bases,* and s is the equilibrium constant for adding each successive base pair to an existing helix. Again, s is taken to be 1 at T_m^0 (T_m of the poly A·poly U helix). In Fig. 58, the loop fraction vs temperature for $A_6C_mU_6$ ($m = 4, 5, 6$ and 8) in 1 M salt solution is shown; it should be noted that these transitions are independent of oligomer concentration. The temperature

* The symbol γ_m is equivalent to $\gamma(x)$ used by Scheffler *et al.* [102–104] with m equivalent to x. Equation (19) is equivalent to Eq. (17), with $f/(1 - f) = C_{1h}/C_0$ and $\sigma s^n = \gamma_m s^{n-1}$.

dependence of s and γ_m can be described by $\Delta\bar{H}^\circ$ (enthalpy per base pair in helix propagation) and by ΔH_i° (enthalpy of loop formation) respectively, through the following relationships:

$$s(T) = \exp\left[-(\Delta\bar{H}^\circ/R)\left(\frac{1}{T} - \frac{1}{T_m^\infty}\right)\right]$$

$$\gamma_m(T_2) = \gamma_m(T_1)\exp\left[-(\Delta H_i^\circ/R)(1/T_2 - 1/T_1)\right]$$

(20)

Furthermore, applying the van't Hoff equation, the relationship between the T_m of the loop and the thermodynamic parameters can be established by the following equation:

$$\frac{1}{T_m} = \frac{1}{T_m^\infty} + \frac{R}{(N-1)\,\Delta\bar{H}_i^\circ}\ln\gamma_m(T_m)$$

(21)

where T_m^∞ is taken to be 78°C, and $\Delta\bar{H}^\circ$ to be -8.0 kcal in 1 M NaCl as derived from thermodynamic studies on poly A · poly U. With these values, γ_m can be calculated from Eq. (20) at its T_m. The enthalpy of initiation, ΔH_i, for these oligomers can be obtained by subtracting the $(N-1)(\Delta\bar{H}^\circ)$ or -40 kcal/mole from the enthalpy of the total reaction given in Eq. (18). The ΔH_i° can then be used to calculate γ_m at 25°C from γ_m at T_m by Eq. (20). All these thermodynamic parameters are reported in Table XIX.

It is important to note that, while the total ΔH° and the total Δs° are negative, the ΔH_i° and Δs_i° in initiation are both positive. As expected, the

TABLE XIX

EXPERIMENTAL THERMODYNAMIC PARAMETERS FOR THE REACTION $A_6C_mU_6$ (SINGLE STRAND) TO $A_6C_mU_6$ (LOOP), AND CALCULATED THERMODYNAMIC PARAMETERS FOR THE INITIAL BASE PAIRING IN THE LOOP FORMATION [a,b]

Loop	T_m(°C)	$H°$ (total)[c] (kcal)	$S°$ (total)[d] (eu)	γ_m[b,e]	H_i° (kcal)[b,e]	S_i° (eu)[b,e]	G_i° (kcal)[b,e]
				(10^{-5})			
$A_6C_4U_6$	2	-12	-44	0.7	$+28$	$+70$	$+7.1$
$A_6C_5U_6$	13.4	-13	-45	1.6	$+27$	$+69$	$+6.6$
$A_6C_6U_6$	21.0	-16	-54	2.6	$+24$	$+59$	$+6.3$
$A_6C_8U_6$	13.4	-14	-51	1.4	$+26$	$+65$	$+6.6$

[a] From Uhlenbeck *et al.* [120].

[b] The equilibrium constant for the initial base pair formation in loop closure is γ_m with ΔH_i°, ΔS_i°, ΔG_i° as the enthalpy, entropy, and free energy of initiations.

[c] From Eq. (18).

[d] ΔS° (total) $= \Delta H^\circ$ (total)$/T_m$.

[e] γ_m was calculated from Eq. (20) and (21); $\Delta G_i^\circ(T) = -RT\ln\gamma_m(T)$. $\Delta\bar{H}_0$ and T_m were taken to -8.0 kcal and 351°K.

entropic effect of a loop on complementary sequences, by forcing them to be close to one another, did much to stabilize the helix. For instance, the T_m of A_6U_6 is estimated to be $-13°C$ at 10^{-4} M strand concentration, while the T_m of the $A_6C_nU_6$ loops are in the range of $+10°C$ at all strand concentrations. The six-membered loop was the most stable, smaller loops were strained and large loops showed the expected decrease in stability due to the lowering in probability of the meeting of A_6 and U_6. As for the conformation of the residues in the loop, the CD studies suggest that the average conformation of the C residues in the unstrained loops ($A_6C_6U_6$ and $A_6C_8U_6$) is not significantly different from the average single-strand conformation of oligo C residues. In the strained loop (short sequence of C) the average conformation of C residues in the loop $A_6C_5U_6$ apparently is different from $A_6C_6U_6$, $A_6C_8U_6$ and oligo C, since the calculated spectrum is considerably different.

III. Interaction between Oligonucleotides and Polynucleotides

Research on oligonucleotide–polynucleotide interaction has yielded much useful information about nucleic acid chemistry and biology. For instance, experiments on the binding of specific oligonucleotides to ribosomes together with the labeled aminoacyl–tRNA have provided the data for the unambiguous assignment of the amino acid codon triplets in the messenger RNA [121,122], a topic which will be discussed further in the chapter on nucleic acid in Volume IV.

Binding of oligoadenylic acids to poly U and oligoguanylic acids to poly C was first reported by Lipsett and co-workers [123,124]. Subsequently, Michelson and Monny have collected a considerable amount of data on several oligomer–polymer interaction systems [125]. In a recent book, data from the literature on the oligomer–polymer interaction are tabulated [126]. The quantitative analyses of the helix-coil transition of the oligomer-polymer system began after the publication of the papers by Damle [127] and by Crothers [128]. These studies all concern the binding of homooligo-nucleotides to the complementary homopolynucleotides.

A. Association of Oligoinosinate with Polycytidylic Acid—Formation of 1:1 Duplex

The CD spectra of $I_9 \cdot$ poly C, $I_{11} \cdot$ poly C, and poly I $\cdot$ poly C at $0°C$ are shown in Fig. 59 [129]. They are distinctly different from their addition spectra, but are identical to each other within experimental error, showing maxima at 245 and 275 nm and minima at 209 and 261 nm. This is also the case for the CD spectra of complexes of poly C with I_6, I_7, I_8, and I_{10}, which are all nearly the same as that of poly I $\cdot$ poly C. Similarly, the UV spectra of

$I_{6-11} \cdot$ poly C measured at $-4°C$ are nearly identical with that of poly I $\cdot$ poly C, although they are very different from their respective addition spectra. Thus, oligo I $\cdot$ poly C complexes have the same optical properties as those of poly I $\cdot$ poly C. This observation indicates that these complexes have the same stoichiometry and the same conformation as those of poly I $\cdot$ poly C complex.

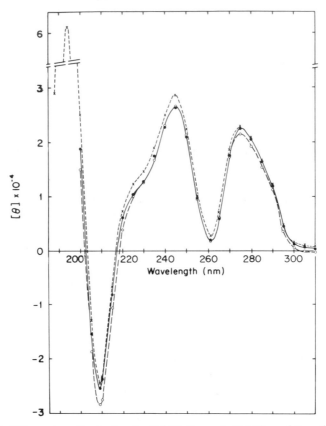

Fig. 59. CD spectra of poly I $\cdot$ poly C ($\bigcirc$), $I_{11} \cdot$ poly C ($\bullet$), and $I_9 \cdot$ poly C ($\times$), at $0°C$ in 0.01 M NaPO$_4$ (pH 7.0)–0.01 M MgCl$_2$. (From Tazawa *et al.* [129].)

This conclusion is supported by the mixing curve for $I_7 +$ poly C, and $I_{11} +$ poly C at $0°C$ (Fig. 60). The absorption spectrum between 220 and 350 nm was measured for every mixture, revealing two isobestic points for each series of mixture, one at 255 ∼ 256 nm and the other at near 262 nm. Every mixing curve at these wavelengths has an intersection at 49 ∼ 50 mole % of poly C. These data, together with the mixing curves at 248 nm which also have intersections around 50 mole % of poly C, clearly indicate that the stoichiometry of the complex is $1I \cdot 1C$. In the following studies, all oligo I's were safely assumed to form only 1:1 complexes with poly C at all nucleotide concentrations used. Indeed, no indication of either 1:2 or 2:1 complex formation was detected from the profile of the melting curves of all oligo I · poly C complexes studied at 1×10^{-5}, 1×10^{-3} M oligo I concentration. The well-defined one-step transition in the melting profiles indicates that oligo I · poly C complexes melt directly to single-stranded oligo I and poly C.

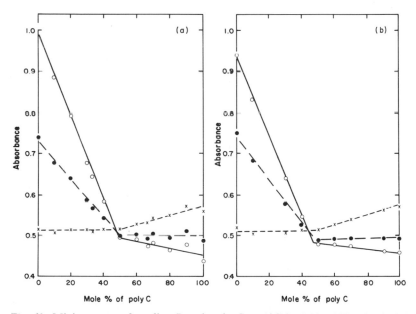

Fig. 60. Mixing curves for oligo I and poly C at 0°C in 0.01 M NaPO$_4$ (pH 7.0)– 0.01 M MgCl$_2$. The total nucleotide concentration is 1×10^{-4} M. (a) I$_7$ + poly C at 248 nm (○), 256.5 nm (●), and 262.3 nm (×). (b) I$_{11}$ + poly C at 248 nm (○), 255.3 nm (●), and 261.5 nm (×). (From Tazawa *et al.* [129].)

The thermal stability of oligo I · poly C complexes is lower than that of poly I · poly C and is dependent on the chain length of the oligomer and the concentration of the complex. An example is given in Fig. 61. By a statistical thermodynamic analysis of the melting data of these complexes, some of the thermodynamic parameters for the melting process (equilibria) can be obtained. The model used in this analysis is that of Magee *et al.* [130], in which oligomers are allowed to bind to the polymers only in an all-or-none fashion. Such a model should be reasonable when the chain lengths of the oligomers are not large [95]. Damle [127] and Crothers [128] both have developed useful equations for oligonucleotide–polynucleotide interaction relating experiment to theory. Since the treatment of Damle and that of Crothers are basically the same, equations from both papers are used in this study depending on which equation is more convenient in application. Symbols from both papers are adopted here and are so identified.

The model of a binding process of an oligomer to a polymer is depicted in Fig. 62. Step 1 represents a nucleation reaction at the end of a preformed

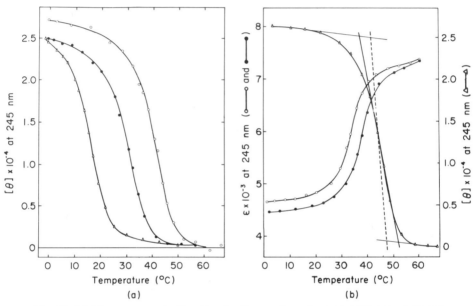

Fig. 61. Melting curves of oligo I·poly C complexes in 0.01 M NaPO$_4$ (pH 7.0)–0.01 M MgCl$_2$ as measured by either CD or UV. (a) I$_6$·poly C ($\triangle$), T$_m$ = 16.8°C; I$_8$·poly C ($\bullet$), T$_m$ = 30.5°C; I$_{10}$·poly C ($\bigcirc$), T$_m$ = 41.1°C at oligomer strand concentration of 1 × 10^{-4} M. (b) I$_{11}$·poly C at oligomer strand concentration of 1.82 × 10^{-6} M ($\bigcirc$), T$_m$ = 34.0°C; 9.1 × 10^{-6} M($\bullet$), T$_m$ = 37.7°C; and 1 × 10^{-4} M ($\triangle$), T$_m$ = 44.5°C. The real slope for I$_{11}$·poly C transition (1 × 10^{-4} M) at T$_m$ is shown as a solid line, and the imaginary slope for a ΔH value of −6.2 kcal/mole for the same transition is shown as a broken line. (From Tazawa *et al.* [129].)

base pair of the nearest oligomer bound to the polymer with an equilibrium constant χs (or $\beta_2 s$ in Crothers' paper). Step 2 represents a growth reaction with an equilibrium constant s, which is assumed to be the same for all subsequent growth reactions. χ is assumed to be independent of both chain length of the oligomer and temperature and represents the entropy consideration and end effects, while s is assumed to be independent of chain length but temperature-dependent. In order to describe the actual binding process, another nucleation reaction has to be considered. This is a nucleation reaction independent of the preformed base pair with the equilibrium constant βs [128], which is not shown in Fig. 62, but has been described in the preceding section.

The dependence of T_m on the oligomer·polymer complexes is given by Damle [127] by the following equation:

$$\ln \chi + \ln C_m = \frac{n \cdot \Delta H}{R}\left(\frac{1}{T_m} - \frac{1}{T_c}\right) \qquad (22)$$

The constant χ is the ratio of the equilibrium constant for forming the first base pair adjacent to an existing base pair versus that for forming subsequent

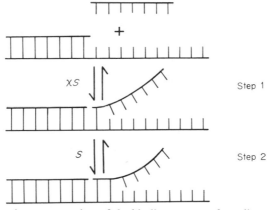

Fig. 62. Schematic representation of the binding process of an oligomer to a polymer. Step 1: a nucleation reaction at the end of preformed base pairs, with an equilibrium constant χs. Step 2: a growth reaction, with an equilibrium constant s. (From Tazawa *et al.* [129].)

base pairs described in Fig. 62. C_m is the strand concentration of the free oligomer at the melting temperature of the complex, T_m. n is the chain length of the oligomer (or χ in Crothers' paper). ΔH is the heat for forming the complex per mole of base pair. T_c is a characteristic temperature at which the equilibrium constant for the growth reaction, s, has a value of unity. Hence $T_c = \Delta H/\Delta S$. The form of Eq. (22) is very similar to that of Eqs. (21) and (13). From Eq. (22), a plot of $1/T_c$ against $\ln C_m$ is predicted to be linear with a slope equal to $R/(n \cdot \Delta H)$. Figure 63 shows the $1/T_m$ vs $\ln C_m$ plot of $I_{6-11} \cdot$ poly C complex. The ΔH value for each complex was calculated from the relationship $\Delta H = R/(n \cdot \text{slope})$ and is listed in Table XX. The ΔH values for these complexes appear to be constant with respect to chain length, with the mean value of -6.2 ± 0.30 kcal/mole of base pair.

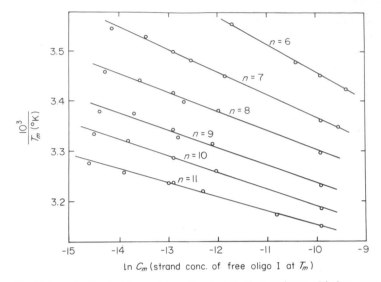

Fig. 63. $1/T_m$ vs $\ln C_m$ plot for various oligo I·poly C complexes, C being strand concentration of the free oligomer at the T_m.

Besides the T_m dependence on concentration, the slope of the thermal melting, especially at the midpoint of the transition, is related to the ΔH. When β/β_2 is very small, and $d\theta/d \ln s'$ is much greater than 1, the following equation has been given by Crothers [128]:

$$\frac{d\theta}{dT} = \frac{n \cdot \Delta H}{2R} \frac{1}{T^2} \tag{23}$$

TABLE XX

$-\Delta H$ FOR OLIGO I·POLY C COMPLEX FORMATION AS CALCULATED FROM THE SLOPES OF LINES IN FIG. 63

Complex	$-\Delta H$ (kcal/mole)
$I_6 \cdot$ poly C	5.87
$I_7 \cdot$ poly C	6.23
$I_8 \cdot$ poly C	6.52
$I_9 \cdot$ poly C	6.35
$I_{10} \cdot$ poly C	6.01
$I_{11} \cdot$ poly C	6.44

where θ is the fraction of polymer base hydrogen-bonded (or f in Damle's paper), $s' = \beta_2 C_m s^n$, and T is the melting temperature. Values of ΔH have been calculated from the slopes $(d\theta/dT)$ at $\theta = 0.5$ or at T_m of the thermal melting profiles of the oligo I·poly C complexes by Eq. (23). An example of the slope measurement (solid line) is shown in Fig. 61 (right) for the CD melting curve of the I_{11}·poly C complex at 10^{-4} M oligomer strand concentration and at T_m 44.5°C. The ΔH values so calculated have a considerable variation, mostly in the range of 3–4 kcal/mole of base pairs [129]. The ΔH values determined by Eq. (23) are definitely smaller than those determined by Eq. (22), and this difference is beyond the experimental error of the slope determination. This point is illustrated in Fig. 61. The broken line passing through the midpoint of the I_{11}·poly C transition (measured by CD) represents the imaginary slope for a ΔH value of -6.2 kcal. The difference between the solid line (real slope) and the broken line (imaginary slope) is too large to be accounted for by experimental error. This phenomenon has been investigated and a solution to this problem in helix-coil transition theory has been proposed by Springgate and Poland [130a].

Calorimetric studies of the poly I·poly C complex formation or dissociation have been reported by Ross and Scruggs [131], and by Heinz, Haar, and Ackermann [132], With a batch microcalorimeter, Ross and Scruggs determined the heat of poly I·poly C complex formation by mixing the two separate polynucleotide solutions at 20° and 37°C and at salt concentration ranging from 0.02 to 0.40 M NaCl. They obtained a ΔH value of -5.6 kcal/mole of base pairs, and found this value to be ionic strength and temperature-independent within the range tested. Heinz *et al.* [132] measured the heat content during the whole transition process with a scanning calorimeter, under conditions of 0.06–1.0 M NaCl with the T_m ranging from 54°–74°C. Under these conditions, they found that the ΔH values vary between -6.5 and 8.4 kcal/mole from low ionic strength (0.06 M) and T_m (54°C) to high ionic strength (1 M) and high T_m (74°C). From the linear extrapolation of the dependence of ΔH on T_m, one would obtain a ΔH value of about 5–6 kcal at T_m of 40°–50°C. Thus, the apparent disagreement between these two laboratories in the ΔH values of poly I·poly C complex could be due to differences in experimental conditions, such as measurements made at different temperature. The ΔH value (average: -6.2 kcal/mole) for the oligo I·poly C complexes measured by T_m dependence on oligomer concentration is between the two values for the ΔH of poly I·poly C complex, i.e., -5.6 kcal (measured at 20° and 37°C) and -6.4 kcal (at T_m 54°C) reported by Ross and Scruggs [131] and by Heinz *et al.* [132], respectively. The T_m's of the oligo I·poly C complexes are between 8° and 45°C. It appears, therefore, that the ΔH values so obtained are nearly identical to those measured calorimetrically for the poly I·poly C complex. The ΔS value for the oligo I·poly C complex determined in this study (-17.4 cal/deg-mole) is, from definition of s and T_c, for the growth reactions and independent of oligomer chain length. This value is identical to that determined for poly I·poly C by Ross and Scruggs (-17.5 cal/deg-mole) and is also comparable to that obtained for poly I·poly C by Heinz *et al.* (-18.6 cal/deg-mole when corrected for ΔH value of -6 kcal/mole).

The stability of the oligo I·poly C complex is markedly lower than that of the oligo A·2 poly U complex. Since poly I·poly C has almost the same T_m as poly A·poly U under similar conditions, the large difference in stability of the oligo I·poly C complex compared with the oligo A·2 poly U complex is probably not due to the difference in base pairs involved. It appears, therefore, that the difference in the strandedness is mainly responsible for the difference in stabilities of these two oligomer–polymer interacting systems. In the case of the polymer–polymer interaction this difference in strandedness does not affect the stability of the complex significantly. The effect of increase in ΔH of the triple-stranded complex compared with that of the double-stranded complex is cancelled out by the same degree of increase in ΔS ($\Delta H = -12 \pm 1.5$ kcal/mole, $\Delta S = -35 \pm 4$ cal/deg-mole for poly A·2 poly U; $\Delta H = -8 \pm 1$ kcal/mole, $\Delta S = -23 \pm 3$ cal/deg-mole for poly A·poly U) resulting in no large difference in T_m. In the oligomer–polymer system, however, the increase in ΔS of the triple-stranded complex (one oligomer strand + two polymer strands) compared with that of the double-stranded complex (one oligomer strand + one polymer strand) is probably less than the increase in ΔH, leading to considerably more stability for the triple-stranded complex compared with the double-stranded complex. The reason is that, while the values of ΔH of the oligomer– (or monomer–) polymer complexes (both double- and triple-stranded) are the same as those of polymer–polymer complexes, the ΔS values of the oligomer–polymer complexes are considerably larger than those of polymer–polymer complexes (ΔS for adenosine·2 poly U is -42 cal/deg-mole), resulting in less stability for the former complex than the latter. This larger ΔS reflects the larger entropy loss of the oligomer (or monomer) compared with the polymer. This effect of the oligomer in increasing ΔS should be more profound for the double-stranded oligomer–polymer complex than for the triple-stranded one, which will lead to less stability for the double-stranded (oligomer· polymer) complex compared with the triple-stranded (oligomer·2 polymer) complex.

The chain length dependence of T_m is also described by Eq. (22). Figure 64 shows $1/T_m$ vs $1/n$ plots. The predicted linearity was indeed observed for both values of C_m, two straight lines extrapolating to the same value, 85.7°C ($T_c = 358.7$°K). From the relation $T_c = \Delta H / \Delta S$, ΔS was calculated to be

-17.4 ± 0.6 (cal/deg-mole). It might be expected that T_m of poly (I)·poly (C) is the same as T_c. However, T_m of poly (I)·poly (C) was found to be 74.4°C. Equation (22) is basically the same as Eq. (14).

From the slope of the lines χ can be obtained, since ΔH is already known and furthermore it is constant. Calculated values of χ are 6.9×10^{-2} liter/mole (from $C_m = 2.5 \times 10^{-6}\ M$), and 7.6×10^{-2} liter/mole (from $C_m = 5 \times 10^{-5}\ M$).

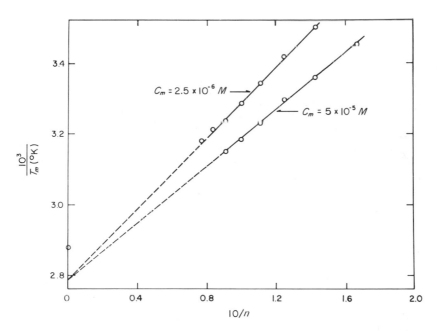

Fig. 64. $1/T_m$ vs $1/n$ plot at two different oligomer strand concentrations.

It should be noted that Eq. (22) from Damle [127] is essentially the same as that given by Crothers [128] in a differential form:

$$\left(\frac{\partial(1/T_m)}{\partial(1/\chi)}\right)a_m = \frac{R \ln (a_m\beta_2)}{\Delta H} \tag{24}$$

where χ, a_m, and β_2 defined in Eq. (24) are equivalent, respectively, to n, C_m, and χ, defined in Eq. (22). Therefore, the same values of ΔH and χ (or β_2) can be obtained from both formulations. However the approach of Crothers also leads to the experimental evaluation of s_m by

$$\left(\frac{\partial \ln a_m}{\partial \chi}\right)T_m = -\ln s_m$$

where s_m is the equilibrium constant for the growth reaction at T_m. From data presented in Fig. 63, various lines in a plot of ln c_m vs n at T were obtained from the range of 285.7°K (or 12.7°C) to 307.7°K (34.7°C). The s_m values at various T_m, which are the slopes of these lines, vary from 9.3 at 12.7°C to 4.8 at 34°C. As may be expected, the s_m values are progressively smaller at higher T_m. The values of β_2 can also be calculated from s_m [128]; varying from 3 to 9×10^{-2} M^{-1}, they are in reasonable agreement with those (7–7.5×10^{-2} M^{-1}) determined by Eq. (24). In the paper by Craig, Crothers, and Doty [119] described in the preceding section, the βs value for the dimeric $A \cdot U$ oligomer complex was found to be around 1×10^{-3} M^{-1}, where βs is the equilibrium constant of nucleation reaction independent of the preformed base pair. The $\beta_2 s$ determined here for 1:1 poly (C) oligo (I) complex is about 0.7 M^{-1} at 285°K. The effect of difference in base pairs ($A \cdot U$ vs $I \cdot C$) on these values is probably not large; therefore this difference between $\beta_2 s$ and βs of nearly three orders of magnitude represents the stabilization due to stacking next to an existing base pair.

The extrapolated T_m (~ 86°C) for a complex of poly (C) with oligo (I) of infinite chain length ($1/n = 0$) is significantly higher than that measured experimentally for poly (I)·poly (C) (~ 75°C). The deviation from linearity in the $1/T_m$ vs $1/n$ plot actually can be observed starting from $I_{12} \cdot$ poly (C) and $I_{13} \cdot$ poly (C) complexes. Among various possible causes, one would be the self-interaction of the inosine residues in the single-stranded poly (I) or oligo (I) of long chain length. Such a self-interaction would reduce the activity of the oligo (I) or poly (I) for complex formation and would reduce the value of T_m.

It should be noted that, while there are many similarities in chemical properties (the main difference being stability) between the oligo I·poly C complex and the poly I·poly C complex, biologically these complexes can be quite different. The poly I·poly C complex is an effective interferon inducer to human cells, while the helical oligo I·poly C complex is not [133].

B. ASSOCIATION OF OLIGOADENYLATE WITH POLYURIDYLIC ACID—FORMATION OF 1:2 TRIPLEX

While the oligo A·poly U association has been studied earlier [123,125], quantitative analyses of the helix–coil transition data began with the thesis of Uhlenbeck [134] and the treatment by Damle [127]. Cantor and Chin [135] showed that the CD spectra of the 2:1 complex of $A_{2-5} \cdot$ poly U are identical to that of poly A·2 poly U, indicating that the conformation of the oligo A·2 poly U complex is the same as that of the poly A·2 poly U complex.

Study of the salt effect on melting [134] showed that the T_m of $(Ap)_4 A \cdot$ poly U and of $(Ap)_9 A \cdot$ poly U are linearly proportional to log concentration of NaCl up from 0.01 to 2 M. However, the slopes of the T_m vs log [NaCl] lines increase as the chain length increases. Thus, the slope varies from 22.7°/log [NaCl] for $A_5 \cdot$ poly U complex, to 25.2 for $A_{10} \cdot$ poly U to 26.0 for poly A $\cdot$ poly U. This slight increase of slope with chain length probably reflects the change in phosphate to base ratio from 0.8 for $(Ap)_4A$ to 1.0 for poly A. In addition, there is a higher concentration of negative charges within the volume occupied by the polymer. The charge effect is also reflected by the observation that the $(Ap)_4A$ > p·poly U complex does have a lower T_m (by 7.6°C at 0.25 M NaCl) than the $(Ap)_4A \cdot$ poly U complex.

For the thermodynamic analysis of the three-stranded oligomer–polymer complex (1:2), Damle [127] stated that "In the case of the three-stranded complex, two possibilities arise regarding the state of the nonbonded segments of the polymer strands: (1) they already exist in a rigid helical structure and undergo very little conformational change upon binding of oligomers, and (2) they exist as random chains or rings and assume rigid helical structure only when bonded to oligomers." The first case is obviously similar to the two-stranded complex in that only the nearest neighbor interactions exist, and hence all the equations derived for the two-stranded complex would apply to this case. It should be noted that while poly U does not have a great tendency to form a hairpin helix, the T_m of the poly U helix is lower than that of the oligo A poly·U complex. For the second case, the situation is more complex, and detailed equations have been given both by Damle [127] and by Crothers [128].

Uhlenbeck [134] had obtained linear plots of $1/T_m$ versus log total nucleotide concentration for complexes of poly U with A_4, A_5, A_6, and A_8 in 0.25 M NaCl–0.02 M Na phosphate, pH 7.1. In applying Eq. (22) [or Eq. (13)] for the analysis of the slopes of these lines in the plot, the value of the ΔH per mole of base pair was found to be 13–13.5 kcal, while the value of ΔH obtained by calorimetry on the poly A·2 poly U complex was -12.7 ± 0.3 kcal/mole at the same salt concentration [136]. This agreement appears to justify the application of Eq. (22) which was formulated originally for the two-stranded complex. A linear $1/T_m$ vs $1/n$ plot at constant strand concentration was also obtained by Uhlenbeck [134]. In addition, he pointed out that the reason for the previous failure of Michelson and Monny [125] in obtaining a linear plot of $1/T_m$ vs $1/n$ was due to their use of constant residue concentration (or constant optical density) instead of constant strand concentration as the basis of handling the data. With the ΔH value taken as 13.5 kcal, the χ value [Eq.(22)] was calculated to be 2 liters/mole. This χ value is considerably larger than that (0.07 liter/mole) observed for the two-stranded oligo I· poly C complex.

The published data on the oligo A·poly U complex obtained by Lipsett *et al.* [123] and by Michelson and Monny [125] were analyzed by Damle [127].

The ΔH value calculated for the oligo A·2 poly U complex in 0.15 M Na$^+$, 0.01 M Mg^{2+}, pH 7 [125] was -12.4 ± 2.5 kcal/mole with a χ value $5 \times 10^{1\pm1}$ liters/mole. Calculation on other sets of data afforded similar results.

Cassani and Bollum [137] have investigated the oligodeoxythymidylate· polydeoxyadenylate (dT·poly dA) and the oligodeoxyadenylate·poly-deoxythymidylate interaction (poly dT·dA) systems. When oligo T and poly dA are mixed in equal proportions, stable double-stranded complexes are rapidly formed. When the poly dA is mixed with a double amount of oligo T, triple-stranded complexes containing 1A and 2T are also formed at a permissive temperature. The triple-stranded complex is less stable than the double-stranded complex. It is interesting to note that the stoichiometry of the interaction is affected by the backbone of the polynucleotide. For instance, in 1 M LiCl–0.01 M cacodylate buffer, pH 6.8, 23°C, the mixing curves showed that d(pT)$_{10}$ forms a double-stranded complex with poly dA, but a triple-stranded complex with poly rA (2T·1A). On the other hand, interaction of oligo dA with poly T always affords a triple-stranded complex (1A:2T) even in a 1:1 mixture under the experimental conditions used [137]. A linear plot of $1/T_m$ versus log concentration of the free oligo T at T_m (or half of the total strand concentration) was obtained either with a fixed stoichiometry (1:1) and varying poly dA concentration or with a fixed poly dA concentration and varying stoichiometry. The calculation on the slope of this line yielded a ΔH value of -9.1 ± 0.1 kcal/mole of base pair. This value is to be compared with the ΔH of -8 ± 1 kcal/mole obtained from calorimetrical measurement of poly A·poly U complex [136].

C. Effects of the Noncomplementary Bases on the Oligomer–Polymer Association

The effect of the noncomplementary bases on the formation of the oligomer–polymer complex was investigated extensively by Uhlenbeck [134] and some of these results were published earlier [138]. The system investigated was the formation of the oligo A·2 poly U complexes with noncomplementary bases (U, G, and C) attached to the termini or inserted at the center of the oligo A.

We shall first consider the influence of the terminal noncomplementary residues. Experiments from the mixing curve indicate that the stoichiometry of 1A to 2U is preserved in the complex formation of poly U with the oligo A containing the terminal, noncomplementary residues. For instance, the stoichiometry for the complex of poly U with CpG(pA)$_3$, CpG(pA)$_4$, (Ap)$_4$A(pU)$_2$, and (Ap)$_4$A(pU)$_3$ was found to be 0.53, 0.56, 0.59, and 0.56, respectively, as expected from the ratio of A residue in oligomer vs U residue in poly U. Similarly, the stoichiometry for the poly U complex with $(Ap)_4N$ or $N(pA)_4$ (N = C,G,U) was found to be 0.60–0.64, in accordance with the theoretical value of 0.615. This observation implies that all the U residues in poly U only form base pairs with A residues in the oligomer and were not occupied by the terminal, noncomplementary bases. This conclusion is also supported by the observation that the extent of maximal hypochromicity per residue of poly U due to the complex formation is also constant, regardless of whether the oligo A contains terminal unmatching bases. The result again indicates that *all* the U residues of the polymer are involved in the helix formation. It may be concluded, therefore, that the terminal, noncomplementary residues attached to the oligo A are looped out upon the complex formation between the A residues in the oligomer with all the U residues in poly U.

Bautz and Bautz [139] were the first to report that the 3'-terminal, unmatched residues significantly lower the melting temperature of the oligo A·poly U complex. The data obtained by Uhlenbeck [134] are presented in Table XXI. Addition of a -pU residue to $(Ap)_4A$ lowers the T_m to about the same extent as the addition of a doubly ionized phosphate group, but more than the addition of a singly ionized cyclic phosphate group. This effect is reduced in high salt. Addition of more -pU residues has a lesser effect, since these additional residues presumably turn away from the helix. It is interesting to note that this destabilization effect of the noncomplementary residue is base specific, and that this base specificity, in turn, is dependent on the location of the unmatched residue. For the unmatched residues located at the 3'- terminus, the order of the destabilization effect is U > G > C, while for those located at 5'- terminus, the order is G ≥ U > C. The data in Table XXI also indicate that the 3'- terminal unmatched residue has a larger destabilization than the 5'- terminal, unmatched residue. The slope in the $1/T_m$ vs the log [Conc] plot for $(Ap)_3A$ was found to be virtually the same as that for $(Ap)_4U$ at 0.27 or 1.0 M salt, indicating that the enthalpy of the helix formation has not been significantly affected by the terminal noncomplementary residue. However, the slope in the $1/n$ vs $1/T_m$ plot for $(Ap)_{n-1}A$ is higher than that for $(Ap)_nU$. These results suggest that the destabilization may be due to entropy or to the reduction in χ value which describes the nucleation reaction at the end of a preformed pair of the nearest oligomer bound to the polymer (Fig. 62).

The effect of the internal, noncomplementary residue was investigated with the $(Ap)_nU(pA)_m$ series and a few of $(Ap)_nG(pA)_m$ series [134]. The stoichiometry data from mixing curves and the hypochromicity data from the maximal complex formation also indicate that the internal, unmatched residue is looped out and does not form a base pair with poly U.

The most direct way of comparing the effect on T_m of the *internal*, noncomplementary residue to that of the *terminal*, noncomplementary residue is illustrated in Fig. 65, cited from the thesis of Uhlenbeck [134]. In this figure, the T_m's of a series of "position isomers" are shown as a function of the position of unmatched U residue inside the oligomers for four different chain lengths. The destabilizing influence of the mismatched U residue is the

TABLE XXI

EFFECTS OF TERMINAL, NONCOMPLEMENTARY RESIDUES
ON THE T_m OF OLIGOMER–2 POLY U
COMPLEX [a]

Oligomer	T_m in 0.27 M NaCl (°C)	T_m in 1.0 M NaCl (°C)
(Ap)₄A	27.8	40.0
(Ap)₄A > p	25.8	39.7
(Ap)₄Ap	20.2	36.0
(Ap)₄ApU	20.5	34.9
(Ap)₄ApUpU	17.0	31.4
(Ap)₄ApUpUpU	15.7	30.5
(Ap)₃A	19.9	31.8
(Ap)₄U	12.1	22.8
(Ap)₄G	13.2	23.7
(Ap)₄C	16.3	27.1
(Ap)₄A	27.8	40.0
(Ap)₄A > p	25.8	39.7
(Ap)₄ApU	20.5	34.9
(Ap)₄ApG	21.7	—
(Ap)₄ApC	23.9	—
U(pA)₄	15.2	—
G(pA)₄	14.4	—
C(pA)₄	20.2	—
U(pA)₅	23.6	—
G(pA)₅	23.0	—
C(pA)₅	28.6	—

[a] From Uhlenbeck [134].

greatest when it is located two residues away from the 5'- terminus or one residue away from the 3'- terminus. This is most clearly illustrated in the U(pA)₅–(Ap)₅U series. When the oligo A contains as many as seven A residues, the position-effect becomes less important, even though the U(Ap)₇ or (Ap)₇U has a T_m value a few degrees lower than (Ap)₆A. The destabilizing effect of G as an internal, unmatched residue is still slightly less than that of U, similar to the base specificity observed for the situation of the terminal unmatched residue. It should be remembered that the situation discussed here is that for a three-stranded complex. The destabilizing effect of the noncomplementary residue is expected to be much greater for a double-stranded complex.

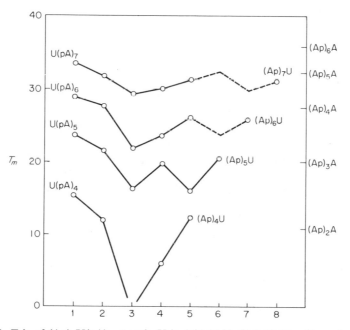

Fig. 65. T_m's of $(Ap)_nU(pA)_m$ + poly U in 0.25 M NaCl–0.02 M sodium phosphate, pH 7.1, as "function" of the position of the U residue in the oligomer. The T_m's of $(Ap)_5UpA$, $(Ap)_6UpA$, and $(Ap)_5U(pA)_2$ (in dotted line) were estimated from the other $(Ap)_nUpA$'s and $(Ap)_nU(pA)_2$'s. (From Uhlenbeck [134].)

D. Association of Complementary Oligonucleotides with Nucleic Acid—Noncooperative Binding

The discussion in the previous three sections mainly concerns the *cooperative* binding of the homooligonucleotides to the homopolynucleotides. In our study [140] on the $(Ap)_6A \cdot$ poly U system, it was found that the formation of the $A_7 \cdot 2$ poly U complex with poly U in excess of 10- to 20-fold is definitely not completely cooperative, but could be partially cooperative. However, the stoichiometry of 1A to 2U always holds regardless of the ratio of the oligo A to the poly U in solution. The results suggest that there is an average of two A_7 oligomers in a stack for an input A/U ratio of 0.05 and three A_7 oligomers in a stack for the input ratio A/U of 0.1. The research discussed in this section concerns the binding of oligonucleotides (usually not homooligomers) with synthetic heteropolynucleotides and nucleic acids in excess. There is little cooperativity between the binding of one oligomer and another, since binding often takes place with separate nucleic acid molecules. At present, the main purpose of this approach is not to investigate the physical chemistry of nucleic acid (such as the experiments described in the above three sections), but to use this procedure as a probe to describe the conformation of nucleic acid, especially RNA. This technique of complementary oligonucleotide binding has been developed rather successfully as a means of identifying unshielded, single-stranded regions of a tRNA or 5 S RNA molecule.

The most important requirement in carrying out this investigative procedure is the availability of a large number of highly radioactive oligonucleotides (trimer to pentamer) of desired sequences. This was made possible by the use of primer-dependent polynucleotide phosphorylase in an addition reaction, in which the radioactive, ribosyl nucleoside diphosphates of high specific activity are added to the di- and trinucleotides isolated from RNA digestion (usually obtained commercially). Following the initial interest in the codon synthesis [122], such technique was redeveloped by Thach in Doty's laboratory [141,142] and later perfected there by Uhlenbeck and Lewis [134,143–146]. The binding has been measured mainly by equilibrium dialysis in a very small volume (100–200 μl). In order to increase the binding constant and to suppress the Donan effect, the dialysis was usually carried out at low temperature (0°–5°C) and in strong salt (such as 1.0 M NaCl, 10 mM MgCl$_2$, and 10 mM Na$_2$ HPO$_4$, pH 7.0). The molar association constant, K, of an oligomer to a single, complementary site on the RNA molecule is defined as

$$K = \frac{[\text{oligomer}]_{\text{bound}}}{[\text{oligomer}]_{\text{free}} \cdot [\text{RNA}]_{\text{free}}}$$

If RNA concentration used is in great excess of the oligomer concentration (say, 100-fold), then K is simply related to the ratio (R) by the following equation [143], where

$$R = \frac{[\text{oligomer}]_{\text{free}} + [\text{oligomer}]_{\text{bound}}}{[\text{oligomer}]_{\text{free}}}$$

which is the ratio of the oligomer concentration in the chamber with the RNA to that in the chamber without RNA,

$$R = 1 + K[\text{RNA}]$$

If an oligomer binds to more than one site on the RNA, the observed K will be a sum of the K values of the various sites. The number of sites per RNA molecule can be ascertained through a conventional Scatchard plot with the data extrapolated to the saturation condition of these sites. The observed value of K for an oligomer can also be affected by the presence of a competing oligomer. In this case, the measured association constant, K_{app}, is related to the association constant of the competitor, K_c, the free competitor concentration [C], and the actual constant of the oligomer K_0 by the expression [145],

$$K_{app} = \frac{K_0}{1 + K_c [C]}$$

1. *Binding to Synthetic Polynucleotides*

In the thesis of Lewis [146], some basic experiments on the binding of oligomers to the synthetic polynucleotides were described. These experiments provide some valuable information about the specificity of the binding process. The following values of K (M^{-1}) were obtained at 0°C, 1 M NaCl, 5 mM EDTA and 10 mM cacodylate, pH 7.2:4,2, and 20,400 for the binding of A_4 to poly A, poly C and poly U, respectively; 4, 2, 2, and 28,900 for C_4 to poly A, poly C, poly U, and poly ($U_{56\%}$, G) respectively; 12, 3, 100, and about 2500, for G_2 to poly A, poly U, poly C and poly ($U_{56\%}$, G), respectively; 2, 2, 60, 250, and 353 for U_5 to poly C, poly U, poly ($U_{62\%}$, C), poly ($U_{56\%}$, G) and poly A, respectively; and 49 for U_4 to poly A. In the calculation of the K values for the binding to mixed polynucleotides, such as poly ($U_{56\%}$, G) and poly ($U_{62\%}$, C) the fraction (f) of the residues in the mixed polymer involved in the unique site sequence has to be determined. It has been computed on the assumption that the mixed polymer has a random sequence and there is only one binding site consisting of a unique sequence. For example, the fraction so calculated for the binding of C_4 to poly ($U_{56\%}$, G) is 0.189, which represents the fraction of -GGGGG- sequence. Similarly, the fraction computed for the binding of U_5 to the poly ($U_{62\%}$, C) is 0.092, representing -UUUUU- sequence, to poly ($U_{56\%}$, G) is 0.016, representing -GGGGG- sequence. The validity of these assumptions remains to be tested and some of the computation appears to be arbitrary. For instance, -UUUUU- sequence was taken as the binding site for U_5 in poly ($U_{62\%}$, C) but -GGGGG- sequence was taken as binding site for U_5 in poly ($U_{56\%}$, G). This uncertainty is the most serious drawback of this approach.

The binding of A_4 to poly U was found to be dependent on polymer/ oligomer ratio, the K value was dropped from 20,400 to 3000 when the polymer/oligomer ratio increased by 20-fold. This observation indicated that the stacking effect between the nearest-neighbor was still in operation for the oligo A–poly U binding experiments, resulting in the high K value observed. On the other hand, the binding of GG to poly C was found to be independent of polymer/oligomer ratio with a low K value of 100 (M^{-1}). It should be noted that at low temperatures and in high salt, poly U forms a double-stranded hairpin helix; therefore, this may be the reason the binding process of U_5 to poly U is lower than to poly ($U_{62\%}$, C) which does not form a helix. Similarly, this may explain why the binding of G_2 to poly U is lower than to poly ($U_{56\%}$, G). Also, in the binding of G_2 to poly ($U_{56\%}$, G), the -GG- sequence in the mixed polymer may also serve as binding site, since G–G interaction has been well documented (Section II,B,2). The strong interactions observed in the binding are those of the A–U and G–C base pair, besides the possible G–C pair.

Lewis [146] has surveyed the effects of polynucleotide composition and oligonucleotide chain length on binding (Table XXII). Binding of GG to poly (A, C) is affected by the ratio of A/C in the polymer. Decrease of C in the composition brings out only a small drop in the observed (R) value; however, as a result of the decrease in the calculated (f) value, the binding constant (K) becomes much larger for poly (A, C) which is rich in A and poor in C (Table XXII). The binding of GG to poly (A, C) was found to be also significantly higher than to poly (U, C). At present, these observations are not well understood, and should be regarded as a complexity in the binding studies. The effect of oligonucleotide chain length on binding is also shown in Table XXII. As expected, both the observed R values and the binding constants are substantially increased by the addition of a complementary residue to the oligonucleotide. In general, the addition of such a residue increases the K value by 10- to 20-fold. On the other hand, addition of one or two noncomplementary residues (A, C, U) at the termini of GG, did not cause a large decrease in K values; in fact, often a small increase was observed (Table XXII). Part of the reason for this increase is probably due to a lowering of GG self-aggregation upon the addition of the noncomplementary residue (especially C or U). Thus, the activity of the oligomer available for the binding process becomes correspondingly higher. The above examples represent the various complexities of the binding process to these synthetic polynucleotides which have been surveyed by Lewis [146]. Based on these observations, he set up arbitrary levels of binding constants at 0°C, 1 M salt and at pH 7.0 to decide whether the binding of complementary oligomers to a specific site in nucleic acid has taken place or not. The lower limit of the binding of a trimer with the capability of forming two G–C pairs and the binding of a tetramer with one G–C pair was set to be $K = 1000$, and the lower limit of a tetramer with two G–C pairs was set to be $K = 3000$.

2. *Binding to Nucleic Acid*

As mentioned earlier in the introduction of this section, this technique of complementary oligonucleotide binding has been used to characterize the unshielded, single-stranded regions of tRNA and 5 S RNA.

TABLE XXII

EFFECTS OF POLYNUCLEOTIDE COMPOSITION AND OLIGONUCLEOTIDE
CHAIN LENGTH, ON THE BINDING [a]

Oligonucleotides	Polynucleotides	R [b]	f [c]	K
GG	$(A_{14\%}, C_{86\%})$	3.77	0.741	400
	$(A_{55\%}, C_{45\%})$	2.68	0.429	2,380
	$(A_{82\%}, C_{18\%})$	2.86	0.033	5,770
	$(U_{28\%}, C_{72\%})$	2.10	0.514	242
	$(U_{62\%}, C_{38\%})$	1.17	0.146	770
AAA	$(U_{62\%}, G_{38\%})$	3.48	0.236	3,850
AAAA	$(U_{62\%}, G_{38\%})$	15.26	0.146	36,200
CC	$(U_{56\%}, G_{44\%})$	1.04	0.189	43
CCC	$(U_{56\%}, G_{44\%})$	1.83	0.085	1,940
CCCC	$(U_{56\%}, G_{44\%})$	7.04	0.037	28,900
UUUU	$(A_{100\%})$	1.27	1.00	49
UUUUU	$(A_{100\%})$	2.83	1.00	353
GG	$(C_{100\%})$	—	1.00	100
GGG	$(C_{100\%})$	3.4–5.7	1.00	500–1,000
AGG, CGG, UGG	$(C_{100\%})$	—	1.00	40–100
GG	$(U_{56\%}, G_{44\%})$	3.4	0.189	2,440
GGG	$(U_{56\%}, G_{44\%})$	5.10	0.085	21,100
GG	$(A_{14\%}, C_{86\%})$	3.77	0.741	400
GGG	$(A_{14\%}, C_{86\%})$	7.44	0.638	2,550
AGG, GGA, CGG, GGC	$(A_{14\%}, C_{86\%})$	—	0.741	240–500
GG	$(A_{82\%}, C_{18\%})$	2.86	0.033	5,770
GGG	$(A_{82\%}, C_{18\%})$	7.97	0.006	452,000
AGG, GGA, AAGG, GGAA	$(A_{82\%}, C_{18\%})$	—	0.033	3,600–6,000
CGG, GGC, GGCC, AGGG	$(A_{82\%}, C_{18\%})$	—	0.033	3,000–8,000
GG	$(U_{28\%}, C_{72\%})$	2.10	0.514	240
GGG	$(U_{28\%}, C_{72\%})$	6.58	0.369	5,000
CGG, GGC, GGU, UGG	$(U_{28\%}, C_{72\%})$	—	0.514	240–980
GG	$(U_{62\%}, C_{38\%})$	1.71	0.146	720
GGG	$(U_{62\%}, C_{38\%})$	6.05	0.056	31,400
CGG, GGC, GGCC, UGGC	$(U_{62\%}, C_{38\%})$	—	0.146	700–2,600

[a] From Lewis [146]. In 1 M NaCl, 5 mM EDTA, 10 mM cacodylate, pH 7.2, at 0°C.

[b] R = observed ratio of the oligomer concentration in the chamber with the RNA to that in the chamber without RNA.

[c] f = calculated fraction of the residues in the mixed polymer involved in the unique site sequence.

In a preliminary experiment, Uhlenbeck *et al.* [143] identified the sequence of -UACU- in the anticodon loop of *E. coli* fMet–tRNA as an unshielded region available for binding. In a subsequent paper, Uhlenbeck [145] reported a systematic investigation on tRNA^{tyr} and tRNA^{fMet} from *E. coli* by this technique. The linear dependence of R values versus tRNA concentration used in the binding experiment was established. At 0°C and in 1 mM EDTA, 10 mM Na$_2$HPO$_4$, pH 7.0, the binding constant (K) of UACA oligomer to tRNA^{tyr} increased from 50,000 to 100,000 when the NaCl concentration increased from 0.2 M to 1.0 M. A Scatchard plot for the binding GGU

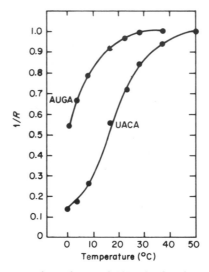

Fig. 66. The temperature dependence of $1/R$, the fraction of free oligomers in the binding of AUGA to tRNA^{fMet} and of UACA to tRNA^{tyr} at 61 μM tRNA, 1.0 M NaCl, 10 mM MgCl$_2$, 10 mM Na$_2$HPO$_4$, pH 7.0. (From Uhlenbeck [145].)

and GGC was made, and the analyses indicated that there is only one binding site for each GGU and GGC in the tRNA^{tyr} with the K value of 40,000 and 83,000, respectively. These values determined from the slope of the Scatchard plot agreed well with the values of 48,000 and 100,000 determined at very low oligomer concentrations. The effect of temperature on the binding of AUGC to tRNA^{fMet} and UACA to tRNA^{tyr} is shown in Fig. 66. The enthalpies of the binding, calculated from the van't Hoff equation, are 22.4 and 21.0 kcal/mole, respectively, for UACA and AUGC interaction. After surveying the binding of over 60 oligonucleotides, the unshielded regions of tRNA^{tyr} were mapped. Similar experiments were done with the tRNA^{fMet} and 5 S RNA [144,146]. It is interesting to note that when the tRNA^{isoleu} is complexed with the tRNA^{isoleu} synthetase, oligomers cannot be hybridized to either the anticodon section or the CCA sequence at the 3′ terminus [147]. This suggests that both the anticodon region and the 3′- terminus are covered or shielded in the same way by the interaction with the enzyme. This type of experiment with yeast and *E. coli* tRNA^{fMet} and with yeast tRNA^{phe} has also been done by others [148–150].

Binding of trinucleotide UUC to the yeast tRNA^{phe} has been measured by a fluorescence assay based on the quenching of the emission of Y base upon interaction [151]. The data on the dependence of binding on temperature yielded the values of $-(15 \pm 1.5)$, $-(13 \pm 1.5)$, and $-(3.7 \pm 0.2)$ kcal/mole, respectively, for ΔH, $T \Delta S$ and ΔF. The relatively weak binding of UUC to tRNA^{phe} is due to the large entropy term in opposing the binding process. This reasoning is supported by the study [152] on the interaction between tRNA^{phe} (anticodon GAA) and tRNA^{glu} (anticodon UUC). The thermodynamic parameters of this tRNA–tRNA interaction through the base pairing of the two anticodon trinucleotide residues were found to be -5.2, $+2.0$, and -7.2 kcal/mole, respectively, for ΔH, $T \Delta S$, and ΔF. The large ΔF value $(-7.2$ kcal$)$ in the RNA (GAA)–RNA (UUC) interaction $(-3.7$ kcal$)$ is due to the change of the entropy term $(T \Delta S)$ from -10.6 kcal to $+2.0$ kcal (a gain of 13 kcal) in spite of the loss of ΔH term from -14 kcal down to -5.2 kcal (a loss of 9 kcal). The result of this investigation shows the importance of the steric factor in the interaction as expressed in the entropy term.

References

1. M. Sundaralingam, *Biopolymers* **7**, 821 (1969).
2. S. Arnott, *Prog. Biophys. Mol. Biol.* **21**, 265 (1970).
3. A. V. Lakshminarayanan and V. Sasisekharan, *Biopolymers* **8**, 475 and 489 (1969).
3a. W. K. Olson and P. J. Flory, *Biopolymers* **11**, 1 (1972).
4. S. Arnott, S. D. Dover, and A. J. Wonacott, *Acta Crystallogr., Sect. B* **25**, 2192 (1969).
5. S. Arnott and D. W. L. Hukins, *Nature (London)* **224**, 886 (1969).
6. V. Sasisekharan and A. V. Lakshminarayanan, *Biopolymers* **8**, 505 (1969).
6a. M. Sundaralingam, *in* "Conformation of Biological Molecules and Polymers" (E. D. Bergmann and B. Pullman, eds.), p. 417. Academic Press, New York, 1973.
7. E. Shefter, M. Barlow, R. Sparks, and K. Trueblood, *J. Amer. Chem. Soc.* **86**, 1872 (1964).
8. E. Shefter, M. Barlow, R. Sparks, and K. Trueblood, *Acta Crystallogr., Sect. B* **25**, 895 (1969).
9. N. C. Seeman, J. L. Sussman, H. M. Berman, and S. H. Kim, *Nature (London) New Biol.* **233**, 90 (1971); J. L. Sussman, N. C. Seeman, S. H. Kim and H. M. Berman, *J. Mol. Biol.* **66**, 403 (1972).
10. J. Rubin, T. Brennan, and M. Sundaralingam, *Science* **174**, 1020 (1971).
10a. R. O. Day, N. C. Seeman, J. M. Rosenberg, and A. Rich. *Proc. Nat. Acad. Sci.* **70**, 849 (1973).

10b. J. M. Rosenberg, N. C. Seeman, J. J. P. Kim, F. L. Suddath, H. B. Nicholas, and A. Rich. *Nature (London)* **243**, 150 (1973).

11. P. O. P. Ts'o, *in* "Fine Structure of Biological Macromolecules" (G. Fasman and S. Timasheff, eds.), p. 49. Dekker, New York. 1970.

12. F. E. Hruska, A. A. Grey, and I. C. P. Smith, *J. Amer. Chem. Soc.* **92**, 214 and 4088 (1970).

13. B. J. Blackburn, A. A. Grey, and I. C. P. Smith, *Can. J. Chem.* **48**, 2867 (1970).

14. H. Dugas, B. J. Blackburn, R. K. Robins, R. Deslauriers, and I. C. P. Smith, *J. Amer. Chem. Soc.* **93**, 3468 (1971).

15. A. A. Grey, I. C. P. Smith, and F. E. Hruska, *J. Amer. Chem. Soc.* **93**, 1765 (1971).

16. F. E. Hruska, A. A. Smith, and J. G. Dalton, *J. Amer. Chem. Soc.* **93**, 4334 (1971).

17. T. Schleich, B. J. Blackburn, R. D. Lapper, and I. C. P. Smith, *Biochemistry* **11**, 137 (1972).

18. D. B. Davies and S. S. Danyluk, *Can. J. Chem.* **48**, 3112 (1970).

19. M. Kainosho, A. Nakamura, and M. Tsuboi, *Bull. Chem. Soc. Jap.* **42**, 1713 (1969).

20. M. Tsuboi, M. Kainosho, and A. Nakamura, *in* "Recent Developments of Magnetic Resonance in Biological Systems" (S. Fujiwara and I. H. Piette, eds.), p. 43. Hirokawa Co., Tokyo, 1968.

21. M. Tsuboi, S. Takahashi, Y. Kyogoku, H. Hayatsu, T. Ukita, and M. Kainosho, *Science* **166**, 1504 (1969).

21a. H. Simpkins and E. G. Richards, *Biochemistry* **6**, 2513 (1967).

22. M. M. Warshaw and I. Tinoco, Jr., *J. Mol. Biol.* **19**, 29 (1966).

23. D. Glaubiger, D. A. Lloyd, and I. Tinoco, Jr., *Biopolymers* **6**, 409 (1968).

24. N. S. Kondo, H. M. Holmes, L. M. Stempel, and P. O. P. Ts'o, *Biochemistry* **9**, 3479 (1970).

25. S. Takashima, *Biopolymers* **8**, 199 (1969).

26. H. DeVoe and I. Tinoco, Jr., *J. Mol. Biol.* **4**, 518 (1962).

27. C. R. Cantor, M. M. Warshaw, and H. Shapiro, *Biopolymers* **9**, 1059 (1970).

28. M. M. Warshaw and C. R. Cantor, *Biopolymers* **9**, 1079 (1970).

29. J. Brahms, J. C. Maurizot, and A. M. Michelson, *J. Mol. Biol.* **25**, 481 (1967).

30. R. C. Davis and I. Tinoco, Jr., *Biopolymers* **6**, 223 (1968).

31. W. C. Johnson, Jr. and I. Tinoco, Jr., *Biopolymers* **8**, 715 (1969).

32. C. A. Bush and I. Tinoco, Jr., *J. Mol. Biol.* **23**, 601 (1967).

33. S. I. Chan and J. H. Nelson, *J. Amer. Chem. Soc.* **91**, 168 (1969).

34. P. O. P. Ts'o, N. S. Kondo, M. P. Schweizer, and D. P. Hollis, *Biochemistry* **8**, 997 (1969).

35. M. P. Schweizer, S. I. Chan, G. K. Helmkamp, and P. O. P. Ts'o, *J. Amer. Chem. Soc.* **86**, 696 (1964).

36. K. N. Fang, N. S. Kondo, P. S. Miller, and P. O. P. Ts'o, *J. Amer. Chem. Soc.* **93**, 6647 (1971).

37. G. K. Helmkamp and P. O. P. Ts'o, *J. Amer. Chem. Soc.* **83**, 138 (1961).

38. M. P. Schweizer, A. D. Broom, P. O. P. Ts'o, and D. P. Hollis, *J. Amer. Chem. Soc.* **90**, 1042 (1968).

39. J. T. Powell, E. G. Richards, and W. B. Gratzer, *Biopolymers* **11**, 235 (1972).

40. B. W. Bangerter and S. I. Chan, *J. Amer. Chem. Soc.* **91**, 3910 (1969).

41. S. I. Chan, B. W. Bangerter, and H. H. Peter, *Proc. Nat. Acad. Sci. U.S.* **55**, 720 (1966).

42. T. Förster and K. Kasper, *Z. Elektrochem.* **59**, 976 (1955).

43. J. Eisinger and R. G. Shulman, *Science* **161**, 1311 (1968).

44. J. Eisinger, M. Gueron, R. G. Shulman, and T. Yamane, *Proc. Nat. Acad. Sci. U.S.* **55**, 1015 (1966).
45. D. T. Browne, J. Eisinger, and N. J. Leonard, *J. Amer. Chem. Soc.* **90**, 7302 (1968).
46. N. J. Leonard, H. Iwamura, and J. Eisinger, *Proc. Nat. Acad. Sci. U.S.* **64**, 352 (1969).
47. H. Iwamura, N. J. Leonard, and J. Eisinger, *Proc. Nat. Acad. Sci. U.S.* **65**, 1025 (1970).
48. I. Tazawa, S. Tazawa, L. M. Stempel, and P. O. P. Ts'o, *Biochemistry* **9**, 3499 (1970).
49. S. Tazawa, I. Tazawa, J. L. Alderfer, and P. O. P. Ts'o, *Biochemistry* **11**, 3544 (1972).
50. P. S. Miller, K. N. Fang, N. S. Kondo, and P. O. P. Ts'o, *J. Amer. Chem. Soc.* **93**, 6657 (1971).
51. C. E. Johnson, Jr. and F. A. Bovey, *J. Chem. Phys.* **29**, 1012 (1958).
52. J. S. Waugh and R. W. Fessenden, *J. Amer. Chem. Soc.* **79**, 846 (1957).
53. C. Giessner-Prettre and B. Pullman, *J. Theor. Biol.* **27**, 87 and 341 (1970).
54. N. S. Kondo, K. N. Fang, P. S. Miller, and P. O. P. Ts'o, *Biochemistry* **11**, 1991 (1972).
55. M. Smith and C. D. Jardetzky, *J. Mol. Spectrosc.* **28**, 70 (1968).
56. M. P. Schweizer, R. Thedford, and J. Slamer, *Biochim. Biophys. Acta* **232**, 217 (1971).
57. S. K. Podder, *Biochemistry* **10**, 2415 (1971).
58. R. L. Scruggs and P. D. Ross, *J. Mol. Biol.* **47**, 29 (1970).
59. C. R. Cantor and W. W. Chin, *Biopolymers* **6**, 1745 (1968).
60. M. Ikehara, S. Uesugi, and M. Yasumoto, *J. Amer. Chem. Soc.* **92**, 4735 (1970).
61. S. Uesugi, M. Yasumoto, M. Ikehara, K. N. Fang, and P. O. P. Ts'o, *J. Amer. Chem. Soc.* **94**, 5480 (1972).
62. C. R. Cantor, R. H. Fairclough, and R. A. Newmark, *Biochemistry* **8**, 3610 (1969).
63. H. G. Khorana, and W. J. Conners, *Biochem. Prep.* **11**, 113 (1966).
63a. C. Giessner-Pettre and B. Pullman, *C. R. Acad. Sci.* **261**, 2531 (1965).
64. L. S. Kan, J. C. Barrett, P. S. Miller, and P. O. P. Ts'o, *Biopolymers* **12**, 2225 (1973).
65. F. M. Schabel, Jr., *Chemotherapy* **13**, 321 (1938).
66. A. Inagaki, T. Nakamura, and G. Wakisaka, *Cancer Res.* **29**, 2169 (1969).
67. J. C. Maurizot, J. Brahms, and F. Eckstein, *Nature (London)* **222**, 559 (1969).
68. A. M. Bobst, F. Rottman, and P. A. Cerutti, *J. Amer. Chem. Soc.* **91**, 4603 (1969).
68a. J. L. Alderfer, I. Tazawa, S. Tazawa, and P. O. P. Ts'o, *Biochemistry* (1974) (in press).
68b. J. L. Alderfer, I. Tazawa, S. Tazawa, and P. O. P. Ts'o, in preparation.
68c. A. Adler, L. Grossman, and G. D. Fasman, *Biochemistry* **8**, 3846 (1969).
69. C. A. Bush and H. A. Scheraga, *Biopolymers* **7**, 395 (1969).
70. A. J. Adler, L. Grossman, and G. D. Fasman, *Biochemistry* **7**, 3836 (1968).
71. N. S. Kondo and P. O. P. Ts'o, unpublished data
72. J. Brahms, J. C. Maurizot, and J. Pilet, *Biochim. Biophys. Acta* **186**, 110 (1969).
73. J. Brahms, A. M. Michelson, and K. E. Van Holde, *J. Mol. Biol.* **15**, 467 (1966).
74. E. Elson and T. Jovin, *Anal. Biochem.* **27**, 193 (1969).
75. H. Simpkins and E. G. Richards, *J. Mol. Biol.* **29**, 349 (1967).
76. G. R. Cassani and F. J. Bollum, *Biochemistry* **8**, 3928 (1969).
77. A. Adler, L. Grossman, and G. D. Fasman, *Proc. Nat. Acad. Sci. U.S.* **57**, 423 (1967). (1967).
78. W. C. Johnson, M. S. Itzkowitz, and I. Tinoco, Jr., *Biopolymers* **11**, 225 (1972).
79. F. Pochon and A. M. Michelson, *C. R. Acad. Sci., Ser. D* **270**, 1879 (1970).

80. W. Voelter, R. Records, E. Bunnenberg, and C. Djerassi, *J. Amer. Chem. Soc.* **90**, 6163 (1968).
80a. Y. Inoue, and K. Satoh, *Biochem. J.* **113**, 843 (1969).
81. S. S. Danyluk and F. E. Hruska, *Biochemistry* **7**, 1038.
82. J. L. Alderfer, I. Tazawa, S. Tazawa, and P. O. P. Ts'o, *Biophys. Soc. Abstr.* **12**, 168a (1972).
82a. L. S. Kan, J. C. Barrett, and P. O. P. Ts'o, *Biopolymers* **12**, 2409 (1973).
83. G. DeBoer, P. S. Miller, and P. O. P. Ts'o, *Biochemistry* **12**, 720 (1973).
84. Y. Kyogoku, R. C. Lord, and A. Rich, *J. Amer. Chem. Soc.* **89**, 496 (1967).
85. Y. Kyogoku, R. C. Lord, and A. Rich, *Proc. Nat. Acad. Sci. U.S.* **57**, 250 (1967).
86. G. M. Nogel and S. Hanlon, *Biochemistry* **11**, 816 (1972).
87. M. N. Lipsett, *Proc. Nat. Acad. Sci. U.S.* **46**, 445 (1960).
88. W. Szer, *J. Mol. Biol.* **16**, 585 (1966).
89. J. C. Thrierr, M. Dourlent, and M. Leng, *J. Mol. Biol.* **58**, 815 (1971).
90. M. Dourlent, J. C. Thrierr, F. Brun, and M. Leng, *Biochem. Biophys. Res. Commun.* **41**, 1590 (1970).
91. M. N. Lipsett, *J. Biol Chem.* **239**, 1250 (1964).
92. S. K. Podder, *Indian J. Biochem. Biophys.* **8**, 239 (1971).
93. M. Eigen and D. Pörschke, *J. Mol. Biol.* **53**, 123 (1970).
94. A. M. Michelson and C. Monny, *Biochim. Biophys. Acta* **149**, 107 (1967).
95. J. Applequist and V. Damle, *J. Amer. Chem. Soc.* **87**, 1450 (1965).
96. D. Pörschke and M. Eigen, *J. Mol. Biol.* **62**, 361 (1971).
97. E. O. Akinrimisi, C. Sander, and P. O. P. Ts'o, *Biochemistry* **2**, 340 (1963).
98. J. Brahms, J. C. Maurizot, and A. M. Michelson, *J. Mol. Biol.* **25**, 465 (1967).
99. J. C. Maurizot, J. Blicharski, and J. Brahms, *Biopolymers* **10**, 1429 (1971).
100. P. O. P. Ts'o, S. A. Rapaport, and F. J. Bollum, *Biochemistry* **5**, 4153 (1966).
101. I. E. Scheffler, E. L. Elson, and R. L. Baldwin, *J. Mol. Biol.* **36**, 291 (1968).
102. I. E. Scheffler, E. L. Elson, and R. L. Baldwin, *J. Mol. Biol.* **48**, 145 (1970).
103. E. L. Elson, I. E. Scheffler, and R. L. Baldwin, *J. Mol. Biol.* **54**, 401 (1970).
104. R. L. Baldwin, *Accounts Chem. Res.* **4**, 265 (1971).
105. B. M. Olivera and I. R. Lehman, *Proc. Nat. Acad. Sci. U.S.* **57**, 1426 (1967).
106. B. M. Olivera, I. E. Scheffler, and I. R. Lehman, *J. Mol. Biol.* **36**, 291 (1968).
107. P. J. Flory and J. A. Semlyen, *J. Amer. Chem. Soc.* **88**, 3209 (1966).
108. I. E. Scheffler and J. M. Sturtevant, *J. Mol. Biol.* **42**, 577 (1969).
109. C. Delisi and D. M. Crothers, *Biopolymers* **10**, 2323 (1971).
110. S. R. Jaskunas, C. R. Cantor, and I. Tinoco, Jr., *Biochemistry* **7**, 3164 (1968).
111. R. B. Gennis and C. R. Cantor, *Biochemistry* **9**, 4714 (1970).
112. D. Pörschke, *Biopolymers* **10**, 1989 (1971),
113. A. D. Cross and D. M. Crothers, *Biochemistry* **10**, 4015 (1971).
114. F. H. Martin, O. C. Uhlenbeck, and P. Doty, *J. Mol. Biol.* **57**, 201 (1971).
115. H. Krakauer and J. M. Sturtevant, *Biopolymers* **6**, 491 (1968).
116. O. C. Uhlenbeck, F. H. Martin, and P. Doty, *J. Mol. Biol.* **57**, 217 (1971).
117. J. Marmur and P. Doty, *Nature (London)* **183**, 1427 (1959).
118. N. Kallenbach, *J. Mol. Biol.* **37**, 455 (1968).
119. M. E. Craig, D. M. Crothers, and P. Doty, *J. Mol. Biol.* **62**, 383 (1971).
120. O. C. Uhlenbeck, P. N. Borer, B. Dengler, and I. Tinoco, Jr., *J. Mol. Biol.* **73**, 483 (1973).
121. M. R. Bernfield and M. W. Nirenberg, *Science* **147**, 479 (1965).
122. "The Genetic Codes," *Cold Spring Harbor Symp. Quant. Biol.* **31**, (1966).
123. M. N. Lipsett, L. A. Heppel, and D. F. Bradley, *Biochim. Biophys. Acta* **41**, 175 (1960); *J. Biol. Chem.* **236**, 857 (1961).

124. M. N. Lipsett, *J. Biol. Chem.* **239**, 1256 (1964).
125. A. M. Michelson and C. Monny, *Biochim. Biophys. Acta* **149**, 107 (1967).
126. B. Janik, "Physicochemical Characteristics of Oligonucleotides and Polynucleotides." Plenum, New York, 1971.
127. V. N. Damle, *Biopolymers* **9**, 353 (1970).
128. D. M. Crothers, *Biopolymers* **10**, 2147 (1971).
129. I. Tazawa, S. Tazawa, and P. O. P. Ts'o, *J. Mol. Biol.* **66**, 115 (1972).
130. W. S. Magee, Jr., J. H. Gibbs, and B. H. Zimm, *Biopolymers* **1**, 133 (1963).
130a. M. W. Springgate and D. Poland, *Biopolymers* **12**, 2241 (1973).
131. P. D. Ross and R. L. Scruggs, *J. Mol. Biol.* **45**, 567 (1969).
132. H. J. Heinz, W. Haar, and T. Ackermann, *Biolpolymers* **9**, 923 (1970).
133. W. A. Carter, P. M. Pitha, L. W. Marshall, I. Tazawa, S. Tazawa, and P. O. P. Ts'o, *J. Mol. Biol.* **70**, 567, (1972).
134. O. C. Uhlenbeck, Ph.D. Thesis, Dept. Biochemistry, Harvard University, Cambridge, Massachusetts, 1969.
135. C. R. Cantor and W. W. Chin, *Biopolymers* **6**, 1745 (1968).
136. H. Krakauer and J. M. Sturtevant, *Biopolymers* **6**, 491 (1968).
137. G. R. Cassani and F. J. Bollum, *Biochemistry* **8**, 3928 (1969).
138. O. Uhlenbeck, R. Harrison, and P. Doty, *in* "Molecular Associations in Biology" (B. Pullman, ed.), p. 107. Academic Press, New York, 1968.
139. E. K. F. Bautz and F. A. Bautz, *Proc. Nat. Acad. Sci. U.S.* **52**, 1476 (1964).
140. P. M. Pitha and P. O. P. Ts'o, *Biochemistry* **8**, 5206 (1969).
141. R. E. Thach, Ph.D. Thesis, Chemistry Dept., Harvard University, Cambridge, Massachusetts, 1963; R. E. Thach and P. Doty, *Science* **148**, 632 (1965).
142. R. E. Thach, *in* "Procedures in Nucleic Acid Research" (G. L. Cantoni and P. R. Davies, eds.), p. 520. Harper, New York, 1966.
143. O. C. Uhlenbeck, J. Baller, and P. Doty, *Nature (London)* **225**, 510 (1970).
144. J. B. Lewis and P. Doty, *Nature (London)* **225**, 510 (1970).
145. O. C. Uhlenbeck, *J. Mol. Biol.* **65**, 25 (1972).
146. J. B. Lewis, Ph.D. Thesis, Dept. of Biochemistry, Harvard University, Cambridge, Massachusetts, 1971.
147. P. R. Schimmel, O. C. Uhlenbeck, J. B. Lewis, L. A. Dickson, E. W. Eldred, and A. A. Schreier, *Biochemistry* **11**, 642 (1972).
148. G. Högenauer, *Eur. J. Biochem.* **12**, 527 (1970).
149. G. Högenauer, F. Turnowsky, and F. M. Unger, *Biochem. Biophys. Res. Commun.* **46**, 2100 (1972).
150. O. Pongs, E. Reinwald, and K. Stamp, *FEBS Lett.* **16**, 275 (1971).
151. J. Eisinger, B. Feuer, and T. Yamane, *Nature (London) New Biol.* **231**, 126 (1971).
152. J. Eisinger, *Biochim. Biophys. Res. Commun.* **43**, 854 (1971).

AUTHOR INDEX

Numbers in parentheses are reference numbers and indicate that an author's work is referred to although his name is not cited in the text. Numbers in italics show the page on which the complete reference is listed.

A

Abel, J.-P., *90*

Abell, C. W., 56(370), *88*

Abraham, D. J., 174(9), 251(9), 253(9), *255*

Achter, E. K., 254, *264*

Ackermann, T., 451, *469*

Adler, A. J., 119(77), 138(77), 145(130), 147 (130), 164(164), *167, 168, 169,* 377(68c, 70), 378(70), 392, 393(77), *467*

Adman, R., 16(84), *81*

Agarwal, M. K., 35(218), *84*

Agarwal, S. C., 28(161), *83*

Aira, M., 22(121), *82*

Akasaka, K., 250(373), *264*

Akinrimisi, E. O., 427(97), *468*

Albert, A., 3(4a), *79*

Alderfer, J. L., 342(49), 346(49), 377 (68a, b), 389(49), 390(49), 392(49), 393 (49), 395(49), 396(49), 397(49), 398(49), 399(49), 400(49), 402(49), 403(49), 404 (49), 405(49), 406(49), 408(49), 410(82), *467, 468*

Alegria, A. H., 26(149), *82*

Aleinikova, T. L., 236(279), *261*

Alexandrowicz, Z., 192(93), 196(93), 197, 198(123), 199(117, 118), 200(118), 208, 236, *257, 258, 259*

Alfrey, T., Jr., 197, 199(115), *257*

Allen, F. S., 156(149a), *169*

Allen, F. W., 42(262), *85*

Allison, W. S., 25(141), *82*

Altman, S., 5(19), *79*

Ames, B. N., 66(407), 67(412, 412a), *89*

Anderson, D. L., 299(68), *303*

Anderson, S. M., 36(223), *84*

Ando, T., 294(46), *303*

Andre, J. M., 116(65), 117(65), *167*

Andre, M. C., 116(65), 117(65), *167*

Andreev, V. M., 220(195a), *259*

Aoyagi, S., 135(108), 136(112), 137(108), *168*

Applequist, J., 119(74), *167*, 418, 420, 421 (95), 422, 447, *468*

Archer, B. G., 238(301), *262*

Arikawa, S., 53(355), *87*

Armstong, R. W., 174, 196(6), 197(6), 216 (6), 220(97), 248, *255, 259, 263*

Arnott, S., 70(425), 71(425), 77(425), 78 (425), *89*, 157(150), *169*, 198(121), 251 (397), 252(404), 253(397), *258, 264*, 307, 308, 309(2), 313(2, 4), 342(2), *465*

Aryo, S. R., 228(230), *260*

Aubel-Sadron, G., 236, *261*

Aubertin, A. M., 134(103), 136(103), 137 (103), *168*

Auer, H. E., 192(93), 196(93), 236, *257*

Augerer, L. M., 248(363), *263*

Augusti-Tocco, G., 30(174), *83*

Axelrod, V. D., 26(148), *82*

B

Bach, D., 239, *262*

Baird, K. M., 56(374), *88*

Baird, S. L., 241(334), *263*

Bakradze, N. G., 230(234), *260*

Balbe, G., 15(78), *80*

Baldwin, R. L., 30(180), 53(358), 58(180), 61(180), *83, 87*, 189(86), 206, 226, 253 (86), *257, 258, 260*, 430, 431(101), 432, 433(102–104), 434(102), 443(102–104), *468*

Balis, M. E., 43(269), *85*

Ball, J. K., 244(338), *263*

Baller, J., 77(452), 78(452), *90*, 460(143), 464(143), *469*

Bamann, E., 43(274, 279), *85*

Bancroft, F. C., 208(153), 213, 214(153), *258, 259*

471

SUBJECT INDEX

A

Absorbance, 92

Absorption bands of bases and nucleosides, 113

Absorption coefficient, 96

Absorption spectrum
  of A-T and G-C base pairs, 120
  bases, 111
  effect of sugar, 111
  electronic origin of transitions, 112, 113
  nucleosides, 111-115
  nucleotides, 111-115

2-Acetaminofluorene, 33

N-Acetoxy-N_2-acetylaminofluorene, binding to DNA, 244

N-Acetoxy-2-fluorenylacetamide
  reaction with guanine, 34, 35
  with proteins, 34

N_4-Acetylcytidine, reaction with sodium borohydride, 15

Acridine
  binding to circular DNA, 289
  formula, 245
  frame-shift mutagen, 67
  intercalating reagents, 66

Acridine dyes, interaction with DNA, 241

Acridine orange
  binding with DNA, 207
  cation formula, 245

Acrylonitrile
  base modification, 24
  phosphate alkylation, 55
  reaction with tRNA, 71, 77

Actinomycin D
  binding to circular DNA, 289
  to DNA, 241, 244
  crystalline complex with deoxyguanosine, 246
  formula, 245

$A_6C_mU_{6'}$, 442

N_2-Acylguanine, formation with periodate, 27

Acylhydrazides, reaction rate with native DNA, 58

Adenine
  absorption spectra, 112
  alkylation of N-1, 21, 63
  of N-3 by dimethyl sulfate, 23
  arabinosyl nucleoside, 369
  bromination, 36
  hydrolysis to hypoxanthine, 3
  reaction with Kethoxal, 27
  with monoperphthalic, 33
  sterochemistry of N-oxidation, 61
  thermal degradation of N-3 alkyl, 53
  uv absorption bands, 113

Adenine oligomers, see Oligoadenylate

Adenosine
  alkylation of N-7, 20
  binding to poly U, 119
  CD, calculation, 125
  conformation in $A_2{}'p_{s'}U$, 311
  Cotton effects of α vs β anomers, 128
  deamination, 30
  N-oxide stability, 33
  reaction with N-acetoxy-2-fluorenylacetamide, 35
    with 7-bromomethylbenz[a]anthracene, 21
    with chloroacetaldehyde, 28
    with diazomethane, 22
    with formaldehyde, 25
    with hydroxylamine, 7
    with nitrous acid, 30
  uv absorption bands, 113

Adenosine-N_1-oxide, synthesis from reaction of hydroxylamine with adenosine, 7

Adenosine pentafuranosides, CD, 125

493

O

V

W

X

Y

Z

A 4
B 5
C 6
D 7
E 8
F 9
G 0
H 1
I 2
J 3